Parasitoid Wasps of South East Asia

Buntika A. Butcher and Donald L.J. Quicke

CABI is a trading name of CAB International

CABI
Nosworthy Way
Wallingford
Oxfordshire OX10 8DE
UK

CABI
200 Portland Street
Boston
MA 02114
USA

Tel: +44 (0)1491 832111
E-mail: info@cabi.org
Website: www.cabi.org

Tel: +1 (617)682-9015
E-mail: cabi-nao@cabi.org

© Buntika A. Butcher and Donald L.J. Quicke 2023. All rights reserved. No part of this publication may be reproduced in any form or by any means, electronically, mechanically, by photocopying, recording or otherwise, without the prior permission of the copyright owners.

The views expressed in this publication are those of the author(s) and do not necessarily represent those of, and should not be attributed to, CAB International (CABI). Any images, figures and tables not otherwise attributed are the author(s)' own. References to internet websites (URLs) were accurate at the time of writing.

CAB International and, where different, the copyright owner shall not be liable for technical or other errors or omissions contained herein. The information is supplied without obligation and on the understanding that any person who acts upon it, or otherwise changes their position in reliance thereon, does so entirely at their own risk. Information supplied is neither intended nor implied to be a substitute for professional advice. The reader/user accepts all risks and responsibility for losses, damages, costs and other consequences resulting directly or indirectly from using this information.

CABI's Terms and Conditions, including its full disclaimer, may be found at https://www.cabi.org/terms-and-conditions/.

A catalogue record for this book is available from the British Library, London, UK.

ISBN-13: 9781800620599 (hardback)
       9781800620582 (ePDF)
       9781800620605 (ePub)

DOI: 10.1079/9781800620605.0000

Commissioning Editor: Ward Cooper
Editorial Assistant: Lauren Davies
Production Editor: James Bishop

Typeset by Straive, Pondicherry, India
Printed and bound in the USA by Integrated Books International, Dulles, Virginia

Contents

Preface	xvii
Acknowledgements	xxi
1. Introduction	**1**
When is a Parasitoid Not a Parasitoid?	2
The Aims of this Book	2
Ecological Role and Importance	3
Utility in Pest Management (Biocontrol)	3
Conservation of natural enemies	3
Classical biological control	3
Augmentative biocontrol	4
Integrative pest management (IPM)	4
The demand for organic produce	4
Importance of Taxonomy	5
History of Parasitoid Wasp Taxonomy in S.E. Asia	5
Early taxonomic work on S.E. Asian parasitoid wasps	6
References	7
2. Biology	**9**
Major Life History Strategies	9
The Idiobiont/Koinobiont Spectrum	10
Idiobionts	10
Koinobionts	11
Adult lifespan, host feeding and synovigeny	11
Concealed versus exposed hosts	11
Developmental Features	12
Polyembryony	12
Teratocytes	12
Larval Instars	12
Superparasitism	12
Gregarious Development	13
Larval Feeding and External Feeding Phase	13
Parasitoid Food Conversion Efficiency	14
Ovoparasitism	14

Sex Determination and Mating System ... 15
   Sex Ratio ... 15
   Sex Ratio Distortion – Thelytoky ... 16
     Wolbachia ... 16
Host Manipulation by Venoms and Viruses ... 16
   Polydnaviruses ... 17
The Fig Ecosystem ... 17
References ... 18

3. Behaviour ... 21
Sex, Courtship and Mating ... 21
   Dispersal and Phoretic Copulation ... 23
Host Location and Assessment ... 23
   Tritrophic Interactions ... 24
   Learning and Associative Learning ... 24
   Locating Xylophagous and Other Concealed Hosts ... 24
Vibrational Sounding, a Form of Echolocation ... 25
References ... 26

4. Parasitoid Diversity, with Special Reference to S.E. Asia ... 29
Surveying Parasitoid Diversity in S.E. Asia ... 30
   What Do Surveys in S.E. Asia Find? ... 31
Convention on Biological Diversity (CBD) and Permitting Issues ... 32
   The Collecting Permit Problem ... 33
   What is Needed for Complete Inventory of a Group? ... 34
   Is There a Type Specimen Impediment? ... 35
DNA Barcoding – The Way Forward ... 36
   Does Barcoding Do a Good Job of Separating Species of Parasitoid Wasp? ... 37
   The Need to Speed Up Taxonomy ... 37
     Turbotaxonomy ... 38
     Barcoding and the Minimalism Movement ... 39
Tropical Parasitoid Diversity and Biology ... 39
   Anomalous Diversity ... 40
   Do We Actually Know Enough to Trust Trends? ... 41
References ... 41

5. Classification and Phylogeny ... 46
Classification Overview ... 46
Hymenoptera Phylogeny Research ... 47
   Morphological Studies ... 47
   Molecular Studies (to 2017) ... 48
   Blaimer *et al.* (2023) ... 48
Conclusions ... 50
References ... 53

6. Morphology ... 55
Body Regions and the Wasp Waist ... 55
   Why the Wasp Waist and Why is it Where it is? ... 56
The Head and Antennae ... 57
   Basic Head Parts ... 57
   Antennae ... 58
   Mouthparts ... 58
Mesosoma (the Middle Body Region Bearing the Legs) ... 59

Legs	61
Trochanter and Trochantellus	62
Wings	63
Hamuli and Wing Coupling	63
The Hymenopteran Wing Venation Ground Plan	64
Vein Naming Systems	65
Pterostigma and parastigma	66
The rest of the venation	67
Where is vein 1rs-m?	68
Problems with the system	68
Fold and Flexion Lines	68
Cell Naming Systems	68
Metasoma	69
Ovipositor	71
Male Genitalia	73
Sculpture	74
References	75
7. Recognition of Major Groups	**79**
How Do Experts Identify Major Groups?	79
Keys to Major Groups for the Majority of Species of Parasitoid Hymenoptera	80
Key A — Is it a Hymenopteran?	80
Key B — Is it a Parasitoid Wasp?	81
Key C — Major Groups of Winged Female Parasitoid and Aculeate Hymenoptera	83
Key D — Taxa with Costal Cell Enclosed by Pair of Tubular Veins	84
Key E — Taxa Lacking Entire Enclosed Fore Wing Costal Cell	85
Key F — Ichneumonidae or Braconidae?	87
References	88
8. Orussoidea	**89**
References	91
9. Braconidae	**92**
Phylogenetic Studies	92
Morphological Terminology	94
Cyclostome	94
Lateral Pronope (or Subpronope)	95
Pronope (or Median Pronope)	95
Precoxal Sulcus	95
Prepectal Carina	95
Anterior Subalar Depression	95
Dorsal, Dorso-Lateral and Lateral Carinae of 1st Metasomal Tergite	95
Dorsope and Laterope	96
Wings	96
Subfamily Identification	96
Key to Females of Common S.E. Asian Subfamilies of Braconidae	97
Cyclostomes	101
Aphidioid Subcomplex	101
Aphidiinae	101
Masoninae	103
Mainline Cyclostomes	103
Braconinae	103
Doryctinae	104

Hormiinae (including former Lysiterminae)	106
Cedriini	106
Hormius group (Hormiini)	107
Lysitermini	107
Pentatermini	107
Pambolinae	107
Rhysipolinae	107
Rhyssalinae	108
Rogadinae	109
Alysioid Subcomplex	111
Alysiinae	112
Exothecinae	113
Telengaiinae (=Gnamptodontinae)	113
Opiinae	114
Non-Cyclostomes	115
The Sigalphoid Complex	116
Agathidinae	116
Sigalphinae	118
The Helconoid Complex	118
Acampsohelconinae	118
Helconoid subcomplex	118
Brachistinae	118
Helconinae	120
Macrocentroid subcomplex	120
Charmontinae	120
Homobolinae	121
Macrocentrinae	121
Orgilinae	122
Xiphozelinae	123
The Euphoroid Complex	123
Cenocoeliinae	123
Euphorinae (including Meteorinae)	123
The Microgastroid Complex	125
Cardiochilinae	126
Cheloninae (including Adeliinae)	128
Dirrhopinae	128
Ichneutinae	128
Microgastrinae	129
Miracinae	132
Proteropinae	132
Unplaced	133
Meteorideinae	133
References	133
10. Ichneumonidae – Darwin Wasps	**147**
Morphological Terminology	150
Tyloids	150
Epomia	150
Epicnemial Carina	151
Sternaulus	151
Episternal Scrobe	152
Postpectal Carina	152

Propodeal Carinae	152
Wings	152
Areolet	152
Wing bullae	154
Glymmae	155
Gastrocoeli and Thyridea	155
Highly Simplified Key to Females (and a Few Males) of Relatively Common or Distinctive S.E. Asian Subfamilies	157
Phylogeny and Informal Subfamily Groupings	160
The Brachycyrtiformes	161
Brachycyrtinae	161
Eucerotinae	161
The Ichneumoniformes	165
Key to Separate Female Ateleutinae, Cryptinae, Ichneumoninae and Phygadeuontinae	166
Agriotypinae	167
Alomyinae	167
Ateleutinae	167
Cryptinae	168
Ichneumoninae	170
Microleptinae	171
Phygadeuontinae	172
The Labeniformes	172
Labeninae	172
The Ophioniformes	172
Anomaloninae	173
Banchinae	174
Key to female tribes of Banchinae relevant to S.E. Asia	175
Atrophini (= Lissonotini)	175
Banchini	175
Glyptini	175
Townesionini	175
Campopleginae (= Porizontinae)	175
Cremastinae	177
Ctenopelmatinae	178
Lycorininae	180
Mesochorinae	180
Metopiinae	182
Nesomesochorinae	184
Ophioninae	185
Oxytorinae	186
Sisyrostolinae (= Brachyscleromatinae)	187
Stilbopinae	187
Tersilochinae	187
Tryphoninae	188
Oedemopsini	189
Phytodietini	190
Sphinctini	190
Tryphonini	190
The Pimpliformes	191
Acaenitinae	191
Cylloceriinae	193

Diplazontinae	193
Orthocentrinae (includes Helictinae, Microleptinae and Oxytorinae of Authors)	195
Pimplinae	196
Key tribes and some groups of Pimplinae (females only) in S.E. Asia	196
Pimplini (= Ephialtini *sensu* Townes)	197
Xanthopimpla	198
Theronia group (= Therioniini *sensu* Townes in part)	198
Delomeristini (= Therioniini *sensu* Townes in part)	199
Ephialtini (= Pimplini *sensu* Townes)	199
Poemeniinae	202
Rhyssinae	204
The Xoridiformes	206
Xoridinae	206
References	207

11. Stephanoidea (Crown Wasps) — 219

Wing Venation	220
Recognition	220
Identification	222
S.E. Asian Genera	222
Foenatopus	222
Megischus	222
Parastephanellus	222
Pseudomegischus	222
Stephanus	222
References	222

12. Evanioidea — 224

Key to the Families of Evanioidea	225
Aulacidae	225
Evaniidae (Ensign Wasps)	226
Key to Evaniidae Genera of S.E. Asia (modified from Deans and Huben, 2003)	227
Gasteruptiidae (Carrot Wasps)	227
References	230

13. Ceraphronoidea — 233

Key to the Families of Ceraphronoidea	235
Ceraphronidae	235
Megaspilidae	237
References	237

14. Megalyroidea — 239

Megalyridae	239
References	241

15. Trigonalyoidea — 242

Morphology	243
S.E. Asian Trigonalyid Fauna	243
Simplified Key to the Genera of Trigonalyidae Occurring in S.E. Asia	245
References	245

16. Parasitoid Aculeates – Chrysidoidea — 247

Key to the Families of Chrysidoidea in S.E. Asia	248
Bethylidae (Flat Wasps)	248

Key to Subfamilies of Fully Winged Bethylidae	249
Bethylinae	250
Epyrinae	250
Mesitiinae	251
Pristocerinae (= Afgoiogfinae)	252
Scleroderminae	252
Chrysididae	252
Simplified Key to Chrysidid Subfamilies Occurring in S.E. Asia	253
Amiseginae	253
Chrysidinae	253
Cleptinae	256
Loboscelidiinae	256
Dryinidae (Pincer Wasps)	258
Embolemidae	263
Sclerogibbidae	264
Scolebythidae	265
References	265

17. Parasitoid Aculeates — Vespoidea *sensu lato* 270

Simplified Key to the Fully Winged and Some Wingless Vespoidea (*s.l.*) in S.E. Asia	272
Pompiloidea	273
Mutillidae (Velvet Ants)	273
Myrmosidae	277
Pompilidae (Spider-hunting Wasps)	277
Sapygidae	278
Scolioidea	279
Bradynobaenidae (extralimital)	279
Scoliidae (Digger Wasps, Mammoth Wasps)	279
Thynnoidea and Tiphioidea (Flower Wasps)	281
Thynnoidea (Thynnidae)	281
Tiphioidea (Tiphiidae)	282
Vespoidea	283
Rhopalosomatidae	283
References	284

18. Platygastroidea 290

Morphological Terminology	291
Mesepimeral Sulcus	291
Skaphion	291
Metasomal Horn	291
Key to S.E. Asian Families of Platygastroidea	291
Neuroscelionidae	292
Nixoniidae	292
Platygastridae	293
Key to Subfamilies of Platygastridae	293
Scelionidae	294
Key to Clusters of Scelionidae	295
Scelioninae *sensu* Authors	296
'Teleasinae'	298
'Telenominae'	299
Sparasionidae	300
References	301

19. Cynipoidea – Gall Wasps and Their Kin — 306
Morphological Terminology — 306
Phylogeny — 308
Simplified Key to Females of Major Groups of S.E. Asian Cynipoidea — 308
Cynipidae — 309
    Cynipini — 309
    Synergini — 310
Figitidae — 310
    Anacharitinae — 311
    Aspicerinae — 312
    Charipinae — 313
    Emargininae — 313
    Eucoilinae — 313
    Figitinae — 314
Ibaliidae — 314
Liopteridae — 315
References — 316

20. Proctotrupoidea — 320
Simplified Key to S.E. Asian Families of Proctotrupoidea — 320
Heloridae — 321
Proctotrupidae — 321
Roproniidae — 323
Vanhorniidae — 323
References — 325

21. Diaprioidea — 327
Key to Separate Families of Diaprioidea in S.E. Asia and Papua New Guinea — 327
Diapriidae — 328
    Simplified Key to the Subfamilies of Diapriidae — 328
    Ambositrinae — 329
    Belytinae — 329
    Diapriinae — 330
        Simplified key to tribes of Diapriinae occurring in S.E. Asia — 330
Ismaridae — 332
Monomachidae — 332
References — 333

22. Chalcidoidea and Mymarommatoidea — 335
Morphological Terminology — 336
    Antenna — 338
    Thoracic Morphology — 338
        Prepectus, Axillae and Axillulae — 338
        Frenum and frenal area — 342
        Wings and venation — 342
        Filum spinosum and linea calva — 342
    Metasoma — 342
Identification — 342
Highly Simplified, Partial Family Key to Female Fully Winged Species — 343
Agaonidae (Fig Wasps) *sensu lato* — 346
    Key to Separate True Agaonidae from Fig-associated Pteromalidae in S.E. Asia — 346
Aphelinidae (and Calesidae) — 349
Azotidae — 350

Chalcididae	351
Chalcidinae (Including Brachymeriinae)	352
Cratocentrinae	352
Dirhininae	352
Epitraninae	353
Haltichellinae	353
Smicromorphinae	354
Chrysolampidae	354
Cynipencyrtidae	354
Encyrtidae	354
Epichrysomallidae	357
Eriaporidae	357
Eucharitidae and Perilampidae	357
Eucharitidae	358
Eucharitinae	358
Gollumiellinae	358
Oraseminae	359
Eulophidae	360
Simplified Key to Groups of Eulophidae (Winged Females)	361
Entedoninae	365
Entiinae (= Euderinae)	365
Eulophinae	365
Opheliminae	366
Tetrastichinae	366
Eupelmidae and *Eopelma*	367
Key to S.E. Asian Subfamilies of Eupelmidae	367
Calosotinae	367
Eupelminae	368
Eopelma	369
Eurytomidae	370
Leucospidae	370
Megastigmidae	371
Metapelmatidae	371
Mymaridae (Fairy Flies)	372
Neanastatidae	374
Ormyridae	375
Perilampidae	375
Former Pteromalidae	376
Asaphesidae (Chalcidoidea *incertae sedis*)	378
Cerocephalidae	378
Chalcedectidae	378
Cleonymidae	378
Diparidae	379
Ditropinotellinae (Chalcidoidea *incertae sedis*)	379
Epichrysomallidae	380
Eunotidae	380
Herbertiidae	381
Heydeniidae	381
Louricinae (Chalcidoidea *incertae sedis*)	381
Lyciscidae	381
Macromesidae	382
Moranilidae	382
Ooderidae	382

Parasaphodinae	382
Pelecinellidae (= Leptofoeninae)	383
Pirenidae (= Eriaporidae)	383
Pteromalidae *sensu stricto*	384
Colotrechninae	384
Miscogasterinae	384
Pachyneurinae	385
Pteromalinae	385
Otitesellini (= former Otitisellinae; = Sycoecinae; = Sycoryctinae)	385
Pteromalini	386
Sycophaginae	387
Trigonoderinae	388
Spalangiidae	389
Storeyinae (Chalcidoidea: *incertae sedis*)	389
Systasidae	389
Signiphoridae	390
Tanaostigmatidae	391
Tetracampidae	391
Torymidae	392
Monodontomerinae	392
Podagrioninae	392
Toryminae	393
Trichogrammatidae	393
References	395
Mymarommatoidea (False Fairy Wasps)	407
References	408

23. Collection, Preservation and Rearing   410

Collecting methods	411
Aerial or Butterfly Nets	411
Sweep Nets and Insect Separators	412
Malaise Traps	413
Yellow Pan Traps	416
Light Traps	418
Canopy Fogging	419
Vacuum Cleaners (Suction Traps)	420
Lesser-used Methods	420
Conclusions	420
Specimen Preparation	420
A Note on Essential Equipment and Consumables	420
Fine forceps	420
Fine forceps – dexterity is a godsend	421
Sorting Wet Samples	422
Drying Small, Fragile Specimens	422
Chemical drying – the AXA method	423
Just amyl acetate	423
Hexamethylisilazane (HMDS)	423
Mounting	423
Direct pinning	424
Stage pinning	424

Adhesives and Gluing Techniques	425
Side gluing	425
Card points and rectangles	425
Store Boxes	426
Microscope Slide Preparations	427
Genitalia preparations	427
Rearing Parasitoids from Wild-collected Hosts	427
Culturing Parasitoids	429
References	430
Appendix 1. Taxa, Authors and Dates	**433**
Index	**455**

Preface

This book is largely for workers who are familiar with entomology and perhaps some familiar with Hymenoptera but who wish to know more about less well-studied parasitoid groups. Such workers might include enthusiastic entomologists who see some remarkable parasitoids in their travels and wish to know more, educators who teach entomology at university level and professional entomologists in agriculture, horticulture and forestry. The parasitoid wasps are without doubt most fascinating insects, many being economically important and probably most being ecologically important due to their high trophic position. All, to some extent, have sophisticated interactions with their hosts, ranging from paralysing venom to subtle modifications of host endocrinology.

They are also, taxonomically, one of the most poorly known groups of arthropods. Even in the north temperate region, there are still many thousands of species awaiting discovery and description. When it comes to the three main tropical regions of the world, the situation is far different. The parasitoid wasp fauna of South East Asia, which we here define to include Cambodia, Indonesia, Laos, Malaysia, Myanmar, the Philippines, Singapore, Thailand, Vietnam and little East Timor, is perhaps the poorest-known. Indeed, there are numerous large cosmopolitan groups that certainly occur here but none have yet been published on, the main reason for this being the lack of accessible relevant literature. Even for those taxa recorded from the region, their distributions are frequently based on records from a single country and in one case in a single botanical garden, because of the lack of collecting and local expertise. Many ecological surveys, including some that aim to provide useful information for agroforestry, simply stop at superfamily level. Frankly, such studies are not of much value.

In surveying the parasitoid Hymenopteran literature dealing with the S.E. Asia region, it is remarkable and depressing that so few biological observations have been made. It is never that easy to rear parasitoids and perhaps it is even harder in the tropics because of such things as humidity, disease, availability of host plants, etc. But it is not that difficult either, and so the almost complete lack of host records apart from those for a few economically important species is something we would like to overcome.

In Chapters 2 and 3, we provide a brief overview of key aspects of parasitoid biology and behaviour such as sex determination, host and mate location, learning, and tritrophic interactions. In Chapter 4, we cover quite a diverse range of topics related to the study of parasitoid diversity. We particularly highlight the problem that taxonomists faced in trying to collect and export materials for study. Such problems are particularly acute for those scientists attempting to carrying out phylogenetic and biogeographical investigations and research programmes, because essentially they need to study material from all around the world if they are dealing with cosmopolitan groups. Even for more endemic taxa perhaps more or less limited to S.E. Asia, it is necessary to examine material from across the whole area in order to determine how many species there are and how they may be distinguished. We examine the taxonomic impediment as experienced in biodiverse countries in which a large proportion of species are as yet to be described. We discuss the potential of DNA barcoding, turbotaxonomy and minimalist approaches to overcome much of this but highlight some of the problems. In Chapter 5, we provide an overview of hymenopteran classification, culminating in 2023 in a new genomic phylogeny by Blaimer et al. that provides a very well-supported backbone of superfamily relationship, which we follow here. Most interestingly, and in agreement with some previous studies, the huge cosmopolitan superfamily Ichneumonoidea (which is our own particular research focus) was recovered as sister group to all the remaining Apocritan parasitoid wasps, including the Aculeates.

Chapter 6 provides a fairly detailed description of the external morphology of Hymenoptera, introducing many of the terms used later in this book. Particular attention is given to the naming of wing veins and cells, because historically several different systems have been employed and the reader is bound to encounter these when consulting other literature. Chapter 7 presents a series of simplified identification keys to allow recognition of whether the specimen is likely to be a parasitoid wasp, all the way through to the identification of each superfamily.

Chapters 8 to 22, the main part of this book, cover each of the superfamilies in turn. For each we describe their biology, often presenting identification keys to their larger subgroups, and we direct readers to the recent taxonomic literature concerning S.E. Asia genera that can be used to take their identification further. Obviously, the different groups of parasitoids have received varying level of taxonomic treatment and biological studies. In nearly all of these chapters we present a labelled diagram of fore wing venation, colour coded to show the wing vein terminology used here. Nearly every group mentioned is illustrated by photographs of mounted or in some cases living specimens. Please note that for purposes of illustration only, we have mounted some specimens in inappropriate ways, such as card-pointing some rather (too) large specimens; in Chapter 23 we explain how to do this properly. Choosing which groups to go into more detail about has been difficult. Whilst we have generally tried to write about and illustrate all the groups that are most likely to be encountered, what an entomologist finds will depend on their collecting or rearing methods; for example, some species are very common but are not often collected in Malaise traps or by light trapping. Rather few hymenopterists use pitfall trapping or concentrate on the edaphic (soil/leaf litter) samples, yet these are the home environments of many groups.

In this book, we attempt to remedy some of the above by providing detailed coverage of all the groups of parasitoid wasps that are known from the region, or are strongly suspected to occur there. We guide the readers with simple identification keys that we hope will enable the great majority of the specimens to be identified correctly to family and in some cases subfamily level. Of course, there is a trade-off, because there are readily available public keys that might provide much closer to 100% correct identification, but necessarily they are far more complicated and often run to more than twice the number of couplets that we have presented here. This is because the other keys are intended to allow for nearly every species, no matter how aberrant, to be identified. Quite often these will not occur in S.E. Asia and are certainly unlikely to be collected.

Separate chapters are devoted to the large families, Braconidae and Ichneumonidae (Ichneumonoidea), as well as to all the other remaining superfamilies. In choosing which species to illustrate we have aimed for a compromise. Many groups are dominated by rather dull reddish brown to black specimens and we have shown some of these; however, to increase interest we have also where possible included some rather more colourful, cute or spectacular members.

In Chapter 23, we describe most of the regularly used techniques for collecting parasitoid wasps and give pointers towards using them most successfully. Then we describe the processes involved in going from freshly collected (wet) specimens through to having nicely mounted ones in which all diagnostic characters can be seen easily. Finally, we discuss the factors that need to be taken into consideration when trying to rear parasitoids and their hosts successfully. To make the book more readable, we have omitted names and dates of authors for taxonomic groups in the text but present these in the Appendix.

We sincerely hope that this book will encourage an upsurge in research on S.E. Asian parasitoid wasps and will be of interest and use to a wide range of biologists.

Buntika A. Butcher and Donald L.J. Quicke

Acknowledgements

Many people have helped with the production of this book and we are exceedingly grateful for their time and generosity.

Photographs of groups that we have not managed to collect ourselves, and also of some collecting apparatus, were kindly provided for use here by the following: Matthew Buffington (Systematic Entomology Laboratory, USDA/ARS c/o NMNH, Smithsonian Institution, Washington, DC, USA), Jochen Drescher, (University of Goettingen, Germany), Ian Dugdale (Baanmaka Nature Lodge, Thailand), Mostafa Ghafouri Moghaddam (Chulalongkorn University, Thailand), Alex Gumovsky (Schmalhausen Institute of Zoology, National Academy of Sciences of Ukraine, Ukraine), Jean-Yves Rasplus (INRAE – Centre de Biologie pour la Gestion des Populations, Montferrier-sur-Lez, France), Alexey Reshchikov (Institute of Eastern-Himalaya Biodiversity Research, Dali University, Yunnan, PR China), Simon van Noort (Iziko Museums of South Africa, Cape Town, South Africa), Michael Sharkey (Hymenoptera Institute, California, USA), Keizo Takasuka (Institute for Advanced Biosciences, Keio University, Yamagata, Japan), Kees van Achterberg (Naturalis Biodiversity Center, Leiden, Netherlands), Wendy Wang (Lee Kong Chian Natural History Museum, University of Singapore, Singapore).

We are very grateful to the following for variously providing, confirming, enhancing or correcting our identifications and often providing valuable new insights: Celso Oliveira Azevedo (Universidade Federal do Espírito Santo, Vitória, Brazil), Gavin Broad (Natural History Museum, London), Denis J. Brothers (University of KwaZulu-Natal, Scottsville (Pietermaritzburg), South Africa), Matthew Buffington (Systematic Entomology Laboratory, USDA/ARS c/o NMNH, Smithsonian Institution, Washington, DC, USA), Minoo EhEl (Chulalongkorn University, Thailand), Michael Gates (Systematic Entomology Laboratory, USDA/ARS c/o NMNH, Smithsonian Institution, Washington, DC, USA), Vincenzo Gentile (via Hymenopterist's Forum), Mostafa Ghafouri Moghaddam (Chulalongkorn University, Thailand), Gary Gibson (Canadian National Collection of Insects, Arachnids and Nematodes, Ottawa, Canada), Alex Gumovsky (Schmalhausen Institute of Zoology, National Academy of Sciences of Ukraine, Ukraine), Christer Hansson (University of Lund, Sweden), John Heraty (University of California Riverside, Riverside, USA), John T. Huber (Canadian National Collection of Insects, Arachnids and Nematodes, Ottawa, Canada), Petr Janšta (Charles University, Prague, Czech Republic), Rajmonana Keloth (ENVIS Centre on Faunal Diversity, Zoological Survey of India, Kolkata, India), Lynn S. Kimsey (University of California, Davis, USA), Abigail Martens (Insect Biodiversity Lab, South Dakota State University,

South Dakota, USA), Mircea-Dan Mitroiu (Alexandru Ioan Cuza University, Iași, Romania), John Noyes (Natural History Museum, London), Jeong Jae You and Christopher Darling (Royal Ontario Museum, Toronto, Canada), Thi Nhi Pham (Institute of Ecology and Biological Resources, Vietnam Academy of Science and Technology, Hanoi, Vietnam), Andrew Polaszek (Natural History Museum, London), A.P. Ranjith (Ashoka Trust for Research in Ecology and the Environment (ATREE), Bangalore, India), Jean-Yves Rasplus (INRAE – Centre de Biologie pour la Gestion des Populations, Montferrier-sur-Lez, France), Alexey Reshchikov (Institute of Eastern-Himalaya Biodiversity Research, Dali University, Yunnan, PR China), Matthias Riedel (Zoologische Staatssammlung München, Germany), Julia Stigenberg (Naturhistoriska Riksmuseet, Stockholm, Sweden) and Lars Vilhelmsen (Natural History Museum of Denmark, København, Denmark).

Max Barclay (Natural History Museum, London), Denis Brothers (University of KwaZulu-Natal, Pietermaritzburg), and George Waldren (Houston, Texas) kindly helped with information on authorships of some taxa.

The following kindly sent us copies of scientific literature that we were unable to access ourselves: Andrew Austin (University of Adelaide, Australia), Gavin R. Broad (Natural History Museum, London), Andrew Davis (J.F. Blumenbach Institute of Zoology and Anthropology, University of Goettingen, Goettingen, Germany), Hassan Ghahari (Islamic Azad University, Tehran, Iran), Norman Johnson (The Ohio State University, Columbus, Ohio, USA), Thi Nhi Pham (Institute of Ecology and Biological Resources, Vietnam Academy of Science and Technology, Hanoi, Vietnam), A.P. Ranjith (Ashoka Trust for Research in Ecology and the Environment (ATREE), Bangalore, India), Matthias Riedel (Zoologische Staatssammlung München, München, Germany), Mark R. Shaw (National Museums of Scotland, Edinburgh, Scotland), Kees van Achterberg (Naturalis Biodiversity Center, Leiden, Netherlands), Cecilia Waichert (Universidade de Brasília, Brasília, Brazil) and Ashleigh Whiffin (National Museums Collection Centre, Edinburgh).

Various specimens were kindly provided by: Worapong Atsawasiramanee (Chulalongkorn University), Yves Basset (Smithsonian Tropical Research Institute, Balboa, Republic of Panama), Kittiphum Chansri (Chulalongkorn University), Weeyawat Jaitrong (Natural History Museum of the National Science Museum, Thailand) and Alexey Reshchikov (Institute of Eastern-Himalaya Biodiversity Research, Dali University, Yunnan, PR China).

Roger Burks (University of California Riverside, Riverside, USA) kindly clarified some aspects of Chalcidoidea classification.

We are grateful to Dr George Beccaloni (Director, the A. R. Wallace Correspondence Project), Dr Adrian Plant and Prof. Mike Sharkey for interesting and thoughtful comments and discussions. Professor Andrew Austin (University of Adelaide), Dr Denis Brothers (University of KwaZulu-Natal, Pietermaritzburg), Dr Istvan Mikó (Department of Entomology, North Carolina State University) and Dr Andrew Polaszek (Natural History Museum, London), each kindly read through chapters of their expertise and offered useful comments, corrections and additions.

Any remaining errors are entirely of our own making.

1

Introduction

Abstract

This chapter discusses what the parasitoid wasps are, how they differ from predators and true parasites and how they may be used in biological control programmes of various types, including integrated pest management (IPM). The early history of parasitoid collecting and taxonomy with special reference to S.E. Asia is described and the importance of taxonomy is discussed. The problems faced by early taxonomists in terms of access to materials and brevity of their descriptions is described.

Insect parasitoids have larvae that feed and develop within or on the bodies of other animals. Each parasitoid larva develops on a single host, although there may be more than one parasitoid larva per host, and they always eventually kill it. Most parasitoids belong to the Hymenoptera; however, in addition to the parasitoid wasps that are the subject of this book, some flies such as Tachinidae and a small number of beetles, moths, lacewings, and even one caddisfly species, *Orthotrichia muscari*, have evolved to be parasitoids (Godfray, 1994; Quicke, 1997).

Historically parasitoid wasps were often referred to simply as parasitic wasps (e.g. Askew, 1971; Quicke, 1997). This refers to the fact that the parasitoid does not kill its host instantly but both remain alive, coexisting for a while. The single injection of venom from the female wasp might actually be lethal in its own right in the long term, but that is not always the case. If a parasitised host's immune system can kill the parasitoid before it has done much damage, again the host may live and ultimately reproduce. On the other hand, most true parasites, especially internal ones but also fleas and lice, generally benefit by keeping their host alive, and in a reasonably healthy state, for as long as possible – the longer it lives, the more reproduction the parasite can accomplish. Obviously, it does not always pan out quite like that, but in principle that is what evolution will favour.

The intimacy of the interaction creates an evolutionary arms race with the host evolving strategies to avoid parasitism (easier to say than 'parasitoidism') in the first place or to kill the parasitoid if that fails; and with the parasitoid being selected for exactly the opposite. Across a million or so species in each category, this evolutionary struggle has led to many amazing adaptations.

The one (or many) to one relationship between parasitoid individuals and those of their hosts means that parasitoid–host systems have different population dynamics than predator–prey relationships. A predator in

© Buntika A. Butcher and Donald L.J. Quicke 2023. *Parasitoid Wasps of South East Asia* (B.A. Butcher and D.L.J. Quicke)
DOI: 10.1079/9781800620605.0001

its lifetime may kill many prey. Consider for example the king of terrestrial predators, the African lion. Individuals live in the wild for about 15 years, for most of which they are nearly full grown. They feed approximately once every three days, consuming some 5–7 kg of meat per day on average, which means that the pride in which they live probably kills a prey animal every day or two – depending on size of pride or size of prey. So each individual lion is responsible for the death of somewhere in the region of 5000 antelope, zebra and buffalo in its lifetime. The lifetime meat consumption of a lion is thus about 35 tonnes. The developing parasitoid wasp will kill at most one host. Mathematical modelling of host–parasitoid interactions is an enormous field of research in its own right.

When Is a Parasitoid Not a Parasitoid?

The definition of parasitoid, when it comes to the Hymenoptera, depends on whose viewpoint you look from. Among all insects the Hymenoptera are almost unique in evolving not only sociality but also, before that, caring for their young by provisioning them with one or more insect food items. Many Sphecidae and solitary Vespidae which construct nests for their offspring provision them with one or more of the preferred, nicely paralysed prey type such as caterpillars, aphids, spiders, etc. In some cases, the prey item is sufficiently large for the complete development of the egg that she will lay on the food; in others, several prey will be needed to feed her larva. In the former the population dynamics is identical to that of a parasitoid; in the latter it is like that of a predator (Weseloh and Hare, 2009). In either case we do not refer to this type of wasp as being a parasitoid, because the female wasp has done much more than just find a host and lay an egg on or in it.

What we and others refer to as a parasitoid is a wasp or fly that locates a host and oviposits on it where the host was initially found. The female does not physically manipulate the host by carrying it to a specially constructed nest cell or to a safer place, or render it immobile by means other than simple envenomation

Sometimes there is a fine dividing line between what is a parasitoid and what is not. For example, one braconid wasp from Australia has been observed to carry the case-bearing beetle host (Quicke and Marshall, 2011). Nearly a century ago, Wheeler noted that some bethylids moved their paralysed hosts to slightly more protected positions before ovipositing on them (Wheeler, 1928) – something that has hardly been followed up. Similarly, at least a few Scoliidae apparently dig a chamber underneath their already subterranean host beetle larva.

For hymenopterologists, there is also a category of 'honorary' parasitoid wasps: wasps that are very similar to and very closely related to proper parasitoid wasps, but which have evolved a different dietary strategy. Many chalcidoids and some braconid wasps, for example, have evolved to be phytophagous, feeding on galled plant tissue, sometimes causing the gall tissue to form themselves, sometimes consuming gall tissue induced by a true gall former – a strategy called inquilinism. A few are known to develop simply as predators of multiple gall inducers within a plant gall (Ranjith et al., 2022)

The Aims of this Book

We hope to provide a basic introduction to the biology, morphology and phylogeny of parasitoid wasps, and obviously point the way to further reading on particular topics. We then provide a basic key to help beginners recognise what is a parasitoid wasp and to identify their specimens to the superfamilies we cover later on. The main part is devoted to treating all those superfamilies and families, and in some cases subfamilies, in each case covering the following topics:

- recognition and identification of subdivisions;
- biological overview;
- use in biological control programmes;
- internal phylogeny;

- recent taxonomic work on S.E. Asian taxa; and
- identification resources.

We provide beginner-level keys to most groups to enable our readers to get started and have a moderately good chance of identifying specimens to family and sometimes subfamily. These are not research-level keys and no one should use the identifications they have made for publication purposes. In writing this book we have tried to identify a very large number of specimens of groups we are not familiar with, using other published research-level keys, and we have made many mistakes. This is normal. The parasitoid Hymenoptera are no exception to the rule that evolution was not directed at making identification easy. There is an enormous amount of convergent evolution – character state syndromes associated with particular ways of life.

In the last chapter (Chapter 23, this volume) we provide a broad overview of collecting methods, specimen sorting and preparation and very basic principles concerning the rearing of live material.

Ecological Role and Importance

Parasitoid insects are of great ecological importance in that they regulate the population densities of their hosts, most of which are phytophagous. Although there are a lot of parasitoid flies, mostly belonging to the family Tachinidae, they are outnumbered manyfold by the parasitoid wasps. It is estimated that worldwide there could easily be more than a million species of Ichneumonoidea alone.

Parasitoid wasps are primarily associated with endopterygote hosts, and although some exopterygotes can also be hosts often in the egg stage, this is mostly when some particular aspect of their biology makes them most suitable for exploitation (Shaw, 1997).

Utility in Pest Management (Biocontrol)

In agroforestry, parasitoid wasps either incidentally or through targeted control programmes contribute to the regulation of various pest insect populations. Biological pest control with parasitoids usually helps the farmer in two ways. The individual hosts attacked are inevitably killed by the developing parasitoid and thus there are fewer pest individuals that can breed. In many cases, the presence of the parasitoid inside a host stunts the latter's growth so that they consume less and so cause less damage (e.g. Syme and Green, 1972; Rohlfs and Mack, 1983; Sato et al., 1986; Schopf and Steinberger, 1996).

It is quite hard to get an overview of progress in biological control, because publications are particularly widely spread over journals and often also government department reports. However, Novita et al. (2023) recently summarised progress in the use of biological control in Indonesia and noted that the trend is upwards.

There are three basic types of biological control:

- conservation of existing natural enemies;
- classical biological control – introduction of non-native, new natural enemies which hopefully establish; and
- augmentative biocontrol — mass rearing and release biocontrol agent, often on a seasonal basis or inundatively.

Conservation of natural enemies

Many pests, native and introduced, may be vulnerable to parasitoids already present in the region. However, crop monocultures and mass spraying with insecticides are generally not conducive to maintaining good populations of natural enemies. Land management actions might be made more favourable for natural enemies by such things as intercropping, providing floral borders as a source of nectar, leaving untreated patches to maintain a reserve parasitoid population, carefully selecting which pesticide and when and how to apply it.

Classical biological control

S.E. Asia has experienced several accidental introductions of potentially serious pests, particularly from the New World. Sometimes, with good fortune, it appears that the local entomofauna can adapt to utilising these as prey/hosts and so stop them becoming pestilential. For example, the Jack Beardsley mealybug, *Pseudococcus jackbeardsleyi* from the neotropics, often a pest of various vegetable, fruit and ornamental crops, was reported from Singapore (1958), Malaysia (1969), Indonesia (1973), Philippines (1975), Thailand (1987), Vietnam (1994) and Cambodia (2010) (Williams, 2004; Muniappan *et al.*, 2009).

The cassava mealybug, *Phenacoccus manihoti*, a major pest of manioc/cassava, was reported from Thailand, Cambodia and Laos in 2009, and from Indonesia in 2010 (Muniappan, *et al.*, 2009; Winotai *et al.*, 2010). In Africa it was controlled successfully by introduction of the encyrtid parasitoid *Anagyrus* (= *Apoanagyrus*; = *Epidinocarsis*) *lopezi* from South America, the pest's native home (Neuenschwander, 2001; Wyckhuys *et al.*, 2021). A stock-rearing colony of the parasitoid *A. lopezi* was imported from the International Institute of Tropical Agriculture (IITA) Benin into Thailand in 2009 and field released in the east, north-east and central parts of the country, but whether this has been successful in controlling the pest is not yet known.

Augmentative biocontrol

Augmentative biocontrol is when large numbers of the control agent are bred for release at times when the 'wild' population is not being as successful at pest control as desired. Both native and introduced control agents can be augmented in principle. The numbers of parasitoids needed for successful augmentation is potentially very large and so the method is most applicable to those that have short developmental periods, and whose hosts (or surrogate hosts) can be produced in large numbers too.

Parasitoids may be reared commercially for release against pests; this is most commonly practised in confined greenhouse situations and greenhouses are not generally needed in warm climates.

In large continental land masses, introduced parasitoids can potentially spread over large areas by themselves (flying or transported by the wind), especially if the crop patches concerned are not too widely separated. This could be an issue with much of S.E. Asia, especially Indonesia, which is a huge collection of islands of various sizes.

Integrative pest management (IPM)

This is essentially when both chemical and biological control methods are combined such that they do not negatively interfere with each other to any significant extent. It requires considerable knowledge about the biological characteristics of all the species involved; for example, the timings of the life cycle of each, their seasonality, their dietary needs, alternative hosts, susceptibilities at various life stages to other control agents, persistence in the environment, etc. When it works it is generally very environmentally sensitive.

One aspect that is very important here is what is the pest species? To untrained eyes a lot of insects look the same – one brown plant hopper can look exceedingly like a dozen others each with different parasitoids and different susceptibilities. This applies not only to the pest, of course, but also to any beneficials. And that means taxonomy, i.e. ways and means of getting things identified.

There are many interesting stories about failed pest management attempts that ultimately turn out to be because of incorrect identification.

The demand for organic produce

A potentially large reason for wanting to increase the role of parasitoids in pest control is to reduce or eliminate the use of chemical pesticides, because real organic products are more exportable to more developed countries as well as being desired by an increasing proportion of local consumers. And there are obvious environmental benefits.

In relation to IPM (above), an important issue is: when does a pest become a real pest? Organic produce carries a premium market value, and probably most shoppers (one hopes) are aware that organic fruit might have a few more blemishes than an inorganic one. Just because a farmer notices some insect-caused damage does not necessarily mean that it is time to open up the tin of neonicotinoids and spray everything.

An important decision in IPM is: what level of pest-caused damage is tolerable? What is the threshold level at which chemical intervention becomes the most profitable strategy?

Importance of Taxonomy

As mentioned above, for any applied work, getting the name of the taxa correct is paramount. The differences in biology between very closely related taxa can be quite major, such as requiring completely different hosts or having very different fecundities. Yet the morphological differences to separate them may be very slight, such that only real experts, with sufficient material for study, can separate them. That is looking on the bright side, because many published identifications are simply wrong, and indeed maybe half of all published host–parasitoid records may be wrong, with either host or parasitoid or both misidentified (Shaw, 1994, 2002). In some cases what was originally believed to be a polyphagous species of parasitoid has turned out to be a complex of numerous far more host-specific species (e.g. Smith *et al.*, 2006). Such errors can lead to completely wrong conclusions about host ranges and their evolution, as well as biological control potential.

Two very well-known, cosmopolitan parasitoid wasps, both important in pest control and both frequently used as metaphoric laboratory rats, the braconid *Habrobracon hebetor* and the chalcidoid *Anisopteromalus calandrae* (Howard), have recently been shown actually to be pairs of morphologically highly similar but biologically quite different taxa (Baur *et al.*, 2014; Kittel and Maeto, 2019).

Misidentifications have sometimes come to light through failed introductions of parasitoids as potential biological control agents. What was thought to be the correct species of parasitoid or host turned out not to be.

History of Parasitoid Wasp Taxonomy in S.E. Asia

Probably the first significant western collector of parasitoid wasps in the region was Alfred Russel Wallace (born 8 January 1823; died 7 November 1913), evolutionary thinker (Wallace, 1855) and co-discoverer of natural selection as the mechanism underlying evolution (Darwin and Wallace, 1858). His epiphany apparently happened whilst suffering from a fever in the village of Dodinga on the island of Halmahera (Beccaloni, 2022). His ideas were stimulated by his great experience in the tropics, starting in South America in the company of another great naturalist, Henry Bates, the later discoverer of now eponymous Batesian mimicry. Wallace, like Charles Darwin, was inspired by the differences he observed between closely related races and species on different but not too distant islands. Wallace was a combination of naturalist, thinker and professional collector, making his living by sending collected specimens of many animal groups back to the United Kingdom for sale. Among the ones that we are aware of were 3069 hymenopterans. His specimens included various Ichneumonoidea, Pompilidae, Scoliidae, Mutillidae, Tiphioidea, Thynnoidea, Chrysididae, Stephanioidea, Platygastroidea, Evaniidae, Chalcidoidea, Cynipoidea, Trigonalyoidea and Diaprioidea, in addition to many ants, bees and social wasps. Van Achterberg and O'Toole (1993) provided a flow chart of what happened to most of Wallace's specimens after they reached his agent, Samuel Stevens, who had premises near the original British Museum in Bloomsbury, London, where the Natural History collections were held until the opening of the new natural history building in South Kensington, London. The majority ended up in the Natural History Museum, London and to a lesser extent in Oxford University Museum, mostly via various private collectors or dealers who had purchased them; however, the British Museum did purchase 7758 specimens directly from S. Stevens and 920 of these were hymenopterans. Sadly, some specimens are lost. A significant proportion of Wallace's material became type specimens, and two of the parasitoid wasps he collected were named after him (*Braunsia wallacei* (Braconidae) and *Spilomicrus wallacei* (Diapriidae)). Wallace wrote up his experience in S.E. Asia in his wonderful book *The Malay Archipelago* (Wallace, 1869).

A less famous but still important early collector was Robert Walter Campbell Shelford (born 1872 in Singapore; died 1912, Margate, England) who was an anthropologist and zoologist, having studied natural history at Cambridge University. He went to Borneo in 1897 to take up the position of curator of the Sarawak Museum and then returned to England in 1905 to be an assistant curator at Oxford University Museum where much of his material now is. He made numerous collecting expeditions which he described in two publications (Shelford, 1900, 1916).

Early taxonomic work on S.E. Asian parasitoid wasps

Not surprisingly, much of the early taxonomic study of parasitoid wasps from the region focused on the larger, more spectacular species such as large braconid and ichneumonid wasps (see Chapter 4). During the last few years of the 19th century and the early part of the 20th, these and various other parasitoid wasp groups were largely reported on by three European workers, Peter Cameron, Gyötö Szépligeti and Günther Enderlein. Cameron (born 1847, died 1 December 1912 in New Mills, Derbyshire) was an amateur British entomologist. Szépligeti (born 21 August 1855 Zirc, died 24 March 1915 in Budapest, Hungary) was a Hungarian but published mostly in the German language, and Enderlein (from Leipzig and Berlin, 7 July 1872, died 11 August 1968 near Hamburg) was a German. Therefore, having a passable knowledge of German is a great help in trying to understand their work – indeed this (and because many type specimens were behind the Berlin Wall) was an important stimulus to one of us to study it. Regarding Peter Cameron, Claude Morley's (1913) obituary briefly summed up his massive output as follows:

> What can we say of his life? Nothing; for it concerns us in no way. What shall we say of his work? Much; for it is entirely ours, and will go down to posterity as probably the most prolific and chaotic output of any individual for many years past.

Enderlein's type specimens are all housed in the Museum and Institute of Zoology, formerly Muzeum i Instytut Zoologii Polskiej Akademii Nauk, Warzsawa [Warsaw], Poland. Szépligeti's are mostly housed in the Hungarian Natural History Museum, Budapest, but with substantial numbers also in Berlin (the Museum für Naturkunde Berlin, formerly called Zoologisches Museum der Humboldt-Universität zu Berlin). Cameron's are more dispersed, but much of his material ended up at the Natural History Museum, London, and Oxford University Museum, whilst some is in Kuching (Malaysia, Sarawak), and Berlin. Although present-day Indonesia was colonised by the Dutch for many years (and academic links still continue) with quite a lot of specimens being located in Amsterdam and Leiden, it was not until the 1970s that there were any major parasitoid wasp researchers in the Netherlands, so there are rather few historic types in those collections, although Cameron borrowed (and returned some) material from the Instituut voor Taxonomische Zoologie, Zoölogisch Museum, Amsterdam (van Achterberg, 1980). The Swedish entomologist Per Abraham Roman (born in Tjos, Westergötland 16 November 1872; died in Stockholm 23 December 1943) was less prolific but also made a significant contribution to the study of S.E. Asian Ichneumonoidea.

Another important (predominantly Hymenoptera) taxonomist active in the first quarter of the 20th century was Jean-Jacques Kieffer (born 1857 in Guinkirchen; died 1925 in Bitche), France who described many species of non-ichneumonoids from the region. Unlike Peter Cameron, with whom he collaborated, he worked on museum material, especially that of the Muséum national d'Histoire naturelle in Paris, rather than having his own collection. His output included works on Proctotrupidae, Platygasteridae, Ceraphronidae, Diapriidae, Scelionidae, Bethylidae, Dryinidae and Embolemidae – and also cecidomyiid flies.

There are big advantages to having type material from many countries largely housed in a small number of large collections in well curated museums, and from where specialists could relatively easily arrange loans. Certainly, in the past this greatly facilitated taxonomic revisions which have largely been carried out by western entomologists. An obvious problem is that it is far harder for taxonomists in poorer S.E. Asian countries to make the expensive research trips to museums in Europe and North America, and this is certainly a huge impediment. But there are also legal issues. Relatively few European and North American entomologists actually make (or made) extensive study visits to museums in other countries; rather they would mostly borrow specimens from museums. This did not used to be a problem as long as the specimens were dead and not CITES listed species. However, we have had an unfortunate incident in Thailand with the postal customs incinerating a parcel of loaned mounted insects because of lack of proper paperwork. Another impediment.

References

Askew, R.R. (1971) *Parasitic Insects*. Heinemann, London, 316 pp.

Baur, H., Kranz-Baltensperger, Y., Cruaud, A., Rasplus, J.-Y., Timokhov, A.V. and Gokhman, V.E. (2014) Morphometric analysis and taxonomic revision of Anisopteromalus Ruschka (Hymenoptera: Chalcidoidea: Pteromalidae) – an integrative approach. *Systematic Entomology* 39, 691–709. doi: 10.1111/syen.12081

Beccaloni, G. (2022) The 'Letter from Ternate': what happened to Wallace's legendary 1858 letter and Darwin's reply to it? *The Linnean* 38, 16–24.

Darwin, C. and Wallace, A.R. (1858) On the tendency of species to form varieties; and on the perpetuation of varieties and species by natural means of selection. *Zoological Journal of the Linnean Society* 3(9), 46–62.

Godfray, H.C.J. (1994) *Parasitoids: Behavioral and Evolutionary Ecology*. Princeton University Press, Princeton, New Jersey.

Kittel, R.N. and Maeto, K. (2019) Revalidation of *Habrobracon brevicornis* stat. rest. (Hymenoptera: Braconidae) based on the CO1, 16S, and 28S gene fragments, *Journal of Economic Entomology* 112, 906–911. doi: 10.1093/jee/toy368

Morley, C. (1913) Peter Cameron. *The Entomologist* 46, 24.

Muniappan, R., Shepard, B.M., Watson, G.W., Carner, G.R., Rauf, A., Sartiami, D., Hidayat, P., Afun, J.C.K., Goergen, G. and Rahman, A.K.M.Z. (2009) New records of invasive insects (Hemiptera: Sternorrhyncha) in southern Asia and West Africa. *Journal of Agricultural and Urban Entomology* 26, 167–174.

Neuenschwander, P. (2001) Biological control of the cassava mealybug in Africa: a review. *Biological Control* 21, 214–229.

Novita, R., Buchori, D., Istiaji, B. and Seminar, A.U. (2023) Mapping biological control research: a systematic review of 20 years of research in Indonesia. In: *IOP Conference Series: Earth and Environmental Science*, Vol. 1133, No. 1, p. 012028. International Conference on Modern and Sustainable Agriculture 2022 02/08/2922 – 03/08/2022. IOP Publishing, Bristol, UK. doi: 10.1088/1755-1315/1133/1/012928

Quicke, D.L.J. (1997) *Parasitic Wasps*. Chapman & Hall, London, 470 pp.

Quicke, D.L.J. and Marshall, S. (2011) Unusual host-carrying behaviour by a parasitoid wasp (Braconidae: Braconinae: *Pycnobraconoides*). *Journal of Hymenoptera Research* 29, 77–79.

Ranjith, A.P., Quicke, D.L.J., Manjusha, K., Butcher, B.A. and Nasser, M. (2022) Hunting parasitoid wasp: first report of mite predation by a 'parasitoid wasp'. *Scientific Reports* 12, 1747.

Rohlfs, W.M. III and Mack, T.P. (1983) Effect of parasitization by *Ophion* 'avidus' Brullé (Hymenoptera: Ichneumonidae) on consumption and utilization of a pinto bean diet by fall armyworm (Lepidoptera: Noctuidae). *Environmental Entomology* 12, 1257–1259.

Sato, Y, Tagawa, J. and Hidaka, T. (1986) Effects of the gregarious parasitoids, *Apanteles ruficrus* and *A. kariyai*, on host growth and development. *Journal of Insect Physiology* 12, 269–274.

Schopf, A. and Steinberger, P. (1996) The influence of endoparasitic wasp *Glyptapanteles liparidis* (Hymenoptera: Braconidae) on the growth, food consumption, and food utilization of its host larva, *Lymantria dispar* (Lepidoptera: Lymantriidae). *European Journal of Entomology* 93, 555–568.

Shaw, M.R. (1994) Parasitoid host ranges. In: Hawkins, B.A. and Sheehan, W. (eds) *Parasitoid Community Ecology*, Oxford University Press, Oxford, pp. 111–144.

Shaw, M.R. (2002) Host ranges of *Aleiodes* species and an evolutionary hypothesis. In: Melika, G. and Thuróczy (eds) *Parasitic Wasps: Evolution, Systematics, Biodiversity and Biological Control*. Agroinform Kiadó, Budapest, pp. 321–327.

Shaw, S.R (1997) Subfamily Euphorinae. In: Wharton, R.A., Marsh, P.M. and Sharkey, M.J. (eds) *Manual of the New World Genera of the Family Braconidae (Hymenoptera)*. Special Publication No. 1 of the International Society of Hymenopterists, Washington, DC, pp. 235–254.

Shelford, R.W.C. (1900) A trip to Mount Penrissen. *Journal of the Straits Branch of the Royal Asiatic Society* 33, 1–26.

Shelford, R.W.C. (1916) *A Naturalist in Borneo. Edited with a biographical introduction by Edward B. Poulton*. T. Fisher Unwin, London, 331 pp.

Smith, M.A., Woodley, N.E., Janzen, D.H., Hallwachs, W. and Hebert, P.D.N. (2006) DNA barcodes reveal cryptic host-specificity within the presumed polyphagous members of a genus of parasitoid flies (Diptera: Tachinidae). *Proceedings of the National Academy of Sciences of the United States of America* 103, 3657–3662. doi: 10.1073/pnas.0511318103

Syme, P.D. and Green, G.W. (1972) The effect of *Orgilus obscurator* (Hymenoptera: Braconidae) on development of the European pine shoot moth (Lepidoptera: Olethreutidae). *The Canadian Entomologist* 104, 523–350.

van Achterberg, C. (1980) The Cameron types of Braconidae in the Netherlands (Hymenoptera, Ichneumonoidea). *Bulletin Zoologisch Museum Universiteit van Amsterdam* 7, 209–214.

van Achterberg, C. and O'Toole, C. (1993) Annotated catalogue of the types of Braconidae (Hymenoptera) in the Oxford University Museum. *Zoologische Verhandelingen, Leiden* 287, 1–48.

Wallace, A.R. (1855) On the Law which has regulated the introduction of new species. *Annals and Magazine of Natural History* 16, 184–196.

Wallace, A.R. (1869) *The Malay Archipelago: The Land of the Orang-utan, and the Bird of Paradise. A Narrative of Travel, with Studies of Man and Nature*. Volumes 1 and 2. Macmillan and Co., London.

Weseloh, R.M. and Hare, J.D. (2009) Predation/predatory insects. In: Resh, V.H. and Cardé, R.T. (eds) *Encyclopedia of Insects*, 2nd Edition. Academic Press, San Diego, California, pp. 837–839.

Wheeler, W.M. (1928) *The Social Insects: Their Origin and Evolution*. Harcourt, Brace and Co., New York, 446 pp.

Winotai, A., Goergen, G., Tamò, M. and Neuenschwander, P. (2010) Cassava mealybug has reached Asia. *Biocontrol News and Information* 31, 10N–11N.

Williams, D.J. (2004) Mealybugs of Southers Asia. The Natural History Museum, Kuala Lumpur, Southdene SDN, 896 pp.

Wyckhuys, K.A.G., Orankanok, W., Ketelaar, J.W., Rauf, A., Goergen, G. and Neuenschwander, P. (2021) Biological control: cornerstone of area-wide integrated pest management for the cassava mealybug in tropical Asia. In: Hendrichs, J., Pereira, R. and Vreysen, J.B. (eds) *Area-Wide Integrated Pest Management: Development and Field Application*. CRC Press, Boca Raton, Florida, pp. 17–32.

2

Biology

Abstract
The major life-history strategies, idiobiont and koinobiont, are described, along with the many biological attributes that they are correlated with them such as, adult lifespan, host feeding and egg production. Various aspects of development biology are described, for example polyembryony, teratocytes, superparasitism and gregarious development. A brief introduction to hymenopteran sex determination mechanisms and their consequences is presented. Sex ratio distortion by *Wolbachia* is briefly explored. The enormously high food conversion efficiency of parasitoid wasps is highlighted. The fig ecosystem is also described briefly.

Although the parasitoid wasp lineage comprises a million or more species, not all of the descendants of their common ancestor are still parasitoids. A few lineages have evolved to be entomophytophagous (e.g. Ranjith *et al.*, 2016), purely predatory within egg masses of other arthropods or inside galls (e.g. Ranjith *et al.*, 2022) or purely phytophagous gall formers (Austin and Dangerfield, 1998) including inside figs (Wiebes, 1979; Kjellberg *et al.*, 2022) or seed predators (Macêdo and Monteiro, 1989; Flores *et al.*, 2006). All of these are usually treated as 'honorary' parasitoid wasps even though, biologically speaking, they are no such thing. However, by far the biggest biological shift occurred in the lineages of 'aculeate' wasps that became the ants, social bees and social wasps (see Chapter 5, this volume).

Some are parasitoids of other parasitoids (hyperparasitoids; = secondary parasitoids), and there are even some hyper-hyperparasitoids.

Major Life History Strategies

Probably the most obvious division of the biology is whether the wasp is an ecto- or an endoparasitoid. This dichotomy is pretty much absolute for early stages of all parasitoid wasps, though in some groups the larvae start off as endoparasitoids but exit the host before fully completing development and continue feeding from the outside (Fig. 2.1).

The challenges facing endo- and ectoparasitoids are very different. The former are immersed in fluid (haemolymph) or wet tissue, and so respiration with an open tracheal system is not possible. They are exposed to host defensive cells (haemocytes) and humoral defensive attack. Ectoparasitoids, on the other hand, are potentially liable to being dislodged or crushed by a mobile host and therefore only rarely feed on mobile host stages (exceptions occur notably among eulophid chalcidoids, and tryphonine and '*Polysphincta* group' Ichneumonidae). In these unusual cases the parasitoids have evolved specialised mechanisms to protect themselves.

Whereas some endoparasitoids pupate within the host's remains (some mummifying the remains; see, for example, Braconidae: Aphidiinae and Rogadinae), many actually exit the host and pupate externally. This is probably related to the fact that the parasitoids will not have consumed the entire soft-tissue contents of the host and once they have pupated they might be vulnerable to putrefaction of the remaining host tissues and liquors.

The Idiobiont/Koinobiont Spectrum

Haeselbarth (1979) suggested another important trait: functional groups differing in their closeness of host associations, whether the parasitoid allows the host to continue to develop after the female has oviposited into it (koinobiont strategy) or further development is stopped (idiobiont strategy). The importance of this distinction was recognised and it was well explained by Askew and Shaw (1986). Table 2.1 summarises most of the important life history characteristics that are associated with the idiobiont/koinobiont spectrum. As with all biology, there are exceptions to everything, but nevertheless this distinction is extremely useful for understanding parasitoid wasp biology. Also, because these strategies are rather well conserved in evolutionary terms, they enable one to make predictions about what biology will be displayed by species yet to be specifically studied.

Idiobionts

Idiobionts prevent further host development after attacking them, therefore parasitoid larvae could have a reliable and immobile source of food when hatching from eggs. They usually attack hosts that are concealed in plant tissues (e.g., leaf rolls, wood) or exposed hosts that possess other kinds of physical protections (e.g. eggs). Females of the former usually have long ovipositors with apico-ventral serrations, that allow them to pierce through these barriers, and are mainly ectoparasitoids. Idiobiont egg-parasitoids and endophagous and have particularly fine and sharp-tipped ovipositors.

For ectoparasitoids, permanent paralysis of their concealed hosts is important so the hosts cannot try to remove them by using their mandibles or scraping them off by crushing them against the substrate. Paralysis of exposed hosts, especially for ectoparasitoids, would make both more conspicuous to

Table 2.1. Main different strategies of idio- and koinobiont parasitoids (modified from Quicke, 2015)

Idibionts	Koinobionts
Ectoparasitoids	Mainly endoparasitoids
Mostly with concealed host	Mostly with exposed hosts
Generalists	Specialists
After oviposition, hosts permanently paralysed	After oviposition, hosts temporarily paralysed
Host development arrested	Host development continues
Rapid larval development	Prolonged larval development, especially during the first instar
Large eggs with bigger yolk	Small eggs with little or no yolk
Females with few mature eggs (lower fecundities)	Females with many mature eggs (higher fecundities)
Synovigenic	Pro-ovigenic
Host feeding more common	Host feeding less common
Long adult lifespan	Short adult lifespan
Attack host stage bigger than parasitoid	Attack host stage smaller than parasitoid
If small hosts, male eggs often laid	Offspring sex is not normally impacted by host size
Sexual size dimorphism is strong	No sexual size dimorphism
Usually diurnal	Diurnal/nocturnal

predators (Quicke, 2015). Idiobionts must develop rather quickly as their hosts cannot gain more resources and their quality declines with time.

Koinobionts

Nearly all koinobionts are endoparasitoid and lay their eggs directly inside the host (mainly exposed hosts but also concealed ones). Koinobionts allow the host to continue its development after attack. The parasitoid larva(e) usually remain in the 1st instar, causing little damage, until the host has reached the right size or state, when they then feed rapidly and eventually kill it. In some species, the larvae continue feeding after chewing their way out of the host's body.

Female venom components are largely responsible for their complex and specific interactions with the host's immune system. Due to the close physiological interactions between host and parasitoid, often koinobionts tend to be rather more host specialists than idiobionts.

Adult lifespan, host feeding and synovigeny

Idiobionts, particularly relatively larger bodied ones (except for egg parasitoids), may have long adult life spans for a combination of reasons. Their hosts tend to be large and mature and this is often associated with them being more dispersed and therefore harder to locate; this in turn means that it takes a while for them to locate a 'sufficient' number. Their larvae have to develop relatively rapidly as their host, once attacked, is only going to decrease in quality over time, and thus they have evolved to produce large yolky (lecithal) eggs so their 1st instar larvae are well-developed. Obviously, there is a physical limit on how many large eggs a female wasp can carry inside her at any one time, and so many have evolved to be synovigenic, which means that they progressively nourish and mature more eggs through their adult lives. Where do the food resources for this come from? In some it comes from fat body reserves, but in many cases, the wasps take in protein-rich nutrients, specifically host haemolymph – something called host feeding (Flanders, 1950; Jervis and Kidd, 1986; Le Ralec, 1995; Quicke, 1997, 2015). Haemolymph is consumed from a wound the wasp makes using either her ovipositor or her mandibles, depending on species.

There are two modes of host feeding: concurrent and destructive. In the former, the host is the same individual that the wasp has or will oviposit on; in the latter she selects a host individual just for feeding from and usually she causes far more damage to it, usually killing it in the process, and so, not surprisingly, seldom actually oviposits on that individual. Indeed, in destructive host feeding, the individual fed upon is often unsuitable to act as a host due to such factors as size (Leius, 1961).

Some larger parasitoids with very long ovipositors whose hosts are deeply concealed (maybe by many centimetres of wood) are clearly unable to host-feed, and yet they are mostly synovigenic idiobionts, and they probably have long adult lifespans. This makes one wonder whether they use other sources of adult food.

Nothing is known at all about the adult lifespan of any S.E. Asian parasitoid species in the wild.

Concealed versus exposed hosts

The evolutionary transition from being an ectoparasitoid of concealed hosts and an endoparasitoid of exposed/concealed hosts was discussed by Quicke *et al.* (2000). There is clear advantage with the host being permanently paralysed, in the case of egg and larval development being external. If not, the host can get rid of them by scraping them off with the mandibles, other substrates or squash them (Quicke, 2015). There is no evolutionary advantage to paralysing the host in the case of the egg laid internally, although many endoparasitoids of exposed hosts do induce temporary host paralysis using a quick sting, thus subduing the host to allow for successful, and sometimes very precise, egg placement.

In the case of exposed hosts, there is a distinct disadvantage if the host is permanently paralysed, not being able to move, feed and grow. It could be vulnerable to attack by a predator. For exposed hosts there are four

Many parasitoids start their development as endoparasitoids, but then egress from the host to pupate. There can be numerous potential evolutionary reasons for this.

- If the innards of a host are not fully consumed when the parasitoid completes development, the host tissue would be expected to putrefy and this will probably make it an unhealthy place to metamorphose.
- The dead host remains might attract scavengers which could consume them and any parasitoid within it.
- The parasitoid pupa may have to enter diapause to survive some period of unsuitable climate, winter in the temperate zone, probably a dry season in the tropics.

Not all parasitoid larvae that egress from a host pupate immediately, but rather they shift from internal feeding to a period of feeding on the host from the outside (Fig. 2.1). This external feeding phase is usually essential for their satisfactory development (e.g. Kuriachan et al., 2011) and may be an adaptation to avoid increasing hostile internal conditions of host immunity or toxicity/ or availability of oxygen, for example.

Parasitoid Food Conversion Efficiency

Parasitoid Hymenoptera have very high food conversion efficiencies (Howell and Fisher, 1977; Harvey et al., 2009). Efficiencies vary according to the type of parasitism, host stage attacked, etc., but values are typically between 50% and 90%. Larvae of some groups of parasitoids spin elaborate, protective and often costly silk cocoons in which they pupate. This is the case in the Ichneumonoidea. Therefore, in such cases, a large part of the food consumed does not lead directly to the production of a larger wasp potentially with more eggs.

Ovoparasitism

One of the major adaptive niches in parasitoid Hymenoptera is egg parasitism, a

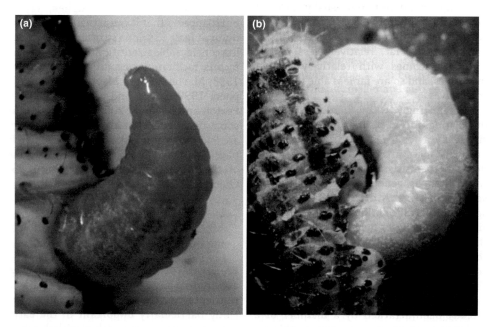

Fig. 2.1. Final instar larva of the extralimital cardiochiline braconid *Toxoneuron nigriceps*, **(a)** exiting from its host caterpillar, and then **(b)** continuing to feed on its tissues externally (external feeding phase). (Source: from Kuriachan et al. (2011) reproduced under terms of Creative Commons licence CC-By 3.0.)

strategy shown by a huge number of species of Platygastroidea and Chalcidoidea. Interestingly, the eggs of slightly fewer than half of insect orders are known to be hosts of hymenopteran parasitoids, which begs the question: why are some insect groups able to escape attack? Fatouros et al. (2020) examined this problem and found that there seemed to be a link with the host order including herbivorous species, although there are some notable exceptions such as the eggs of Odonata. Most of the eggs attacked are laid on plants. It is well known that plants may send out chemical signals to attract predators and parasitoids of their herbivores when the plant suffers herbivore damage. There is also evidence that plants respond to the deposition of herbivore eggs onto them in the same way (see *Tritrophic Interactions*, Chapter 3, this volume).

Sex Determination and Mating System

Hymenoptera are all haplodiploid insects which means that unfertilised eggs develop into males and fertilised ones normally into females. This system is called arrhenotoky. Thus, the trigger for development of unfertilised eggs must be through the act of oviposition (physical egg distortion) or physicochemical signals from the host.

At least two mechanisms appear to be involved in hymenopteran sex determination and the type has important consequences for mating system. It was long known that for some species of parasitoid wasp being maintained in culture, the proportion of males rapidly increases (Crozier, 1971). In these cases, the mechanism was worked out to be what is known as complementary sex determination (CSD). In its simplest form this involves a single gene locus (single locus CSD), but it might involve a small number of loci (Heimpel and de Boer, 2008). There has been a lot of effort put in to understanding the details of the mechanism and mapping the gene loci involved (Matthey-Doret et al., 2019). The alleles at this/these loci are heterozygous in the population and so their fertilised eggs will very frequently be heterozygous for the sex-determination locus allele, and this means the egg will develop into a female wasp. Unfertilised eggs, however, can only be homozygous at a given gene locus and so develop into males. Inbreeding leads to loss of genetic diversity (something very important in zoos breeding endangered species) and so the proportion of fertilised eggs that are homozygous increases. The colonies will finally go extinct when it is no longer possible to have heterozygous females.

The above mechanism cannot be the only one, because there are also many species in which mating between siblings is essentially the rule, for example many gregarious chalcidoids. In these there is neither CSD nor any marked cost of inbreeding, although some outbreeding probably does occur from time to time.

Sex Ratio

Female parasitoid wasps usually have considerable control either over the fertilisation of every individual egg that they lay, or the general sex ratio if they are laying many in quick succession. To produce a son they simply have to withhold sperm from contacting it as it passes the spermatheca. As Godfray (1994) put it, 'Sex ratio is a major preoccupation of parasitoid behavioural ecologists'.

The key thing to understand is that in haplodiploid insects, there is no great restoring evolutionary (Fisherian) selection to produce equal numbers of sons and daughters (see Godfray, 1994, pp. 156–158). In fact, the evolutionary value of a daughter to a haplodiploid parasitoid is greater than that of a son as long as she has a good chance of getting mated.

A consequence of the asymmetry in the value of offspring gender is that for idiobionts, a large daughter who can carry and lay more eggs is of greater value than a small daughter, and since offspring size in solitary parasitoids is determined by host size, a wasp benefits by selectively laying female (fertilised) eggs on larger hosts, and male (unfertilised) eggs on smaller ones. Here the key is that although a small male may not be able to produce as many sperm as a larger individual, the number produced may still be quite sufficient to fertilise all the eggs of even a large female.

Sex Ratio Distortion – Thelytoky

Since Hymenoptera are haplodiploid one does not expect the typical 50:50 sex ratio of most other animals, but instead usually a female-biased one. However, it is not uncommon to encounter populations, and even entire species, in which males are unknown or astonishingly rare, including from broods produced by virgin females.

There is more than one system that can lead to all-female lineages, including intrinsic genetic mechanisms and male-killing bacterial endosymbionts, the best-known of which is a group of bacteria in the genus *Wolbachia*. Apomixis does not always involve any bacterial endosymbionts (Du et al., 2023).

Wolbachia

Wolbachia are obligate endosymbionts which are vertically transmitted from mother to offspring inside the cytoplasm of the egg in many arthropods and nematodes (Gottlieb and Zchori-Fein, 2001; Kaur et al., 2021). *Wolbachia* are known to cause thelytokous (asexual) reproduction in many species of parasitic Hymenoptera. However, whilst being the commonest they are not the only 'male-killing' endosymbionts in parasitic Hymenoptera (Adachi-Hagimori et al., 2008), and there are numerous bacterial endosymbionts that are not male-killers. In addition to sex-ratio distortion, *Wolbachia* infection can induce reproductive incompatibility in many insects, including parasitoid wasps (Quicke, 2015). In this case individuals of a species may contain different *Wolbachia* strains and these act as selfish genetic elements, that make sexual reproduction only possible if the male and female carry the same bacterial strain.

Wolbachia infection is very frequent among parasitoid wasps in the wild (e.g. Betelman et al., 2017), and some species host multiple *Wolbachia* strains (Betelman et al., 2017). Pradana et al. (2019) showed that *Wolbachia* infection is widespread among parasitoids in Indonesia. Many cases are known (principally in species investigated due to their economic importance) where within a species there are both sexual and thelytokous *Wolbachia*-bearing strains (Starý, 1999).

Depending on how long a species or population has been thelytokous as a result of *Wolbachia* infection, it may be possible to cure them. The term 'cure' is something of a misnomer here but it means that either by feeding antibiotic to an infected female or by rearing them at high temperature, the *Wolbachia* can be killed, and the unmated wasp can then lay unfertilised eggs that will develop into functional males. If the thelytoky has been going on for a very long time though, genes necessary for making males may have become lost or non-functional, and so even if the *Wolbachia* are eliminated, no males can result, or if they do, then they may be non-functional. In one chalcidoid that had been asexual for many thousands of generations, males may not be able to produce mature sperm, females may be reluctant to mate, and a major muscle can be absent from the spermatheca reducing its functionality even had there been mature functional sperm inside it (Gottlieb and Zchori-Fein, 2001). Not only *Wolbachia* but also other species of Rickettsiales endosymbionts produce similar effects in parasitoid wasps (Adachi-Hagimori and Miura, 2020). These genetic blocks have been hypothesised to be a possible route to sympatric speciation (Adachi-Hagimori et al., 2011).

Host Manipulation by Venoms and Viruses

Most female parasitic wasps envenomate their host before they lay their eggs in or on it. The envenomation and oviposition may occur in a single injection, as is the case with most endoparasitoids, or the female may first envenomate the host, sometimes causing temporary host paralysis, and only after the host is subdued for a while will she lay her egg(s). Sometimes the effects of the venom on host development may be delayed, as shown by Shaw (1981) with a nice experiment. Quicke and Butcher (2021) surveyed the roles and chemistries of venoms of wasps that do not also produce polydnaviruses (see below).

Venoms of those idiobiont ectoparasitoids of larval hosts (and also those from some pupal endoparasitoids) induce long-term paralysis, and usually this will ultimately be fatal. In addition, both paralysing and non-paralysing venoms usually have effects on various aspects of the host's immune system and in the latter case often on its development.

Polydnaviruses

Although this section only applies to microgastroid Braconidae (see Chapter 9, this volume) and campoplegine and banchine Ichneumonidae (see Chapter 10, this volume), the number of species and economic importance of many of these wasps warrant some discussion. Braconid polydnaviruses are called bracoviruses, ichneumonoid ones, ichnoviruses. Most insects, including parasitoid wasps, experience infections by members of a range of viral families. Sometimes these cause major pathologies and sometimes there seems to be a very large degree of tolerance. During the evolution of the above-mentioned three groups, the genomes (or part thereof) of one of their regular viruses became incorporated into the genome of the wasp. This happens with virus and retroelement genomes often in all organisms, but in these particular instances the endogenised viral genomes evolved to become expressed in the female wasp reproductive tract – specifically in a region called the calyx posterior of each lateral oviduct. The early stages of this evolutionary viral capture and utilisation have not been observed but are short, the wasp produces millions of 'virus' particles in the calyx gland, each particle containing a number of circular DNA molecules which are also replicated from the wasp's genome, hence the family name Polydnaviridae. The 'virus' particles are injected into the host at the time of oviposition in very large numbers. We put the word 'virus' in inverted commas because these polydnaviruses do not replicate at all – they are evolutionary dead ends. But they do infect various specific types of host cell where their enclosed DNA is expressed. There is also usually an important interaction between the wasp's true venom from its venom glands and the polydnaviruses.

The Fig Ecosystem

Fig trees are all classified in the genus *Ficus*, family Moraceae, with the edible fig, *F. carica*, as type species. With more than 800 species occurring throughout the tropics and subtropics and because they provide fruit as a major food source for a whole range of animals, they are considered keystone species. Fig wasps in the family Agaonidae *sensu stricto* (Chalcidoidea) (see Chapter 22, this volume) are the sole pollinators of fig trees (Cook and Rasplus, 2003; Herre *et al.*, 2008). All fig wasps are bound to figs as larvae, and their specialised diets are restricted to fig embryos (Bouček, 1988). Figs and their pollinator fig wasps are almost all species-specific, and this system has provided a good model for the study of co-evolution, co-speciation and insect–plant interactions. Flowers of fig trees are unique because they are completely concealed within the fig, an enclosed inflorescence, with the hundreds of tiny florets lining the inside of a central cavity, called a syconium. Fig wasps consequently show many morphological and behavioural adaptations for this microcosm. In general they display extreme host specificity, and life cycles that are tightly synchronised with fig phenology (Wiebes, 1979). Sexual dimorphism is extreme which reflects the different activities of males and females. Female wasps or foundresses have functional wings and eyes and are responsible for dispersal and colonising the new hosts (Weiblen, 2002). These females are also characterised by modifications of the head which is especially flattened with mandibular appendages with rows of teeth (see Fig. 22.6) that push against the inflorescence bracts lining the fig opening (or ostiole) and antennae in response to the shape of the inflorescence opening (van Noort and Compton, 1996) and by the evolution of pollen transport mechanisms in most (Ramirez, 1978). Their antennal scapes fit into a deep groove on the dorsal surface of the head, and the third segment bears a spine that serves as a hook for prying at the outer bracts and also as a point of detachment for the distal segments on contact with the inner bracts (Galil and Neeman, 1977).

When figs are ready to be pollinated, the foundress enters through the terminal narrow ostiole and claws her way into the lumen, then she spreads pollen onto the flower styles whilst ovipositing. The pollination activity of the female wasp is essential for the fig to continue development rather than being aborted by the tree. Once the female has oviposited, the fig wasp larvae develop within a single ovary which forms soft, nutritious gall tissue around the developing wasp and which the wasp larva consumes. Therefore, for mutualism between the fig tree and the fig wasp, the wasp must pollinate and also leave some flowers ungalled for seed production (Dunn, 2020).

During the oviposition, the foundress inserts her ovipositor down the style of an individual flower and deposits an egg into the ovary (Jansen-Gonzalez et al., 2012). She will also inject chemicals directly into an ovary, inducing the ovary to form a gall (Martinson et al., 2014).

Adaptive character matching has been observed between figs and fig wasps (Weiblen, 2004), with a strong correlation between ovipositor length in wasps and style length in figs (Chen et al., 2021). The foundresses are short-lived and not feeding; evolution acts to favour ones with a high ability to locate the specific fig tree using chemical cues for a long distance and olfactory and visual cues for a short distance.

Interestingly, a few non-pollinating wasps belonging to the Pteromalidae are also called true fig wasps, because their presence prevents the fig from being aborted and instead grow to maturity. However, these do not perform a pollination role.

The fig microcosm also supports a wide diversity of other parasitoid wasps, some as inquilines within galled ovaries, some as parasitoids, and probably some as predators. Most of these belong to the Chalcidoidea but a few Braconidae (Braconinae) are also frequently reared from figs, but no details of their biology are known.

References

Adachi-Hagimori, T. and Miura, K. (2020) Limited mating ability of a wasp strain with *Rickettsia*-induced thelytoky. *Annals of the Entomological Society of America* 113, 355–358.

Adachi-Hagimori, T., Miura, K. and Stouthamer, R. (2008) A new cytogenetic mechanism for bacterial endosymbiont-induced parthenogenesis in Hymenoptera. *Proceedings of the Royal Society of London, Series B, Biological Sciences* 275, 2667–2673.

Adachi-Hagimori, T., Miura, K. and Abe, Y. (2011) Gene flow between sexual and asexual strains of parasitic wasps: a possible case of sympatric speciation caused by a parthenogenesis-inducing bacterium. *Journal of Evolutionary Biology* 24, 1254–1262.

Askew, R.R. and Shaw, M.R. (1986) Parasitoid communities: their size, structure and development. In: Waage, J. [K.] and Greathead, D. [J.] (eds) *Insect Parasitoids*. London, Academic Press, pp. 225–264.

Austin, A.D. and Dangerfield, P.C. (1998) Biology of *Mesostoa kerri* Austin and Wharton (Insecta: Hymenoptera: Braconidae: Mesostoinae), an endemic Australian wasp that causes stem galls on *Banksia marginata* Cav. *Australian Journal of Botany* 46, 559–569.

Betelman, K., Caspi-Fluger, A., Shamir, M. and Chiel, E. (2017) Identification and characterization of bacterial symbionts in three species of filth fly parasitoids. *FEMS Microbiology Ecology* 93(9), p.fix107.

Boivin, G. and van Baaren, J. (2000) The role of larval aggression and mobility in the transition between solitary and gregarious development in parasitoid wasps. *Ecology Letters* 3, 469–474.

Boucek, Z. (1988) Australasian Chalcidoidea (Hymenoptera). CAB International, Wallingford, UK, 832 pp.

Chen, L., Segar, S.T., Chantarasuwan, B., Wong, D.M., Wang, R., Chen, X. and Yu, H. (2021) Adaptation of fig wasps (Agaondinae) to their host revealed by large-scale transcriptomic data. *Insects* 12, 815.

Cook, J.M. and Rasplus, J.Y. (2003) Mutualists with attitude: coevolving fig wasps and figs. *Trends in Ecology and Evolution* 18, 241–248.

Crozier, R.H. (1971) Heterozygosity and sex determination in haplo-diploidy. *The American Naturalist* 105, 399–412.

Du, S., Ye, F., Xu, S., Liang, Y., Wan, F., Guo, J. and Liu, W. (2023) Apomixis for no bacteria- induced thelytoky in *Diglyphus wani* (Hymenoptera: Eulophidae). *Frontiers in Genetics* 13, 1061100.

Dunn, D.W. (2020) Stability in fig tree–fig wasp mutualisms: how to be a cooperative fig wasp. *Biological Journal of the Linnean Society* 130, 1–17.

Fatouros, N.E., Cusumano, A., Bin, F., Polaszek, A. and van Lenteren, J.C. (2020) How to escape from insect egg parasitoids: a review of potential factors explaining parasitoid absence across the Insecta. *Proceedings of the Royal Society, Series B* 287, 20200344.

Flanders, S.E. (1950) Regulation of ovulation and egg disposal in the parasitic Hymenoptera. *Canadian Entomologist* 82, 134–140.

Flores, S., Nassar, J. M. and Quicke, D.L.J. (2006) Reproductive phenology and pre-dispersal seed predation in *Protium tovarense* (Burseraceae), with description of the first known phytophagous '*Bracon*' species (Hymenoptera: Braconidae: Braconinae). *Journal of Natural History* 39, 3663–3685.

Galil, J. and Neeman, G. (1977) Pollen transfer and pollination in the common fig (*Ficus carica* L.). *New Phytologist* 79, 163–171.

Godfray, H.C.J. (1994) *Parasitoids: Behavioral and Evolutionary Ecology*. Princeton University Press, Princeton, New Jersey.

Gottlieb, Y. and Zchori-Fein, E. (2001) Irreversible thelytokous reproduction in *Muscidifurax uniraptor*. *Entomologia Experimentalis et Applicata* 100, 271–278.

Haeselbarth, E. (1979) Zur Parasitierung der Puppen von Forleule (*Panolis flammea* [Schiff.]), Kiefernspanner (*Bupalus piniarius* [L.]) und Heidelbeerspanner (*Boarmia bistortana* [Goeze]) in bayerischen Keifernwäldern. *Zeitschrift für Angewandte Entomologie* 87, 186–202.

Harvey, J.A., Wagenaar, R. and Bezemer, T.M. (2009). Interactions to the fifth trophic level: secondary and tertiary parasitoid wasps show extraordinary efficiency in utilizing host resources. *Journal of Animal Ecology* 78, 686–692.

Heimpel, G.E., de Boer, J.G. (2008) Sex determination in the Hymenoptera. *Annual Review of Entomology* 53, 209–230.

Herre, E.A., Jandér, K.C. and Machado, C.A. (2008) Evolutionary ecology of figs and their associates: recent progress and outstanding puzzles. *Annual Review of Ecology, Evolution, and Systematics* 39, 439–458.

Howell, J. and Fisher, R.C. (1977) Food conversion efficiency of a parasitic wasp, *Nemeritis canescens*. *Ecological Entomology* 2, 143–151.

Jansen-González, S., Teixeira, S.D. and Pereira, R.A.S. (2012) Mutualism from the inside: coordinated development of plant and insect in an active pollinating fig wasp. *Arthropod–Plant Interactions* 6, 601–609.

Jervis, M.A. and Kidd, N.A.C. (1986) Host-feeding strategies in Hymenopteran parasitoids. *Biological Reviews* 61, 395–434.

Kaur, R., Shropshire, J.D., Cross, K.L., Leigh, B., Mansueto, A.J., Stewart, V., Bordenstein, S.R. and Bordenstein, S.R. (2021) Living in the endosymbiotic world of *Wolbachia*: a centennial review. *Cell Host & Microbe* 29, 879–893.

Kjellberg, F., van Noort, S. and Rasplus, J.Y. (2022) Fig wasps and pollination. In: Sarkhosh, A., Yavari, A. and Ferguson, L. (eds) *The Fig. Botany, Production and Uses*. CAB International, Wallingford, UK, pp. 231–254.

Kuriachan, I., Henderson, R., Laca, R. and Vinson, S.B. (2011) Post-egression host tissue feeding is another strategy of host regulation by the koinobiont wasp, *Toxoneuron nigriceps*. *Journal of Insect Science* 11(1), Article 3 (11 pp.). doi: 10.1673/031.011.0103

Le Ralec, A. (1995) Egg contents in relation to host-feeding in some parasitic Hymenoptera. *Entomophaga* 40, 87–93.

Leius, K. (1961) Influence of food on fecundity and longevity of adults of *Itoplectis conquisitor* (Say) (Hymenoptera: Ichneumonidae). *Canadian Entomologist* 93, 771–780.

Macêdo, M.V. de and Monteiro, R.F. (1989) Seed predation by a braconid wasp, *Allorhogas* sp. (Hymenoptera). *Journal of the New York Entomological Society* 97, 358–362.

Martinson, E.O., Jandér, K.C., Peng, Y.Q., Chen, H.H., Machado, C.A., Arnold, A.E. and Herre, E.A. (2014) Relative investment in egg load and poison sac in fig wasps: implications for physiological mechanisms underlying seed and wasp production in figs. *Acta Oecologica* 57, 58–66.

Matthey-Doret, M., van der Kooi, C.J., Jeffries, D.L., Bast, J., Dennis, A.B., Vorburger, C. and Schwander, T. (2019) Mapping of multiple complementary sex determination loci in a parasitoid wasp. *Genome Biology and Evolution* 11, 2954–2962.

Mayhew, P.J. (1998) The evolution of gregariousness in parasitoid wasps. *Proceedings of the Royal Society, London, Series B* 265, 383–389.

Mayhew, P.J. and van Alphen, J.J.M. (1999) Gregarious development in alysiine parasitoids evolved through a reduction in larval aggression. *Animal Behavior* 58, 131–141.

Montoya, P., Pérez-Lachaud, G. and Liedo, P. (2012) Superparasitism in the fruit fly parasitoid *Diachasmimorpha longicaudata* (Hymenoptera: Braconidae) and the implications for mass rearing and augmentative release. *Insects* 3, 900–911.

Pexton, J.J. and Mayhew, P.J. (2001) Immobility: the key to family harmony? *TRENDS in Ecology & Evolution* 16, 7–9.

Pexton, J.J. and Mayhew, P.J. (2002) Siblicide and life-history evolution in parasitoids. *Behavioural Ecology* 13, 690–695.

Pexton, J.J. and Mayhew, P.J. (2004) Competitive interactions between parasitoid larvae and the evolution of gregarious development. *Oecologia* 141, 179–190.

Pradana, M.G., Giyanto, G., Furukawa, S., Nakamura, S. and Buchori, D. (2019) Detection of *Wolbachia* endosymbiont in several agriculturally important insect parasitoids in Bogor, Indonesia. *AGRIVITA, Journal of Agricultural Science* 41, 364–371.

Quicke, D.L.J. (1997) *Parasitic Wasps*. Chapman & Hall, London, 470 pp.

Quicke, D.L.J. (2005) Biology and immature stages of *Panteles schnetzeanus* (Hymenoptera: Ichneumonidae) a parasitoid of *Lampronia fuscatella* (Lepidoptera: Incurvariidae). *Journal of Natural History* 39, 431–443.

Quicke, D.L.J. (2015) *The Braconid and Ichneumonid Parasitic Wasps: Biology, Systematics, Evolution and Ecology*. Wiley Blackwell, Oxford, 688 pp.

Quicke, D.L.J. and Butcher, B.A. (2021) Review of venoms of non-polydnavirus carrying ichneumonoid wasps. *Biology (MDPI)* 10, 50. doi: 10.3390/biology10010050

Quicke, D.L.J., LeRalec, A. and Vilhelmsen, L. (2000) Ovipositor structure and function in the parasitic Hymenoptera with an exploration of new hypotheses. *Rendiconti* 47, 197–239.

Ramirez, W.B. (1978) Evolution of mechanisms to carry pollen in Agaonidae (Hymenoptera Chalcidoidea). *Tijdschrift voor Entomologie* 121, 279–293

Ranjith A.P., Quicke, D.L.J., Saleem, U.K.A., Butcher, B.A., Zaldivar-Riverón, A. and Nasser, M. (2016) Entomophytophagy in an Indian braconid 'parasitoid' wasp (Hymenoptera): specialized larval morphology, biology and description of a new species. *PLoS ONE*, 11, e0156997.

Ranjith, A.P., Quicke, D.L.J., Manjusha, K., Butcher, B.A. and Nasser, M. (2022) Hunting parasitoid wasp: first report of mite predation by a 'parasitoid wasp'. *Scientific Reports* 12, 1747.

Rosenheim, J.A. (1993) Single-sex broods and the evolution of nonsiblicidal parasitoid wasps. *The American Naturalist* 14(1), 90–104.

Shaw, M.R. (1981) Delayed inhibition of host development by the non paralysing venoms of parasitic wasps. *Journal of Invertebrate Pathology* 37, 215–21.

Starý, P. (1999) Biology and distribution of microbe-associated thelytokous populations of aphid parasitoids (Hym., Braconidae, Aphidiinae). *Journal of Applied Entomology* 123, 231–236.

van Noort, S. and Compton, S.G. (1996) Convergent evolution of agaonine and sycoecine (Agaonidae, Chalcidoidea) head shape in response to the constraints of host fig morphology. *Journal of Biogeography* 23, 415–424.

Vinson, S.B. (1976) Host selection by insect parasitoids *Annual Review of Entomology* 21, 109–133.

Vinson, S.B. and Hegazi, E.M. (1998) A possible mechanism for the physiological suppression of conspecific eggs and larvae following superparasitism by solitary endoparasitoids. *Journal of Insect Physiology* 44, 703–712.

Weiblen, G.D. (2002) How to be a fig wasp. *Annual Review of Entomology* 47, 299–330.

Weiblen, G.D. (2004) Correlated Evolution in Fig Pollination. *Systematic Biology* 53, 128–139.

Wiebes, J.T. (1979) Co-evolution of figs and their insect pollinators. *Annual Review of Ecology and Systematics* 10, 1–12.

Williams, D.J. (1988) The distribution of the neotropical mealybug *Pseudococcus elisae* Borchsenius in the Pacific region and Southern Asia (Hem.-Hom., Pseudococcidae). *Entomologist's Monthly Magazine* 124, 123–124.

3

Behaviour

Abstract
The two most important behaviours that a female parasitoid wasp needs to accomplish are to have sex and to find hosts. Both involve a small insect finding another small insect in an incredibly large world. Some of the most important features that have evolved to achieve those goals are explored. These include innate and learning host cues, associative learning, the role of volatiles, and tritrophic interactions. The amazing techniques of vibration sounding, a form of echolocation evolved independently by several groups of parasitoid wasps, is described.

In this chapter we discuss two very important aspects of parasitoid behaviour. The first involves a generally tiny insect first finding another tiny insect in a very large world, i.e., finding a mate, and what then proceeds. The second also starts with a generally tiny insect finding another tiny insect in a very large world: it is locating, assessing and accessing a host.

Sex, Courtship and Mating

If we ignore for the moment that some species are thelytokous, that is they have female offspring without mating (see Chapter 2, this volume), then females of the rest of the species, the great majority, have to mate before they start finding hosts. There is a slight caveat in that while females will attempt to find mates, their unfertilised eggs can still produce male offspring, and in some cases if they have failed to mate, they may just lay unfertilised eggs and the resulting sons may yet pass on the female's genes.

What happens after a female ecloses will then depend very much on the species. Factors involved include its sex determination mechanism, whether the host species may be multigenerational on the same plant or in the same substrate, and so on.

For species with complementary sex determination (CSD) (see *Sex Determination and Mating System*, Chapter 2, this volume), especially single locus CSD, outbreeding is very important for fitness so the females will usually disperse away from their emergence site to reduce the risk of mating with close relatives. In the laboratory with far higher chance of inbreeding, allelic diversity decreases and this means that some fertilised eggs will be homozygous for the CSD locus and therefore they develop into diploid males if they make it through to adulthood. Diploid males have zero or greatly decreased genetic fitness. For such taxa, the need to avoid mating between siblings is high. Since unmated females can still produce male offspring, their fitness is not absolutely zero, but fitness is reduced because their sons will on average have a lower probability of locating a virgin female to mate with.

Most female parasitoid wasps release volatile pheromones to initiate courtship which will last only a short time (Quicke, 1997; McClure et al., 2007). It may involve wing fanning (Bredlau and Kester, 2019), male–male struggles over access to females (Eggleton, 1991), or the transfer of pheromones from the antennae of the male to those of the female (Steiner et al., 2010). For parasitoids of wood-boring insects such as Pimplinae, Rhyssinae and Xoridinae, several males are waiting to be mated with a single female (Shaw et al., 2021; Chansri et al. (in press)), however, only a single male has ever been observed to achieve copulation at any one time, even though females are sometimes receptive to repeated matings with the same or a different male for a short time afterwards (Quicke, 2015; M.R. Shaw, observations). This behaviour is also reported by Chansri et al. (2023), the first record of *Cyrtorhyssa moellerii* males emerging before females and aggregating around where the females would emerge. Some males showed aggressive guarding behaviour for their marked location (by tergal stroking). Larger males inserted their metasomas into the female's chewed tunnel before completely emerging, while the smaller males did not show this behaviour but waited nearby and tried to mate with the females when they exited the tree (Chansri et al., 2023).

Mating successfully is crucial for any sexual species and in many species the olfactory system is a key factor responsible for this behaviour. Recently molecular techniques have advanced to the point where olfactory genes can be studied without resorting to sticking electrodes in the wasp to record responses to volatiles wafted over their antennae. Olfactory receptor co-receptor (Orco) genes are the most conserved olfactory receptor genes, and form an essential part of the pathway, transducing olfactory chemical detection to nervous impulses through forming a ligand-gated ion channel complex with conventional ligand-binding odorant receptors. Zhang et al. (2023) studied Orco genes of four *Drosphila* parasitoids belonging to the genus *Leptopilina* (Cynipoidea) and the results showed that these genes are highly conserved with a typical antennae-biased tissue expression pattern. Orco-deficient male parasitoid wasps lost their ability to mate, while female Orco-deficient parasitoid wasps could mate successfully, but with defective host-searching performance and reduced oviposition rate.

In some species, when a male parasitoid detects a female close by, he will perform wing-fanning behaviour to waft female pheromone over him (Quicke, 2015). In some species this wing-fanning is to do with the male signalling the female via vibration (Villagra et al., 2011). There are some variations in whether the wasps are ready to mate immediately after emerging or require a pre-mating time (a day or so, or 1 h in *Habrobracon hebetor*) (Quicke, 2015). In some species female pheromones play a crucial part in mating behaviour with both short- and long-distance pheromones (McClure et al., 2007) modulating attraction and courtship behaviour (McNeil and Brodeur, 1995; Marchand and McNeil, 2000), such as the aphid parasitoid, *Aphidius ervi*. McClure et al. (2007) studied courtship behaviour and sex pheromones of this *braconid* and found that short-distance pheromones worked effectively in the morning, therefore successful mating was significantly higher in the morning than in the afternoon, probably due to cooler temperature and more humidity.

In some parasitoid species, antennae play an important role during courtship. *Trichopria drosophilae* (Diapriidae) is a pupal parasitoid of the common fruit fly *Drosophila melanogaster* (Drosophilidae). During courtship, the male's fourth flagellomeres come into contact with the two apical female flagellomeres, then secrete sex pheromone and spread onto the female's receptor. Therefore, antennal contact between male and female is essential (Romani et al., 2008). The male's antennal gland on the fourth flagellomere is crucial for mating behaviour (Sacchetti et al., 1999). Male antennal glands are a common character in several superfamilies of Hymenoptera and in a few other insect orders (Belcari and Kozánek, 2006).

Dispersal and Phoretic Copulation

Mating position can be a male on top of a female facing the same direction; this position is commonly found in parasitoids. Other positions can be found, such as male and female joined end-to-end during the mating and facing in opposite directions or both sexes standing side-by-side or even copulating in the air (Camarao and Morallo-Rejesus, 2003). Several groups with flightless females, such as Mutillidae and Tiphiidae, rely on phoretic copulation, that is, the much larger male having located a mate and started copulating then literally flies away with her, and so is responsible for dispersing the female to new sites where hosts may be found.

Host Location and Assessment

Competition among parasitoids involves a mixture of both extrinsic (among free-living adults) and intrinsic (among developing immatures) competition (Ode et al., 2022). Parasitoids, when compared with predators, often display higher levels of intraspecific and interspecific competition, because many species are specific to their hosts while predators are more generalist. Moreover, suitable hosts for parasitoids are distributed in the natural habitats, leading to strong pressure on the adult female wasps to locate the hosts (extrinsic). In addition, the total mass of adult parasitoid offspring is less than that of the host, therefore the parasitoid larvae developing inside the same hosts must compete for resources (intrinsic) (Ode et al., 2022).

Parasitoids and their hosts have been in an evolutionary arms race for a long period of time, starting from immune responses to a variety of behavioural strategies (Hoedjes, et al. 2011; Dicke et al., 2020). When the herbivorous insect larvae (caterpillars) start feeding, they have already given up their hiding places, because feeding would trigger the release of herbivore-induced plant volatiles (HIPVs), which are used by the parasitoid wasps to locate their hosts (Bernays, 1997). The composition of HIPVs depends on both biotic and abiotic factors (Haverkamp and Smid, 2020) which poses a challenge to inexperienced parasitoid wasps to differentiate plants infested by hosts from plants infested by non-hosts. Once a parasitoid has landed on an infested leaf, it will start searching for cues of potential suitable hosts (i.e. caterpillar frass or silk) by drumming its antennae on the leaf surface and probing with the ovipositor. If the parasitoid finds host-derived cues, the wasp will be aroused, leading to increased sensitivity of the sensory systems (Bleeker et al., 2006; Turlings et al., 1993) and intense searching behaviour. These cues are detected by gustatory neurons on the antennae and the ovipositor which associate with HIPVs and will last for 24 h. However, if the wasp does oviposit, this experience will last longer (Takasu and Lewis, 2003).

For hosts, even very slight movements nearby, such as a parasitoid flying or walking close to them, will cause vibrations that can be detected by specific mechanosensory hairs on their body (Taylor and Yack, 2019).

Eschbach et al. (2011) studied host-searching behaviour of the parasitoid *Leptopilina* and its host *Drosophila*; they found that hosts can learn to avoid odours associated with specific sound frequencies. In general, when a caterpillar encounters a parasitoid, it will try to defend itself by rapid and powerful head-strikes and oral secretion, resulting in repelling or even killing of the parasitoid (Nofemela, 2017; Potting et al., 1999). If the parasitoid wasp overcomes the host's defence behaviour, she inserts her ovipositor into the host and quickly oviposits her eggs.

When the hosts are scarce, parasitoids express aggressive behaviour solely to consume the host by chasing other females away or sometimes physical combat. Brood guarding and fighting behaviours are normally found in most idiobiont parasitoids, especially in Bethylidae, Mymaridae and Scelionidae. Females guard the parasitised hosts until their offspring have hatched; sometimes the mothers will even attack and kill other females that attempt to superparasitise hosts

(Hardy and Blackburn, 1991; Batchelor et al., 2005; Guerra-Grenier et al., 2020).

Vinson (1976) reviewed most of the essential points about the challenges faced by a newly emerged female parasitoid. Obviously, finding a mate is a high priority, but after that she needs to locate a host. It is generally accepted that the process involves a hierarchical series of stages. A typical host will be in a given habitat, will be associated with a particular range of plant species and will be associated with a particular part of the plant (root, leaf, stem, flower, fruit, etc.). Exactly how many stages are involved depends on the parasitoid, but generally four are recognised:

- habitat location (assuming not already in it);
- host location within habitat;
- host acceptance (as suitable species); and
- host suitable state (host suitability assessment).

The last of these may involve decisions about host size, health and previous possible parasitisation. For example, for obligate hyperparasitoids such as mesochorine ichneumonids, finding a caterpillar of a suitable species is not enough: the wasp must then assess whether that caterpillar contains the larva of a suitable species of primary parasitoid, and then, during oviposition, the mesochorine must place her egg within the developing parasitoid within the host – quite a task, but mesochorines are exceedingly successful at it and abundant in S.E. Asia. Many idiobionts preferentially lay unfertilised male eggs on lower-quality hosts because body size is less important for fitness in male wasps. Chemical cues used at a distance have to be volatile compounds, but once physical contact is made, sense organs on the female wasp's ovipositor, feet, antennae, etc., may detect non-volatile, host-associated compounds such as proteins (Quicke, 2015, pp. 130–135).

Tritrophic Interactions

A fascinating aspect of parasitoid–host–plant interactions is that many plants have evolved to send out chemical signals (kairomones) when attacked by a herbivore that attract parasitoids – sometimes even herbivore-specific ones. Experiments often show that parasitoids respond to odours given off by plants damaged by herbivores such as caterpillars and mealy bugs, but not to undamaged plants or to the host herbivore in isolation (Nadel and van Alphen, 1987; Turlings et al., 1990; Turlings and Tumlinson, 1991, 1992; Godfray, 1994).

It is now widely known that the host plants of host herbivores are often far from passive. Whilst parasitoids may have evolved to smell host-damaged plants as a cue for host location, the plants themselves may produce and emit volatiles that specifically attract the parasitoid and predators of the given herbivore that is feeding upon them (Turlings and Wäckers, 2004).

Learning and Associative Learning

When a female parasitoid emerges as an adult she may experience many cues that will help her to go on to find her own hosts. These are usually chemical cues and she will learn these. For example, there may be lingering odours of a damaged host plant, of slowly decaying wood, of host frass, and so on. One or more of these may be of great potential use in her subsequent foraging for hosts.

In many cases, however, the site of emergence of a female parasitoid might not contain many remaining cues at all and the wasp at first has to rely on innate preferences. Nevertheless, parasitoids generally emerge at approximately the same time of year that their hosts will be available, and so even the general smell of the local environment will generally be helpful. Having found a suitable host, the associated cues may help refine her search strategy (Godfray, 1994, pp. 43–48). The new cues learned at the time of successfully locating a host may be chemical or visual and possibly tactile

Locating Xylophagous and Other Concealed Hosts

Many parasitoids are unable to make direct contact with a host except when their ovipositor touches it, because the host is concealed, and therefore they have evolved to use other types of host-location cues.

Many parasitoids attack leaf-mining hosts of various insect orders. The mines themselves undoubtedly provide a mix of chemical and visual cues. Parasitoids of such hosts rely a lot on vibrations created by the activity of the host larva within the leaf mine to home in on it, although obtaining definitive experimental evidence of the actual mechanisms involved is technically challenging (Meyhöfer and Casas, 1999). However, evolution will have acted upon the host therein to minimise the chance of the parasitoid successfully attacking it. They can usually take evasive action; indeed, hosts that have already pupated are still capable of detecting vibrations caused by the parasitoid wasp walking around over the leaf surface (Bacher et al., 1997).

What signals might beetle larvae living deep in wood give off to allow a parasitoid to locate them? Locating a deeply concealed host is never going to be easy but there are many species of wasps that achieve this – see for example the long-ovipositored braconids and ichneumonids illustrated here (see Figs 9.8 g, 10.46 b). The hosts of such wasps are far too deeply concealed in a probably rather heterogeneous wood matrix for vibrational sounding (see below) to be effective. Not surprisingly, there is some evidence that such wasps really cannot locate their hosts precisely. Hocking (1968) used X-rays to determine where the ovipositor of the well-known temperate rhyssine wasp *Rhyssa persuasoria* (an idiobiont parasitoid of siricid wood wasp larvae) actually went in the wood, and found that many oviposition attempts missed host borings. Since it takes such wasps at least an hour to 'drill' into the substrate to an appropriate depth, it makes one wonder how long their adult lifespans must be. No such data are available for any tropical (let alone S.E. Asian) rhyssines or similar.

Vibrational Sounding, a Form of Echolocation

Parasitoids of deeply concealed hosts that cannot be seen or touched obviously have additional problems to overcome. For example, for a parasitoid of a wood-boring beetle larva, the host (when it is at the right stage of development) might be several centimetres below the wood's surface and perhaps tens of centimetres away from the site where the mother beetle laid her egg. How can the wasp locate it? Chemical signals are unlikely to be accurate enough, so that leaves two possibilities: listening for a host's movement or feeding noises (neither being applicable to host pupal stages, which are also attacked) or using echolocation. It is now known that a number of parasitoid wasps have evolved a version of echolocation that solves this problem. Technically, because the vibrations are borne through the solid substrate rather than through the air, it is called vibrational sounding.

The wasp generates vibrations in the substrate using highly modified terminal flagellomeres called antennal hammers. In many ichneumonids that have evolved this method of host location, the antennal hammers appear to have evolved from sensilla which lost their sensory function and became solid cuticular knobs and then often fused together (Fig. 3.1) (Laurenne et al., 2009; Laurenne and Quicke, 2010). The behaviour and adaptations are the only mechanism used by Orussidae which have the whole of the apex of the terminal flagellomere modified (Vilhelmsen et al., 2001)

The ears (hearing organs) of parasitoid wasps are in their legs, specifically the tibiae, and these detect vibrations via the wasp's feet (tarsi) – this is exactly how snakes hear, via vibrations transmitted from the ground via their jaws. In the case of insects, the 'ear' is called more precisely a subgenual organ (name from Latin, meaning 'behind the knee') and comprises enlarged trachea, tympanic membrane and a cluster of vibration-sensitive scolapidial units (Vilhelmsen et al., 2001; Otten et al., 2002). One can infer whether a parasitoid wasp uses echolocation by the combination of having some modified antennal structures and the females in particular having swollen fore tibiae to accommodate an enlarged, more sensitive subgenual organ.

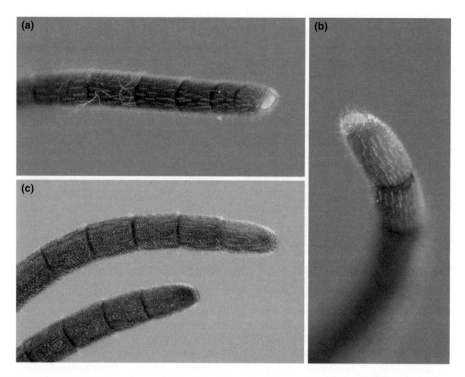

Fig. 3.1. Apical parts of the antennae of females of cryptine Ichneumonidae showing the flattened apex of the terminal flagellomeres which are used to hammer the substrate, creating vibrations which may be reflected back to the feet and then ears from hidden objects such as hosts: (**a, c**) gen. sp., indet.; (**b**) *Arrhytis* sp.

This mode of host location relies on the host substrate being relatively homogeneous apart from the host's tunnels and the host within them. Therefore, wasps that employ vibrational sounding are almost always associated with older wood that has lost its bark, or even wooden telegraph poles. Quicke *et al.* (2003) described how one large Afrotropical cryptine ichneumonid walked to-and-fro over the wood surface rapidly tapping the surface with the tips of both antennae, which were moved inwards in arcs.

References

Bacher, S., Casas, J., Wäckers, F. and Dorn, S. (1997) Substrate vibrations elicit defensive behaviour in leafminer pupae. *Journal of Insect Physiology* 43, 885–978

Batchelor, T.P., Hardy, I.C.W., Barrera, J.F. and Pérez-Lachaud G. (2005) Insect gladiators II: competitive interactions within and between bethylid parasitoid species of the coffee berry borer, *Hypothenemus hampei* (Coleoptera: Scolytidae). *Biological Control* 33, 194–202.

Belcari, A. and Kozánek, M. (2006) Secretory material from antennal organs and its possible role in mating behaviour of Pipunculidae (Diptera). *Canadian Journal of Zoology* 84, 1727–1732.

Bernays, E.A. (1997) Feeding by lepidopteran larvae is dangerous. *Ecological Entomology* 22, 121–123.

Bleeker, M.A.K., Smid, H.M., Steidle, J.L.M., Kruidhof, H.M., van Loon, J.J.A. and Vet, L.E.M. (2006) Differences in memory dynamics between two closely related parasitoid wasp species. *Animal Behaviour* 71, 1343–1350.

Bredlau, J.P. and Kester, K.M. (2019) Evolutionary relationships of courtship songs in the parasitic wasp genus, *Cotesia* (Hymenoptera: Braconidae). *PLoS ONE* 14(1), e0210249.

Camarao, G.C. and Morallo-Rejesus, B. (2003) Parasitoids of the Asian corn borer, *Ostrinia furnacalis* (Guenee), and their biological attributes. *Philippine Agricultural Scientist* 86, 17–26.

Chansri, K., Somsiri, K., Quicke, D.L.J. and Butcher, B.A. (2023) First confirmed parasitism of pleasing fungus beetles by a tropical rhyssine ichneumonid, and first record for *Cyrtorhyssa moellerii* B., 1898 from Thailand. *Journal of Hymenoptera Research*

Dicke, M., Cusumano, A. and Poelman, E.H. (2020) Microbial symbionts of parasitoids. *Annual Review of Entomology* 65, 171–190.

Eggleton, P. (1991) Patterns in male mating strategies of the Rhyssini: a holophyletic group of parasitoid wasps (Hymenoptera: Ichneumonidae). *Animal Behaviour* 41, 829–838.

Eschbach, C., Cano, C., Haberkern, H., Schraut, K., Guan, C., Triphan, T. and Gerber, B. (2011) Associative learning between odorants and mechanosensory punishment in larval *Drosophila*. *Journal of Experimental Biology* 214, 3897–3905.

Godfray, H.C.J. (1994) *Parasitoids: Behavioral and Evolutionary Ecology*. Princeton University Press, Princeton, New Jersey.

Guerra-Grenier, F., Abram, P.K. and Brodeur, J. (2020) Asymmetries affecting aggressive contests between solitary parasitoids: the effect of host species. *Behavioral Ecology* 31, 1391–1400.

Hardy, I.C.W. and Blackburn, T.M. (1991) Brood guarding in a bethylid wasp. *Ecological Entomology* 16, 55–62.

Haverkamp, A. and Smid, H.M. (2020) A neuronal arms race: the role of learning in parasitoid–host interactions. *Current Opinion in Insect Science* 42, 47–54.

Hocking, H. (1968) Studies on the biology of *Rhyssa persuasoria* (L.) [Hymenoptera: Ichneumonidae] incorporating an X-ray technique. *Journal of the Australian Entomological Society* 7, 1–5.

Hoedjes, K.M., Kruidhof, H.M., Huigens, M.E., Dicke, M., Vet, L.E.M. and Smid, H.M. (2011) Natural variation in learning rate and memory dynamics in parasitoid wasps: opportunities for converging ecology and neuroscience. *Proceedings of the Royal Society B: Biological Sciences* 278, 889–897.

Laurenne, N. and Quicke, D.L.J. (2010) Antennal hammers: echos of sensillae past. In: Pontarotti, P. (ed.) *Evolutionary Biology – Concepts, Molecular and Morphological Evolution*. Springer, Cham, Switzerland, pp. 271–282.

Laurenne, N.M., Karatolos, N. and Quicke, D.L.J. (2009) Hammering homoplasy: multiple gains and losses of vibrational sounding in cryptine wasps (Insecta: Hymenoptera: Ichneumonidae). *Biological Journal of the Linnean Society* 96, 82–102.

Marchand, D. and McNeil, J.N. (2000) Effects of wind speed and atmospheric pressure on mate searching behaviour in the aphid parasitoid *Aphidius nigripes* (Hymenoptera: Aphidiidae). *Journal of Insect Behaviour* 13, 187–199.

McClure, M., Whistlecraft, J. and McNeil, J.N. (2007) Courtship behaviour in relation to the female sex pheromone in the parasitoid, *Aphidius ervi* (Hymenoptera: Braconidae). *Journal of Chemical Ecology* 33(10), 1946–1959.

McNeil, J.N. and Brodeur, J. (1995) Pheromone-mediated mating in the aphid parasitoid, *Aphidius nigripes* (Hymenoptera: Aphidiidae). *Journal of Chemical Ecology* 21, 959–972.

Meyhöfer, R. and Casas, J. (1999) Vibratory stimuli in host location by parasitic wasps. *Journal of Insect Physiology* 45, 967–971.

Nadel, H. and van Alphen, J.J.M. (1987) The role of host plant odours in the attraction of a parasitoid, *Epidinocarsus lopezi*, to the habitat of its host, the cassava mealybug, *Phaenococcus manihoti*. *Entomologia Experimentalis et Applicata* 45, 181–186.

Nofemela, R.S. (2017) Strong active defensive reaction of late instar *Plutella xylostella* (L.) (Lepidoptera: Plutellidae) towards *Cotesia vestalis* (Haliday) (Hymenoptera: Braconidae) correlates with its low suitability for parasitism. *African Entomology* 25, 454–461.

Ode, P.J., Vyas, D.K. and Harvey, J.A. (2022) Extrinsic inter- and intraspecific competition in parasitoid wasps. *Annual Review of Entomology* 67, 305–328.

Otten, H., Wäckers, F.L., Isidoro, N., Romani, R. and Dorn, S. (2002) The subgenual organ in *Pimpla turionellae* L. (Hymenoptera: Ichneumonidae): ultrastructure and behavioural evidence for its involvement in vibrational sounding. *Redia* 85, 61–76.

Potting, R.P.J., Vermeulen, N.E. and Conlong DE (1999) Active defence of herbivorous hosts against parasitism: adult parasitoid mortality risk involved in attacking a concealed stemboring host. *Entomologia Experimentalis et Applicata* 91, 143–148.

Quicke, D.L.J. (1997) *Parasitic Wasps*. New York City, Springer, 470 pp.

Quicke, D.L.J. (2015) *The Braconid and Ichneumonid Parasitic Wasps: Biology, Systematics, Evolution and Ecology*. Wiley Blackwell, Oxford, 688 pp.

Quicke, D.L.J., Laurenne, N.M., Broad, G.R. and Barclay, M. (2003) Host location behaviour and a new host record for *Gabunia* aff. *togoensis* Krieger (Hymenoptera: Ichneumonidae: Cryptinae) in Kibale Forest National Park, West Uganda. *African Entomology* 11, 308–310.

Romani, R., Rosi, M.C.m Isidoro, N. and Bin, F. (2008) The role of the antennae during courtship behaviour in the parasitic wasp *Trichopria drosophilae*. *Journal of Experimental Biology* 211, 2486–2491.

Sacchetti, P., Belcari, A., Romani, R., Isidoro, N. and Bin, F. (1999) External morphology and ultrastructure of male antennal glands in two diapriids (Hymenoptera: Diapriidae). *Entomology Problems* 30, 63–71.

Shaw, M.R., Giannotta, M., Herrera-Flórez, A.F. and Klopfstein, S. (2021) Two males, one female: triplet-style mating behaviour in the Darwin wasp *Xorides ater* (Gravenhorst, 1829) (Hymenoptera, Ichneumonidae, Xoridinae) in the Swiss Alps. *Alpine Entomology* 5, 15–22. https://doi.org/10.3897/alpento.5.64803

Steiner, S.M., Kropf, C., Graber, W., Nentwig, W. and Klopfstein, S. (2010) Antennal courtship and functional morphology of tyloids in the parasitoid wasp *Syrphoctonus tarsatorius* (Hymenoptera: Ichneumonidae: Diplazontinae). *Arthropod Structure and Development* 39, 33–40.

Takasu, K. and Lewis, W.J. (2003) Learning of host searching cues by the larval parasitoid *Microplitis croceipes*. *Entomologia Experimentalis et Applicata* 108, 77–86.

Taylor, C.J. and Yack, J.E. (2019) Hearing in caterpillars of the monarch butterfly (*Danaus plexippus*). *Journal of Experimental Biology* 222, jeb211862.

Turlings, T.C.J. and Tumlinson, J.H. (1991) Do parasitoids use herbivore-induced plant chemical defences to locate hosts? *Florida Entomologist* 74, 42–50.

Turlings, T.C.J. and Tumlinson, J.H. (1992) Systematic release of chemical signals by herbivore-injured corn. *Proceedings of the National Academy of Science of the U.S.A.* 89, 8399–8402.

Turlings, T.C.J. and Wäckers, F. (2004). Recruitment of predators and parasitoids by herbivore-injured plants. *Advances in Insect Chemical Ecology* 2, 21–75.

Turlings, T.C.J., Tumlinson, J.H. and Lewis, W.J. (1990) Exploitation of herbivore-induced plant odours by host-seeking parasitic wasps. *Science* 250, 1251–1253.

Turlings, T.C.J., Wackers, F.L., Vet, L.E.M., Lewis, W.J. and Tumlinson, J.H. (1993) Learning of host-finding cues by hymenopterous parasitoids. In: Papaj, D.R. and Lewis, A.C. (eds) *Insect Learning: Ecological and Evolutionary Perspectives*. Chapman & Hall, New York, pp. 51–78.

Vilhelmsen, L., Isidoro, N., Romani, R. Basibuyuk, H.H. and Quicke, D.L.J. (2001) Host location and oviposition in a basal group of parasitic wasps: the subgenual organ, ovipositor apparatus and associated structures in the Orussidae (Hymenoptera, Insecta). *Zoomorphology* 121, 63–84.

Villagra, C.A., Pinto, C.F., Penna, M. and Niemeyer, H.M. (2011) Male wing fanning by the aphid parasitoid *Aphidius ervi* (Hymenoptera: Braconidae) produces a courtship song. *Bulletin of Entomological Research* 101, 573–579.

Vinson, S.B. (1976) Host selection by insect parasitoids *Annual Review of Entomology* 21, 109–133.

Zhang, Q., Chen, K., Wang, Y., Lu, Y., Dong, Z., Shi, W., Pang, L., Ren, S., Chen, X. and Huang, J. (2023) The odorant receptor co-receptor gene contributes to mating and host-searching behaviors in parasitoid wasps. *Pest Management Science* 79, 454–463.

4

Parasitoid Diversity, with Special Reference to S.E. Asia

Abstract
Despite S.E. Asia being a vast and biologically diverse area, only a small proportion of its parasitoid wasp diversity has been described, and this is probably significantly less than the proportion in Africa or South America. The very large number of the islands suggests that there is likely to be much endemism but far too little is known to draw any conclusion. Although there have been quite a few published surveys specifically aimed at parasitoid diversity, predominantly in agroecosystems in the regions, the poor taxonomic state of the group means that most specimens are only identified to superfamily or family level, which is not very informative. The largely negative impact on the application of the Convention on Biological Diversity (CBD), and of permitting issues, massively handicaps taxonomic progress on all groups, not just parasitoid wasps. Overcoming the taxonomic impediment by rapid means such as DNA barcoding is discussed. Global trends in parasitoid biology are considered given the paucity of taxonomic information.

It is estimated that there are about 2.3 million valid described extant species on the planet recognised by taxonomists at the present time (Bánki et al., 2023). Every year, new species are being discovered and described in all major groups of organisms (Songvorawit et al., 2021). However, the vast majority of species remain unknown to science and are waiting to be discovered, especially in the tropical regions where at most only two-thirds of the world's floral and faunal diversity have been discovered (Bierregaard et al., 1992; DeFries et al., 2005). Even for the relatively well-studied Lepidoptera it seems that only half, or maybe less than half, of the world's fauna has been described (Kristensen et al., 2007). For parasitoid wasps we suspect that rather fewer than 10% have been described. It was long believed that Coleoptera were the most speciose order of insects, but that was largely due to the large amateur enthusiast contribution to their taxonomy (Forbes et al., 2018); now it seems to be more of a race between Hymenoptera and Diptera (Chimeno et al., 2022).

S.E. Asia is one of the world's three major biodiversity regions, and contains six of the world's 25 biodiversity hotspots (Myers, 1988; Cincotta et al., 2000). Indo-Burma, Sundaland, the Philippines, and Wallacea, contain 20% of the world's vertebrate and plant species. However, compared with other hotspots, S.E. Asia is far more densely populated, with a human population of over 800 million – more than the Amazon and Congo combined. Most countries in this region have

© Buntika A. Butcher and Donald L.J. Quicke 2023. *Parasitoid Wasps of South East Asia*
(B.A. Butcher and D.L.J. Quicke)
DOI: 10.1079/9781800620605.0004

low mean incomes, and low per capita GDP. As a consequence, natural habitats such as forest, had been cut down for economic purposes i.e. agricultural areas, buildings, etc. This region also has the highest proportion of threatened vascular plants, reptiles, birds, and mammal species compared to the other tropical regions. Unfortunately, deforestation and habitat loss due to economic development, i.e. clear-cutting for food production (cash crops and agriculture) combined with poor natural resource management, poverty and corruption means that many species are threatened and probably some species have gone extinct before ever being recorded. Lack of adequate conservation funding creates alarming challenges for Southeast Asian conservationists (Sodhi et al., 2004; Giam et al., 2010; Webb et al., 2010).

Furthermore, not only is S.E. Asia's annual deforestation rate as a percentage of total land area the highest in the tropics, it has also increased between the periods 1990–2000 and 2000–2005. It has been estimated that over 40% of the region's biodiversity may disappear by 2100 (Sodhi et al., 2004). Insects are the most diverse groups of invertebrates in the terrestrial ecosystems, and their high diversity is linked directly to the ecosystem services they provide (Schwartz et al., 2000). Therefore, studying and measuring insect diversity has a great impact on human activities in natural, agricultural or forested environments (Mazón, 2016). To be able to do that, appropriate surrogates with a well-known taxonomy and higher taxa should be selected for biodiversity monitoring (Vieira et al., 2012). Using surrogates can be very useful when sampling species-rich areas such as tropical regions or in those studies where time and human resources are limited (Lovell et al., 2007; Moreno et al., 2008). Good surrogates should be a group with well-known taxonomy and ecology and should be cosmopolitan (Caro and O'Doherty, 1999).

Parasitoid wasps are a highly diverse set of insects in the Hymenoptera. Due to their life histories as parasitoids, they play a crucial role in most terrestrial ecosystems by regulating populations of their insect hosts (Quicke, 2015; Songvorawit et al., 2021). Moreover, some species are used as natural enemies to control insect pest population in biological control and integrated pest management (IPM) programmes to limit damage caused by insect pests both in agricultural areas and in stored products, with more success rate when compared with other insect groups as natural enemies. Parasitoid wasps are important for both ecology and economy, but knowledge about species composition and diversity of parasitoid wasps in the tropics is still limited (Sann et al., 2018). The number of ichneumonoid species in this region includes 1294 species of Braconidae and 710 species of Ichneumonidae which had been recorded in the Taxapad interactive database up until 2012 (Yu et al., 2016) – much lower than extrapolated numbers (Jones et al., 2009). This indicates that taxonomic work on this superfamily is far from complete and requires a great deal more taxonomic effort. Indeed, the difference becomes really stark when one compares the ichneumonid number with those of the United Kingdom with a land area of c. 245,000 km^2, where approximately 2500 species of Ichneumonidae are known, with barely half that number for tropical S.E. Asia whose combined land area of 4,489,324 km^2 is more than 18 times greater.

Most described parasitoid species are recorded from temperate areas, where a far higher proportion of the fauna has been described when compared with tropical areas (Gaston, 1993). Although numbers of described parasitoid wasp species are rather high, the total number of extant species is still unclear. The Chalcidoidea alone comprises 22,500 described species, but their overall diversity is estimated at less than 500,000 species (Heraty, 2017; Noyes, 2019). Similarly, the Ichneumonidae and the Braconidae currently have approximately 23,000 and 18,000 described species, respectively, but these numbers are definitely underestimates (Quicke, 2015).

Surveying Parasitoid Diversity in S.E. Asia

Several studies have tried to compare the parasitoid diversities of various managed agro-forestry habitats with less disturbed ones, though almost always not with actual pristine primary forest. Although not parasitoids, ants have also featured highly in such comparative studies, because their collection and identification are less challenging. Frighteningly, most studies have drawn completely invalid conclusions. This is actually an interesting topic in its own right. The question may be asked: why, if you survey/sample species from, say, an oil palm plantation, and use the same methods to collect in a natural habitat, might the differences not be valid? Indeed, not just that the comparison might not be exact because of some biases, but actually pointing to completely the opposite trend than really exists.

The reason is that the species abundance distribution relationships will be different, and often extremely different, between natural and disturbed/managed habitats. Although all field biologists and taxonomists realise this, it is seldom allowed for in the actual surveying. Disturbed habitats and agricultural areas often have quite a few moderately abundant species which can all be sampled fairly quickly, but the total number is quite small. Small samples/surveys will probably include almost of the common ones. Primary forests have a few commoner species, but many uncommon ones, so a small sample is likely to pick up the few common ones but not very many from the long tail of increasingly rare ones.

What is now clear is that it is nearly impossible to compile a complete inventory of a diverse group of insects such as parasitoid Hymenoptera in any area. This was exemplified by a fine study carried out in New Zealand, a country with a naturally very small fauna (Saunders and Ward, 2018). Running various numbers of Malaise traps (see Chapter 23, this volume) they concluded that use of only a few traps, or using them for only a very short time, severely limits the estimates of richness because:

- fewer individuals are caught, leading to a greater number of singletons; and
- the considerable variation of individual traps means some traps will contribute few or no individuals.

The combination of the difficulty in identifying parasitoid wasps to major groups, let alone to genera and morphospecies, with the large numbers of individuals often collected, plus the need for sampling over long periods with many traps, makes it exceedingly hard to obtain meaningful data.

There have been quite a few studies of parasitoid wasp diversity in S.E. Asia, many concerned with diversity in agroforestry systems. Nearly all of these only identify the parasitoids to family (or even superfamily) level. The reason is not surprising, it is part of the purpose of this book to correct that.

Putra (2018) studied parasitoid diversity in banana plantations at Yogyakarta, Indonesia, from April to June 2018. The specimens were collected every week using two sampling methods: direct (collecting pests from three banana plants at the point); and indirect (aerial net and yellow pan traps). The results showed that Chalcidoidea were the most highly diverse and the highest number of individuals were Scelionidae.

What Do Surveys in S.E. Asia Find?

Ikhsan and Hamid (2020) studied diversity and abundance of Hymenoptera on tidal swamp rice fields in Indragiri Hilir District, Riau Province, Indonesia, employing four different collecting methods (aerial net, Malaise trap, pitfall trap and yellow pan trap). Their sampling was carried out at three sub-districts of rice production centres: Batang Tuaka, Keritang and Reteh, from February to October 2018. In total, there were 4701 individuals consisting of 39 families and 319 species of Hymenoptera at the three studied areas. Formicidae, Braconidae, Ichneumonidae and Scelionidae had the highest number of species, and Formicidae, Scelionidae, Diapriidae and Braconidae had the highest number of individuals.

Azhar *et al.* (2022) studied effects of rainforest conversion to cash crops on the abundance, biomass and species richness of parasitoid wasps in Sumatra, Indonesia. Parasitoid wasps play a crucial role in terrestrial ecosystems by significantly reducing herbivorous populations, including agricultural pests (Quicke, 2015), leading to increased plant growth (Dyer *et al.*, 1993), which will promote animal biodiversity (Bascompte and Jordano, 2007). Two important agricultural crops in S.E. Asia are natural rubber (*Hevea brasiliensis*) and oil palm (*Elaeis guineensis*) (Azhar *et al.*, 2022), especially in Indonesia (Azhar *et al.*, 2022); plantations of these two crops cause decreasing ecosystem functions and biodiversity. Abundance and morphospecies richness of six families of parasitoid wasps (Braconidae, Ceraphronidae, Encyrtidae, Eulophidae, Scelionidae and Platygastridae) were compared between four land-use systems. Four potential host orders (Coleoptera, Diptera, Hemiptera and Lepidoptera) from the same sites were also recorded. The Braconidae had been the focus of this study due to the family's highly diversity. Insect samples were collected once during dry (July–October, 2013) and rainy seasons (November–March, 2013–2014) using canopy fogging. The results showed that oil palm and rubber plantations had just over 50% of all morphospecies; moreover, the measured abundances of parasitoid wasps were significantly affected by land use, with those of rainforest almost two-and-a-half times higher than those in rubber and more than four times as high as in oil palm. This study showed that rainforest transformation to monocultures of rubber and oil palm reduces abundance, richness and biomass of parasitoid wasps and affects their community composition across seasons (Azhar *et al.*, 2022).

Alternatively, sampling may be done to assess the (relative) abundance of particular groups of parasitoids that are more easily recognised. For example, Sann *et al.* (2018) investigated the rice fields at Laguna province, Luzon, Philippines concentrating only on four important egg-parasitoid genera (*Anagrus*, *Oligosita*, *Gonatocerus*, and *Paracentrobia*) of relevance to the control of rice hopper pests, such as *Nilaparvata lugens* and *Nephotettix* spp. In this case they supplemented traditional morphological studies with DNA sequencing of the 28S rRNA and the barcoding part of the *cytochrome c oxidase* (COI) genes.

Convention on Biological Diversity (CBD) and Permitting Issues

Without doubt, the study of S.E. Asian parasitoid wasps has been, and continues to be, enormously harmed by a combination of anti-imperialism and the Convention on Biological Diversity (CBD). In 2010, relevant parties adopted a set of rules and guidelines about the fair and equitable sharing of benefits arising from the utilisation of countries' biodiversity genetic resources. The Nagoya Protocol came into force in October 2014 (Secretariat of the Convention on Biological Diversity, 2011; Kiehn *et al.*, 2021), and has been ratified by the EU plus 113 other UN member states (De Prins, 2022). Although the Nagoya Protocol actually exempts taxonomy from its remit, there is widespread belief and experience that the ways that signatories have implemented changes in their legislations has severely hampered taxonomic research in many of the most biologically diverse and understudied regions of the world. Making it really difficult to carry out taxonomic research in these areas actually impedes the potential development of countries' genetic resources – if you do not know what it is and how to recognise it, how can you make use of it? Instead of fostering collaborations between 'users' and 'providers' of resources, the implementations seem in practice to have had quite the opposite effect (Deplazes-Zemp *et al.*, 2018; Prathapan *et al.*, 2018).

One of the requirements of the CBD is that signatory parties should simplify measures for access for non-commercial research purposes (Kiehn *et al.*, 2021). They also point out that many of the impediments to fundamental research (e.g. taxonomy, inventory) will have negative impacts on human development and biodiversity. This is of course excellent, the problem being that the implementation does not always live up to the ideal. Kiehn *et al.* (2021) pointed out that some difficulties arise from lack of clear definitions of terms, e.g. commercial versus non-commercial (Tran *et al.*, 2019). There is often a grey area and administrators tend to be cautious. In many countries there are issues of demarcation of responsibilities, imprecisely defined terms, and gaps in the law (Morgera *et al.*, 2015; Nguyen and Tran, 2018) and therefore a lot more to be done before things are likely to run as smoothly as intended.

The difficult issues relate not just to specimens but also to DNA sequence data, or at least might do (Hammond, 2017; Lyal, 2019). Do DNA sequences from a specimen belong to the country the specimen came from? Very clearly, if the samples have been processed to obtain information about genes whose products or derivatives could have commercial implications, for example, in the pharmaceuticals domain, then yes, that might seem reasonable. But what about DNA barcoding in which the *cytochrome oxidase c subunit* 1 sequence is almost certainly incapable of being exploited to make a profit whilst potentially being of great scientific value to the source country as well as the wider world's scientific community. Should we expect permit administrators to have enough biological background knowledge to make scientifically valid decisions in each case?

The Collecting Permit Problem

Permits are certainly a significant issue, especially when it comes to the involvement of foreigners, because of fears that it might lead to biopiracy and because it might seem to be disrespecting local talent.

A summary of collecting permit requirements for many countries is provided by the website www.theskepticalmoth.com/collecting-permits/ [accessed 24 January, 2023]. Here is what it says about countries in S.E. Asia, and then a selection of others where biodiversity is better known.

- CAMBODIA: Collecting permits required, make contacts with the Cambodian Entomology Initiative for further information.
- INDONESIA: Permits required and difficult to obtain. Permits going by the acronym 'LIPI' are issued by the Indonesian Institute of Sciences, but they have a great website for this entire process. Probably there is a strict quota per species, regardless of CITES status.
- MALAYSIA: Permits are issued through the Wildlife Department, and it looks like an extensive process. Needs at least six months for approval.
- PHILIPPINES: Permits necessary and require a collaboration of a local scientist.
- PAPUA NEW GUINEA: Research permits are available through the National Research Institute for Au$41.
- FRANCE No permit needed in France, amateur entomologists are welcome.
- MALTA: Collecting allowed outside of national parks and private land.
- POLAND: Collecting permitted outside of protected areas.
- ITALY: Very collector friendly, permits required only for national parks.
- JAPAN: Collecting allowed outside of protected areas.

A good question might be: who worked out all the insect taxonomy in European countries? The answer is of course a mixture, but dominated by amateurs with some input from people working in museum collections, and really very few in academic university situations. Although there may be some amateur butterfly and moth enthusiasts (and a few for other groups of prettier insects) in S.E. Asia, there really is no tradition in amateur entomology. Many factors underlie this and some of the more important ones include:

- insufficient disposable income and free time; and
- lack of peer interest groups that can foster interest even at an early age, such as the UK's Amateur Entomological Society.

Many of the people making decisions are bureaucrats, often with little or no practical biological experience such as having done a university degree course in biology, and usually very little time spent in the field in the company of naturalists who could explain things. Nearly all collecting permit and export permit application forms are not designed with bulk insect sample in mind, nor do they reflect the reasons behind the desire to collect or export. They almost always ask for a list of species, when getting the list of species (and authoritatively identified voucher specimens) from an overseas expert is the main point of the exercise.

How many specimens do you want to collect/export? This question cannot be answered honestly in most cases if there is to be any hope of being issued with a permit. For a taxonomist the answer is probably 'as many as possible'. But just how many? There are actually rather few data and the numbers have huge variance according to trap design, location, season, etc. One study, in New York state, USA, rather than in the tropics, by Matthews and Matthews (1970) gave the following answer.

- Four traps operated between early June and late July yielded a total of 40,348 insects.
- Daily catch per trap ranged from 36 to 749 individuals.
- Diptera were the most frequently collected order.
- Hymenoptera constituted nearly 15% of individuals overall (but this was largely due to one trap catching an enormous number of Plectoptera, which are uncommon in the tropics).

We think that the range of individuals per trap day is probably quite representative. Of course, very many of these will be extremely small dipterans and hymenopterans. They are nevertheless part of the catch that one is asking to collect.

Now let us consider how much sampling is needed to get even somewhere close to inventorying the small Hymenoptera of an area of a given habitat in the tropics.

What is Needed for Complete* Inventory of a Group?

(* close to)
We will just concentrate on a programme of Malaise trapping, such as the TIGER programme in Thailand. Malaise traps of course only collect species that are amenable to being caught in Malaise traps, which is only a subset of all the species. Many other groups spend most of their time crawling around in leaf litter, or even if they do fly, perhaps only rarely disperse at Malaise trap height. Parasitoid wasps associated with aquatic habitat such as egg parasitoids of Gerridae (Hemiptera) or dragonflies (Odonata) seldom stray far from water. In Chapter 23 of this volume, we describe various other collecting methods. So, discounting those, what is needed? Four factors spring to mind:

- Different habitats have different taxa inhabiting them. Further, species abundance curves may vary enormously between different habitats, for example between dry dipterocarp forest, montane evergreen forest, mangrove, tropical rain forest, savannah.
- Species turnover with distance (β-diversity) can be considerable across tranches of apparently similar habit and between hotspots of diversity or endemism. Sampling would need to reflect these patterns at various scales.
- Year-on-year differences: many species can be found in one year but not the next. They do not emerge regularly each year and their presence or absence and their abundance can vary. You need to trap for more than one year. (This phenological separation is a potential mechanism of sympatric divergence.)
- There is enormous variance in what individual, apparently similarly set up traps in the same habitat, even quite close to one another, will catch.

What can we conclude?

- From the first factor: that you need to sample each type of habitat.
- From the second: that one needs to sample each habitat across as broad a geographical range as possible.
- From the third: that sampling must be conducted over multiple years.
- From the fourth: that you need a lot of traps at one site, not one or two, or five, but many more, maybe even a hundred.

Satisfying the above involves not just the initial cost of purchasing and servicing the traps, but also the subsequent storage and sorting. It should be obvious that even on a cut-down scale, every effort should be made not to waste the valuable by-catch. Just because a project is, say, only concerned with Ichneumonoidea, the by-catch of beetles, flies and other creatures will be of great scientific value to many individuals around the world. A friend said, 'Taxonomists cherry pick what they work on.' This is true but it is the only way, because the diversity of insects and other arthropods is so vast that any individual taxonomist can only have real expertise in a very limited subset of taxa, typically in the case of parasitoid Hymenoptera just one family or even subfamily if the group is large and diverse.

To conclude, any inventory of parasitoid Hymenoptera requires a veritable army of taxonomists with their individual specialisations, and no one country can support this. It is financially and educationally impossible. Long, long ago, great international natural history museums, such as the one in London, did try to have expertise covering almost all big groups, but that attitude simply became impossible as collections grew and finances were limited. The essential expertise is international.

Is There a Type Specimen Impediment?

The situation is reached in fairly species-rich genera where X number of species are described in a revision of material from country A, then another Y species from nearby country B, and C species from another nearby region. Add to this a spattering of individual descriptions dating back maybe 100 years or more, which are essentially always inadequate for species recognition, and the genus become unworkable. The reasons are some combination of the following.

- Descriptions and/or associated illustrations are totally inadequate for species recognition.
- Types from at least one study are inaccessible.
- Attempts at specific keys are based on previous descriptions and often rely heavily on coloration.
- Many species descriptions are based on either males or females, but not both.

It is a simple historic fact that the vast bulk of research carried out on tropical faunas and floras over the past 200 years was carried out by western scientists, with the material they were working on transported to and curated almost entirely in museums in European capital cities. This is where most type specimens of parasitoid wasps are housed. More recently, over the past 100 years, North American researchers have also amassed tropical collections and published taxonomic works on them, although they have been far more involved with the Neotropical fauna because of its more direct relevance to that of their own countries.

In all groups of insects, a relatively greater proportion of larger-bodied species were described in the early years of their taxonomic investigation (Blackburn and Gaston, 1994). Thus, large-bodied Ichneumonoidea were the most well-known group of parasitoid wasps because they attracted the attention of local collectors and predominantly European taxonomists, such as those mentioned in Chapter 1 (this volume). Further, many taxonomic entomologists during the first part of the 20th century had broad taxonomic interests, so the depth of knowledge that some had in any particular group was limited. A contributing factor is that mostly these workers only had access to relatively small numbers of individuals and species from tropical regions, either commercially collected specimens or ones resulting from various broad scientific expeditions, plus some material collected by amateurs, often ministers of religion or military officers, posted abroad.

There is no doubt that there was a real and significant taxonomic impediment, not only for workers in S.E. Asia, but everywhere. Practicalities such as time to pack loaned items properly, cost of shipping and risk of damage in transit (especially by seamail) meant, we think generally rightly so, that big museums in Europe or North America would be unlikely to loan material, especially type specimens, to workers in small tropical countries. This was also a 'Catch-22' situation, because, without access to types, workers in overseas countries were not able to publish as much and so it was harder to build up a good academic reputation. We have had the dreadful experience of a shipment being destroyed by Customs authorities without permission to have the parcel returned to the sender. Such incidents are not that uncommon and certainly greatly reduce the possibility of international loans even if all the necessary paperwork seems complete.

There is still an extremely great need for taxonomists to have access to type specimens. Years ago, for example, in many works by Cameron and Szépligeti (see Chapter 1, this volume) descriptions of new species, and sometimes of new genera, were exceedingly brief and not illustrated. One example, certainly not the shortest, is the description of the braconid *Ipobracon parvispeculum* from Sumatra by Enderlein (1920):

> "Unterscheidet sich von I. basispeculum nur durch folgendes: Basalfeld des 2. Tergites sehr klein und nur bis zum Ende des 1. Viertels reichend; Basalhälfte der Flügel braun oder hellbraun. Pubescenz der Legescheide sehr kurz.
> Körperlänge 6½—9 mm, Vorderflügellänge 6½—9½ mm,
> Bohrerlänge 4½—6½ mm."

This translates as: 'Differs from *I. basispeculum* only in the following: basal area of the 2nd tergite very small and reaching only to the end of the 1st quarter; basal half of wings brown or light brown. Pubescence of the sheath very short.' Even when you are familiar with the group, this is not terribly helpful because many species will comfortably fit this description. Therefore, there is a huge need for all the available type specimens to be photographed from several angles and the images made publicly available. There are some initiatives in this direction. For example, Museum für Naturkunde, Berlin, is making high-definition images of all of its type specimens available via its *Zoosphere* project. Such images will go a long way towards at least being able to exclude species that one is trying to identify if not making a definite match possible.

There is still an impediment (but it is getting less), not because major museums are more willing to loan specimens, and not really because some countries are investing in their own museum facilities, but because of technology. All major museum insect collections now have excellent specimen imaging systems, and so in principle, a curator or technician can take relevant, high-quality photographs of type specimens and characters and send them to the requestor. Some are even uploading digital images of all their types on the World Wide Web (www). The impediment now is largely the cost of the time, and availability of time, of the curators in the major collections. It should be noted that having a good habitus photo of a type specimen on the web may be sufficient to confirm its genus (but often not) however it may be sufficient to say whether someone has a different species or has something very similar. One or a few images will almost always be inadequate to make definite species matches.

DNA Barcoding – The Way Forward

Probably the majority of taxonomists understand the important role that molecular DNA methods can have in the discovery and delimitation of species. Good traditional taxonomy is a skilled but laborious and time-consuming, and therefore costly, activity. There are simply too few taxonomists out there to describe and to write identification keys to all the species of animals, and probably even plants. This is especially true for hyperdiverse groups of organisms such as parasitoid Hymenoptera. The issues are confounded by many being rather small, which almost inevitably leads to a reduction in the number of characters available to permit visual recognition (Polilov, 2017). Indeed, neuronal nuclei in adult brains have even been lost in some tiny chalcidoids (Polilov *et al.*, 2023). Carried out by experts, given enough resources including the necessary time, good (expensive) equipment such as photo-microscopes, ability to access the bulk of material in all relevant museum collections, new field work, collecting permits, DNA sequencing, and so on, the products can be excellent, user-friendly treatments. A recent example is Polaszek *et al*.'s (2022) recent revision of the trichogrammatid chalcidoid genus *Megaphragma* (see Chapter 22, this volume). This is a small, cosmopolitan and economically important genus of small wasps (indeed the smallest known insects belong to it). We inquired how long the work had actually taken in researcher hours, and the answer was at least six months. Given that the entomologists involved are diligent and probably many work long hours, that means at least 1000 scientist hours (we suspect more). What was the product? Well, the revision includes formal original descriptions of 22 species and provides a key to all 32 valid species. Broadly speaking, therefore, approximately 30 scientist hours per species or 50 scientist hours per new species. Whilst *Metaphragma* is not a particularly easy genus, there are many others of similar difficulty and complexity.

Given the above, and that estimates of numbers of undescribed insect species range between a minimum of approximately 5.5 million (García-Robledo et al., 2020) and up to about 30 million (Erwin, 1991), and that parasitoid Hymenoptera are one of the groups with the highest proportion of undescribed species, we are talking about numbers in the order of 100 million scientist hours being needed before all parasitoid wasps are included in thorough revisionary studies using traditional methods. It is not going to happen. Sadly, the massive Anthropocene decline in insect numbers must surely mean that huge numbers of these are going extinct as we write (Montgomery et al., 2019; Wagner, 2020).

A number of studies applying barcoding to understudied tropical parasitoid insect faunas, particularly that of Costa Rica, have shown an enormous previously unsuspected diversity among morphologically very similar forms (Smith et al., 2006, 2007, 2008).

Does Barcoding Do a Good Job of Separating Species of Parasitoid Wasp?

It depends on the group and in particular how fast their sequences evolve relative to the actual process of biological speciation, i.e., the virtual or total cessation of gene flow between populations. Obviously, this is not the only factor but if a group is speciating rapidly then there is unlikely to be a really good signal in just the 650 or so base pair barcoding gene region. On the other hand, it can work very well. Fig. 4.1 shows an example from the family Stephanidae. These wasps are really conservative in their morphology and coloration, and very few species have been described from the region. Even generic identification is not that easy and some of the characters used in keys are rather indistinct. Nevertheless, the molecular data separated the three genera that were collected. We have only been able to identify two species with any degree of confidence based on existing keys, but neither appears to be a single entity. We cannot guarantee as yet that the different barcodes do not reflect incomplete mitochondrial lineage sorting; the magnitude of differences really suggests strongly that each cluster represents a distinct, yet morphologically cryptic, species. This is quite typical, and finally resolving such issues beyond any reasonable doubt would require an integrative taxonomic approach. However, we strongly suspect that these clusters are really previously unrecognised cryptic species. Now the problem is, which ones – if any – are the same species as the named holotypes?

The Need to Speed Up Taxonomy

One of us recalls the time when world experts in museums might spend almost their entire working lifetime on some major taxonomic revision. It is hardly a joke that some of these mammoth tomes were never completed because the by-then ageing taxonomist got run over by a bus or similar. Like it or not, both museum specialists and university academics are no longer able to follow such careful, plodding paths. Short-termism is the key. Research grants are typically for three-year periods and one or, usually, more good scientific papers have to be published from each in order to both secure further funding and ensure the worker keeps their job. And even if money can be obtained to support taxonomy it is still necessary to demonstrate research products every three years or so.

The above causes workers to avoid tackling really big important genera or to seek short cuts, nowadays mostly molecular shortcuts. It seems highly unlikely that there will ever be a workable morphological key to *Bracon* (Braconidae: Braconinae) species of any large region. There are probably thousands of mostly undescribed species, and globally, concepts of subgenera or species groups tend to become very blurred. Molecules are probably the only

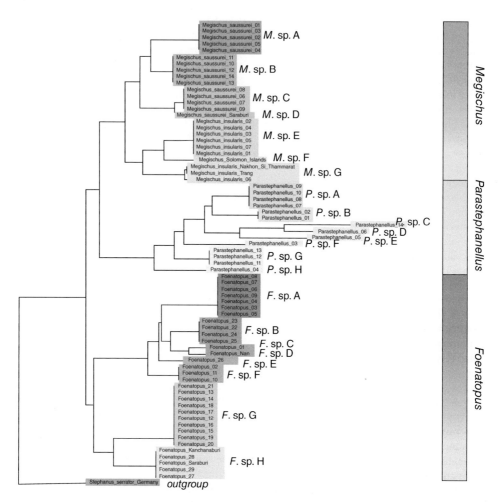

Fig. 4.1. Maximum likelihood phylogenetic tree of stephanid wasps based on the barcoding (COI) gene region, mostly collected from one dry forest site, in Thailand. (Source: data kindly provided by Kittiphum Chansri.)

hope, even if, currently, most people do not have access to a pocket-sized, affordable DNA barcoding machine. With technology advancing so quickly this may soon be a reality.

Turbotaxonomy

In 2012, Butcher *et al.* described 179 new species of only a single genus, *Aleiodes* (Braconidae: Rogadinae) from Thailand, using a combination of traditional morphology but with considerable input from the results of DNA barcoding. To put this in context, prior to their study, only three species had been recorded from the country. This represents a 60-fold increase in known species for the country. Their work was based mainly on the copious material that resulted from Mike Sharkey's TIGER (Thailand Insect Group for Entomological Research) project, a USA National Science Foundation (NSF) funded biodiversity inventory survey of the country which utilised 3605 Malaise traps, serviced at 7-day intervals, at 559 separate sites spread across more than 30 National Parks and reserves between 2006 and 2009, and run for a year with the assistance of numerous local assistants (Sharkey and Clutts, 2010; Plant *et al.*, 2012). The holotypes and much other material from this project were deposited in the modern collection facility at Queen Sirikit Botanic Garden (Chiang Mai), Thailand. The scientific (ecological, taxonomic, phylogenetic) benefit of running similar projects in other S.E. Asian countries, perhaps especially the islands of Indonesia (where local endemism may be high despite the islands of Sundaland having been a largely contiguous land mass from about 110,000-12,000 years ago), would be enormous. With developments in DNA sequencing technology over the past 15 years, molecular data for a far larger proportion of specimens could now be obtained. A start was made in Mt Halimun National Park, W. Java, in Indonesia by Cancian de Araujo et al. (2018), but only a tiny amount of material (4531 specimens) was processed for DNA barcoding, with approximately a 50% success rate. Nevertheless, these represented 1195 exclusive BINs (barcode index numbers) or putative species and 96% of these had never before been sequenced.

Butcher et al. (2012) coined the term 'Turbo taxonomy' for their approach because each species description was kept fairly short, and many characters were illustrated photographically, thus obviating the need for lengthy verbal descriptions. A picture paints a thousand words, as they say. They also made considerable use of DNA barcode data to help sort morphologically similar specimens into putative species clusters. Having assessed morphological and molecular evidence they decided on where to draw the boundaries between species, and presented a key to species that had a backbone based largely on morphology, but in places also used DNA base characters. Of course, amateurs at home (if there ever were to be such people working on parasitoid wasps in the tropics) would not be able to separate and identify all those cryptic species for which only DNA provides an answer, but they could make potentially correct identifications for those keyed using morphology.

Barcoding and the Minimalism Movement

Mike Sharkey and collaborators pioneered a far more revolutionary approach to formally naming large numbers of almost certainly undescribed tropical parasitoid wasps, starting with the agathidine braconid genus *Zelomorpha* (Meierotto et al., 2019) and subsequently other groups of Braconidae from Costa Rica (Sharkey et al., 2021a, b). The reason for the material all being from Costa Rica was because they were all (or nearly all) reared as part of Dan Janzen and Winnie Hallwach's enormous, multi-year caterpillar rearing programme in the Area de Conservation Guanacaste, Costa Rica, in the north-west of that country, and so the majority of specimens had host Lepidoptera information and usually host plant information, even though many hosts were also recognised on the basis of barcodes and caterpillar characters. This minimalist approach relied largely on a combination of the host information and the barcode clusters identified as barcode index numbers (BINs) that the specimens fell into. No description of the adult wasp's morphology is given, and just the representative barcode presented, but each species was illustrated photographically at least by a habitus image (Fig. 4.2). There is no reason, apart from time or publication costs, why many more photos could not be included for each described species, but who knows exactly what body parts would need to be illustrated in order to differentiate the new species from others that may be discovered subsequently?

Needless to say, this approach was quickly, and often aggressively, criticised by a number of authors who did not like the departure from classical descriptive methodology (Zamani et al., 2020, 2022; Meier et al., 2022). Heated debate followed (e.g. Sharkey et al., 2022). Firstly, the minimalist approach was never intended to be applied to well-known groups or for well-studied faunas; it is not applicable to butterflies, birds, beetles, or for the Holarctic. We think the shortest and best explanation of Sharkey's rationale is given in the introduction of Sharkey et al. (2021c).

Tropical Parasitoid Diversity and Biology

For most groups of parasitoid Hymenoptera there are only fairly bland statements, such as, 'they are very diverse but poorly studied in the tropics' or occasionally, 'the group appears to be principally temperate in distribution'. It is certainly the case that in temperate habitats and at high elevations in the tropics, ichneumonids predominate in terms of the numbers of individuals encountered (trapped) and the

Chelonus osvaldoespinozai Sharkey, sp. nov.
http://zoobank.org/1CF9C9DC-F60A-4925-B275-B132541DDF38
Figure 144

Diagnostics. BOLD:ABU8005. Consensus barcode. TGTATTATATTTTATTTTTG-
GTATATGATCTGGGATATTAGGTTTATCACTAAGTATATTAATTCGAATA-
GAATTAAGTTTGGTAGGTAGATTATTAATAAATGATCAGTTATATAATA-
GAATTGTTACTTTACATGCCTTTGTTATAATTTTTTTTATAGTTATAC-
CAATTATAATTGGTGGATTTGGTAATTGATTAATTCCTTTAATATT-
AGGTTTACCTGATATAGCATTTCCACGTATAAATAATATAAGATATT-
GACTATTAATTCCTTCTTTATTTATATTATTATTAAGTGGATTTGT-
TAATATAGGGGTAGGTACTGGATGAACAGTTTATCCTCCTTTATCTT-
TATTGATTGGTCATGGGGGAATTTCTGTAGATTTATCTATTTTTTC-
TYTACATTTAGCTGGAATATCTTCTATTATAGGGGCYATTAATTTTAT-
TACTACTAGATTAAATACTTGAATTAATAATAAGTATATGGATAAATTYC-
CTTTATTTGTTTGGTCAGTATTAATTACTGCTGTTTTATTATTATTGT-
CYTTACCTGTATTGGCAGGTGCTATTACTATATTATTAAGGGATCGAAAT-
TTAAATACCAGATTTTTTGATCCATCTGGTGGGGGGGATCCAGTTTTA-
TATCAGCATTTATTT.

Holotype ♂. Alajuela, Sector Rincon Rain Forest, Sendero Venado, 10.897, -85.27, 420 meters, caterpillar collection date: 24/iv/2011, wasp eclosion date: 21/v/2011. Depository: CNC.

Host data. *Stenoma* Janzen230 (Depressariidae) feeding on *Xylopia frutescens* (Annonaceae).

Caterpillar and holotype voucher codes. 11-SRNP-41837, DHJPAR0042858.
Paratype. Host = *Stenoma* Janzen230: DHJPAR0052184. Depository: CNC.
Etymology. *Chelonus osvaldoespinozai* is named to honor Sr. Osvaldo Espinoza of GDFCF and ACG for his many years as a dedicated inventory parataxonomist for ACG.

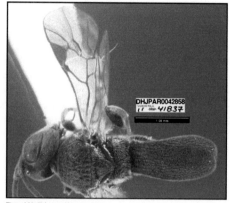

Figure 144. *Chelonus osvaldoespinozai*, holotype.

Caterpillar and holotype voucher codes. 11-SRNP-41837, DHJPAR0042858.
Paratype. Host = *Stenoma* Janzen230: DHJPAR0052184. Depository: CNC.
Etymology. *Chelonus osvaldoespinozai* is named to honor Sr. Osvaldo Espinoza of GDFCF and ACG for his many years as a dedicated inventory parataxonomist for ACG.

Fig. 4.2. An example of a minimalist revision of a parasitoid wasp. (Source: from Sharkey et al. (2021a), reproduced under terms and conditions of Creative Commons Attribution License CC-BY 4.0.)

numbers of species that there appear to be in samples. In contrast, many (but not all) subfamilies of Braconidae appear to be tropicocentric, but they have been studied far less in this respect (see *Anomalous Diversity*, below).

Precise data were absent until the use of Malaise traps became more widespread. Henry Townes was a significant figure here and largely responsible for popularising the use of the Malaise trap device for collecting ichneumonids (Townes, 1962, 1972). These devices allowed efficient, fairly standardised collecting to be carried out all over the world.

Anomalous Diversity

The husband-and-wife researchers Denis and Jennifer Owen ran such Malaise traps in the (rather extensive suburban and mature) back garden of their house in Leicester in England (Owen, 1991, 2010), and also in a garden area in Sierra Leone in West Africa (Jennifer Owen worked at Fourah Bay College, near Freetown, which subsequently became the University of Sierra Leone). Expert identification of the collected Ichneumonidae turned up a surprise result: the collected wasps were far more diverse in the temperate European localities (Owen and Owen, 1974; Owen *et al.*, 1981). This was opposite to what was known for most other groups of animals and plants in which there is a marked increase in diversity towards the (moist) tropics, and became known as 'anomalous diversity'. There are of course several other well-known groups that exhibit anomalous diversity, for example penguins, bumblebees and freshwater planktonic crustaceans. Since Ichneumonidae are a very conspicuous part of the parasitoid wasp fauna everywhere, this attracted considerable theoretical attention and several ecological and physiological hypotheses have been formulated to explain it (Quicke, 2015). The main ones are:

- resource fragmentation hypothesis (Janzen and Pond, 1975);
- a predation hypothesis (Rathke and Price, 1976); and
- 'Nasty Host Hypothesis' (NHH) (Gauld *et al.*, 1992; Gauld and Gaston, 1994).

Sime and Brower (1998) reviewed the evidence for and against each of these and concluded that 'the nasty-host hypothesis is best supported by the limited evidence available while evidence in favour of other hypotheses is either lacking or ambiguous'. But why Ichneumonidae in particular? What about other groups of parasitoid

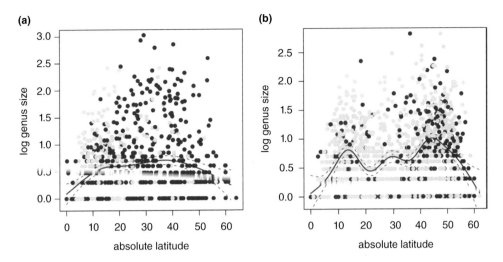

Fig. 4.3. Latitudinal trends in described species of ichneumonoid wasps: **(a)** Braconidae; **(b)** Ichneumonidae. The y-axes are logarithmic base 15. Both plots show as a red solid line the generalised additive model (GAM) fitted to the data, ± (red dashed lines) standard errors, genus data points colour coded for wasp body size (dark blue < 6.5 mm, cyan = 6.6–19.5 mm, magenta = 22.6–33 mm, yellow = > 33.1 mm). (Source: © Donald L.J. Quicke, reproduced from Quicke, 2012.)

wasps? Most, if not all, of the above hypotheses put forward to explain the anomalous diversity of ichneumonids ought to apply to all other groups of parasitoids. We do not have an answer unless that it is because, overall, Ichneumonidae are just not very good at coping with toxic environments around their developing larvae. Maybe investigation of gene families involved in detoxification in genera across a range of parasitoid hymenopteran groups might throw some light on the situation.

Do We Actually Know Enough to Trust Trends?

Quicke (2012) examined the latitudinal trends in the numbers of described species of Braconidae and Ichneumonidae in relation to biology and body size. The two graphs in Fig. 4.3 show several things but, most importantly for both families, there is a massive difference in the body sizes of taxa described between temperate and tropical latitudes. In each case, a far larger proportion of species described from low (tropical) latitudes are relatively large-bodied insects. Unless (which would be exceedingly odd) there are markedly fewer small-bodied tropical species, the data show the typical pattern – large conspicuous species get described before smaller inconspicuous ones. The described tropical faunas of both families are therefore likely to be heavily biased to large conspicuous species and so we simply do not yet have the data needed to be able to say anything about latitudinal trends with confidence.

References

Azhar, A., Hartke, T.R., Böttges, L., Lang, T., Larasati, A., Novianti, N., Tawakkal, I., Hidayat, P., Buchori, D., Scheu, S. and Drescher, J. (2022) Rainforest conversion to cash crops reduces abundance, biomass and species richness of parasitoid wasps in Sumatra, Indonesia. *Agricultural and Forest Entomology* 24, 506–515.

Bánki, O., Roskov, Y., Döring, M., Ower, G., Vandepitte, L., Hobern, D., Remsen, D., Schalk, P., DeWalt, R. E., Keping, M., Miller, J., Orrell, T., Aalbu, R., Abbott, J., Adlard, R., Adriaenssens, E. M., Aedo, C., Aescht, E., Akkari, N. *et al.* (2023). Catalogue of Life Checklist (Version 2023-01-12). Available at https://www.catalogueoflife.org. doi: 10.48580/dfqz

Bascompte, J. and Jordano, P. (2007) Plant–animal mutualistic networks: the architecture of biodiversity. *Annual Review of Ecology, Evolution, and Systematics* 38, 567–593.

Bierregaard, R.O. Jr, Lovejoy, T.E., Kapos, V., dos Santos, A.A. and Hutchings, R.W. (1992) The biological dynamics of tropical rainforest fragments. *BioScience* 42, 859–866.

Blackburn, T.M. and Gaston, K.J. (1994) Animal body size distributions change as more species are described. *Proceedings of the Royal Society of London, series B* 257, 293–297.

Butcher, B.A., Smith, M.A., Sharkey, M.J. and Quicke, D.L.J. (2012) A turbo-taxonomic study of Thai *Aleiodes (Aleiodes)* and *Aleiodes (Arcaleiodes)* (Hymenoptera: Braconidae: Rogadinae) based largely on COI bar-coded specimens, with rapid descriptions of 179 new species. *Zootaxa* 3457, 1–232.

Cancian de Araujo, B., Schmidt, S., Schmidt, O., von Rintelen, T., Ubaidillah, R. and Balke, M. (2018) The Mt Halimun-Salak Malaise Trap project – releasing the most species rich DNA Barcode library for Indonesia. *Biodiversity Data Journal* 6, e29927. doi: 10.3897/BDJ.6.e29927

Caro, T.M. and O'Doherty, G. (1999) On the use of surrogate species in conservation biology. *Conservation Biology* 13, 805–814.

Chimeno, C., Hausmann, A., Schmidt, S., Raupach, M.J., Doczkal, D., Baranov, V., Hübner, J., Höcherl, A., Albrecht, R., Jaschhof, M., Haszprunar, G. and Hebert, P.D.N. (2022) Peering into the darkness: DNA barcoding reveals surprisingly high diversity of unknown species of Diptera (Insecta) in Germany. *MDPI Insects* 13(1), 82.

Cincotta, R., Wisnewski, J. and Engelman, R. (2000) Human population in the biodiversity hotspots. *Nature* 404, 990–992.

DeFries, R., Hansen, A., Newton, A.C. and Hansen, M.C. (2005) Increasing isolation of protected areas in tropical forests over the past twenty years. *Ecological Applications* 15, 19–26.

Deplazes-Zemp, A., Abiven, S., Schaber, P., Schaepman, M., Schaepman-Strub, G., Schmid, B., Shimizu, K.K. and Altermatt, F. (2018) The Nagoya Protocol could backfire on the Global South. *Nature Ecology & Evolution* 2, 917–919.

De Prins, J. (2022) Editorial. *The Nagoya Protocol and Taxonomy. Phegea* 50, 122.

Dyer, M.I., Turner, C.L. and Seastedt, T.R. (1993) Herbivory and its consequences. *Ecological Applications* 3, 10–16.

Enderlein, G. (1920) Zur Kenntnis außereuropäischer Braconiden. *Archiv Naturgeschichte* 84A, 51–224.

Erwin, T.L. (1982) Tropical forests: their richness in Coleoptera and other arthropod species. *Coleopterist's Bulletin* 36, 74–75.

Forbes, A.A., Bagley, R.K., Beer, M.A., Hippee, A.C. and Widmayer, H.A. (2018) Quantifying the unquantifiable: why Hymenoptera, not Coleoptera, is the most speciose animal order. *BMC Ecology* 18(1), 21.

García-Robledo, C., Kuprewicz, E.K., Baer, C.S., Clifton, E., Hernández, G.G. and Wagner, D.L. (2020) The Erwin equation of biodiversity: from little steps to quantum leaps in the discovery of tropical insect diversity. *Biotropica*, 52, 590–597. https://doi.org/10.1111/btp.12811.

Gaston, K.J. (1993) Spatial patterns in the description and richness of the Hymenoptera. In: LaSalle, J. and Gauld, I.D. (eds) *Hymenoptera and Biodiversity*. CAB International, Wallingford, UK, pp. 277–293.

Gauld, I.D. and Gaston, K.J. (1994) The taste of enemy-free space: parasitoids and nasty hosts. In: Hawkins, B.A. and Sheehan, W. (eds) *Parasitoid Community Ecology*. Oxford University Press, Oxford, pp. 279–299.

Gauld, I.D., Gaston, K.J. and Janzen, D.H. (1992) Plant allochemicals, tritrophic interactions and the anomalous diversity of tropical parasitoids: The "nasty" host hypothesis. *Oikos* 65, 353–357.

Giam, X., Ng, T.H., Yap, V.B. and Tan, H.T. (2010) The extent of undiscovered species in Southeast Asia. *Biodiversity and Conservation* 19, 943–954.

Hammond, E. (2017) *Gene Sequences and Biopiracy: Protecting Benefit Sharing as Synthetic Biology Changes Access to Genetic Resources*. Third World Network Briefing Paper 93, 1–4. Third World Network, Penang, Malaysia.

Heraty, J. (2017) Parasitoid biodiversity and insect pest management. In: Foottit, R.G. and Adler, P.H. (eds) *Insect Biodiversity: Science and Society*. Wiley-Blackwell, Oxford, pp. 603–625.

Ikhsan, Z., Hidrayani, Yaherwandi, Y. and Hamid, H. (2020) The diversity and abundance of Hymenoptera insects on tidal swamp rice field in Indragiri Hilir District, Indonesia. *Biodiversitas* 21, 1020–1026.

Janzen, D.H. and Pond, C.M. (1975) A comparison by sweep-net sampling of the arthropod fauna of secondary vegetation in Michigan, England and Costa Rica. *Transactions of the Royal Entomological Society, London* 127, 33–50.

Jones, O.R., Purvis, A., Baumgart, E. and Quicke, D.L.J. (2009) Using taxonomic revision data to estimate the geographic and taxonomic distribution of undescribed species richness in the Braconidae (Hymenoptera: Ichneumonoidea). *Insect Conservation & Diversity* 2, 204–212.

Kiehn, M., Fischer, F. and Smith, P.P. (2021) The Nagoya Protocol and Access and Benefit Sharing regulations of the Convention on Biological Diversity (CBD) and its impacts on botanic gardens' collections and research. CABI Reviews, CABI International, Wallingford, UK. doi: 10.1079/PAVSNNR202116034.

Kristensen, N.P., Scoble, M.J. and Karsholt, O. (2007) Lepidoptera phylogeny and systematics: the state of inventorying moth and butterfly diversity. *Zootaxa* 1668, 699–747.

Lovell, S., Hamer, M., Slotow, R. and Herbert, D. (2007) Assessment of congruency across invertebrate taxa and taxonomic levels to identify potential surrogates. *Biological Conservation* 139, 113–125.

Lyal, C.H.C. (2019) Current situation on Digital Sequence Information (DSI). In: Chege Kamau, E. (ed.) *Implementation of the Nagoya Protocol. Fulfilling new obligations among emerging issues*. BfN-Skripten 564. Bonn, Germany, Bundesamt für Naturschutz, pp. 119–124.

Matthews, R.W. and Matthews, J.R. (1970) Malaise trap studies of flying insects in a New York mesic forest I. Ordinal composition and seasonal abundance. *Journal of the New York Entomological Society* 78, 52–59.

Mazón, M. (2016) Taking shortcuts to measure species diversity: parasitoid Hymenoptera subfamilies as surrogates of species richness. *Biodiversity and Conservation* 25, 67–76.

Meier, R., Blaimer, B., Buenaventura, E., Hartop, E., von Rintelen, T., Srivathsan, A. and Yeo, D. (2022) A re-analysis of the data in Sharkey et al.'s (2021) minimalist revision reveals that BINs do not deserve names, but BOLD Systems needs a stronger commitment to open science. *Cladistics* 38, 264–275.

Meierotto, S., Sharkey, M.J., Janzen, D.H., Hallwachs, W., Hebert, P.D.N., Chapman, E.G. and Smith, M.A. (2019) A revolutionary protocol to describe understudied hyper-diverse taxa and overcome the taxonomic impediment. *Deutsche Entomologische Zeitschrift* 66, 119–145.

Montgomery, G.A., Dunn, R.R., Fox, R., Jongejans, E., Leather, S.R., Saunders, M.E., Shortall, C.R., Tingley, M.W. and Wagner, D.L. (2019) Is the insect apocalypse upon us? How to find out. *Biological Conservation* 233, 332–333.

Moreno, C.E., Guevara, R., Sánchez-Rojas, G., Téllez, D. and Verdú, J.R. (2008) Community level patterns in diverse systems: a case study of litter fauna in a Mexican pine-oak forest using higher taxa surrogates and resampling methods. *Acta Oecologica* 33, 73–84.

Morgera, E., Tsioumani, E. and Buck, M. (2015) *Unraveling the Nagoya Protocol: A Commentary on the Nagoya Protocol on Access and Benefit-Sharing to the Convention on Biological Diversity*. Brill, Leiden, The Netherlands.

Myers, N. (1988) Threatened biotas: hotspots in tropical forests. *Environmentalist* 8, 178–208.

Nguyen, D.T.C. and Tran, T.H.T. (2018) Vietnam's legislation on access to genetic resources and benefit sharing (ABS): achievements, remained weakness and solutions. *Journal of Korean Law* 18, 139–155.

Noyes, J.S. (2019) Universal Chalcidoidea Database. World Wide Web electronic publication. 1304. Available at http://www.nhm.ac.uk/chalcidoids (accessed 12 July 2023).

Owen, D.F. and Owen, J. (1974) Species diversity in temperate and tropical Ichneumonidae. *Nature* 249, 583–584.

Owen, J. (1991) *The Ecology of a Garden: The First Fifteen Years*. Cambridge University Press, Cambridge, UK.

Owen, J. (2010) *Wildlife of a Garden: A Thirty-Year Study*. Royal Horticultural Society, London.

Owen, J., Townes, H. and Townes, M. (1981) Species diversity of Ichneumonidae and Serphidae (Hymenoptera) in an English suburban garden. *Biological Journal of the Linnean Society* 16, 315–36.

Plant, A.R., Surin, C., Saokhod, R. and Srisuka, W. (2012) Higher taxon diversity, abundance and community structure of Empididae, Hybotidae and Brachystomatidae (Diptera: Empidoidea) in tropical forests – results of mass-sampling in Thailand. *Studia Dipterologica* 18, 121–149.

Polaszek, A., Fusu, L., Viggiani, G., Hall, A., Hanson, P. and Polilov, A.A. (2022) Revision of the World Species of *Megaphragma* Timberlake (Hymenoptera: Trichogrammatidae). *Insects* 13, 561.

Polilov, A.A. (2017) Anatomy of adult *Megaphragma* (Hymenoptera: Trichogrammatidae), one of the smallest insects, and new insight into insect miniaturization. *PLoS ONE* 12, e0175566.

Polilov, A.A., Hakimi, K.D. and Makarova, A.A. (2023) Extremely small wasps independently lost the nuclei in the brain neurons of at least two lineages. *Scientific Reports* 13, 4320. doi.org/10.1038/s41598-023-31529-4

Prathapan, K.D., Pethiyagoda, R., Bawa, K.S., Raven, P.H., Rajan, P.D., Acosta, L.E. [plus full list of signatories provided in the supplementary materials to the paper] (2018) When the cure kills – CBD limits biodiversity research. *Science* 360(6396), 1405–1406.

Putra, I.L.I. (2018) Diversity of parasitoid Hymenoptera in banana germplasm plantation Yogyakarta. *Jurnal Biologi Udayana* 23, 26–33.

Quicke, D.L.J. (2012) We know too little about parasitoid wasp distributions to draw any conclusions about latitudinal trends in species richness, body size and biology. *PLoS ONE* 7(2): e32101. doi: 10.1371/journal.pone.0032101

Quicke, D.L.J. (2015) *The Braconid and Ichneumonid Parasitic Wasps: Biology, Systematics, Evolution and Ecology*. Wiley Blackwell, Oxford, 688 pp.

Rathcke, B.J. and Price, P.W. (1976) Anomalous diversity of tropical ichneumonoid parasitoids: a predation hypothesis. *The American Naturalist* 110, 889–893.

Sann, C., Wemheuer, F., Beaurepaire, A., Daniel, R. Erler, S. and Vida, S. (2018) Preliminary investigation of species diversity of rice hopper parasitoids in Southeast Asia. *Insects*, 9: 19.

Saunders, T.E. and Ward, D.F. (2018) Variation in the diversity and richness of parasitoid wasps based on sampling effort. *PeerJ* 6, e4642.

Schwartz, M.W., Brigham, C.A., Hoeksema, J.D., Lyons, K.G., Mills, M.H. and van Mantgem, P.J. (2000) Linking biodiversity to ecosystem function: implications for conservation ecology. *Oecologia* 122, 297–305.

Secretariat of the Convention on Biological Diversity (2011) Nagoya Protocol. Available at: https://www.cbd.int/abs/doc/protocol/nagoya-protocol-en.pdf (accessed 7 March, 2023)

Sharkey, M.J. and Clutts, S.A. (2010) A revision of Thai Agathidinae (Hymenoptera, Braconidae), with descriptions of six new species. *Journal of Hymenoptera Research* 22, 69–132.

Sharkey, M.J., Janzen, D.H., Hallwachs, W., Chapman, E.G., Smith, M.A., Dapkey, T., Brown, A., Ratnasigham, S., Naik, S., Manjunat, R., Perez, K., Milton, M., Hebert, P.D.N., Shaw, S.R., Kittel, R.N., Solis, A., Metz, M., Goldstein, P.Z., Brown, J.W., Quicke, D.L.J., van Achterberg, C., Brown B.V. and Burns, J.M. (2021a) Minimalist revision and description of 411 new species in 11 subfamilies of Costa Rican braconid parasitic wasps, including host records. *Zootaxa* 1013, 1–665.

Sharkey, M.J., Baker, A., McCluskey, K., Smith, M.A., Naik, S., Ratnasingham, S., Manjunath, R., Perez, K., Sones, J., D'Souza, M., Jacques, B.S., Hebert, P., Hallwachs, W. and Janzen, D. (2021b) Addendum to a minimalist revision of Costa Rican Braconidae: 28 new species and 23 host records. *ZooKeys* 1075, 77–136.

Sharkey, M., Brown, B., Baker, A. and Mutanen, M. (2021c) Response to Zamani et al. (2020) The omission of critical data in the pursuit of "revolutionary" methods to accelerate the description of species. *ZooKeys* 1033, 191–201.

Sharkey, M.J., Tucker, E.M., Baker, A., Smith, M.A., Ratnasingham, S., Manjunath, R., Hebert, P., Hallwachs, W. and Janzen, D (2022) More discussion of minimalist species descriptions and clarifying some misconceptions contained in Meier et al. 2021. *ZooKeys* 1110, 135–149.

Sime, K.R. and Brower, A.V.Z. (1998) Explaining the latitudinal gradient anomaly in ichneumonid species richness: evidence from butterflies. *Journal of Animal Ecology* 67, 387–399.

Smith, M.A., Woodley, N.E., Janzen, D.H., Hallwachs, W. and Hebert, P.D.N. (2006) DNA barcodes reveal cryptic host-specificity within the presumed polyphagous members of a genus of parasitoid flies (Diptera: Tachinidae). *Proceedings of the National Academy of Sciences of the United States of America* 103, 3657–3662. doi: 10.1073/pnas.0511318103

Smith, M.A., Wood, D.M., Janzen, D.H., Hallwachs, W. and Hebert, P.D.N. (2007) DNA barcodes affirm that 16 species of apparently generalist tropical parasitoid flies (Diptera, Tachinidae) are not all generalists. *Proceedings of the National Academy of Sciences of the United States of America* 104, 4967–4972. doi: 10.1073/pnas.0700050104

Smith, M.A., Rodriguez, J.J., Whitfield, J.B., Deans, A.R., Janzen, D.H., Hallwachs, W. and Hebert, P.D.N. (2008) Extreme diversity of tropical parasitoid wasps exposed by iterative integration of natural history, DNA barcoding, morphology, and collections. *Proceedings of the National Academy of Sciences of the United States of America* 105, 12359–12364.

Sodhi, N.S., Koh, L.P., Brook, B.W. and Ng, P.K. (2004) Southeast Asian biodiversity: an impending disaster. *Trends in Ecology & Evolution* 19, 654–660.

Songvorawit, N., Quicke, D.L.J. and Butcher, B.A. (2021) Taxonomic progress and diversity of ichneumonoid wasps (Hymenoptera: Ichneumonoidea) in Southeast Asia. *Tropical Natural History* 21, 79–93.

Townes, H.K. (1962) Design for a Malaise trap. *Proceedings of the Entomological Society of Washington* 64, 253–262.

Townes, H.K. (1972) A light weight Malaise trap. *Entomological News* 83, 239–247.

Tran, T.H.T., Nguyen, B.T. and Nguyen, D.T. (2019) Viet Nam: New ABS legislation and practice, compliance with the Nagoya Protocol. In: Chege Kamau, E. (ed.) *Implementation of the Nagoya Protocol. Fulfilling New Obligations Among Emerging Issues. BfN-Skripten 564*. Bundesamt für Naturschutz, Bonn, Germany, pp. 15–21.

Vieira, L.C., Oliveira, N.G., Brewster, C.C. and Gayubo, S.T. (2012) Using higher taxa as surrogates of species level data in three Portuguese protected areas: a case study of Spheciformes (Hymenoptera). *Biodiversity and Conservation* 21, 3467–3486.

Wagner, D.L. (2020) Insect declines in the Anthropocene. *Annual Review of Entomology* 65, 457–480.

Webb, C.O., Slik, J.F. and Triono, T. (2010) Biodiversity inventory and informatics in Southeast Asia. *Biodiversity and Conservation* 19, 955–972.

Yu, D.S., van Achterberg, C, and Dorstmann, K. (2016) *World Ichnemonoidea. Taxonomy, Biology, Morphology and Distribution*. Nepean, Ottawa, Canada.

Zamani, A., Vahtera, V., Sääksjärvi, I.E. and Scherz, M.D. (2020) The omission of critical data in the pursuit of "revolutionary" methods to accelerate the description of species. *Systematic Entomology* 46, 1–4.

Zamani, A., Fric, Z.F., Gante, H.F., Hopkins, T., Orfinger, A.B., Scherz, M.D., Bartoňová, A.S. and Pos, D.D. (2022) DNA barcodes on their own are not enough to describe a species. *Systematic Entomology* 47, 385–389.

5

Classification and Phylogeny

Abstract

The history of Hymenoptera classification and particularly that of parasitoid wasps is discussed, starting with general opinions about groups, through formal morphological phylogenetic analyses, followed by analysis of data from a few specific genes, to the latest genomic research. The latest study, Blaimer *et al.* (2023), forms the basis for the arrangement used here. It appears now to be well supported that the Ichneumonoidea are the sister group to all the rest of Parasitica+Aculeata. The latter comprises two large groups: (1) the Proctotrupomorpha and (2) the Evaniomorpha containing the superfamilies, Stephanoidea, Evanioidea, Megalyroidea, Trigonalyoidea, as well as the Aculeata.

The order Hymenoptera comprises the familiar sawflies, bees, social wasps and ants together with a diverse array of families of parasitoid wasps and their kin, which constitute by far the greater part of the species richness. The most basal Hymenoptera are the phytophagous sawflies and woodwasps, a paraphyletic grade taxon, often referred to as the 'Symphyta'. From these, the parasitoid Hymenoptera evolved, and subsequently from within the latter, the monophyletic Aculeata, or 'stinging wasps' evolved, although many aculeate wasps are also parasitoids.

Here we briefly summarise progress towards current understanding of the evolutionary relationships between major hymenopteran groups, with particular emphasis on the traditional 'Parasitica' families, although parasitoid biology is also obligate for many families of aculeate wasps.

Classification Overview

There have been relatively few major changes in the overall classification of the order over the past 100 or more years, until the advent of DNA. Whilst the relationships between non-aculeate parasitoid groups appear to have settled into stability at superfamily level, there are some differences of opinion when it comes to the aculeates. Aguiar *et al.* (2013) provided a comprehensive and annotated list of all Hymenoptera superfamilies and families, extinct and extant, together with their approximate numbers of genera and species. It does not, however, recognise the several aculeate superfamilies that have been split (or reinstated) from Vespoidea *s.l.*, i.e. Formicoidea, Pompiloidea, Scolioidea, Thynnoidea, as recognised by Branstetter *et al.* (2017) and Peters *et al.* (2017).

The phytophagous sawflies and woodwasps were long believed to be basal, and, because they are easily recognised by their lack of a wasp waist and have generally more complete venation and (with one rare exception) are not parasitoids, these were lumped into the traditional suborder 'Symphyta'. We place the name in inverted commas because the group is not monophyletic.

The wasp-waisted wasps traditionally were treated as two further suborders: the 'Parasitica' comprising the majority of species with parasitoid biology, and the Aculeata in which all the wasps possessing a stinger were placed, together with various related, non-stinging parasitoids. Again 'Parasitica', since it is quite a useful term, is placed in inverted commas because it is not monophyletic, but Aculeata is not because they are monophyletic. Nowadays, suborders are not recognised although Russian authors in particular, especially palaeoentomologists, use a number of infraorder names.

The 'one rare exception' referred to is the Orussidae (Orussoidea), which lack a wasp waist and are generally sawfly-like except for their internalised thin flexible ovipositor, but are parasitoids. All evidence placed them as the sister group to the 'Parasitica' + Aculeata, a very early evolution of parasitoidism. The most basal extant sawflies are the Xyelidae and their very complete wing venation helps in the interpretation of veins in other wasp groups.

The extant species are highly derived, but various fossils of related extinct families have a more typical parasitoid appearance. The 'Parasitica' + Aculeata are collectively called the Apocrita, referring to their wasp waists, whereas the combination Orussoidea + 'Parasitica' + Aculeata is sometimes referred to as the Vespina (e.g. Nyman et al., 2019).

Hymenoptera Phylogeny Research

Morphological Studies

For many years most workers considered that the Ichneumonoidea were the sister group to the Aculeata, based on several shared features, and especially that they both possess flap-like articulated structures from the lower ovipositor/stinger valves, called valvilli, that project into the egg/venom canal (Quicke et al., 1992).

Major morphological phylogenetic investigations are:

- Ronquist et al. (1999), who formally analysed a dataset constructed by the famous Russian palaeoentomologist Alexander Pavlovich Rasnitsyn;
- Sharkey and Roy (2002), who re-analysed Ronquist et al.'s data and showed that the recovered relationships among the Apocrita were strongly influenced by losses of wing veins;
- Sharkey (2007), who summarised previous studies and presented an intuitive phylogeny, which, with major exceptions of the position of the Ichneumonoidea and Ceraphronoidea, quite closely resembles the tree generated by Peters et al. (2017) (see below);
- Vilhelmsen et al. (2010), based on many mesosomal characters; and
- Sharkey et al. (2012), who presented a comprehensive combined morphological analysis including representatives of all extant superfamilies, based on 392 morphological characters, and sequence data for four genes (18S, 28S, COI and EF1-α).

Hardly surprisingly, given their better state of classification, there have been quite a number of morphological cladistic analyses of the aculeate Hymenoptera as a whole (e.g. Brothers, 1975) and of particular groups such as the Chrysidoidea (Brothers and Carpenter, 1993).

It is worth mentioning that in the days of morphological cladistic studies of parasitoid evolution, there was a very strong dogma that the ancestral biology of the Parasitica and within most superfamilies, was idiobiont parasitoidism of deeply concealed hosts, because this is how the parasitoid life history strategy seems to have evolved from woodwasp-like ancestors (Quicke, 2015). It was also deemed that it would be evolutionarily far easier to evolve from an idiobiont ectoparasitoid to an endoparasitoid and thence to a koinobiont, than from koinobiont to idiobiont. The advent of molecular phylogenetics has shown that such reversals have happened probably numerous times, especially in the Chalcidoidea.

Molecular Studies (to 2017)

Peters *et al.* (2017) employed a bioinformatics pipeline to retrieve and process more than 120,000 DNA sequences representing more than 1100 species and 80,000 informative sites and obtained the relationship Stephanoidea + (Ichneumonoidea + (Proctotrupomorpha + (Evanioidea + Aculeata))). The problem with these analyses is that there is inevitably very uneven gene coverage.

Klopfstein *et al.* (2013) analysed objectively aligned 18S and 28S rDNA data together with sequences from one mitochondrial and four nuclear protein-coding genes and also carried out combined molecular and morphological analyses based on the data matrix of Sharkey *et al.* (2012). Their analyses supported monophyly of Unicalcarida (the Apocrita plus the symphytan group Cephoidea, Siricoidea, Xiphidrioidea and Orussoidea), Vespina, Apocrita, Proctotrupomorpha and core Proctotrupomorpha, and the protein-coding sequences additionally indicated that the Aculeata are nested within a paraphyletic Evaniomorpha (Fig 5.1).

Many workers are following the phylogeny from Peters *et al.* (2017) probably because it was based on data from 3256 protein-coding genes from 173 insect species, without doubt a very formidable dataset. Whilst the traditional 'Proctotrupomorpha' was recovered, the sister-group relationship between the Ichneumonoidea and the Ceraphronoidea makes very little morphological sense. In the same year, Branstetter *et al.* (2017) carried out a similarly large analysis with greater emphasis on the aculeates but also including a good representation of the 'Parasitica'. Their tree is shown in Fig. 5.3.

Blaimer *et al.* (2023)

Following the molecular studies published in 2017 (Branstetter *et al.*, 2017; Peters *et al.*, 2017) there was a bit of a gap. As we were about to submit the manuscript of this volume in March 2023, a new and very impressive study had just been published. Blaimer *et al.* (2023) assembled an ultraconserved elements (UCEs) dataset for 771 species representing 94 out of 109 recognised extant families and belonging to all 22 recognised superfamilies. Some criticisms levelled at the earlier molecular studies of having rather incomplete taxon sampling (see e.g. Brothers, 2021) are thus negated. There is thankfully considerable similarity in the recovered relationships to those found by the earlier molecular investigations. However, previous studies had varied in the position of the Ichneumonoidea. Branstetter *et al.* (2017) had recovered them as the sister group to all other Apocrita whereas Klopfstein *et al.* (2013) and Peters *et al.* (2017) had both recovered them in rather derived positions (Figs 5.1–5.3).

The new arrangement of higher groups (Fig. 5.4) thus includes: (i) a monophyletic Apocrita; (ii) a monophyletic basal Ichneumonoidea; (iii) a monophyletic Proctotrupomorpha; (iv) a monophyletic Aculeata; and (v) a monophyletic 'Evaniomorpha'+Aculeata, but the former informal Evaniomorpha are paraphyletic. At the base of the Aculeata, nodes generally have less support and the Chrysidoidea was not recovered as monophyletic, and their paraphyly with respect to the other aculeates should probably not be given any great weight at this time.

In addition to generating a phylogeny (Fig. 5.4), Blaimer *et al.* (2023) included fossil calibration points enabling them to estimate the times of origins of major clades (Fig. 5.5) which they then went on to use to see what factors might have been associated with increased rates of diversification. In particular they examined the wasp waist of Apocrita, the evolution of parasitoid lifestyle, the loss of use of the ovipositor for oviposition (i.e. the evolution of a stinger in the Aculeata), and the evolution of secondary phytophagy, including pollen feeding in Apoidea and plant galling in Cynipoidea and Chalcidoidea.

As was previously obvious, parasitoidism has been the dominant strategy since the

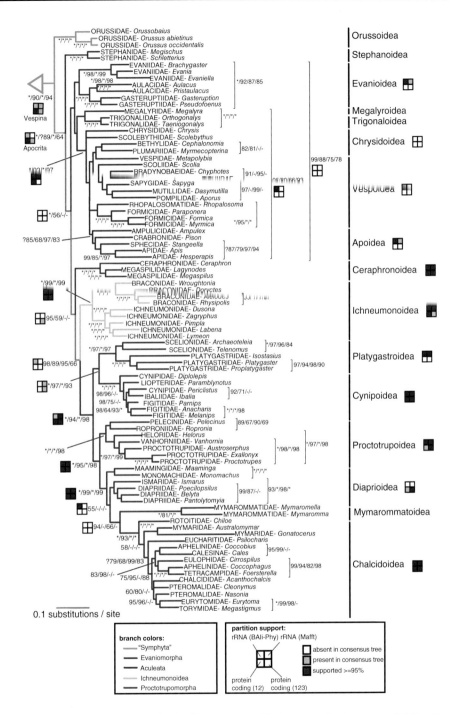

Fig. 5.1. Bayesian phylogenetic tree for the Hymenoptera with strong emphasis on parasitoid families based on combined protein-coding, RNA and morphological data, showing total and partitioned support values. (Source: from Klopfstein *et al.*, 2013, reproduced under terms and conditions of Creative Commons Licence CC.BY 4.0.)

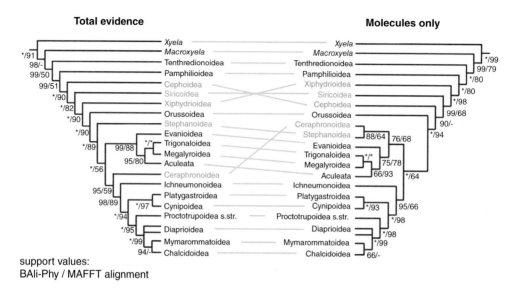

Fig. 5.2. Simplified total-evidence tree based on the combined molecular and morphological data contrasted with the tree obtained from the molecular data alone. Support values are from BAli-Phy-based and MAFFT alignments of the rRNA data; asterisks represent 100% support. As regards parasitoids, note the conflicting placements of Stephanoidea and Ceraphronoidea. (Source: from Klopfstein *et al.* (2013), reproduced under terms and conditions of Creative Commons Licence CC.BY 4.0.)

Late Triassic, some 220 million years ago, but their analysis concluded that it was not an immediate driver of diversification, whereas there was strong evidence that transitions to secondary phytophagy were. Many of the early parasitoid families are extinct but there was quite a family-level diversity in the late Jurassic and early Cretaceous (Rasnitsyn, 2002).

It may seem surprising that parasitoidism did not stand out as a marked promoter of diversification. In trees, based on all fossil groups, it certainly seems to, at least on superficial inspection. We suspect that really parasitoidism is a spectrum. In the earliest parasitoids the mother probably stabbed the host with her ovipositor, so incapacitating it, and then her larvae would feed up on it quickly. This is not dissimilar to a pride of lions feeding over the course of a few days on the carcass of a large prey they have brought down – a vaguely disguised form of predation. From that, it is a very long evolutionary journey to the complicated physiological host–parasitoid interactions of koinobiont parasitoids, perhaps especially those such as microgastrine braconids with their endogenised polydnavirus genomes (see Chapters 2 and 17, this volume).

Conclusions

The message we take away from the recent phylogenetic studies is that morphological conclusions about relationships cannot generally be relied upon. In placing extinct taxa, morphology is the only way of course, but ever-increasing types and volumes of molecular data suggest relationships that are very often incongruent with morphology. Conflicts between molecular stories show that molecular data based on only a small number of genes are not infallible. However, the huge sequence datasets now becoming available seem overwhelming.

Why such conflicts? Evolution is the simple answer. Evolution works to promote successful genotypes and not to make the life of the taxonomist/systematist easy. Homoplasy, i.e. parallel evolutionary acquisition

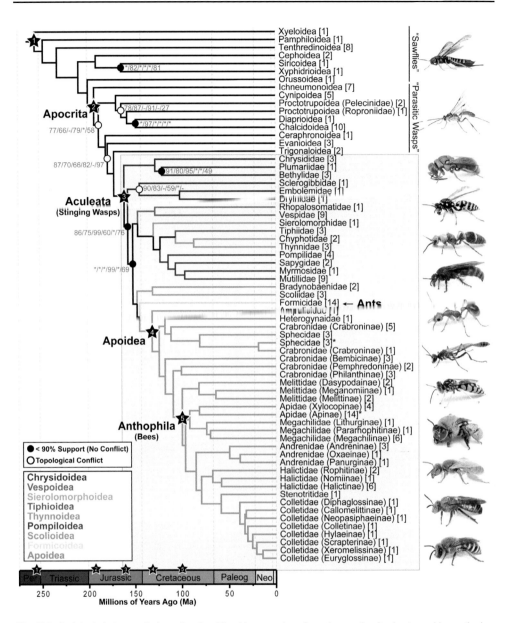

Fig. 5.3. A dated phylogenetic hypothesis of the Hymenoptera focusing on the Aculeata and in particular, the bees (Apoidea). (Source: from Branstetter et al., 2017, reproduced under terms and conditions of Creative Commons licence CC.BY 4.0.)

of similar characters given a common selection pressure (convergent evolution), is so powerful. This includes also parallel losses: when a character state no longer confers a benefit, it will be lost. The loss may be passive, but much more likely it will be subject to passive selection (a random walk), because expressing almost any character incurs some evolutionary cost, no matter how slight.

Finally, while we expect most readers to have some familiarity with the general

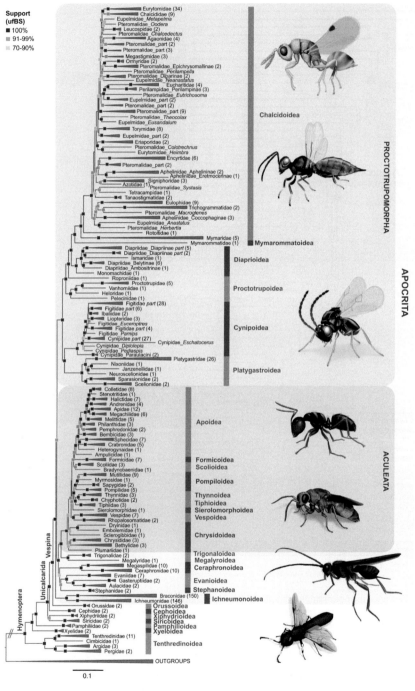

Fig. 5.4. Maximum-likelihood Hymenoptera phylogeny estimate based on 446 UCE loci in the 70% complete nucleotides data subset. Many nodes are shown collapsed, with the numbers of species included in each shown in brackets. Ultrafast bootstrap support values (ufBS) are indicated by coloured squares (note legend top left). (Source: from Blaimer *et al*. (2023), reproduced under terms of Creative Commons Attribution 4.0 International License; to view a copy of this license, visit http://creativecommons.org/ licenses/by/4.0/.)

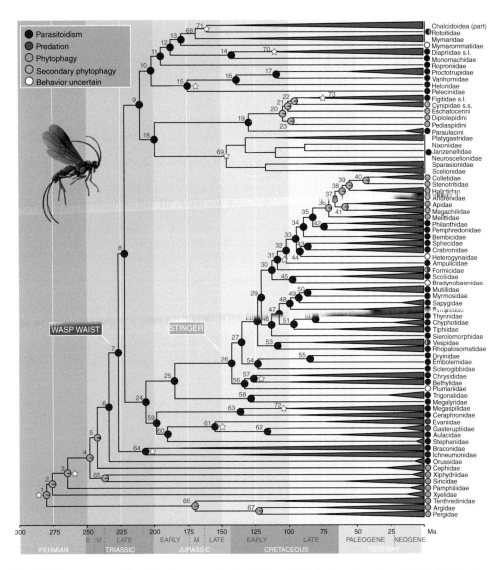

Fig. 5.5. Chronogram estimated using approximate likelihood with the topology shown in Fig. 5.4. Twelve fossil calibration points are indicated by a white star. Pie charts indicate ancestral state reconstructions. (Source: from Blaimer *et al.* (2023), reproduced under terms of Creative Commons Attribution 4.0 International License; to view a copy of this license, visit http://creativecommons.org/licenses/by/4.0/.)

rational of obtaining and analysing DNA sequence data because that has been around for the past 30 years, not everyone may be familiar with ultraconserved element (UCE) studies, which have basically taken over higher-level molecular phylogenetics. We point readers to the excellent review of Zhang *et al.* (2019).

References

Aguiar, A.P., Deans, A.R., Engel, M.S., Forshage, M., Huber, J.T. *et al.* (2013) Order Hymenoptera. In: Zhang, Z.-Q. (ed.) Animal biodiversity: an outline of higher-level classification and survey of taxonomic richness (Addenda 2013). *Zootaxa* 3703, 51–62.

Blaimer, B.B., Santos, B.F., Cruaud, A., Gates, M.W., Kula, R.R., Mikó, I., Rasplus, J.-Y., Smith, D.R., Talamas, E.J., Brady, S.G. and Buffington, M.L. (2023) Key innovations and the diversification of Hymenoptera. *Nature Communications* 14, 1212.

Branstetter, M.G., Danforth, B.N., Pitts, J.P., Faircloth, B.C., Ward, P.S., Buffington, M.L., Gates, M.W., Kula, R.R. and Brady, S.G. (2017) Phylogenomic insights into the evolution of stinging wasps and the origins of ants and bees. *Current Biology* 27, 1019–1025. doi: 10.1016/j.cub.2017.03.027

Brothers, D.J. (1975) Phylogeny and classification of the aculeate Hymenoptera, with special reference to Mutillidae. *University of Kansas Science Bulletin* 50, 483-648.

Brothers, D.J. (2021) Aculeate Hymenoptera: Phylogeny and Classification. In: Starr, C.K. (ed.) *Encyclopedia of Social Insects*. Springer Nature Switzerland AG, Cham, Switzerland, pp. 3–11.

Brothers, D.J. and Carpenter, J.M. (1993) Phylogeny of Aculeata: Chrysidoidea and Vespoidea (Hymenoptera). *Journal of Hymenoptera Research* 2, 227–304.

Klopfstein, S., Vilhelmsen, L., Heraty, J.M., Sharkey, M. and Ronquist, F. (2013) The Hymenopteran tree of life: evidence from protein-coding genes and objectively aligned ribosomal data. *PLoS ONE* 8(8), e69344.

Nyman, T., Onstein, R.E., Silvestro, D., Wutke, S., Taeger, A., Wahlberg, N., Blank, M. and Malm, T. (2019) The early wasp plucks the flower: disparate extant diversity of sawfly superfamilies (Hymenoptera: 'Symphyta') may reflect asynchronous switching to angiosperm hosts. *Biological Journal of the Linnean Society* 128, 1–19.

Quicke, D.L.J. (2015) *The Braconid and Ichneumonid Parasitic Wasps: Biology, Systematics, Evolution and Ecology*. Wiley Blackwell, Oxford, 688 pp.

Quicke, D.L.J., Ingram, S.N. and Fitton, M.G. (1992) Phylogenetic implications of the structure and distribution of ovipositor valvilli in the Hymenoptera. *Journal of Natural History* 26, 587–608.

Peters, R.S., Krogmann, L., Mayer, C., Donath, A., Gunkel, S. et al. (2017) Evolutionary history of the Hymenoptera. *Current Biology* 27, 1013–1018.

Rasnitsyn, A.P. (2002) Superorder Vespidea Laicharting, 1781. Order Hymenoptera Linne, 1758 (= Vespida Laicharting, 1781). In: Rasnitsyn, A.P. and Quicke, D.L.J. (eds) *History of Insects*. Kluwer Academic, Dordrecht, pp. 242–254.

Ronquist, F., Rasnitsyn, A.P., Roy, A., Erikson, K. and Lindgren, M. (1999) Phylogeny of the Hymenoptera: A cladistic reanalysis of Rasnitsyn's (1988) data. *Zoologica Scripta* 28, 13–50.

Sharkey, M.J. (2007) Phylogeny and classification of Hymenoptera. *Zootaxa* 1668, 521–548.

Sharkey, M.J. and Roy, A. (2002) Phylogeny of the Hymenoptera: a reanalysis of the Ronquist et al. (1999) reanalysis, emphasising wing venation and apocritan relationships. *Zoologica Scripta* 31, 57–66.

Sharkey, M.J., Carpenter, J.M., Vilhelmsen, L., Heraty, J., Liljeblad, J., Dowling, A.P.G., Schulmeister, S., Murray, D., Deans, A.R., Ronquist, F., Krogmann, L. and Wheeler, W.C. (2012) Phylogenetic relationships among superfamilies of Hymenoptera. *Cladistics* 27, 1–33.

Vilhelmsen, L., Mikó, I. and Krogmann, L. (2010) Beyond the wasp-waist: structural diversity and phylogenetic significance of the mesosoma in apocritan wasps (Insecta: Hymenoptera). *Zoological Journal of the Linnean Society* 159, 22–194,

Zhang, Y.M., Williams, J.L. and Lucky, A. (2019) Understanding UCEs: a comprehensive primer on using ultraconserved elements for arthropod phylogenomics. *Insect Systematics and Diversity* 3, 5:3. doi: 10.1093/isd/ixz016

6

Morphology

Abstract

Basic body divisions of apocritan parasitoid wasps are illustrated and described. Sections are devoted to the structure of the head (including the antennae and mouthparts), mesosoma, legs, wings, metasoma, the ovipositor, male genitalia and cuticular sculpture terminology. The wing venation and its terminology as used by different authors is compared in detail, various systems are compared in tables to facilitate understanding the diverse literature, and the system used in this book is thoroughly explained.

The morphological terminology needed to understand and use identification keys to the Hymenoptera is particularly challenging. Workers on different groups have often had their own systems. Very old publications often did not appreciate the fact that the wasp waist does not separate the thorax from the abdomen but rather the 1st abdominal segment from the rest. Wing venation is very important for identification at all levels but several different terminologies have been employed, often with workers in Russia, Western Europe and North America each adopting different systems.

A very good and thoroughly illustrated description of the cuticular structures was provided by Huber and Sharkey (1993). Karlsson and Ronquist (2012) presented a detailed description of the whole anatomy of exemplar opiine braconids. Townes' (1969) illustrations of Ichneumonidae morphology, especially of the mesosoma, are widely reproduced (e.g. Bennett *et al.*, 2019), as are van Achterberg's (1979, 1988) of a braconid. Bouček (1988) provided detailed illustrations of body parts important for chalcidoid recognition.

Yoder *et al.* (2010) introduced an extremely useful online resource, *The Hymenoptera Anatomy Ontology Portal* (available at http://portal.hymao.org/projects/32/public/ontology/search, accessed 13 December, 2022) which provides a tree-like set of definitions for basically all hymenopteran anatomical characters.

Body Regions and the Wasp Waist

All insects have three major body regions, sometimes called tagmata:

- head – bearing the antennae, eyes (usually), ocelli (usually) and mouthparts;
- thorax – bearing the wings (usually) and three pairs of legs; and
- abdomen – containing most of the gut, reproductive organs and terminal genitalia and cerci.

It is therefore no wonder that when early entomologists looked at a typical wasp-waisted hymenopteran they termed the three conspicuous body parts head, thorax and abdomen. This usage was normal in the 19th century and you may need to be aware of that when trying to interpret early descriptions. However, it was eventually realised that the narrow waist in the apocritan wasps (i.e. not a sawfly or woodwasp) is actually located between the 1st and 2nd abdominal segments, so the middle body region is composed of the thorax proper (three segments, each with a pair of legs) plus the 1st abdominal segment and this body region was named the mesosoma by Michener (1944) (Fig. 6.1). This is the term used now by most hymenopterists, although ant researchers tend to refer to it as the alitrunk. The 1st abdominal segment which makes up the postero-dorsal part of the mesosoma is now the propodeum. The rest of the abdomen is now mostly termed the metasoma but different terminology (e.g., gaster) is used for some groups.

Why the Wasp Waist and Why Is It Where It Is?

The generally accepted hypothesis is that two things were of great importance in the evolution and success of the Hymenoptera: a highly manoeuvrable ovipositor (later evolving into a stinger) and strong flight. A flexible joint somewhere before the apex of the abdomen involving a narrow, specialised joint

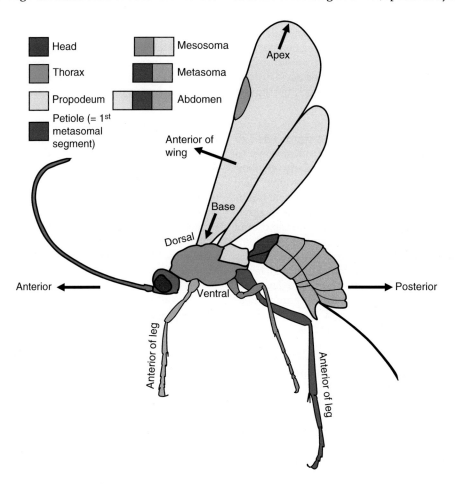

Fig. 6.1. Body regions of an apocritan (wasp-waisted) hymenopteran.

(the wasp waist) evolved under selection pressure for ovipositor manoeuvrability but selection for strong flight involves having a large and powerful main longitudinal flight muscle. This muscle inserts posteriorly onto an internal cuticular projection of the thoracic skeletomuscular apparatus, the metaphragma, which projects backwards into the anterior of the abdomen, allowing for the muscle's length. This would not be possible had the waist evolved between metathorax and abdomen, but instead evolved immediately behind the 1st abdominal segment.

The Head and Antennae

The insect head is embryologically derived from six fused segments, starting with a pre-antennal or ocular segment that lacks appendages, followed by five segments with paired 'appendages'. These are, from anterior to posterior: the antennae, the labral or intercalary segment bearing the flap-like labrum, the mandibles, the maxillary palps and the labial palps. In addition, it has a pair of compound eyes (completely lost in a few ants) and three dorsal ocelli (simple eyes). The area bearing the ocelli, bordered by a sulcus (groove), is called the stemmaticum.

There is considerable variation between hymenopteran groups in the overall shape of the head, ranging from the typical orthognathous form the mouth pointing at right angles (orthogonal) to the long axis of the body, to prognathous in which the mouth parts are directed anteriorly in line with the long axis. A few diapriids and bethylids approach the opisthognathous condition with the mouthparts pointing posteroventrally.

Basic Head Parts

In most groups of parasitoid wasps, looking at the head from the front one sees the mandibles, immediately above which are the clypeus and then the face, the latter being bordered dorsally by the antennal sockets (toruli) (Fig. 6.2 a). The part of the head between the antennal sockets and the ocelli is called the frons, which often has important taxonomic characters. Adjacent to the antennal sockets many parasitoids that pupate within a hard substrate, such as wood, have a pair of deep depressions called scrobes into which the base of the antenna (scapus and pedicellus) can lie protected, probably being important when the wasp emerges (Vilhelmsen, 1997; Vilhelmsen and Turrisi, 2011). The term scrobe usually refers to depressions on the frons, but functionally similar subantennal grooves (ventral or ventrolateral to the antennal sockets) occur in Orussidae and Megalyridae.

The parts of the head behind the ocellae and eyes have names: mediodorsally the area is called the vertex, dorsolaterally the temples, and down the side of the head towards the mandibles is the gena (genae) or cheeks (Fig. 6.2 b, c). An important character in many groups is the occipital carina which, when present, runs around the outside of the back of the head down to join the base of the mandibles. The lower lateral part of the occipital carina is often called the genal carina (Fig. 6.2 c). The back of the head behind the occipital carina is called the occiput, the ventral part of which sometimes being called the postgena. The head connects to the thorax at the small foramen magnum (the same name as the hole in the mammalian skull through which the spinal cord and brain connect). The mouthpart structures are enclosed dorsally by the foramen magnum dorsally and the ventrolateral part of the occiput, i.e. postgena, and the inner margin of the postgena is usually marked by a carina called the hypostomal carina (Fig. 6.2 c, d).

The clypeus and face are usually separated by a groove or a change in sculpture but they may be confluent (e.g. Ichneumonidae: Campopleginae) (see Fig. 10.20 a). This is sometimes referred to as the epistomal suture or epistomal sulcus. In a few groups (e.g. some Ichneumonoidea) or if the mandibles are open, then the labrum (a medial, somewhat triangular plate articulated with the clypeus) may be visible (see Figs 9.1, 9.4, 10.40 e, 10.47 c). Labrum translates as 'upper lip' and should not be confused with the more posteriorly originating labium.

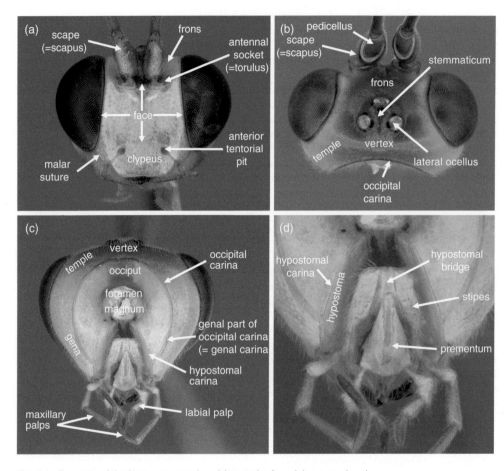

Fig. 6.2. Features of the hymenopteran head (exemplar from Ichneumoninae).

Antennae

All hymenopteran antennae comprise three parts: two basal segments called the scapus and pedicellus (the latter usually quite a lot smaller), and a multisegmented flagellum which may be further subdivided into distinct zones, sometimes with some of the segments fused. The individual segments (or articles) of the flagellum are called flagellomeres. The simplest flagellum has a series of essentially similar cylindrical segments that abut each other; broadly, this is called filiform. In some, especially rather smaller-bodied, groups the individual flagellomeres may be more bead-like and this is called moniliform. Sexual dimorphism in antennal shape is common, and the female flagellum may have a distinct terminal club (or clava) composed of one or a few expanded and fused flagellomeres. The flagellum may also gradually widen distally, also called clavate. In several groups, males may have very serrate (see Figs 13.4 c, 22.22 b) or pectinate (see Fig. 16.2 f) flagella. Some chalcidoids have one or more basal flagellomeres greatly reduced and these are called anelli (see Chapter 22, this volume).

The flagellar segments of some groups bear conspicuous elongate placoid (= multiporous plate) sensillae which are important phylogenetically and taxononomically (Basibuyuk and Quicke, 1999).

Mouthparts

All Hymenoptera have a pair of mandibles, even bees in which other parts of the mouthparts are developed into a long proboscis. The mouthparts other than the mandibles are usually largely concealed from the front but there are many exceptions in which they are enlarged and more protruding; sometimes this is part of a concealed nectar extraction apparatus (Jervis, 1998; see also Quicke, 2015).

The next developmental segment posterior to the mandibles bears a pair of maxillae which have a united basal cardo part from which arises a structure called the stipes. From the apex of each stipes emerges a segmented maxillary palp and an outer galea process and an inner lacinia process, all of which help manipulate food. Posterior to the maxillary segment is the labium, which is attached to the base of the head by a structure called the gula which is concealed except with dissection, then the large mentum (a plate-like structure comprising a basal postmentum and distal prementum) (Fig. 6.2 d). The labial palps arise laterally from the prementum, on either side of the paired glossae.

Mesosoma (the Middle Body Region Bearing the Legs)

Each of the three thoracic segments (prothorax, mesothorax, metathorax) has a dorsal part called the notum, a lateral part called the pleuron and a ventral sternum. In species that have well-developed wings, the mesothorax which bears the fore wings (the main power-generating pair) is by far the largest of the three because of its associated large flight muscles. However, in flightless species (consider worker ants) evolution rapidly adjusts the proportions because legs and walking are much more important. In these cases, the three thoracic segments tend to be of similar size. The first really detailed investigation of the hymenopteran mesosoma was by Snodgrass (1910) and 100 years later Vilhelmsen et al.'s (2010) very thorough work. Other detailed studies have been published for various specific groups: Bethylidae (Lanes et al., 2020), Chalcidoidea (Gibson, 1986; Krogmann and Vilhelmsen, 2006), Mymarommatoidea (Vilhelmsen and Krogmann, 2006), Scelionidae (Mikó et al., 2007), Tiphiidae s.l. (Boni Bartalucci, 2004).

In fully winged, flying taxa, the prothorax (particularly the pronotum as viewed from above) is a short, U-shaped structure; within it essentially are just muscles for the anterior legs. Sometimes it is not visible at all. In apterous and brachypterous species it evolves to be far larger as the importance of the fore legs for locomotion becomes relatively larger (see Figs 10.17 a, 16.10 a, 17.2, 17.3).

The mesothoracic spiracles are usually concealed and the metathoracic always concealed in fully winged species as they are located beneath the bases of the hind wings (Reid, 1941; Fedoseeva, 2017). In apterous species, the metathoracic spiracle is seen in the membrane between the meso- and metapleuron.

The mesoscutum usually displays three parts: a large medial lobe separated in some way from a pair of lateral lobes. Often the separation is in the form of a pair of longitudinal, posteriorly converging grooves called notauli, but these may be less conspicuous; they might be indicated by a change in coloration, setosity, sculpture, or perhaps not indicated at all. Notauli and their development and form are important in the recognition of many taxa. Immediately posterior to the mesoscutum is a differentiated somewhat triangular structure called the scutellum (meaning shield), sometimes called the mesoscutellum. Anteriorly it may be separated from the mesoscutum by a flexible trans-scutal suture but this is often not present, i.e. the mesoscutum and scutellum are completely fused. In addition, there it often has a separate transverse groove close to the anterior scutellar margin, the scutellar sulcus or scutellar groove (Fig. 6.3 a, b, d). Some workers confuse the terms and refer to the scutellar sulcus as the trans-scutal suture.

In Chalcidoidea, in particular, the axillae are often important. These are approximately triangular areas of the dorsal surface of the mesoscutum, lateral to the scutellum (see Fig. 22.17). Do not confuse them with axullae, which are the areas immediately posterior to them. In most groups the axillae terminate anteriorly at the same level as the scutellum (at the trans-scutal suture), but in some taxa they extend as pointed structures lateral to the lateral lobes of the mesoscutum.

The metanotum is also short in most taxa and occasionally, for example in some groups of Bethylidae, completely lost externally. When present, its median part is usually differentiated as a raised, sometimes carinate area which in some groups may be referred to as the metascutellum. Posterior to the metanotum another narrow transverse strip-like sclerite, the metapostnotum, may be visible (Fig. 6.3 c, d) but its development is quite variable (Brothers, 1976; Whitfield *et al.*, 1989); it may be fused to the anterior margin of the propodeum and differentiated from it by a groove. It has some taxonomic and more phylogenetic significance, depending on the group.

There are only two pairs of thoracic spiracles: the mesothoracic and metathoracic (Fig. 6.4). In addition, there is a propodeal spiracle.

The largest structure visible from the side in winged species is the mesosternal+mesopleural complex which is sometimes called the ventral mesopectus because there is often no external indication of the join between pleuron and sternum (Fig. 6.4). The anterior face of the mesosternal+mesopleural complex is called the prepectus which

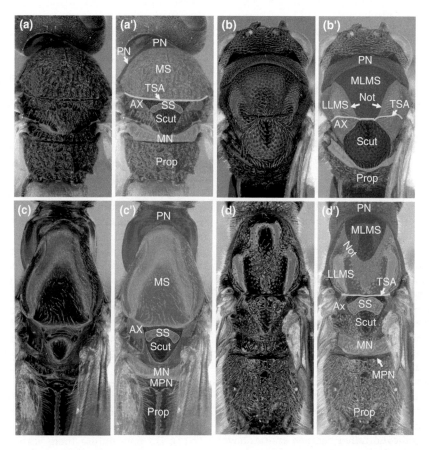

Fig. 6.3. Features of the dorsal mesosoma: (**a, a'**) *Chelonus* (Braconidae: Cheloninae); (**b, b'**) *Brachymeria* (Chalcidoidea: Chalcididae); (**c, c'**) Proctotrupidae gen. sp.; (**d, d'**) *Wroughtonia* (Braconidae: Helconinae). Abbreviations: Ax, axilla; LLMS, lateral lobe of mesoscutum; MLMS, medial lobe of mesoscutum; MN, metanotum; MPNT, metapostnotum; MS, mesoscutum (undifferentiald into distinct lobes); MT, metanotum; Not, notauli; Pron, pronotum; Prop, propodeum, Scut, scutellum; SS, scutellar sulcus; TSA, trans-scutal articulation.

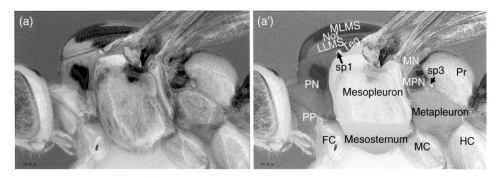

Fig. 6.4. Features of the lateral mesosoma. Abbreviations: FC, fore coxa; HC, hind coxa; LLMS, lateral lobe of mesoscutum; MC, mid coxa; MLMS, middle lobe of mesoscutum; MN, metanotum; MPN, metapostnotum; Not., notaulus; PN, pronotum; PP, propleuron; Pr, propodium; sp, spiracle.

may or may not be differentiated from the larger ventral portion by a carina (prepectal carina = epicnemial carina = epicnemium = acetabular carina). The part of the mesosternum differentiated by the epicnemial carina (or furrow) is often called the epicnemium (e.g. Kimsey and Bohart, 1990) but also sometimes the acetabulum, referring to the fact that it is sometimes a smooth surface posterior to the fore coxae (Vilhelmsen et al., 2010). The midventral part of the mesosternum often has a midlongitudinal groove called the discrimen or mesosternal suture – this may be smooth or crenulate. A small anterodorsal part of the mesopleuron is often differentiated and referred to as the subalar region or acropleuron.

Ventrally from anterior to posterior, most winged hymenopterans show the propleuron whose medial faces abut one another, and nestled between the posteroventral propleura and the fore coxae, a typically rather small prosternum (Fig. 6.5). The coxae are all nestled within smooth coxal cavities sometimes called acetabula in reference to the smooth depressions in the human pelvis where the rounded heads of the femora sit. In some groups the separate prepectus may meet medioventrally behind the fore coxae. In the bethylid example shown in Fig. 6.5 this is very obvious and the anterior half may be seen to be differentiated (anterior prepectal flange). In other groups there is no ventrally connecting part of the prepectus.

The anterior of the mesosternal+mesopleural complex often has a transverse carina. In most groups this is called the epicnemial carina and the part anterior to it is the epicnemoum. However, in some groups it is called the acetabular carina and the anterior part the acetabulum (because it is smooth and abuts the fore coxae). In the Braconidae, it is usually called the prepectal carina.

The ventral mesosternal+mesopleural complex (Fig. 6.5) posteriorly is where the mid coxae insert (mesocoxal cavities) and the metasternum may be visible between the mid coxae. In some groups (some braconids and ichneumonids in particular) there may be a transverse carina in front of the mid coxae, sort of mirroring the prepectal carina.

Legs

Insect legs comprise a basal coxa, followed by the typically small trochanter, a usually rather robust femur which contains most of the intrinsic leg musculature, a usually fairly long and thin tibia, and finally a multi-segmented tarsus (Fig. 6.6). Ancestrally, the tibiae have two apico-ventral spurs on each leg. However, in the informal group Unicalcarida (the Apocrita plus the symphytan group Cephoidea, Siricoidea, Xiphidrioidea and Orussoidea) there is only a single spur – as always there is an exception; there are two fore tibial spurs in the Ceraphronoidea, but

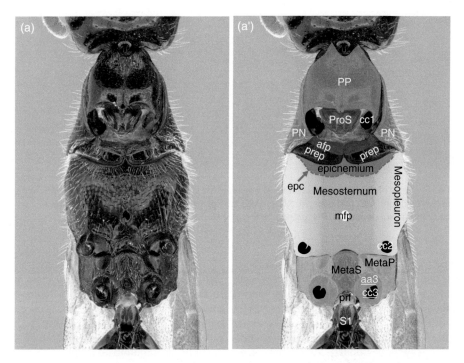

Fig. 6.5. Features of the ventral mesosoma (of a pristocerine Bethylidae). Abbreviations: aa3, acetabular area of metathorax; afp, anterior flange of prepectus; cc, coxal cavity; epc, epicnemial carina; mfp, mesofurcal pit; MetaP, metapleuron; PetaS, metastenum; PN, pronotum; PP, propleuron; prep, prepectus; prf, propodeal foramen; Pros, prosternum; S1, 1st metasomal sternite.

this is most probably an atavism. The single spur of the Unicalcarida is highly modified and is sometimes called the calcar. It operates in conjunction with a modified basal concavity of the fore basitarsus as an antennal cleaner (Basibuyuk and Quicke, 1995).

Primitively there are five tarsal segments (articles) but fewer in some groups. The 1st tarsal segment (or article) is called the basitarsus, and the terminal, claw-bearing segment is the telotarsus. The claws are usually but not always symmetric, and may be simple (that is, with the base only roundly swollen to some extent), or they may have a large pointed basal lobe, or the terminal claw may be divided (sometimes on the inner side) and this is called bifurcate. In addition, the posterior margin has a few thickened spines near its base and these are normally inconspicuous and ignored, but in some taxa these spines may be much enlarged and extend towards the end of the basal lobe or even almost towards the claw proper – this condition is called pectinate. Between the claws is a small sclerite visible from above and this shows some phylogenetically significant variation (Basibuyuk *et al.*, 2000).

Trochanter and Trochantellus

The normal insect leg comprises, from base to apex: coxa, trochanter, femur, tibia and a multisegmented tarsus. In ichneumonoids and many other Apocrita (excluding the Aculeata and Proctotrupoidea), the base of the femur is sharply demarcated by a groove (not an articulation), giving the appearance of a second trochanter, but the correct term for this is trochantellus. However, in much of the literature it is referred to as the second trochanter, for example in the keys to Ichneumonidae by Henry Townes, and some workers have called it the prefemur.

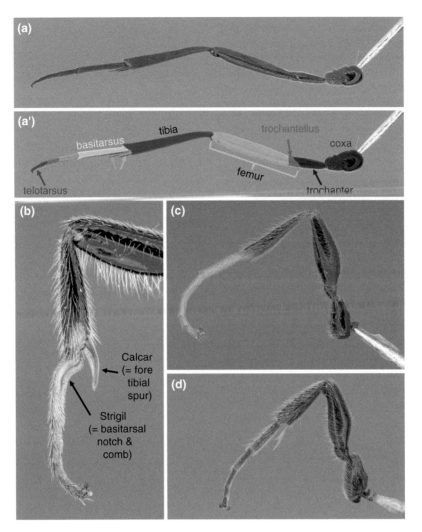

Fig. 6.6. Exemplar parasitoid wasp legs: (**a, a'**) Evaniidae hind leg showing major parts, note that the base of the femur is differentiated from the larger part by a non-flexible groove, the basal part being called the trochantellus, although it is not part of the trochanter; (**b**) part of fore leg of a bethylid showing the antennal cleaner, which is composed of a single curved fore tibial spur, and basally concave basitarsus, the two parts having opposing combs of setae; (**c, d**) hind legs of two aculeates (bethylid and tiphiid, respectively) showing that the femur lacks a differentiated trochantellus.

Wings

Hamuli and Wing Coupling

Hymenoptera have two pairs of wings but in flight they functionally have a single 'wing' surface, because the smaller hind wings are coupled to the fore wings by a system of hooks called hamuli. In the aculeate Hymenoptera, the hamuli are arranged in two groups (Basibuyuk and Quicke, 1997), the distal ones being hook or 'S'-shaped, the more basal ones called secondary hamuli being simpler curved bristles with bifurcate tips. The hamuli system is extremely efficient and is inspiring engineering developments (Eraghi et al., 2021). The fore wings provide nearly all of the power and correspondingly are driven by far larger flight muscles, which explains both the relatively large size of the mesothorax in winged species/forms as well as the wasp waist being located behind the 1st metasomal segment.

Apart from vespid social wasps, a few other groups have species that can fold their wings longitudinally and some are involved in Batesian mimicry of vespids (Danforth and Michener, 1988)

The Hymenopteran Wing Venation Ground Plan

Hymenopteran wing venation, of both fore and hind wings, despite the wide variation, derives from a similar set of longitudinal veins connected by a smaller number of transverse cross-veins. The picture becomes more complicated because 'longitudinal' veins can bend to be physically transverse, can fuse with others over part of their length, and can re-separate, or divide, and parts of longitudinal veins can be entirely lost in some taxa.

We start by showing the basic plan of longitudinal veins starting from the anterior wing margin (Fig. 6.7). These are, in order:

- Costal (abbreviated C)
- Subcostal (abbreviated SC or Sc)
- Radial (abbreviated R)
- Medial (abbreviated M)
- Cubital (abbreviated CU or Cu)
- Anal veins (there are up to three, abbreviated A1, A2 and A3)

Sharkey (1988) argued that the subcosta (SC) is probably lost in all Hymenoptera except for in the basal xyelid sawflies, but accepted the possibility that it is fused indistinguishably with either the costa or radius, and it is traditional to include it in vein names, usually as SC+R or C+SC+R. We also argue that it is lost in all parasitoid wasps but a short stub appears to be present in many sawflies (Fig. 6.8 a, b). The basal and distal true hamuli of xyelid, pamphiliid and xiphydriid sawflies are inserted on the vein at the anterior edge of the hind wing and therefore we consider them as being of costal origin (Basibuyuk and Quicke, 1997). In parasitoids which have only the distal cluster of true, hook-shaped, hamuli they are always indicated as arising from vein R (technically R1), but we believe this is untrue. In sawflies, as shown in Fig. 6.8 b, they definitely seem to arise from the costa, and they never extend distally beyond where vein R1 joins the wing margin. In larger Ichneumonoidea there are also occasional hamuli along

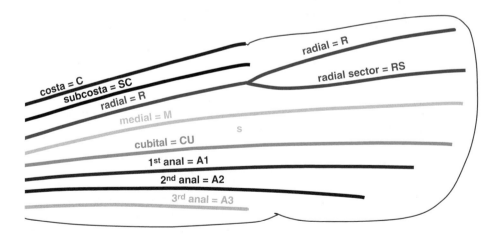

Fig. 6.7. Stylised arrangement of longitudinal veins in a hypothetical ancestral wing. In the system used here, these longitudinal veins are written in capital letters whereas cross-veins (not shown here) are given in lower case.

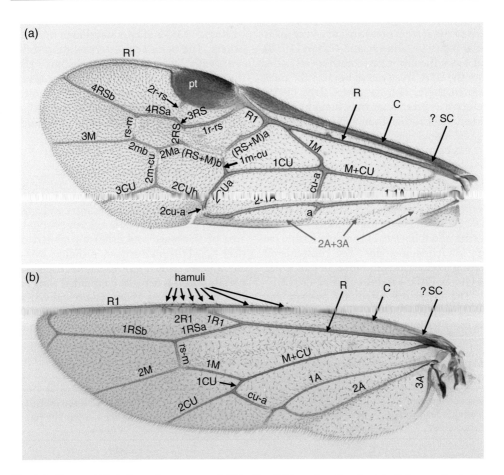

Fig. 6.8. Fore and hind wings of a non-parasitoid tenthredinid sawfly with veins labelled according to the modified Comstock-Needham system employed here. Note that, especially in the hind wing, there is a short thin vein stub adjacent to the costa at the wing base which we consider to be the subcosta.

the anterior hind wing margin after vein C appears to have terminated but before the distal cluster; this seems to be an indication that they are derived only from the costal vein even if a tubular vein is not visible.

Vein Naming Systems

Wing venation has been a quagmire of names for both veins and cells, and this has undoubtedly hindered learning and work, because hymenopteran wing venation provides many of the most important sets of characters for recognising groups at all levels. Therefore, there is no escaping having to learn some terms and also becoming familiar with a number of quite different and sometimes contradictory naming systems (Table 6.1). An additional headache, not only for beginners, arises because particular nomenclatural systems have been informally adopted by different regions as well as by workers on different taxonomic groups.

By far the worst system is the Jurinian, which was proposed more than 200 years ago (Jurine, 1807) and which was developed by Cresson (1887), which gets the identities of the longitudinal veins in the way that we understand them today. This system was the basis of the system presented in a very influential paper by Rohwer and Gahan (1916), and, since the latter three authors all hailed from the USA, this system became the most adopted there. Rohwer and Gahan (1916) preferred a system that they thought helped systematists rather than giving priority to the real homologies of veins. In most groups this has been replaced by various versions of the Comstock–Needham system used here, but its use persists in particular in the aculeate groups (Day, 1988) and by some workers on Ichneumonoids (e.g. Marsh, 1971; He et al., 2000) and more importantly by Henry Townes and followers in their books and papers on Ichneumonidae. For comparisons relating to the ichneumonoids in particular, see Quicke (2015) and Broad et al. (2018).

The Comstock–Needham system employed here, and nowadays by most workers, was specifically applied to the Hymenoptera by Ross (1936) and later developed by many others. The main vein interpretation difference with the Jurine–Cresson system is that the medial and cubital veins in the Comstock–Needham system were called the cubitus and discoidus, respectively. Thus the path of CU after separating from M is completely different.

Pterostigma and parastigma

A prominent landmark on the anterior margin of the fore wing of most larger and a few smaller parasitoid wasps is a nearly triangular area of thickened and often pigmented cuticle, called the pterostigma. Although it usually appears fairly uniform, careful inspection reveals that it is actually a modified wing cell (Fig. 6.9) (Sharkey, 1988). The costal vein of the fore wing, after the flexion break at the base of the pterostigma, carries on forming

Table 6.1. Comparison of fore wing vein name systems based on Ichneumonoidea with complete venation.

Name used here following Sharkey and Wharton (1997)	Goulet and Huber (1993)	van Achterberg (1993)	Jurine–Cresson system, as used by e.g. Henry Townes
C	C		costa
C+SC+R[a]	C+Sc+R	C+Sc+R	costa
SC[b]	C+Sc		costa and subcosta
R	R1	1-R1	metacarpus
r-rs	r-rs	r	1st abscissa of radius
2RS	Rs	2-SR	2nd interrcubitus
3RSa	Rs	3-SR	2nd abscissa of radius
3RSb	Rs	SR1	3rd abscissa of radius
(RS+M)a	Rs+M	1-SR+M	cubitus
(RS+M)b	Rs+M	2-SR+M	
r-m	r-m	r-m	
2M	M	2-M	
M+CU	M+CU	M+Cu1	medius
1m-cu	m-cu	m-cu	recurrent vein
2m-cu	2m-cu	—	2nd recurrent vein
1CU	Cu	1CU1	discoideus
2CUa	Cu	3-CU1	
2CUb	Cu	Cu1a	
1cu-a	1cu-a	cu-a	
2cu-a	2cu-a	CU1b	
1-1A	A	1-1A	

Notes: [a]in several groups C is separate from R+Sc, (or is absent) and a costal cell present; [b]it seems likely that SC is absent in all Apocrita.

the anterior margin of the pterostigma whereas the posterior margin of the pterostigma is formed by R1. For various reasons we think the costal vein probably terminates at the apex of the pterostigma, and the vein that continues along the anterior wing margin after the pterostigma is R1.

The rest of the venation

Things are complicated slightly because the radial vein divides into an anterior branch which runs along the anterior wing margin, in the Hymenoptera forming the pterostigma, and a posterior longitudinal branch called the radial sector (RS). This branch in particular can be confusing because it can run for a short distance retrograde from the pterostigma, before turning towards the wing tip.

Wing venation has been a quagmire of names for both veins and cells, and this has made life awkward for many a person trying to get to grips with Hymenoptera identification.

For groups with relatively complete wing venation, we have adopted the Comstock–Needham system of nomenclature modified by Redtenbacher and then further developed largely by Michael Sharkey, who first presented a review of all previous systems, and a proposal for a revised terminology, in the newsletter *Ichnews* (Sharkey, 1988). Since then, the orthography of Sharkey's system has been substantially revised (Huber and Sharkey 1993; Sharkey and Wharton, 1997). We use it because it is reasonably internally consistent.

In this system the longitudinal veins are abbreviated based on their Anglicized names and start with a capital letter. Cross-veins are named according to which longitudinal veins they run between, from anterior to posterior, and are given in small letters. When a longitudinal vein splits into an anterior and posterior part, the anterior branch is usually designated by the suffix 1 and the posterior ones by 2, 3, etc. However, for parasitoid wasp venation this mostly only concerns the radius (R) and traditionally the more posterior part is called the radial sector (RS).

Longitudinal veins may be free or fused. A free vein is composed only of the vein it is part of. However, in the basic hymenopteran wing, certain longitudinal veins are always fused, and this is indicated by a plus sign (+). Thus, the fused basal section of the medial vein (M) and cubital vein (CU) is designated M+CU.

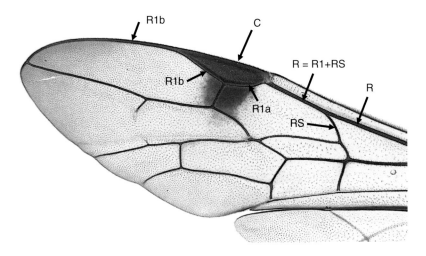

Fig. 6.9. Fore wing of a *Pristaulacus* sp. (Evanioidea: Aulacidae) from Thailand showing the differentiation between the middle part of the pterostigma, which is postulated to be thickened wing membrane, and proposed precise vein nomenclature.

When a section of two fused veins is lost but their origin at either end is still present, one cannot usually know where one vein ends and the other begins (perhaps even whether that question makes sense). In these cases, the combined vein is written with an ampersand (&). Thus in many braconids and all ichneumonids the fused vein (RS+M)a is absent. At the basal end of where it would be are Rs running some short way from the parastigma.

Longitudinal veins are usually divided into two or more separate sections, called abscissae, by their intersections with cross-veins. Each abscissa of free vein created by intersection with a cross-vein running from a more anterior longitudinal vein is numbered in sequence, with the number placed immediately before the vein abbreviation. When a vein is further subdivided by intersection with a cross-vein on its posterior side, the sections (abscissas) are designated by a lower-case letter immediately following the vein name. For example, from base to apex: the medial vein of a typical sawfly or braconid wasp (Fig. 6.8 a), after the cubital vein, separates from the combined M+CU; the first abscissa of the free medial (1M) runs anteriorly to fuse with the radial vein for a while, then more distally becomes free again with the 2nd (2M), 3rd (3M) and sometimes 4th (4M) abscissae forming the posterior margins of the submarginal cells.

Where is vein 1rs-m?

Although in some popular wing venation naming systems the first cross-vein between RS and M, which usually forms the distal border of the 2nd submarginal cell, is called 1rs-m (or 1r-m for short), many workers refer to this as 2rs-m. This particularly affects Ichneumonidae (see Chapter 9, this volume). The explanation is provided by Ross (1936) who noted that in some Xyelidae the medius does not fuse with RS and actually separates from M+CU far more distally than in all other Hymenoptera, and is connected to RS by a short cross-vein at the level of the 1st submarginal cell. So this cross-vein, which is absent in all non-xyelids, is technically 1rs-m.

The consequence of the above is that some workers label the rs-m cross-veins in a way that is homologous to the xyelid case, so the first rs-m cross-vein in non-xyelids is called 2rs-m, whereas other workers number it 1rs-m (or just rs-m or r-m) if there is only one radiomedial cross-vein.

Problems with the system

Whilst veins and probably most cross-veins are homologous throughout the order, the separate abscissae of longitudinal veins are not. When a cross-vein is lost (absent) the numbering of the subdivisions (abscissae) of the longitudinal veins involved shifts back by one. Cross-veins are far more variable (i.e. may move over evolutionary time or be lost). Some workers simply number them starting with the most basal one present being designated; others, as mentioned above, try to preserve some concept of vein homology.

Fold and Flexion Lines

Wings are not static plates in flight but fold dynamically in various ways. Just as with the ground-plan venation, there are a small number of lines of flexibility and these have often been used to try to ascertain vein and cell homologies (Wootton, 1979). In the Ichneumonidae, for example, where one bifurcating fold line splits relative to fore wing vein 2m-cu is very important from the point of view of identification (see Fig. 10.6). However, only a few of these fold lines are given names, the most important exception being the claval furrow which runs just anterior to the anal vein in both fore and hind wings (Figs 6.8–6.10).

Cell Naming Systems

The names of wing cells may present even more of nightmare than venational nomenclature, and again different systems used by various groups (see Table 6.2). In some ways fortunately, many groups of parasitoid

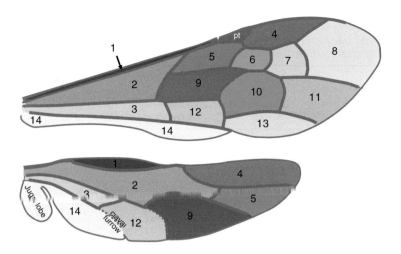

Fig. 6.10. Fore and hind wings of a stylised hymenopteran showing the terminology used here for wing cells. Cell abbreviations: 1, costal; 2, basal; 3, sub-basal; 4, marginal; 5–8, 1st to 4th submarginal cells in order; 9–11, 1st to 3rd discal cells in order; 12, 13, 1st and 2nd subdiscal cells; 14, anal.

Hymenoptera have greatly reduced wing venation and often no closed cells, so the problems are restricted to the rest, notably Ichneumonoidea, parasitic Aculeates, and rare groups such as Orussidae and Trigonalyidae. Few cells have essentially specific names in most systems, the costal and the anal. Huber and Sharkey (1993) in *Hymenoptera of the World* presented a system with names based on the name given to the longitudinal vein forming the anterior margin of the cell (as indeed with the costal and anal cells) but this was at variance with what most workers used. Then series of cells between the same pair of longitudinal veins but separated by cross-veins are usually numbered in sequence from the basal 1st one. For the Braconidae, Sharkey and Wharton (1997) adopted the widely used system that most western workers used (e.g. Shaw and Huddleston, 1991; van Achterberg, 1979, 1993), and this is the system used here (Fig. 6.10).

In the system used by Henry Townes for Ichneumonidae (e.g. Townes, 1969) and followed by Broad *et al.* (2018) for the same group, the names of hind wing cells are distinguished from those of the homologous fore wing ones by using the suffix '-ellan'. Thus, the costal becomes the costellan, the radial becomes the radiellan, the medial becomes the mediellan, and so on.

If this area is formed into a marked lobe, it can be called the claval lobe but it may also be referred to as the 'anal lobe' or 'plical lobe' by various workers. The hind wings of Evaniidae, Vespoidea and Chrysidoidea have a very distinct even more basal lobe called the jugal lobe (Fig. 6.10). This is located posterior to the 1st anal cell (and clavus) which may also be somewhat lobe-shaped because the wing margin is strongly incised where the claval furrow intersects it. Wootton (1979) recommended that the area behind the claval furrow should be referred to as the clavus. Note that some workers refer to this as the anal lobe (e.g. Kimsey, 1991) – potentially very confusing and therefore clear explanations and preferably diagrams should always be provided.

Metasoma

The metasoma (Fig. 6.11) comprises nine segments but some posterior ones are often reduced or lost. The 1st segment is often strongly differentiated from subsequent ones, especially in parasitoid taxa, obviously in ants but also in several Vespoidea and Apoidea, the modification usually being in connection with allowing the posterior

Table 6.2. Various commonly used fore wing cell names for taxa with relatively complete wing venation, such as a braconid (see Fig. 6.10 for system used here). Note that hind wings have few cross-veins in most parasitoid Hymenoptera so there is seldom need for use of 1st, 2nd, etc.

Cell number (Fig. 6.10)	Huber and Sharkey (1993)[a]	Broad et al. (2018)	Various other names used	Names used by Azevedo et al. (2018)	Cell name used here following Sharkey and Wharton (1997)
1	costal	—	costal	costal	costal
2	radial	median	median	radial	basal
3	1st cubital	submedian	cubomedial	1st cubital	subbasal
4	2nd radial	radial	radial	2nd radial	marginal
5	1st radial	1st submarginal	1st radiomedial	1st radial	1st submarginal
6	1st radial sector	areolet[b]	2nd radiomedial	—	2nd submarginal
7	2nd radial sector	3rd cubital	3rd radiomedial	—	3rd submarginal
8	3rd radial sector			—	4th submarginal
9	1st medial	—	1st discoidal	1st medial	1st discal
5+9	—	discocubital[c]	—	—	disco-submarginal (= 1st submarginal + 1st discal)
10	2nd medial	2nd discoidal	2nd discoidal	—	2nd discal
11	3rd medial	3rd discoidal		—	3rd discal
12	2nd cubital	1st brachial	1st brachial	2nd cubital	1st subdiscal
13	3rd cubital	2nd brachial	2nd brachial	—	2nd subdiscal
14	anal	anal	anal	—	anal

[a]Cell names based on the vein forming their anterior margin.
[b]Here we take the ichneumonid areolet to be bordered by 2RS basally (see e.g. Lanham, 1951).
[c]When vein RS+M absent, as in Ichneumonidae.
[d]When vein 2m-cu present.

metasoma greater movability. In many parasitoid groups the 1st segment, which is often called the petiole, is often modified more than simply being narrowed anteriorly, and is often tubular and sometimes petiolate, i.e. narrow and rather tubular basally and widened rather abruptly posteriorly.

Dorsal features generally receive most attention. The sclerotised upper part of each segment is called a metasomal tergite, and is developmentally divided into a central tergum and a pair of laterotergites. The terga of metasomal tergites, particularly the 2nd and 3rd, are often fused to form a syntergum, and the sutures between them may be invisible. In sawflies there are 10 visible tergites, the last one of which bears a pair of piliferous sensory organs called cerci or pygostyles on its postero-lateral margin (Fig. 132 in Richards, 1977). In Chalcidoidea there are only 9 abdominal segments visible (Bouček, 1988; Gibson et al., 1997), the most posterior of these bearing the cerci. Richards wrote: 'Where the tenth segment is membranous or fused with the ninth, the pygostyles may become attached to the latter', so the assumption seems to be that in chalcidoids, and possibly other groups, the 10th tergite is lost. However, many chalcidoids also have a more posterior mediodorsal sclerite which is often small and sometimes described as being like a finger nail, but in some groups (e,g. Torymidae and some Eulophidae) is much larger – it is referred to as the epipygium by some workers, but Gibson et al. (1997) treated it as the 10th abdominal (= 9th metasomal) tergite and thus the cerci would be on the 9th tergite.

Primitively, metasomal segments 1 to 9 have a pair of spiracles but some of these are lost in many groups. The most posterior spiracle of species that 'drill' into hard host substrates are particularly enlarged, as are the posterior metasomal muscles that control the ovipositor system.

In most parasitoid wasps the tergites and sternites are more or less equally well sclerotised

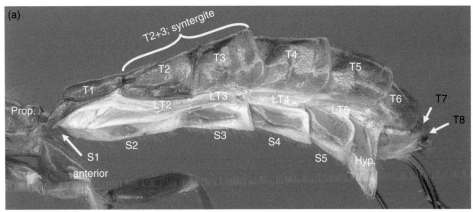

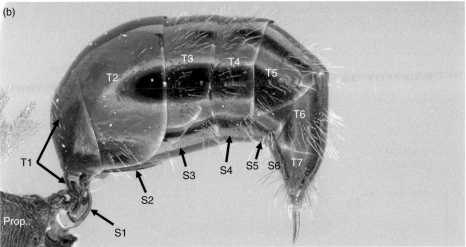

Fig. 6.11. Metasomas of two large-bodied parasitoids: **(a)** *Ischnobracon* (Braconidae: Braconinae) showing that all the tergites are exposed, and the extensive desclerotisation of the sternites; **(b)** a pristocerine bethylid (Chrysidoidea) with reduced number of externally visible metasomal segments and showing equivalent degree of sternal and tergal sclerotisation.

(Fig. 6.11 b) but in the Ichneumonoidea, the sclerotised zones are greatly reduced and surrounded by membrane (Fig. 6.11 a).

The postero-most metasomal (10th abdominal) sternum is modified as a somewhat larger structure called the hypopygium, which in females may help support the base of the ovipositor during oviposition. In males the last visible tergite is called the epipygium.

The most posterior abdominal tergites are strongly modified to form parts of the genitalia, the ovipositor system in females and the 'terminalia' (parameres and aedeagus, functionally equivalent to the penis) of males. It is beyond the scope of the present work to go into much detail, but it is a fascinating area of study and the following references provide useful introductions: Smith (1969), Quicke (1997), Vilhelmsen (2000).

Ovipositor

Evolutionarily one of the most important systems in the order showing many specialised adaptations (Quicke, 2015).

Ovipositors of taxa attacking exposed hosts (or ones that the wasp has physical access to) are typically rather short, seldom longer than the metasoma and often much shorter. Necessarily for wasps whose hosts are more deeply concealed such as wood-borer larvae, the ovipositor has to be longer. In these cases, it may be exposed and sticking out from the end of the metasoma like a tail (as in many Ichneumonoidea and Chalcidoidea) but in other taxa it may be held concealed (coiled or looped) within the metasoma.

The ovipositor (= terebra) in life is protected by a pair of setose ovipositor sheaths (3rd valvulae) except while it is being used. The ovipositor proper comprises three parts: a single dorsal valve (2nd valvula) and a pair of lower valves (1st valvulae). These enclose the egg/venom canal and they are interlocked by a sliding tongue (rhachis) and groove (aulax) mechanism, called the olistheter mechanism (Fig. 6.12). Given the vital importance of the ovipositor it is not surprising that it has been subject to much evolutionary elaboration in various groups (Quicke et al., 1994; Quicke and Fitton, 1995), including its complete loss of oviposition function and redeployment as a stinger in the Vespoidea s.l. and Apoidea.

Understanding the diverse adaptive variants in the parasitoid wasps and in the stinging aculeates has been greatly facilitated by reference to the sometimes less modified 'architecture' of the sawflies (Vilhelmsen, 2000). General overviews of structure, function

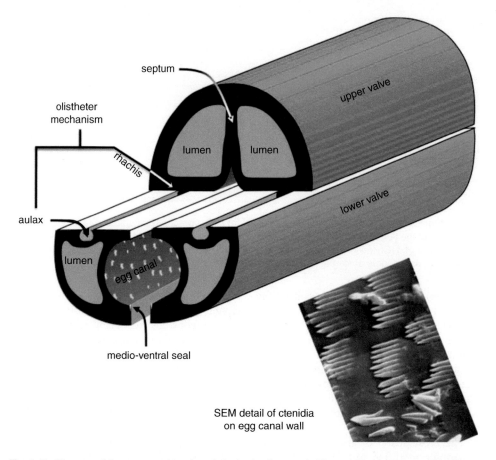

Fig. 6.12. Diagram of the components of a relatively simple parasitoid wasp (e.g. Ichneumonoidea) ovipositor, and inset showing comb-like microsculpture of the egg canal wall that is used to drag the egg chorion down the canal during oviposition. (Source: reproduced from Quicke (2015) with permission of John Wiley and Sons.)

and adaptations in parasitoid wasps are provided by Quicke et al. (1994, 1999). A number of papers describe the ovipositor and associated abdominal sclerites and musculature in considerable detail for various groups, e.g. Ceraphronoidea (Ernst et al., 2013); Chalcidoidea (King, 1962; Copland and King, 1971 a,b, 1972 a-c; Copland et al., 1973); Cynipoidea (Fergusson, 1988); Ichneumonoidea (Eggs et al., 2018); Vespoidea s.l. (Kumpanenko and Gladun, 2018; Kumpanenko et al., 2022). Unfortunately, these works do not all use the same anatomical terminology and the system in general is too complicated to go into detail here.

Within the Ichneumonoidea there is a spectrum of ovipositor apex morphologies associated with function. Taxa which use their ovipositor to 'drill' through hard or tough substrates have the upper valve with a protruding preapical nodus, and the tips of the two lower valves with well-developed serrations. In endoparasitoids that oviposit into medium-sized larval hosts, there is often a pre-apical notch and serrations are very reduced on the lower valves, usually just three tiny ones remaining. Species that attack host eggs or very early instars tend to have very fine, needle-like ovipositor tips. These features can allow the biologies of taxa to be predicted with high accuracy (Belshaw et al., 2003).

Boring et al. (2009) described in considerable detail how such a notched ovipositor works, based on that of a koinobiont caterpillar parasitoid braconid. A similar morphology in which the notch has a more active role as a clip was described by van Lenteren et al. (1998) in eucoiline cynipoids. In these the mechanism interlocks more firmly with the host fly larva cuticle and prevents it from escaping into its food medium until oviposition has been achieved.

Parasitoid wasps do not oviposit the same way as, for example, chickens do, because squeezing eggs down the narrow ovipositor lumen using metasomal pressure would undoubtedly force the ovipositor interlocking mechanism apart. The mechanism was first elucidated by Austin and Browning (1981) who showed that fine comb-like sculpture inside the egg canal is used to drag the egg chorion along the tube. In the case of large yolky (lecithal eggs) the chorion must be capable of considerable distortion (Quicke, 2015).

Male Genitalia

Whether male genitalia morphology is important for generic or species-level taxonomy depends very much on the group. It is hardly used, for example, in the Ichneumonoidea; occasionally it is important in the Chalcidoidea for helping to separate closely related species; but it is of extreme importance in many aculeate groups such as the Bethylidae, Chrysididae Dryinidae, Mutillidae and Scoliidae. Indeed, for some groups the only characters used in key couplets are from the male genitalia. In some of the latter families it provides the only means of recognising genera (e.g. Azevedo et al., 2018). In Chapter 23, this volume, we briefly describe some methods for making microscope slide preparations of male genitalia.

The whole structure is often referred to as the genital capsule. The aedeagus is the analogue of the mammalian penis and has a central position, emerging from between the volsellae (Fig. 6.13). The medio-distal margin of the volsella has an articulated, approximately triangular and usually toothed appendage called the digitus. Lateral to that the volsella usually has a second usually partially articulated and also often toothed structure called the cuspis or cuspidal process. The outer part is formed from the paired parameral processes which some authors call harpes. The parameres may be divided into distinct distal and proximal parts but in other groups they form the continuous outer part of the capsule, such as in Ichneumonoidea (Quicke, 1988).

Unfortunately, for historical reasons, the same structures are sometimes given different names in different groups. For detailed descriptions and terminology see: Snodgrass (1941) (all Hymenoptera); Peck (1937) and Phand and Ahirrao (2015) (Ichneumonidae); Azevedo et al. (2018) and Lanes et al. (2020) (Bethylidae); Mikó et al. (2013) (Ceraphronoidea). Dal Pos et al. (2023) have recently published an ontogeny-based, revised terminology for hymenopteran male genitalia.

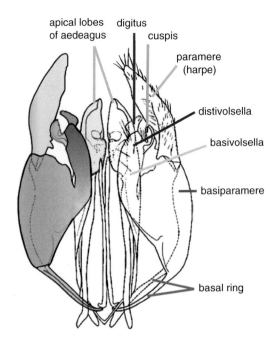

Fig. 6.13. Diagram showing major parts of male genitalia based on an Oriental bethylid. (Source: modified from Liao *et al.* (2022) under terms of Creative Commons Attribution licence 4.0.)

Sculpture

The Hymenoptera collectively display an enormous variety of cuticular sculpture patterns, and a pretty broad range of specialist terms are used in taxonomic works to describe them. Sculpture is important because in various groups it is used in keys for identifying everything from family to species.

In response to various requests for help, Eady (1968) started by presenting clear line drawings and brief descriptions of about 30 sculptural types. Most people nowadays refer to the illustrated descriptions by Harris (1979). The diversity in the highly sculptured Chrysididae prompted Martynova (2017) to present a specialised version which has good illustrations that introduced some new useful terms for when multiple types of sculpture are superimposed, but it did not include several widely used descriptive terms.

The main terms used here with brief descriptions are shown in the table below.

Term	Description
Alutaceous	With a network of very shallow, very fine grooves ('cracks')
Costate	With longitudinal raised ribs or ridges (costae)
Coriaceous	With a network of fine grooves (deeper and wider than alutaceous) giving a leather or skin like appearance
Foveate	With numerous, regular, depressions or pits (foveae)
Nitid	Smooth and shiny
Punctate	With numerous uniform small punctures ('pits') from which setae may or may not originate
Reticulate	With network of ridges defining flat-bottomed pits, each pit often with a central seta
Rugose	With irregular wrinkles
Rugulose	Like rugose but relatively finer
Striate	With parallel, fine, longitudinal impressed lines or grooves
Strigate	With numerous fine subparallel, usually raised, striae

Combined terms such as foveate-rugose indicate a sort of transitional combination.

References

Azevedo, C.O., Alencar, I.D.C.C., Ramos, M.S., Barbosa, D.N., Colombo, W.D., Vargas, J.M.R. and Lim, J. (2018) Global guide of the flat wasps (Hymenoptera, Bethylidae). Zootaxa 4489, 1–294. https://doi.org/10.11646/zootaxa.4489.1.1

Austin, A.D. and Browning, T.O. (1981) A mechanism for the movement of eggs along insect ovipositors. *International Journal of Insect Morphology and Embryology* 10, 93–108.

Basibuyuk, H.H. and Quicke, D.L.J. (1995) Morphology of the antenna cleaner in the Hymenoptera with particular reference to non-aculeate families (Insecta). *Zoologica Scripta* 24, 157–177.

Basibuyuk, H.H. and Quicke, D.L.J. (1997) Hamuli in the Hymenoptera (Insecta) and their phylogenetic implications. *Journal of Natural History* 31, 1563–1585.

Basibuyuk, H.H. and Quicke, D.L.J. (1999) Gross morphology of multiporous plate sensilla in the Hymenoptera (Insecta). *Zoologica Scripta* 28, 51–67.

Basibuyuk, H.H., Quicke, D.L.J., Rasnitsyn, A.P. and Fitton, M.G. (2000) Morphology and sensilla of the orbicula, a sclerite between the tarsal claws, in the Hymenoptera. *Annals of the Entomological Society of America* 93, 625–636.

Belshaw, R., Grafen, A., Quicke, D.L.J. (2003) Inferring life history from ovipositor morphology in parasitoid wasps using phylogenetic regression and discriminant analysis. *Zoological Journal of the Linnean Society* 139, 213–238.

Bennett, A.M.R., Cardinal, S., Gauld, I.D. and Wahl, D.B. (2019) Phylogeny of the subfamilies of Ichneumonidae (Hymenoptera). *Journal of Hymenoptera Research* 71, 1–156.

Boni Durluccl, M. (2004) Tribe-groups of the Myzininae with special regard to the Palaearctic taxa of the tribe Meriini (Hymenoptera, Tiphiidae). *Linzer biologische Beiträge* 36, 1205–1308.

Boring. A.C., Sharkey, M.J. and Nychka, J.A. (2009) Structure and functional morphology of the ovipositor of *Homolobus truncator* (Hymenoptera: Ichneumonoidea: Braconidae). *Journal of Hymenoptera Research* 18, 1–24.

Bouček, Z. (1988) *Australasian Chalcidoidea (Hymenoptera): A Biosystematic Revision of Genera of Fourteen Families, with a Reclassification of Species.* CAB International, Wallingford, UK.

Broad, G.R., Shaw, M.R. and Fitton, M.G. (2018) *The Ichneumonid Wasps of Britain and Ireland (Hymenoptera: Ichneumonidae).* Royal Entomological Society, Telford, UK.

Brothers, D.J. (1976) Modifications of the metapostnotum and origin of the 'propodeal triangle' in Hymenoptera Aculeata. *Systematic Entomology* 1, 177–182.

Copland M.J.W. and King P.E. (1971a) The structure and possible function of the reproductive system in some Eulophidae and Tetracampidae, *Entomologist* 104, 4–28.

Copland M.J.W. and King P.E. (1971b) The structure of the female reproductive system in the Chalcididae. *Entomologist's Monthly Magazine* 107, 230–239.

Copland, M.J.W. and King, P.E. (1972a) The structure of the female reproductive system in the Pteromalidae (Chalcidoidea: Hymenoptera). *Entomologist* 105, 77–96.

Copland, M.J.W. and King, P.E. (1972b) The structure of the female reproductive system in the Eurytomidae (Chalcidoidea: Hymenoptera). *Journal of Zoology (London)* 166, 185–212.

Copland, M.J.W. and King, P.E. (1972c) The structure of the female reproductive system in the Torymidae (Hymenoptera: Chalcidoidea). *Transactions of the Royal Entomological Society of London* 124, 191–212.

Copland, M.J.W., King, P.E. and Hill, D.S. (1973) The structure of the female reproductive system in the Agaonidae (Chalcidoidea, Hymenoptera). *Journal of Entomology (A)* 48, 25–35.

Cresson, E.T. (1887) Synopsis of the families and genera of the Hymenoptera of America, north of Mexico, together with a catalogue of the described species, and bibliography. *Transactions of the American Entomological Society (Supplement)* 13, vii +1–351.

Dal Pos, D., Mikó, I., Talamas, E.J., Vilhelmsen, L. and Sharanowski, B.J. (2023) A revised terminology for male genitalia in Hymenoptera (Insecta), with a special emphasis on Ichneumonoidea. *PeerJ* 11, e15874.

Danforth, B.N. and Michener, C.D. (1988) Wing folding in the Hymenoptera. *Annals of the Entomological Society of America* 81, 342–349.

Day, M.C. (1988) Spider wasps. *Handbooks for the Identification of British Insects* 6(4), 1–60.

Eady R.D. (1968) Some illustrations of microsculpture in the Hymenoptera. *Proceedings of the Royal Entomological Society of London. Series A, General Entomology* 43, 66–72.

Eady, R.D. (1974) The present state of nomenclature of wing venation in the Braconidae (Hymenoptera); its origins and comparison with related groups. *Journal of Entomology Series B, Taxonomy* 43, 63–72.

Eggs, B., Birkhold, A.I., Röhrle, O. and Betz, O. (2018) Structure and function of the musculoskeletal ovipositor system of an ichneumonid wasp. *BMC Zoology* 3, 12.

Eraghi, S.H., Toofani, A., Khaheshi, A., Khorsandi, M., Darvizeh, A., Gorb, S. and Rajabi, H. (2021) Wing coupling in bees and wasps: from the underlying science to bioinspired engineering. *Advanced Science* 2021, 2004383

Ernst, A.F., Mikó, I. and Deans, A.R. (2013) Morphology and function of the ovipositor mechanism in Ceraphronoidea (Hymenoptera, Apocrita). *Journal of Hymenoptera Research* 33, 25–61.

Fedoseeva, E.B. (2017) The metathoracic spiracles in some ants and wasps (Hymenoptera: Formicidae; Vespidae). *Russian Entomological Journal* 26, 49–62.

Fergusson, N.D.M. (1988) A comparative study of the structures of phylogenetic significance of female genitalia of the Cynipoidea (Hymenoptera). *Systematic Entomology* 13, 12–30.

Gibson, G.A.P. (1986) Mesothoracic skeletomusculature and mechanics of flight and jumping in Eupelminae (Hymenoptera, Chalcidoidea: Eupelmidae). *Canadian Entomologist* 118, 691–728. doi: 10.4039/Ent118691-7

Gibson, G.A.P., Huber, J.T. and Woolley, J.B. (eds) (1997) *Annotated Keys to the Genera of Nearctic Chalcidoidea (Hymenoptera)*. Ottawa, Canada, NRC Research Press, 794 pp.

Goulet, H. and Huber, J. (eds) (1993) *Hymenoptera of the World: An Identification Guide to Families* (Vol. vii). Research Branch, Agriculture Canada, Ottawa, Canada, pp. 13–59.

Harris, R.A. (1979) A glossary of surface sculpturing. *Occasional Papers in Entomology* 28, 1–31.

He, J., Chen, X. and Ma, Y. (2000) *Fauna Sinica vol. 18. Hymenoptera, Braconidae*. Science Press, Beijing, China, 757 pp.

Huber, J.T. and Sharkey, M.J. (1993) Structure. In: Goulet, H. and Huber, J. (eds) *Hymenoptera of the World: An Identification Guide to Families (Vol. vii)*. Research Branch, Agriculture Canada, Ottawa, Canada, pp. 13–59.

Jervis, M.A. (1998) Functional and evolutionary aspects of mouthpart structure in parasitoid wasps. *Biological Journal of the Linnean Society* 6, 461–493.

Jurine, L. (1807) Nouvelle méthode de classer les Hyménoptères et les Diptères. Volume 1. J.J. Paschoud, Geneva, 319 pp.

Karlsson, D. and Ronquist, F. (2012) Skeletal morphology of *Opius dissitus* and *Biosteres carbonarius* (Hymenoptera: Braconidae), with a discussion of terminology. *PLoS ONE* 7(4), e32573.

Kimsey, L.S. (1991) Relationships among the tiphiid wasp subfamilies (Hymenoptera). *Systematic Entomology* 16, 427-438. doi: 10.1111/j.1365-3113.1991.tb00677.x

Kimsey, L.S. and Bohart, R.M. [1990] (1991) *The Chrysidid Wasps of the World*. Oxford University Press, 652 pp.

King, P.E. (1962) The muscular structure of the ovipositor and its mode of function in *Nasonia vitripennis* (Walker) (Hymenoptera: Pteromalidae). *Proceedings of the Royal Entomological Society of London, Series A General Entomology* 37, 121–128.

Krogmann, L. and Vilhelmsen, L. (2006) Phylogenetic implications of the mesosomal skeleton in Chalcidoidea (Hymenoptera, Apocrita) – tree searches in a jungle of homoplasy. *Invertebrate Systematics* 20, 615–674.

Kumpanenko, A.S. and Gladun, D.V. (2018) Functional morphology of the sting apparatus of the spider wasp *Cryptocheilus versicolor* (Scopoli, 1763) (Hymenoptera: Pompilidae). *Entomological Science* 21, 124–132.

Kumpanenko, A., Gladun, D. and Vilhelmsen, L. (2022) Morphology of the sting apparatus in velvet ants of the subfamilies Myrmosinae, Dasylabrinae, Myrmillinae and Mutillinae (Hymenoptera: Mutillidae). *Zoomorphology* 141, 81–94.

Lanes, G.O., Kawada, R., Azevedo, C.O. and Brothers, D.J. (2020) Revisited morphology applied for systematics of flat wasps (Hymenoptera, Bethylidae). *Zootaxa* 4752, 1–127.

Lanham, U.N. (1951) Review of the wing venation of the higher Hymenoptera (Suborder Clistogastra), and speculations on the phylogeny of the Hymenoptera. *Annals of the Entomological Society of America* 44, 614–628.

Liao, X.-P., Chen, B. and Li, T.-J. (2022) A taxonomic revision of the subfamily Myzininae from China, with a key to the Chinese species (Hymenoptera: Tiphiidae). *Zootaxa* 5154, 152–174. doi: 10.11646/zootaxa.5154.2.3

Marsh, P.M. (1971) Keys to the Nearctic Genera of the Families Braconidae, Aphidiidae, and Hybrizontidae (Hymenoptera). *Annals of the Entomological Society of America* 64, 841–850.

Martynova, K.V. (2017) Microsculpture of cuckoo wasps (Hymenoptera, Chrysididae): general overview with first attempt of classification. Morphologia. *Український ентомологічний журнал* 1 (12), 7–19.

Michener, C.D. (1944) Comparative external morphology, phylogeny, and a classification of the bees (Hym.). *Bulletin of the American Museum of Natural History* 82, 151–326; figures 1–246.

Mikó, I., Vilhelmsen, L., Johnson, N.F., Masner, L. and Pénzes, Z. (2007) Morphology of Scelionidae (Hymenoptera: Platygastroidea): head and mesosoma. *Zootaxa* 1571, 1–78.

Mikó, I., Masner, L., Johannes, E., Yoder, M. J., & Deans, A. R. (2013) Male terminalia of Ceraphronoidea: morphological diversity in an otherwise monotonous taxon. *Insect Systematics & Evolution* 44, 261–347.

Peck, O. (1937) The male genitalia of Hymenoptera, especially the Ichneumonidae. *Canadian Journal of Research* 15(D) (12), 21–274.

Phand, D.L. and Ahirrao, D.V. (2015) Morphological study of male genitalia of Ichneumonidae (Insecta: Hymenoptera) of subfamily Pimplinae – I. *Journal of Experimental Biology and Agricultural Sciences* 3, 32–36.

Quicke, D.L.J. (1988) Inter-generic variation in the male genitalia of the Braconinae (Insecta, Hymenoptera, Braconidae). *Zoologica Scripta* 17, 399–409.

Quicke, D.L.J. (1997) *Parasitic Wasps*. Springer, New York City, 470 pp.

Quicke, D.L.J. (2015) *The Braconid and Ichneumonid Parasitic Wasps: Biology, Systematics, Evolution and Ecology*. Wiley Blackwell, Oxford, 688 pp.

Quicke, D.L.J. and Fitton, M.G. (1995) Ovipositor steering mechanisms in parasitic wasps of the families Gasteruptiidae and Aulacidae (Hymenoptera). *Proceedings of the Royal Society, London, Series B* 261, 99–103.

Quicke, D.L.J., Fitton, M.G., Tunstead, J., Ingram, S.N. and Gaitens, P.V. (1994) Ovipositor structure and relationships within the Hymenoptera, with special reference to the Ichneumonoidea. *Journal of Natural History* 28, 635–682.

Quicke, D.L.J., LeRalec, A. and Vilhelmsen, L. (1999) Ovipositor structure and function in the parasitic Hymenoptera with an exploration of new hypotheses. *Rendiconti* 17, 107–235.

Reid, J.A. (1941) The thorax of wingless and short winged Hymenoptera. *Transactions of the Royal Entomological Society of London* 91, 361–446.

Richards, O.W. (1977) Hymenoptera. Introduction and key to families (2nd edition). *Handbooks for the Identification of British Insects* 6(1), 1–100.

Rohwer, S.A. and Gahan, A.B. (1916) Horismology of the hymenopterous wing. *Proceedings of the Entomological Society of Washington* 18, 20–76.

Ross, H.H. (1936) The ancestry and wing venation of the Hymenoptera. *Annals of the Entomological Society of America* 29, 99–111.

Sharkey, M.J. (1988) Ichneumonoid wing venation. *Ichnews* 11, 2–12.

Sharkey, M.J. (1994) Another look at wing vein/cell nomenclature. *Ichnews* 14, 2–5.

Sharkey, M.J. and Wharton, R.A. (1997) Morphology and Terminology. In: Wharton, R.A., Marsh, P.M. and Sharkey, M.J. (eds) *Identification Manual to the New World Genera of Braconidae*, Special Publication 1 of the International Society of Hymenopterists, Washington, DC, pp. 19–37.

Sharkey, M.J., Laurenne, N.M., Quicke, D.L.J., Sharanowshi, B. and Murray, D. (2006) Revision of the Agathidinae (Hymenoptera: Braconidae) with comparisons of static and dynamic alignments. *Cladistics* 22, 546–567.

Shaw, M.R. and Huddleston, T. (1991) Classification and biology of braconid wasps (Hymenoptera: Braconidae). *Handbooks for the Identification of British Insects* 7. British Museum (Natural History), London, pp. 1–126.

Smith, E. L. (1969) Evolutionary morphology of external insect genitalia. 1. Origin and relationships to other appendages. *Annals of the Entomological Society of America* 62, 1051–1079.

Snodgrass, R.E. (1910) The thorax of the Hymenoptera. *Proceedings of the United States National Museum* 39, 37–91.

Snodgrass, R.E. (1941) The male genitalia of the Hymenoptera. *Smithsonian Miscellaneous Collections* 99, 1–86.

Townes, H.K. (1969) The genera of Ichneumonidae, Part 1. *Memoirs of the American Entomological Institute* 11, 1–300.

van Achterberg, C. (1979) A revision of the subfamily Zelinae auct. (Hymenoptera, Braconidae). *Tijdschrift voor Entomologie* 122, 241–479.

van Achterberg, C. (1988) Revision of the subfamily Blacinae Foerster (Hymenoptera, Braconidae). *Zoologisches Verhandelingen, Leiden* 249, 1–324.

van Achterberg, C. (1993) Illustrated key to the subfamilies of the Braconidae (Hymenoptera: Ichneumonoidea). *Zoologische Verhandelingen* 283, 1–189.

van Lenteren, J.C., Isidoro, N. and Bin, F. (1998) Functional anatomy of the ovipositor clip in the parasitoid *Leptopilina heterotoma* (Thompson) (Hymenoptera: Eucoilidae), a structure to grip escaping host larvae. *International Journal of Insect Morphology & Embryology* 27, 263–268.

Vilhelmsen, L. (1997) Head capsule concavities accommodating the antennal bases in Hymenoptera pupating in wood: possible emergence-facilitating adaptations. *International Journal of Insect Morphology and Embryology* 26, 129–138.

Vilhelmsen, L. (2000) The ovipositor apparatus of basal Hymenoptera (Insecta): phylogenetic implications and functional morphology. *Zoologica Scripta* 29, 319–345.

Vilhelmsen, L. and Krogmann, L. (2006) Skeletal anatomy of the mesosoma of *Palaeomymar anomalum* (Blood & Kryger, 1922) (Hymenoptera: Mymarommatidae). *Journal of Hymenoptera Research* 15, 290–306.

Vilhelmsen, L. and Turrisi, G.F. (2011) Per arborem ad astra: morphological adaptations to exploiting the woody habitat in the early evolution of Hymenoptera. *Arthropod Structure & Development* 40, 2–20.

Vilhelmsen, L., Mikó, I. and Krogmann, L. (2010) Beyond the wasp-waist: structural diversity and phylogenetic significance of the mesosoma in apocritan wasps (Insecta: Hymenoptera). *Zoological Journal of the Linnean Society* 159, 22–194.

Whitfield, J.B., Johnson, N.F. and Hamerski, M.R. (1989) Identity and phylogenetic significance of the metapostnotum in nonaculeate Hymenoptera. *Annals of the Entomological Society of America* 82, 663–673.

Wootton, R.J. (1979) Function, homology and terminology in insect wings. *Systematic Entomology* 4(1), 81–93.

Yoder, M.J., Mikó, I., Seltmann, K.C., Bertone, M.A. and Deans, A.R. (2010) A gross anatomy ontology for Hymenoptera. *PLoS ONE* 5, e15991.

7

Recognition of Major Groups

Abstract

In this chapter, a series of identification keys is provided. The first is to help beginners be confident that they are dealing with a hymenopteran and then progressively to whether it is a parasitoid wasp (rather tricky when it comes to some of the aculeates) then to the recognition of major groups (superfamilies) of winged female parasitoid and aculeate Hymenoptera. And finally, a section is devoted to distinguishing the two extant Ichneumonoidea, Ichneumonidae and Braconidae, which often gives beginners some trouble.

How Do Experts Identify Major Groups?

The answer is that they usually do not need to use keys. So why do we include keys? The same experts did use the keys available at the time and spent many thousands of hours looking at specimens and trying to run them through keys. This is why they are experts. Many experts will have often seen something very similar to a given specimen and will recall what it was. They will then give it a quick check to make sure that they have not been misled.

Often particular groups have a definite *Gestahlt*. They are the only insects that look remotely like what they do. There is a popular guide to the mammals of Africa that includes a key. The first couplet gives the options, 'an elephant' and 'not an elephant'. This works because everyone knows what an elephant looks like, so there is no need to go into a series of options about trunk, tusks, large ears, five toe nails, etc. Thus, experts only occasionally have to resort to the use of keys to major groups, and even novices reading this book should soon start to get the 'feel' of particular groups. But do not become overconfident. Always check the important confirmatory characters.

Here we present a series of keys to superfamilies and some families that should enable more than 90% of S.E. Asian species to be determined correctly. To present full, virtually foolproof keys would take many pages because there are so many exceptions, an inevitable consequence of the order being so species-rich. To make this point, and emphasise that determinations made using our keys should not be relied upon and certainly never used as the basis for any publication, the recent key to the subfamilies of Ichneumonidae of the British Isles (Broad *et al.*, 2018) comprises 100 couplets and has an accompanying nearly 600 illustrations. The key to the equivalent set of taxa as treated here in *The Hymenoptera of the World* (Mason, 1993) is effectively 84 couplets long.

Here and throughout, we will point readers to more complete, though necessarily more technical, and longer identification keys. In general, female wasps are more easily identifiable than males and we suggest beginners try to avoid trying to identify males. On the other hand, the following keys should serve to make novice readers familiar with some of the most important identification features used in other works.

Always refer to the figures. We generally illustrate the extremes of variation so for most specimens the option to follow should be obvious. If the first character of a key couplet is obvious, there is no need to consider the second. Unfortunately, use of these keys relies on the user having an intact and well-mounted specimen (see Chapter 23, this volume). Also, for the most part, it is far easier to identify female wasps, though in some groups with strong sexual dimorphism of morphology and behaviour, such as some aculeate groups with flightless females, males are by far the most commonly encountered and so the keys accommodate them too.

Keys to Major Groups for the Majority of Species of Parasitoid Hymenoptera

The following keys do not include taxa that have yet to be recorded from S.E. Asia. In using the key, pay attention to the comment '[rare]' that follows the names of some taxa. If you think you have keyed out a rare taxon, the truth is that you almost certainly have not, and you should go back through your answers and compare with illustrations here and in other sources. Alternative more thorough keys (often applicable to the entire world fauna) and which ought to be consulted for any serious studies are:

- Goulet and Huber (1993) – a remarkable very well-illustrated (with line drawings) work that covers all families and in some cases also provides keys to subfamilies. It is now available for free on the World Wide Web.
- Hanson and Gauld (1995) – although dealing with an extralimital fauna, there are very few differences compared with S.E. Asia at superfamily or family level.
- Currently in press is a new book by van Noort and Broad, titled "Wasps of the World: A Guide to Every Family" which will profile every family.

When starting to study any group, it is always a very good idea to use more than one identification key as they invariably have different characters or key groups in a different order.

Key A – Is It a Hymenopteran?

This short key is aimed only at distinguishing apocritan Hymenoptera from insects that might appear superficially similar but are not. We assume the reader can recognise a beetle.

1	Wings present even if reduced (brachypterous)	2
-	Wings absent	4
2(1)	With two pairs of wings (hind wing just a naked protruding vein in some Mymaridae and in Mymarommatoidea – see Chapter 22, Figs 22.27 c, 22.51, this volume)	3
-	With only one pair of wings **AND** hind wing modified into a club-shaped haltere	**Diptera**
3(2)	Hind wing with a row of (usually at least 3) hook-like thickened setae (hamuli) approximately half way along anterior wing margin (Fig. 6.8 b)	go to **Key B**
-	Hind wing without hamuli	**not Hymenoptera**
4(1)	Mandibles present	5
	Mandibles absent **not Hymenoptera** (probably Diptera or Hemiptera)	
5(4)	Mesosoma and metasoma joined at a narrow waist	go to **Key B**
-	Mesosoma and metasoma broadly joined	**not Hymenoptera**

Key B – Is It a Parasitoid Wasp?

Recognising what is or is not a parasitoid wasp can be difficult, though in some cases it is really easy. The problem lies in the fact that parasitoid wasps are not monophyletic. To be sure, when parasitoid wasps evolved from their sawfly/woodwasp ancestors, they were monophyletic. At some point, approximately 230 million years ago (Blaimer et al., 2023), a species evolved that would diversify into another major group of hymenopterans, the aculeates. This huge clade included various very familiar insects – bees, ants, social stinging wasps, spider-hunting wasps, and a lot of others. They are characterised by a change in function of the ovipositor from being an egg-laying organ that also injects venom into hosts, to a specialised stinging organ. In these aculeates, the egg is extruded at the base of the 'stinger' and does not pass down its interior 'egg' canal (now of course called the venom canal). Many aculeates are still parasitoids, and while not typically referred to as 'parasitoid wasps', we include them here because they are biologically the same.

All of the above means that there is no one diagnostic character for a parasitoid wasp. Especially when it comes to the aculeate wasps, recognising the parasitoid members (essentially all the superfamily Chrysidoidea and a few others that used to be classified in the Vespoidea) may involve a process of elimination.

1	Mesosoma and metasoma broadly joined (Fig. 8.1 a) **AND** fore wing with numerous cells enclosed by tubular veins **AND** costal cell always present and margined anteriorly by vein C **AND** fore wing always with an anal vein [sawflies, woodwasps and relatives (Hymenoptera 'Symphyta') though one family (Orussidae) are parasitoids too]	2
–	Mesosoma and metasoma joined at a narrow waist, but note, when the metasoma is broad anteriorly and abutting the mesosoma, the narrow waist may not be obvious, so view specimen from above with illumination also from below	3
2(1)	Antennae inserted on ventral side of head adjacent to mouth, below ventral margin of eye (Fig. 8.1 c); head with a crown or tooth-like projections surrounding the anterior ocellus (Fig. 8.1 a, c); (rare)	**Orussoidea**
–	Antennae inserted on front of head, removed from mouth .... **other Hymenoptera 'Symphyta', not parasitoids**	
3(1)	One or two dorsally protruding, node-like segments present between mesosoma and main part of metasoma (called the gaster) (Fig. 7.1 a) **OR** if male with exactly 11 flagellomeres and with moderate anterior constriction between 1st and 2nd metasomal tergites, then genitalia occupying 0.25–0.5 metasomal length (Fig. 7.1 b) **AND** eyes rather small located well away from top/back of head in lateral view (Fig. 7.1)	an ant, **Vespoidea: Formicidae**
–	Mesosoma and main part of metasoma (gaster) directly connecting without a dorsally protruding node like segment at anterior of metasoma **OR** if with moderate anterior constriction between 1st and 2nd metasomal tergites, female **OR IF** male, then genitalia not so relatively large and/or eyes larger and not conspicuously on anterior (mouth end) of head	4
4(3)	Female with an elongated ovipositor and its sheaths extending behind the apex of metasoma by at least half of metasomal length ....**a parasitoid wasp** (go to Key C)	
–	Male **OR IF** female, without an elongated ovipositor protruding behind the apex of the metasoma, although sometimes with a naked protruding stinger (see Fig. 17.3)	5
5(4)	Fore wing without any cells enclosed by tubular veins..**a parasitoid wasp** (go to Key C)	
–	Fore wing with at least one cell enclosed by tubular veins	6
6(5)	Flagellum with more than 11 segments (i.e. with more than 13 antennal segments ....**a parasitoid wasp** (go to Key C)	
–	Flagellum with 11 or fewer segments (i.e. with fewer than 14 antennal segments	7

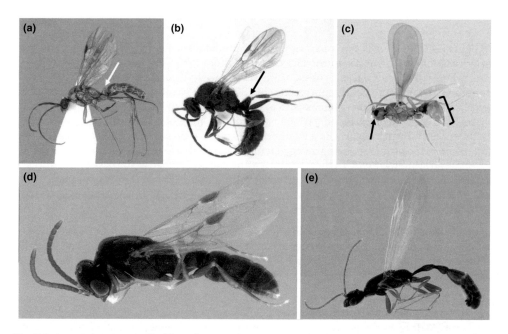

Fig. 7.1. Assorted male ants: **(a, b)** easily recognised as ants (Formicoidea: Formicidae) because of the well-developed node-like 1st segment of metasoma, i.e. separating mesosoma from gaster; **(c)** with only weak constriction between metasomal segments 1 and 2 but relatively huge genital capsule and small anteriorly located eye; **(c, d)** with enhanced constrictions 1st and 2nd and between 2nd and 3rd segments; **(e)** more attenuated form still with relatively large genitalia.

7(6)	Flagellum with exactly 8 segments **AND** hind wing without closed cells ....**a parasitoid wasp** (go to Key C)
-	Flagellum with 10 or 11 segments 8
8(7)	Pronotum widely separated from tegula with a strong postero-dorsal lobe directed somewhat below the tegula (Fig. 7.2); sometimes with many setae, especially near wing base and on propodeum, plumose............................ not a parasitoid, **Apoidea**
-	Pronotum usually reaching to touch tegula and without a prominent lobe; never with plumose setae.. 9

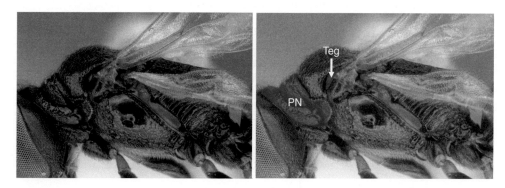

Fig. 7.2. Mesosoma of a sphecid wasp (Apoidea) showing pronotal separation from tegula and pronotal lobe.

9(8) Hind coxae widely separated basally by flat metasternum **OR** mid coxae partly covered basally by lamelliform lobe protruding from mesosternum (see Fig. 17.8 c) **OR** side of metasomal tergite 2 and/or metasomal sternite 2 with a felt line composed of fine adpressed setae (see Fig. 17.2 a) **OR** male and last visible sternite forming a strong medial upcurved prong (see Fig. 17.8 b) **OR** at least two basal flagellomeres with strong dorsal spine-like seta (Fig. 17.9 b); fore wing not longitudinally folded ....**a parasitoid wasp** (go to Key C)

- Not as above .... not a parasitoid **Vespoidea s.l.**

Key C — Major Groups of Winged Female Parasitoid and Aculeate Hymenoptera

1 Metasoma arising from propodeum far above hind coxae (see Fig. 7.3) **AND** costal cell present; (common)..**Evanioidea**

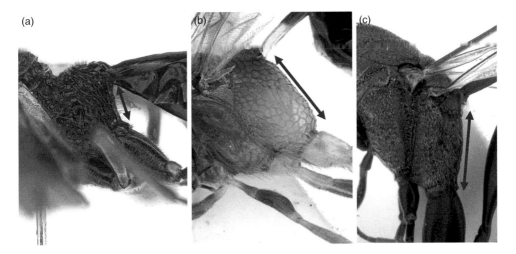

Fig. 7.3. Showing the long gap between insertion of metasoma and hind coxae of Evanioidea.

- Metasoma usually with sternite 1 below level of top of hind coxal insertion, or occasionally slightly above hind coxae IF far above (the rare Braconidae – Cenocoeliinae) then costal cell absent because veins C, Sc and R fused.. 2
2(1) Head with 5 tooth like projections around frons (see Fig. 11.3 c) **AND** hind femur with ventral serrations (see Fig. 11.3 b); body 5 or more mm long; pronotum extended into a long neck (see Fig. 11.3 a, c) **AND** wings with several cells enclosed by tubular veins; (uncommon) ....**Stephanoidea**
- Head without ring of teeth around margin of frons; without the other combination of characters 3
3(2) Lower part of malar region with deep concavity which is sharply bordered dorsally (Fig. 14.1 c) and in life, is where the antenna scape is protected; ovipositor and sheaths exserted beyond apex of metasoma; (very rare) ....**Megalyroidea**
- Lower part of malar region without deep, sharply defined concavity; ovipositor and sheaths variable.. 4
4(3) Fore wing with 9 or 10 closed cells (Fig. 15.1) **AND** antennae with more than 25 segments; (rare)..**Trigonalyoidea**
- Fore wing with no more than 8 closed cells; antennae with often with fewer than 20 flagellomeres 5

5(4)	Hind wing stalked, wing membrane not reaching the base of the wing stalk (Fig. 22.27 b, e) **OR** no wing membrane present (Fig. 22.27 c) and sometimes hind wings apparently absent (although technically a short spur is present (Fig. 22.51 b)); fore wing with very long marginal setae..	6
-	Hind wing with membrane present and reaching the base of the wing; fore wing marginal setae normally shorter	7
6(5)	With two tubular segments between propodeum and remaining swollen part of abdomen (gaster) (see Fig. 22.51); minute, body length < 1 mm; (very rare, mostly collected in pan traps)..**Mymarommatoidea**	
-	With at most one tubular petiole segment (see Fig. 22.27 a); size variable but often small...**Chalcidoidea: Mymaridae**	
7(5)	Head produced into a strong posterior lobe bordered by a flange (Fig. 16.7 c); scape, femurs and tibias with a transparent ventral flange (Figs 16.7 a, d); (uncommon but highly distinctive)..**Chrysidoidea: Lobosceliidae**	
-	Head not as above, without a large posterior lobe bordered by a flange; scape and legs without transparent flanges..	8
8(7)	Basal third or so of fore wing with a pair of anterior veins running separately but closely parallel to each other enclosing a costal cell which is open for its whole length (Fig. 7.4 a), usually also with more posterior tubular vein M+CU .... go to key D	
-	Basal third or so of fore wing with single vein along basal anterior margin (C or C+SC+R) (Fig. 7.4 a) **OR** with a single vein running just posterior to membranous anterior margin (SC+R) (Fig. 7.4 b) at least for most of the distance between wing base and pterostigma if present **OR** essentially with no venation at all (Fig. 18.4 b), i.e., without a pair of veins enclosing a narrow costal vein thus costal cell absent **BUT** sometimes with more posterior tubular vein M+CU and/or 1-1A............... go to key E	

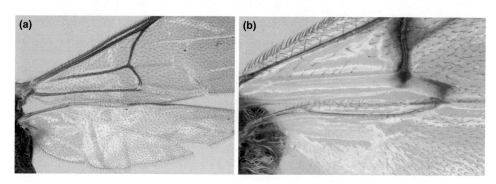

Fig. 7.4. Fore wings showing two or one tubular vein(s) at or near anterior margin and hind wing bases showing lack of cells completely enclosed by tubular veins: **(a)** Pristocerinae (Bethylidae: Chrysidoidea); **(b)** Cynipidae.

Key D — Taxa with Costal Cell Enclosed by Pair of Tubular Veins

1	Flagellum with exactly 8 segments (see Figs 16.8, 16.9 d, 16.11) **AND** hind wing without closed cells (Fig. 7.4 a).. some **Chrysidoidea**	
-	Flagellum with more than 8 flagellomeres (see Fig. 21.2 e); hind wing variable.....	2
2(1)	Antennae inserted on a projecting shelf far above the mouth, their sockets (toruli) more or less pointing upwards (see Fig. 21.2); fore wing usually without pterostigma; trochantelli differentiated (Fig. 21.2) ....a few **Diaprioidea**	

-	Antennae not inserted on a projecting shelf, their sockets pointing forwards or obliquely not directly upwards, and often arising close to the mouthparts (see Figs 16.2 b, e, 17.3); fore wing often with pterostigma; trochantelli usually not differentiated (Figs 6.6 c, d) 3
3(2)	Hind wing with at least one cell enclosed by tubular veins closed cells .... **Vespoidea s.l.** (go to key in Chapter 17, this volume)
-	Hind wing without cells enclosed by tubular veins closed cells (Fig. 7.4 a) 4
4(3)	Antenna with more than 18 flagellomeres (Fig. 16.12); (rare).................................... .... **Chrysidoidea: Sclerogibbidae**
-	Antenna with 14 or fewer flagellomeres... 5
5(4)	Body with metallic lustre (Figs 16.5, 16.6) .... some **Chrysidoidea**
-	Body without metallic colours/lustre 6
6	**IF** path of fore wing vein CU clearly discernible, even if spectral (Figs 16.7 e, 20.1, 20.4) 7
-	**IF** path of fore wing vein CU not discernible... 8
7(6)	Fore wing vein CU continues more or less straight towards end of wing after the junction with m-cu (Figs 20.1, 20.2) or even diverts anteriorly (Fig. 20.4) .... some **Proctotrupoidea**
-	Fore wing vein CU continues angles markedly towards posterior wing margin after vein m-cu (see Fig. 16.12) .... some **Chrysidoidea**
8(6)	Fore wing venation distinctive, usually with well-developed pterostigma, short vein r-rs and RS reaching wing margin only a short distance past pterostigma (Figs 20.2, 20.3 b-d); hind wing without jugal lobe; mandible unidentate; antenna with 11 flagellomeres... some **Proctotrupoidea: Proctotrupidae**
-	Fore wing venation different; hind wing with jugal lobe; mandible with multiple teeth; antenna sometimes with 8 or 10 flagellomeres.............. some **Chrysidoidea**

Key E — Taxa Lacking Entire Enclosed Fore Wing Costal Cell

1	Fore wings with no cells completely enclosed by tubular veins (Figs 22.18 a, 22.45 d, 22.5 f) although sometimes cells indicated by spectral veins) (Figs 22.4 b, 22.23 c) 2
-	Fore wings with at least one cell completely enclosed with tubular veins; (note Cynipoidea keyed both ways) 7
2(1)	Fore wing stigmal vein (r-rs, see Fig. 22.4) long and strongly curved towards wing margin but not reaching it (see Fig. 13.1 b, c), with or without a pterostigma; fore tibia with two spurs (see Fig. 13.1 a) [note: can be hard to see without high magnification]; (small insects, less than 3 mm long) ....**Ceraphronoidea**
-	Fore wing stigmal vein absent or straight (see Figs 21.2 d, 22.6 a, 22.93; fore tibia with only one spur, usually distinctly modified as a calcar (see Fig. 6.6 b)............. 3
3(2)	Antennae inserted on a projecting shelf far above the mouth, their sockets (toruli) more or less pointing upwards (see Fig. 21.2) .... most **Diaprioidea**
-	Antennae not inserted on a projecting shelf, their sockets not pointing upwards, and often arising close to the mouthparts (Figs 16.2, 22.12, 22.30) 4
4(3)	Hind wing with large, clearly defined jugal lobe, separated from clavus by a deep incision (see Fig. 6.10); head prognathous (Fig. 16.2 c-e) ....a few **Bethylidae**
-	Hind wing without jugal lobe; head not usually prognathous................................... 5
5(4)	Antennae not geniculate (i.e. not elbowed between scape and pedicellus (Figs 19.4, 19.6 b); scapus not elongate, less than 3 × longer than wide, usually less than 2 × (Figs 19.4, 19.6); fore wing usually with well-defined, nearly triangular cell formed

by veins r-rs and RS (Figs 19.1 a, 19.4 f, 19.5 f, 19.6 b); metasoma usually laterally compressed.. most **Cynipoidea**

- Antennae geniculate (i.e. elbowed between scape and pedicellus (Figs 18.2, 18.5, 22.18); scapus usually elongate and > 3.5 × longer than preapically wide (Figs 22.1, 22.21); fore wing without cells bordered by tubular veins except for costal cell; metasoma usually dorsoventrally compressed 6

6(5) Pronotum in lateral view not reaching to tegula (see Figs 7.5 a, 22.3, 22.18 b, 22.22 d); prepectus almost always externally visible as a separate sclerite between posterodorsal part of pronotum and anterior of mesopleuron thus separating the pronotum from the tegula (see Figs 7.5, 22.3); basal antennal flagellomeres frequently ring-like (anelli) (see Fig. 22.1); vein R of hind wing complete, defining the posterior margin of costal cell but vein C is not present along the anterior margin (Fig. 7.4 b) .... **Chalcidoidea**

- Pronotum in lateral view reaching to and abutting tegula (Figs 7.5 b, 18.1, 18.9); no separate prepectal sclerite present; basal antennal flagellomeres never highly reduced as ring-like anelli; vein R of hind wing often absent incomplete .... **Platygastroidea**

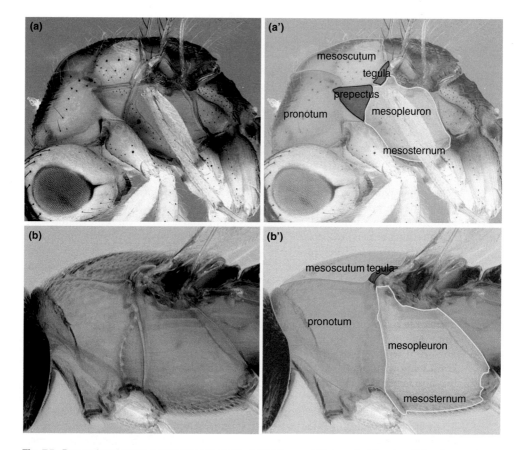

Fig. 7.5. Pronotal and tegular arrangements in Chalcidoidea and Platygastroidea: **(a, a')** *Bootanomyia* sp. (Chalcidoidea: Megastigmidae); **(b, b')** Scelionidae gen. sp.

7(1) Anterior margin of basal third of fore wing without vein C running along the edge, the longitudinal vein present (SC+R) separated from wing margin by wing membrane (see Figs 7.4 b, 19.3 b) .... some **Cynipoidea**
- Anterior margin of basal third of fore wing with vein C+SC+R running along the edge (see Figs 9.5, 10.10 a, 10.31 d, 17.9 b) 8
8(7) Antenna almost always with more than 11 flagellomeres except in small species; hind wing jugal lobe absent (99.99%); basal 2 or more flagellomeres without apical spine; (size variable; very common) .... **Ichneumonoidea** (go to Key F)
- Antennal flagellum with 11 flagellomeres **AND** hind wing with clearly defined jugal lobe; basal 2 or more flagellomeres with apical spine (see Fig. 17.10 b); (large wasps; sometimes common at light; technically the costal cell is just present at extreme base and apex but at first glance can easily appear to lack one) ....**Vespoidea**: **Rhopalosommatidae**

Key F – Ichneumonidae or Braconidae?

1 With well-developed wings 2
- Wingless or brachypterous... 0
2(1) Fore wing with vein 2m-cu absent (Fig. 7.0 a), i.e. without a vein running posteriorly from the middle of 2nd submarginal cell **OR** from near rs-m to the distal part

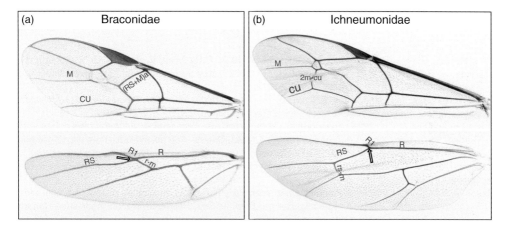

Fig. 7.6. Wing character distinguishing most of Braconidae and Ichneumonidae easily. Black and yellow arrow indicates point of separation of hind wing veins R1 and RS from R.

of vein CU (100%); hind wing with rs-m vein originating basal to the separation of veins (98%); fore wing vein (RS+M)a usually present (> 85%)...............**Braconidae**
- Fore wing with vein 2m-cu present, originating anteriorly from posterior of areolet (2nd submarginal cell) (Fig. 7.6 b) **OR** if areolet absent, then from close to the transverse rs-m vein (99.9%); hind wing with r-m vein originating distal to the separation of veins R1 and RS (98%); fore wing vein (RS+M)a always absent (100%) .... **Ichneumonidae**
3(1) Spiracle of first metasomal tergite situated on anterior third................**Braconidae**
- Spiracle of first metasomal tergite situated posterior to middle....**Ichneumonidae**

Beginners often have difficulty distinguishing between these two families. In S.E. Asia there are very few taxa for which there should be any problem. It is important to locate the free distal part of fore wing vein CU running to the wing margin from the distal end of the 1st subdiscal cell (see cell 12 in Fig. 6.10). The question is whether or not there is a vein, either tubular or spectral, running anteriorly from it to vein M approximately half way along it. In braconids there is no cross-vein present (Fig. 7.6 a) whereas in Ichneumonidae, there is (Fig. 7.6 b) – this vein is 2m-cu. In nearly all ichneumonids it joins vein M at the posterior of the areolet (see Fig. 10.13 b) or close to where the areolet would be if there is only one rs-m cross vein (see Fig. 10.4 b, c, Fig. 10.10 c, 10.34 c). Only in < 0.1% of ichneumonids is 2m-cu reduced (spectral) or absent, and there are few such wasps in the region.

There is another good distinguishing character in the hind wing which concerns where the rs-m vein originates anteriorly. In braconids it originates basal to or at the separation of veins RS and R1 (Fig. 7.6 a) whereas in ichneumonids it originates distally, often by a considerable margin (Fig. 7.6 a). There are a few exceptions but in all those cases the fore wing vein 2m-cu situation should be clear.

If fore wing vein (RS+M)a is present it is a braconid, it is never present in Ichneumonidae although a few have a short vein stub called the ramulus, originating from vein 1m-cu&M (see Fig. 10.4).

Wingless and highly brachypterous taxa are rather few, but essentially in this region, if the 1st metasomal segment is petiolate (narrow basally and strongly widened posteriorly) and the spiracle situated beyond the middle of the tergite, the specimen is probably a phygadeuontine ichneumonid such as *Gelis* (see Fig. 10.17 a).

References

Blaimer, B.B., Santos, B.F., Cruaud, A., Gates, M.W., Kula, R.R., Mikó, I., Rasplus, J.-Y., Smith, D.R., Talamas, E.J., Brady, S.G. and Buffington, M.L (2023) Key innovations and the diversification of Hymenoptera. *Nature Communications* 14, 1212.

Broad, G.R., Shaw, M.R. and Fitton, M.G. (2018) *The Ichneumonid Wasps of Britain and Ireland (Hymenoptera: Ichneumonidae)*. Royal Entomological Society, Telford, UK.

Goulet, H. and Huber, J.T. (1993) *Hymenoptera of the world: an identification guide to families*. Research Branch, Agriculture Canada Publication 1894/E, Ottawa, Ontario, 668 pp. Available from https://esc-sec.ca/wp/wp-content/uploads/2017/03/AAFC_hymenoptera_of_the_world.pdf (and various other sites)

Hanson, P.E. and Gauld, I.D. (eds) (1995) *The Hymenoptera of Costa Rica*. Oxford University Press, Oxford, UK, 893 pp.

Mason, W.R.M. (1993) Chapter 5, Key to superfamilies of Hymenoptera. In: Goulet, H. and J.T. Huber (eds) *Hymenoptera of the World: An identification guide to families*. Research Branch, Agriculture Canada Publication 1894/E, Ottawa, Ontario, pp. 65–100.

van Noort, S. and Broad, G. (in press) *Wasps of the World: A Guide to Every Family*. Princeton University Press, New Jersey, USA, 240 pp. ISBN: 9780691238548

8

Orussoidea

Abstract

This very small, uncommon and rarely collected superfamily is described, including what little is known of their biology. Their classification is discussed and recognition of subfamilies is not recommended. The species occurring in S.E. Asia and the countries where they have been collected are tabulated. Key identification works are referenced. One species from Singapore is illustrated.

These are rarely collected and rarely observed insects that spend most of their adult lives searching for hosts on dead wood, specifically on tree trunks that have lost their bark. They may be thought of as the most derived of the sawflies ('Symphyta'); they are the sister group to the remaining Hymenoptera and are the only 'sawflies' that are parasitoids. As with all other sawflies, but very few Apocrita, they lack a wasp waist (Fig. 8.1 a, b). The sole extant family Orussidae has traditionally been divided into two subfamilies, Ophrynopinae and Orussinae, although Vilhelmsen (2003a) recommended that these be disregarded, given how few genera there are. Further, in the morphological phylogeny of Vilhelmsen *et al.* (2014) the Ophrynopinae were recovered as a derived genus group rather than as sister group to the Orussinae.

Orussids have a very interesting host location mechanism (Vilhelmsen *et al.*, 2001), which explains the highly specialised antennal apex (Fig. 8.1 c) and modified fore legs of females with highly swollen tibia (which contains the 'ear', technically called the subgenual organ) and strengthened tarsus (Fig. 8.1 d). The antennae tap the wood and the leg pick up the vibration. This is called vibrational sounding because the vibrations are transmitted through a solid (wood), compared with aquatic sonar or the echolocation that bats do.

The general biology of Orussidae was described by Cooper (1953), who found that the egg is not laid directly on the host but a distance away from it in the host's gallery. Therefore, host location is completed by the 1st instar orussid larva. Larval anatomy was described in detail by Vilhelmsen (2003b). Unlike all other 'sawflies', the larval head region is greatly reduced, lacking eyes and with tiny antennae, adaptations to the parasitoid way of life in the dark. Many details are still unclear but of note is that the mid and hind guts although highly modified are connected, unlike in most developing larval apocritans, although there may be a muscular closing mechanism.

Vilhelmsen (2003a, b) presented a morphological phylogeny of the family, a key to world genera and also an annotated key to the species of *Orussus*. Vilhelmsen *et al.* (2014) provided a revised and updated key for *Orussus*. Although there are many temperate, African, Australian and especially South American species, rather few orussids have been found in S.E. Asia (Table 8.1).

Table 8.1. Orussidae species recorded in the literature from S.E. Asia.

Species	Distribution	Reference
Mocsarya metallica	Indonesia, Sri Lanka	Vilhelmsen, 2001
Ophrynopus (as *Stirocorsia*) *kohli*	Indonesia (widely distributed, including Sulawesi), Laos, Malaysia (Sarawak), Papua New Guinea, Philippines	Vilhelmsen and Smith, 2002 (see also Vilhelmsen, 2020)
Ophrynopus (as *Stirocorsia*) *maculipennis*	Indonesia (Ambon, Aru, Irya Jaya, Halmahera, Moluccas), Papua New Guinea	Vilhelmsen and Smith, 2002 (see also Vilhelmsen, 2020)
Orussus decoomani[a]	Vietnam	Yasumatsu, 1954
Orussus bensoni	Philippines	Yasumatsu, 1954
Orussus loriae	New Guinea	Yasumatsu, 1954
Orussus punctulatissimus	Malaysia	Vilhelmsen *et al.*, 2014

[a]Vilhelmsen *et al.* (2014) note that this species from Vietnam cannot be interpreted at the moment so they consider it a *species inquirenda*.

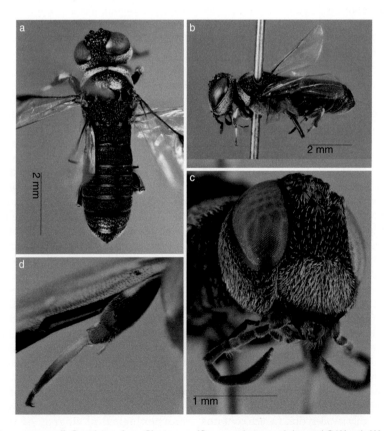

Fig. 8.1. *Orussus* sp., aff. *O. striatus*, from Singapore. (Source: photograph by and © Wendy Wang (Lee Kong Chian Natural History Museum, LKCNHM, National University of Singapore), reproduced with permission.)

References

Cooper, K.W. (1953) Egg gigantism, oviposition and genital anatomy: their bearing on the biology and phylogenetic position of *Orussus* (Hymenoptera: Siricoidea). *Proceedings of the Rochester Academy of Science* 10, 38–68.

Vilhelmsen, L. (2001) Systematic revision of the genera *Chalinus* Konow, 1897 and *Mocsarya* Konow, 1897 (Hymenoptera: Orussidae). *Insect Systematics & Evolution* 32, 361–380.

Vilhelmsen, L (2003a) Phylogeny and classification of the Orussidae (Insecta: Hymenoptera), a basal parasitic wasp taxon. *Zoological Journal of the Linnean Society* 139, 337–418. doi: 10.1046/j.1096-3642.2003.00080.x

Vilhelmsen, L (2003b) Larval anatomy of Orussidae (Hymenoptera). *Journal of Hymenoptera Research* 12, 346–354.

Vilhelmsen, L. (2020) Two new species of *Ophrynopus* Konow, 1007 (Hymenoptera: Orussidae), with a new definition of the genus. *Zootaxa* 4790, 121–137.

Vilhelmsen, L. and Smith, D.R. (2002) Revision of the 'ophrynopine' genera *Argentophrynopus* gen. n., *Guiglia* Benson, *Kulcania* Benson, *Ophrella* Middlekauff, *Ophrynon* Middlekauff, *Ophrynopus* Konow, and *Stirocorsia* Konow (Hymenoptera: Orussidae). *Insect Systematics & Evolution* 33, 387–420.

Vilhelmsen, L., Isidoro, N., Romani, R., Basibuyuk, H.H. and Quicke, D.L.J. (2001) Host location and oviposition in a basal group of parasitic wasps: the subgenual organ, ovipositor apparatus, and associated structures in the Orussidae (Hymenoptera, Insecta). *Zoomorphology* 121, 63–84.

Vilhelmsen, L., Blank, S.M., Liuc, Z. and Smith, D.R. (2014) Discovery of new species confirms Oriental origin of *Orussus* Latreille (Hymenoptera: Orussidae). *Insect Systematics & Evolution* 45, 51–91.

Yasumatsu, K. (1954) *Orussus boninensis*, a new species of Orussidae from the Bonin Islands. *Insecta Matsumurana* 18, 115–118.

9

Braconidae

Abstract
The huge and diverse family Braconidae is described and various specific morphological terms used in their description are illustrated. A simplified identification key to most of the subfamilies occurring in S.E. Asia is provided. The world and regional generic and specific diversity of subfamilies is tabulated. Each major group and each subfamily are then discussed individually, including what is known of their biology, recent taxonomic descriptions and relevant identification works for the S.E. Asian representatives. Representative species and identification characters are illustrated photographically.

Braconidae is a large family of Hymenoptera, second in size in terms of numbers of described species only to its sister family, Ichneumonidae. It is a cosmopolitan group, with more than 1100 genera and more than 21,220 valid described species (Yu et al., 2016). This family has great use as natural enemies in biological control programmes. Because of their high diversity, they provide a good example for studying diversity, evolutionary biology, host–parasitoid interactions and morphological adaptation as well as evolution of polydnaviruses (Sharanowski et al., 2011).

Many of the larger or colourful braconids from tropical regions were described by a handful of European workers during the first two decades of the 20th century, notably the Hungarian Guy Szépligeti and the British Peter Cameron. Considering that borrowing type specimens was almost impossible in those days and that publications were usually without illustrations, we think they did remarkably well. However, Cameron's work leaves a lot to be desired and we recommend that readers should have a look at the one-page (or less) obituary of him by Claude Morley, who also worked at the Natural History Museum, then the British Museum (Natural History) (see Chapter 4, this volume).

Global and regional described braconid diversity is summarised in Table 9.1 together with notes on the biologies of the subfamilies.

The Vietnamese braconids recorded until 2014 were listed by Long and Belokobylskij (2003) and supplemented by Long and van Achterberg (2014), but there are no other checklists for other S.E. Asian countries.

Phylogenetic Studies

The first major attempt to work out phylogenetic relationships of all the subfamilies recognised at the time was by Quicke and van Achterberg (1990), based on morphology. The resulting trees were certainly not to the

Table 9.1. Summary of currently recognised subfamilies of Braconidae recorded from S.E. Asia, their described generic and species richness, biology and hosts. (Source: data updated from various sources, notably Yu et al. (2016), Sharkey et al. (2021a, b) and Gadallah et al. (2022a)

Subfamily	World genera	S.E. Asia genera	World species	Described S.E. Asia species	Life history strategy	Host order(s) in S.E. Asia
Acampsohelconinae	3	1	c.130	11	Egg-larval endo.	C
Agathidinae	61	23	1230	255	Koino, endo.	L
Alysiinae	108	23	2446	71	Koino, endo.	D
Aphidiinae	63	14	657	20	Koino, endo.	Hem
Brachistinae	41	5	794	59	Koino, endo.	C
Braconinae	190	69	3109	455	Idio, ecto, [endo] [pred]	C, L, D, Hem
Cardiochilinae	17	4	220	28	Koino, endo.	L
Cenocoeliinae	6	1	91	4	Koino, endo.	C
Charmontinae	2	1	10	1	Koino, endo.	L
Cheloninae	23	7	1643	63	Koino, endo.	L
Dirrhopinae	1	1	5	1	Koino, endo.	L
Doryctinae	c. 200	44	c. 2000	296	Idio, ecto.	C, L
Euphorinae	59	18	1270	45	Koino+imago, endo.	C, L, H, Hem, P
Exothecinae	9	3	92	13	Idio, ecto.	L
Helconinae	18	3	119	13	Koino, endo.	C
Homolobinae	3	1	73	10	Koino, endo.	L
Hormiinae	c. 25	13	225	52	Idio, ecto.	L, C
Ichneutinae	9	1	86	2	Koino, endo.	L
Macrocentrinae	8	3	278	11	Koino, endo.	L
Meteoredeinae	2	17	1	No named species recorded but several occur	Koino, endo.	L
Microgastrinae	2999	21	81	270	Koino, endo.	L
Miracinae	2	1	47	1	Koino, endo.	L
Opiinae	39	18	2063	128	Koino, endo	D
Orgilinae	13	5	363	33	Koino, endo.	L
Pambolinae	9	1	70	4	Idio, ecto.	C
Proteropinae	6	1	30	2	Koino, endo	H
Rhysipolinae	8	1	55	3	Koino, ecto.	L
Rhyssalinae	14	3	63	3	Idio, ecto.	C, ?L
Rogadinae	57	31	1351	361	Koino, endo.	L
Sigalphinae	8	1	c. 50	3	Egg-larval endo	L
Telengaiinae	4	3	90	4	?Koino	L
Xiphozelinae	2	1 (but probably 2 occur)	6	3	Koino, endo.	L

Abbreviations: C, Coleoptera; D, Diptera; H, Hymenoptera; Hem, Hemiptera; K, kleptoparasitoids; L, Lepidoptera; N, Neuroptera; P, Psocoptera.

taste of many other braconid researchers and soon thereafter a lengthy criticism was published (Wharton et al., 1992) followed by a rebuttal (van Achterberg and Quicke, 1992). However, reanalysis of the dataset by the critics yielded essentially the same result. To cut the story short, morphology was subsequently discovered to be highly influenced by convergent character state sets strongly influenced by biology, and this led to a number of follow-up papers investigating character conflict (e.g., Quicke and Belshaw, 1999).

Recent research where the results are to be trusted far more are based entirely on molecular data and these consistently show that the family is divided into two large informal groups which can largely be recognised by a feature of the mouthparts. The two groups are the cyclostomes and the non-cyclostomes. The former has the lower part of the clypeus (the hypoclypeus) turned internally, revealing a depressed and concave glabrous labrum. The cyclostomes are mostly idiobiont ectoparasitoids but with several independent transitions to endoparasitism, both as idiobionts and koinobionts. However, some of the cyclostomes have secondarily lost the cyclostome mouthparts (e.g. Aphidiinae, Alysiinae, most Opiinae and Telengaiinae). The non-cyclostomes do not have the lower part of the clypeus reflexed inwards, and if the labrum is visible it is flat and setose (Quicke, 2015). All non-cyclostomes are koinobiont endoparasitoids.

The non-cyclostomes include the majority of subfamilies, most of which have been confidently assigned to a number of informal subfamily groups or complexes, specifically the microgastroids, sigalphoids, euphoroids and helconoids. Within the last of these a macrocentroid subcomplex is also recognised (Sharanowski et al., 2011).

Many braconids possess single locus sex determination and therefore breeding colonies rapidly tend towards producing allmale offspring, but this is not true of all, for example in some alysiines (see e.g. Ma et al., 2013).

Morphological Terminology

There is relatively little specialist terminology associated with this family but there are a few, mostly concerning the pronotum and the 1st metasomal tergite, mostly introduced by van Achterberg (1974, 1976a, 1979a, 1994a).

Cyclostome

One of the most important characters for braconid identification, but one which often confuses beginners, is whether or not they are cyclostome. The term cyclostome means round hole and refers to a nearly circular or oval recessed gap between the upper edge of the mandibles and the lower border of the clypeus as seen from the front (Fig. 9.1 a). The cavity formed thus reveals all or nearly all of the labrum, which is usually concave and largely polished (but see Pambolinae). What is being seen as the lower edge of the clypeus is where it turns abruptly inwards and forms the roof of the cavity. The reflexed (turned inward) part of the clypeus is called the hypoclypeus. The dorsal margin of the hypoclypeus is usually separated from the forward-facing part of the clypeus by a

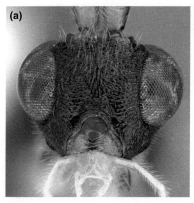

Fig. 9.1. Comparison of mouthparts showing **(a)** cyclostome and **(b)** non-cyclostome conditions.

carina. In non-cyclostomes, the ventral margin of the clypeus is usually convex and more or less abuts to the closed mandibles. If a wasp is clearly cyclostome it belongs to the cyclostome lineage, but cyclostomicity has been secondarily lost in most Opiinae and all Alysiinae.

Lateral Pronope (or Subpronope)

A round pit-like depression in the anterolateral part of the pronotum (van Achterberg, 1994a). These are found in Agathidinae and Sigalphinae and some Ichneutinae.

Pronope (or Median Pronope)

This term was used by van Achterberg (1994a) for a deep pit in the mid-anterior part of the pronotum which is present in all Agathidinae. It is bordered posteriorly by the pronotum, which distinguishes it from the antescutal depression which is between pronotum and anterior of mesonotum.

Precoxal Sulcus

Term introduced by Richards (1956), but much more extensively by van Achterberg and braconidologists in general, for the nearly horizontal longitudinal depression of the mesopleuron in Braconidae. It is located higher than the sternaulus that is found in many Ichneumonidae. We know that these structures are not homologous, because some opiine braconids have both (Fig. 9.2).

Prepectal Carina

This is the widely used term for the transverse carina close to the anterior margin of the mesopleuron+mesosternum which is called the epicnemial carina by most ichneumonidologists.

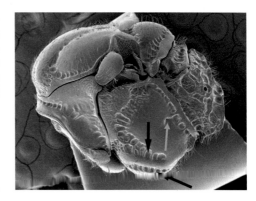

Fig. 9.2. SEM of the mesosoma of a *Sternaulopius* sp. (Braconidae: Opiinae) showing both precoxal sulcus (upper black arrow) and sternaulus (lower black arrow), and additionally indicating the episternal scrobe (blue arrow). (Source: image by and © Robert A. Wharton (Texas A & M University), reproduced with permission.)

Anterior Subalar Depression

This term was introduced by van Achterberg (1994a) and describes a depression above the subalar prominence (Richards, 1977) which may be smooth, crenulate or with a distinct transverse carina. This is used in several of van Achterberg's subfamily keys.

Dorsal, Dorso-Lateral and Lateral Carinae of 1st Metasomal Tergite

Van Achterberg (1974) noted and coined terms for several features of the 1st metasomal tergite (= petiole) in Braconidae. The tergum of the 1st metasomal tergite may have up to three distinct pairs of carinae. The lateral carina when present runs close to the ventral margin of the tergum. Arising from near the base of the lateral carina a second carina may run above and parallel to the lateral carina and enclose a longitudinal groove. Also arising from the anterior of these is often another carina pair (dorsal carinae) which can either run separately to the posterior margin of the tergite, or meet medially, forming a curved transverse carina.

Dorsope and Laterope

Moderate to deep pits at the anterior confluence, respectively, of the dorsal and dorso-lateral carinae, and of the dorso-lateral and lateral carinae. The dorsopes may be so deep that the pair medially meet and are separated by only a thin, often transparent, cuticular septum. These are essentially similar to the glymmae of the 1st tergite in some Ichneumonidae when these are located far anteriorly.

Wings

The majority of braconids have six or seven fore wing cells enclosed by tubular veins (Fig. 9.3). Only the extralimital (Chilean) genus *Apoxyx* has a more or less well-developed vein 2m-cu. However, loss of wing veins is common, almost always affecting more distal parts of the wing, although M+CU can be largely obliterated too in many groups.

The terms antefurcal, interstitial and postfurcal are used to indicate that the anterior junction of a vein (typically fore wing vein 1cu-a) is basal, in line with or distal to another vein (typically 1M) opposite it. Inclivous and reclivous indicate, respectively, whether a transverse wing vein slopes posteriorly towards or away from the wing tip (van Achterberg, 1976a, 1979b).

Subfamily Identification

Braconid identification was aided enormously when Cornelis (Kees) van Achterberg published his preliminary key to the subfamilies of Braconidae (van Achterberg, 1976b). Amazingly, prior to that work, the most recent general key to subfamilies was by Szépligeti (1904).

Fortunately, there are several more up-to-date keys available that can help identification of braconids to subfamily, with various degrees of completeness and relevance to the region. The most comprehensive is by van Achterberg (1993a) but this does not include several recent changes in classification, especially as regards the minor cyclostome subfamilies and the Helconinae s.l. Also, in the same year, a key was published by Sharkey (1993) which is somewhat simpler to use but lumps several smaller groups with a broad concept of Rogadinae. Gadallah *et al.* (2022b) and Sharkey *et al.* (2023) both provided a more up-to-date, photographically illustrated keys to the taxa occurring in the Middle East and the New World but neither include Xiphozelinae.

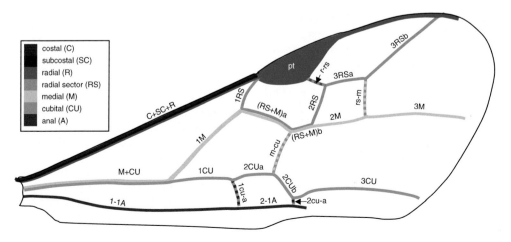

Fig 9.3. Stylised braconid fore wing venation using terminology of Sharkey and Wharton (1997).

Key to Females of Common* S.E. Asian Subfamilies of Braconidae

(* does not include Charmontiinae, Dirrhopinae, Proteropinae, Xiphozelinae)

1 Fore wing vein RS tubular all the way to wing margin where it joins tubular vein R1 (Figs 9.13 a, 9.14 a, b, e) 2
- Fore wing vein RS not tubular all the way to wing margin, usually indicated in part as a nebulous or spectral vein (Figs 9.30 c, 9.32 a) 30

 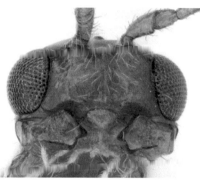

Fig. 9.4. Two examples showing exodont mandibles of Alysiinae.

2(1) Mandibles with three or more conspicuous teeth or apical lobes, usually directed outwards (Fig. 9.4 a) **OR** if directed medially, their tips remain separated from one another (Fig. 9.4 b) .... **Alysiinae**
- Mandibles bidentate apically (sometimes appearing unidentate in anterior view) **AND** when closed (the usual condition in specimens) their tips clearly overlapping (Fig. 9.1) 3
3(2) Fore wing vein RS running in a nearly straight line to wing margin, closely parallel to the distal margin of pterostigma, the two being separated by only a very narrow straight marginal cell which is at least 6.0 x longer than maximally wide (Fig. 9.5);

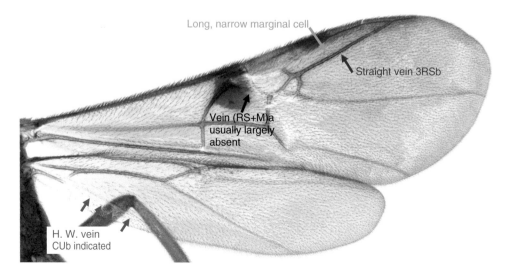

Fig. 9.5. Features of Agathidinae wing venation.

	hind wing vein 2CU2 indicated (Fig. 9.5); fore wing vein (RS+M)a largely or entirely absent (except *Earinus* spp.) (Fig. 9.5) .... most **Agathidinae**
–	Fore wing vein RS diverging from pterostigma making the marginal cell much wider anteriorly than posteriorly (Figs 9.15 c) **OR** vein RS strongly curved (Fig. 9.27 c, d) 4
4(3)	Propodeum with a pair of posteromedial spines (apophyses) (Fig. 9.11 b,d); mouthparts cyclostome (see couplet 7) (Fig. 9.11 c) 5
–	Propodeum without spine; mouthparts variable 6
5(4)	Eyes emarginate opposite antennal sockets **AND** (nearly always) with metasomal tergites 1 and 2 with a midlongitudinal carina and tergite 2 virtually always sculptured .... few **Rogadinae**
–	Eyes not emarginate; metasomal tergites 1 and 2 with a midlongitudinal carina, tergite 2 smooth and shiny ....**Pambolinae**
6(4)	Tip of ovipositor much darker (usually black) contrasting with rest of shaft which is yellow-brown (N.B. not the setose ovipositor sheaths, which may need to be separated or broken to reveal the ovipositor proper); ovipositor tip dorsally with two nodes; fore tibia with a longitudinal row of pegs or thickened spines (can be very difficult to see in small-bodied species); mouthparts cyclostome .... most **Doryctinae**
–	Ovipositor unicolorous; ovipositor tip with one or no nodes; fore tibia without longitudinal row of thickened spines; mouthparts variable 7
7(6)	Mouthparts cyclostome, i.e. the lower part of the clypeus is angled inwards forming a dorsally arched depression and the labrum is thus largely visible, and usually glabrous and concave 8
–	Mouthparts not cyclostome, the clypeus ventrally straight or convex, usually closely abutting mandibles and the labrum not largely visible 17
8(7)	Occipital carina completely absent **AND** hind wing subbasal cell very small, length of vein M+CU less than 0.3 × combined length of M+CU and 1-M; fore wing vein cu-a interstitial or nearly so except in *Euurobracon* which is a very large wasp with body length > 2 cm; hind wing vein m-cu always absent .... **Braconinae**
–	Occipital carina present, at least laterally **OR** if absent, length of vein M+CU more than 0.45 × combined length of M+CU and 1-M; fore wing vein cu-a often postfurcal; hind wing vein m-cu present or absent; hind wing vein m-cu often present 9
9(8)	Metasomal tergites 1–3 forming a complete, strongly sculptured carapace concealing all other tergites (except in some bloated specimens) **AND** prepectal carina present (absent in *Parahormius*); (common) ....**Hormiinae – Lysitermini**
–	Metasoma not formed as a sculptured carapace **OR** if rarely so, prepectal absent 10
10(9)	Metasomal tergites dorsally membranous, usually white, semi-transparent **AND** hypostomal carina joining occipital carina above base of mandible; (common) .... most ....**Hormiinae – Hormiini**
–	Metasomal tergites usually strongly sclerotised dorsally; hypostomal carina variable 11
11(10)	Fore tibia with a longitudinal row of pegs or thickened spines (Fig. 9.6) but these can be very difficult to see in small-bodied species; (common) ....**Doryctinae**
–	Fore tibia without a longitudinal row of pegs or thickened spines 12

Fig. 9.6. Fore tibial row of pegs (chaetobothria) or thickened spines of Doryctinae.

12(11) Prepectal carina present 13
-       Prepectal carina absent 15
13(12) Ventral third of occipital carina well developed and running direct to base of mandible separate from hypostomal carina (Fig. 9.12 b); (uncommon) .... **Rhysipolinae**
-       Ventral third of occipital carina absent **OR** if well developed, then curving towards, and often joining hypostomal carina before the latter reaches the base of the mandible.. 14
14(13) Spiracles of metasomal tergites 2 and 3 in the sclerotized notum; metasomal tergites usually strongly sculptured (foveate, rugulose, striate), 1st and 2nd tergites usually with midlongitudinal carina distinguished from remaining sculpture; fore wing anal vein (2A) never indicated; hind basitarsus shorter than remaining part of tarsus; (common) .... most **Rogadinae**
-       Spiracles of metasomal tergites 2 and 3 in the laterotergites; metasomal tergites 2 onwards usually smooth rarely sculptured striate; fore wing with vein 2a distinctly visible nearly opposite 1cu-a **OR** hind basitarsus elongate, longer than articles 2–5 combined; (rare) .... **Rhyssalinae**
15(12) Metasomal tergite 2 with a raised transverse basal area bordered posteriorly by a groove; occipital carina completely absent; (small wasps, less than 2 mm (uncommon) .... **Telengaiinae** (= Gnamptodontinae)
-       Metasomal tergite 2 without a raised transverse basal area 16
16(15) Occipital carina absent dorsally, usually present laterally (Fig. 9.17); (common) .... some **Opiinae**
-       Occipital carina complete; (rare) .... **Exothecinae**
17(7)   Fore, middle and often hind trochantellus with one or more blunt pegs (Fig. 9.25 a) .... **Macrocentrinae**
-       Trochantelli without pegs 18
18(17) Metasomal tergites 1–3 forming a complete, strongly sculptured carapace concealing all other tergites (except in some bloated specimens) (Figs 9.20, 9.21 a, 9.22 b, 9.29 a, b), with or without transverse crenulate grooves indicating borders between segments 19
-       Metasoma not formed as a carapace 22
19(18) Fore wing 2nd submarginal cell closed, i.e. vein rs-m present (Fig. 9.22 d) 20
-       Fore wing 2nd submarginal cell open, vein rs-m absent (Fig. 9.22 b, c) .... some **Brachistinae** (also *Fornicia*, Microgastrinae, Fig. 9.31 d)
20(19) 1st metasomal tergite flexibly connected (articulated) with 2nd (Fig. 9.20); (rare) .... **Sigalphinae**
-       1st metasomal tergite immovably connected with 2nd the two separated by a crenulate groove (Fig. 9.29 d, e) or without any groove (Figs 9.21 a, 9.29 a) 21

21(20)	Fore wing 2nd submarginal cell closed, i.e. vein rs-m present although usually spectral (Fig. 9.29 a–d); hind wing vein 2CU1 absent (Fig. 9.29 c); (common) .... most **Cheloninae**	
–	Fore wing 2nd submarginal cell open, vein rs-m absent; hind wing vein 2CU1 present (Fig. 9.21 d); (rare) ....**Acampsohelconinae**	
22(18)	Antescutal depression present (a deep crescent-shaped pit between front of middle lobe of mesoscutum and posterior of middle part of pronotum (see Fig. 9.24 c-e); often ophionoid yellow-brown nocturnal species (uncommon) .... **Homolobinae**	
–	Antescutal depression absent; colour and shape variable 23	
23(22)	Metasoma arising from propodeum far above upper margin of hind coxae by considerably more than the basal width of the hind coxa (rare) ....**Cenocoeliinae**	
–	Metasoma arising from propodeum below upper margin of hind coxa, rarely slightly above it.. 24	
24(23)	2nd submarginal cell of fore wing small and triangular, veins 2RS and rs-m touching anteriorly (or even joining), a short vein 2M present posteriorly ....some **Orgilinae**	
–	2nd submarginal cell of fore wing absent or quadrilateral 25	
25(24)	Hind wing vein cu-a present 26	
–	Hind wing vein cu-a absent (small wasps, less than 2 mm body length); (uncommon in the region) ....**Aphidiinae – Ephedrini**	
26(25)	Prepectal carina absent laterally; anterior subalar depression smooth 27	
–	Prepectal carina present laterally; anterior subalar depression with a carina 28	
27(26)	Metasomal tergite 2 with a (raised) transverse basal area bordered posteriorly by a transverse groove (Fig. 9.16); occipital carina usually completely absent; fore wing 2nd submarginal cell always short with vein 3RSa approximately as long as rs-m; fore wing vein M+CU tubular for whole length except extreme base; (uncommon) ....**Telegaiinae** (= Gnamptodontinae)	
–	Metasomal tergite 2 without a differentiated anterior transverse; occipital carina usually present laterally but always absent dorsally; fore wing 2nd submarginal cell often elongate with vein 3RSa much longer than rs-m, and often narrowing distally; fore wing vein M+CU often largely unsclerotized .... most **Opiinae**	
28(26)	Hind femur partially rugose ventrally, sometimes with strong ventral tooth; frons with strong midlongitudinal carina. (medium to large wasps, body length usually 7–12 mm) .... **Helconinae** s.s.	
–	Hind femur not coarsely sculptured ventrally, without ventral tooth; frons usually without midlongitudinal carina 29	
29(28)	Outer apex of hind tibia with a few pegs among the setae (these may be very difficult to see and often require adjusting the angle of illumination and aspect of the specimen)... some **Orgilinae**	
–	Outer apex of hind tibia without pegs 30	
30(29)	Hind wing vein 2-CU present; metasoma laterally compressed; ovipositor concealed within metasoma; (often collected at UV light) .... **Meteorideinae**	
–	Hind wing vein 2-CU absent; metasoma usually not laterally compressed; ovipositor nearly always clearly exserted beyond apex of metasoma .... **Euphorinae** (including Meteorini), **Brachistinae**	
31(1)	Fore wing vein 1-M&1RS strongly curved into parastigma and vein (RS+M)a arising from parastigma (Fig. 9.30 c); veins 2RS and rs-m absent; (uncommon) ....**Ichneutinae**	
–	Fore wing vein 1-M&1RS **NOT** strongly curved into parastigma; vein 2RS present and vein rs-m sometimes present (Fig. 9.32 a) 32	
32(31)	Antenna with exactly 12 flagellomeres; fore wing vein 2RS reaching pterostigma more or less where r-rs emerges from it; (uncommon) .... **Miracinae**	
–	Antenna with more flagellomeres; fore wing vein r-rs meeting 2RS a distance from pterostigma... 33	

33(32)	Antenna with exactly 16 flagellomeres (note that placoid sensillae may be arranged in two rings on each flagellomere giving superficial appearance of more segments **AND** fore wing 2nd submarginal cell absent (Fig. 9. 31 f) **OR** very short (Fig. 9. 31 a, b) **AND** hind wing vein cu-a present; (extremely common) ....**Microgastrinae**
-	Antenna with different number of flagellomeres 34
34(33)	Fore wing vein 3RSb strongly arched at base (Fig. 9.28 c); metasoma not formed into a three-segmented carapace.. **Cardiochilinae**
-	Fore wing vein 3RSb straight or only weakly curved at base (Fig. 9.7 a, 9.27, 9.29 a–e) **OR** veins rs-m and 3RSb absent; metasoma variable 35
35(34)	No tubular venation distal to fore wing vein 1-M&1RS ,,, ... **Agathidinae** (*Aneurobracon*)
-	Tubular veins present in distal part of fore wing 36
36(35)	Fore wing vein 3RS arising directly from pterostigma remote from 2-RS (i.e. r-rs absent); metasomal tergites 1 and 2 fused connected by a crenulate groove; (small wasps less than 2 mm) .... **Cheloninae – Adeliini**
-	Fore wing vein 3RS arising distinct r-rs vein (even if 2-RS absent); metasomal tergites 1 and 2 movably joined 37
37(36)	**Hind** wing vein cu-a, and usually also veins M+CU and 1-M absent; sometimes with pair of prongs arising from hypopygium; (small, less than 4 mm body length; uncommon, mostly at higher altitudes) .... **Aphidiinae**
-	Hind wing veins cu-a, M+CU and 1-m present; never with pair of prongs arising from hypopygium .... some **Euphorinae**

Cyclostomes

Many of the subfamilies included here have been recognised as a group since early times.

Aphidioid Subcomplex

Rather to the surprise of many workers, the Aphidiinae which are not conspicuously morphologically cyclostome, the endemic Australian Maxfischeriinae which is not cyclostome and the Gondwanan Mesostoinae (some of which are cyclostome) form a robustly supported clade as sister group to all the remaining cyclostomes (Zaldivar-Riverón et al., 2006; Sharanowski et al., 2011, 2021; Jasso-Martínez et al., 2022a, b). Of these only the first is known from the region.

Aphidiinae

Aphidiinae (aphid mummy wasps) are solitary koinobiont endoparasitoids exclusively of aphids (Hemiptera) many of which are major insect pests worldwide, therefore members of this subfamily are one of the most effective groups of natural enemies of aphids with high-level host–parasitoid specificity (Žikić et al., 2017). There are about 60 genera and more than 650 species worldwide (Yu et al., 2016). Starý et al. (2010a) studied aphid parasitoids from Thailand, recording a total of 11 species in 10 genera. Later, Martens et al. (2021) studied Thai Aphidiinae and revealed 20 species from 15 genera and erected the new genus and species, *Ishtarella thailandica* with very distinctive mesoscutal horns, collected by light trap. The Thai aphidiine fauna shares numerous species with Bangladesh, China, India, North Korea, South Korea, Malaysia, Japan, Pakistan and Vietnam. *Lipolexis oregmae* is a common species and cosmopolitan.

Because of their economic importance and ease of culture, more is known about aphidiine biology and development than for almost all other braconid groups. Aphidiines have completely or largely alecithal, hydropic eggs and the serosal membrane surrounding the developing embryo dissociates to produce teratocytes. Realised fecundity can be very high – more than 1000. Larval stages have been described for various genera (e.g. Pennacchio and Digilio, 1990; more references in Quicke, 2015).

A number of papers have presented molecular analyses of relationships within the subfamily (Belshaw and Quicke, 1997; Smith *et al.*, 1999; Sanchis *et al.*, 2000; Shi and Chen, 2005). Although the subfamily has a predominantly north temperate distribution these days, Belshaw *et al.* (2000) provided evidence that its origins were in Gondwanaland, which is consistent with its close relationships to the Mesostoinae and Maxfischeriinae. Molecular data support recognition of four tribes, Aphidiini, Ephedrini, Praini and Trioxini, with the Ephedrini, which have complete wing venation, probably being basal.

In members of some genera the female hypopygium is modified into a pair of spines that are used for host restraint during oviposition (Fig. 9.7 c,d).

In S.E. Asia the main aphid family attacked is the Greenideidae, but there are also some species associated with introduced crop pest aphids that originate in more temperate regions. Aphidiine species attacking greenideid aphids are common in Thailand as well as in East and North Asia (Starý *et al.*, 2008, 2010b), for example *Archaphidius greenideae*, *Fissicaudus thailandicus*, *Indaphidius curvicaudatus* (Fig. 9.7 b), and *Parabioxys songbaiensis*.

The key to genera from the Middle East by Gadallah *et al.* (2022c) includes all the common genera recorded from S.E. Asia. However, several additional uncommon ones occur (although most are unpublished). Starý and Ghosh (1983) provided a key to the Indian genera. Another quite useful and quite well illustrated key with most genera is by

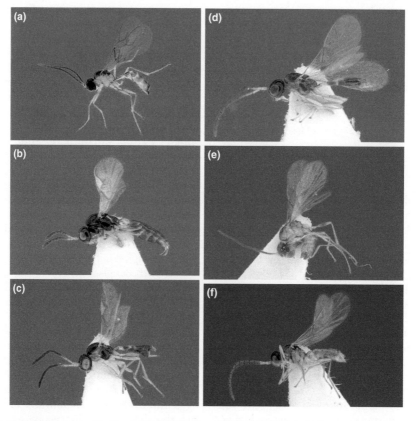

Fig. 9.7. Six aphidiines from Thailand: **(a)** *Aphidius* sp., (Aphidiini); **(b)** *Inaphidius curvicaudus* (unplaced); **(c)** *Binodoxys* nr *indicus* female (Trioxini), showing hypopygidial prongs; **(d)** *Trioxis* sp. (Trioxini); **(e)** *Lipolexis oregmae* (Trioxini); **(f)** *Praon* sp., (Praiini).

Raychaudhuri (1990) for north-east India, although as always with extralimital studies, the species level keys for each genus could give wrong answers for S.E. Asian species.

Masoninae

This is the archetypic enigmatic group with a single extant genus, *Masona*. Very small, they are associated with the ground layer and collected mostly in flight-intercept traps used by coleopterists and by Burlese funnels to separate arthropods from leaf litter. They were originally described as aberrant Braconidae (van Achterberg, 1995b), then molecular analysis of an even more aberrant species from Australia strongly supported them as ichneumonids (Quicke et al., 2019a), but most recently, ultra-conserved element analysis (Jasso-Martínez et al., 2022b) again places them in the Braconidae close to Aphidiinae.

One undescribed species has been seen by one of us (DLJQ) extracted from leaf litter in Cambodia (Quicke et al., 2019b) and we have just discovered another in Thai Malaise trap samples.

Mainline Cyclostomes

Most of the traditional cyclostome subfamilies form a grade leading to the secondarily non-cyclostome Alysiinae+Opiinae which are Diptera parasitoids and which we treat separately below. Most of this grade are idiobiont ectoparasitoids of weakly or deeply concealed hosts.

Braconinae

This is the largest subfamily of Braconidae, with at least 2500 species in approximately 192 genera (Yu et al., 2012; Ranjith et al., 2016a, b; Li et al., 2017, 2021; Quicke et al., 2018), mostly recorded from the tropics. There are approximately 70 genera recorded from S.E. Asia (Yu et al., 2012); however, some of these are based on very old descriptions of taxa whose type material is lost or in terrible condition and are probably misidentified, such as *Vipio* and *Glyptomorpha*, which we do not believe occur in the region.

Braconines are cosmopolitan but highly diverse in tropical regions (Quicke, 2015), especially those of the Old World, including S.E. Asia. In addition to the key characters, most genera have the 1st metasomal tergite flattened laterally (Quicke, 1987).

The subfamily includes *Habrobracon*, of which *H. hebetor* was the first hymenopteran taken into space (see Quicke, 2015). It is still widely used as a laboratory animal, but its heyday has passed. This species, which occurs widely in S.E. Asia in association with stored product pests, provides a strong example of why taxonomy is important. Two species of the genus, *H. hebetor* and *H. brevicornis*, were recognised at one time, but they were synonymised because they are morphologically very similar. They are now separate again as they do not interbreed (e.g. Chomphukhiao et al., 2018) and actually have rather different biological traits, and DNA sequence data conclusively separated them (Kittel and Maeto, 2019). The problem is that many papers were published on '*H. hebetor*' whilst they were treated as synonyms, and since no voucher material was kept and records of the origins of the cultures often lost, it is probably impossible to know which papers refer to which species now.

The genus *Bracon* has more than 1000 described species known worldwide and several species are important natural enemies of various pests. In addition to various *Bracon* species, several genera have species that are occasionally important in controlling pests. Several genera are frequently recorded from stem-borer pests in Asia, especially crambids such as *Chilo*, *Scirpophaga* and *Schoenobius* species. The braconines involved are various *Amyosoma*, *Pseudoshirakia*, *Stenobracon* and *Tropobracon* species.

The biology of this subfamily is diverse. Whilst most are idiobionnt parasitoids; a few species are reported as entomophytophagous (Ranjith et al., 2016b) or phytophagous (Flores et al., 2005; Perioto et al., 2011) or entirely predatory (Ranjith et al., 2022a). In general, braconines, including most *Bracon* species, are idiobiont ectoparasitoids of concealed immature hosts such as gall inducers and leaf rollers or miners, or deeply concealed such as wood-borers (Shaw and Huddleston 1991; Quicke 2015; Ranjith et al., 2016b). The Aspidobraconina and the related *Crinibracon* (which occurs in S.E. Asia) are endoparasi-

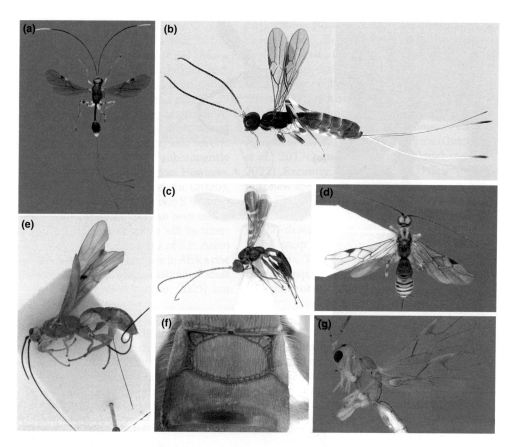

Fig. 9.9. Exemplar doryctines: **(a)** *Spathius* sp.; **(b–d)** gen. spp.; **(e, f)** *Zombrus* sp., habitus and metasomal syntergum 2+4 showing variant of common doryctine groove arrangement; **(g)** gen. sp. showing pseudostigma in hind wing that characterises males of a number of genera and is basically diagnostic of being a doryctine.

Hormiinae (including former Lysiterminae)

This group has been the most subject to taxonomic and nomenclatural changes of any other group within the family (Wharton, 1993). Whitfield and Wharton (1997) and various other workers used the name as a catch-all for several groups of small cyclostome braconids including Rhysipolinae, Mesostoinae, Rhyssalinae and Exothecinae, although most of these workers recognised that they are heterogeneous. This reflects the small number of distinct morphological characters that can be used to separate them.

It is a tribute to traditional taxonomists that they long ago recognised Hormiinae *sensu stricto* with their highly desclerotised metasomal tergites, and the Lysiterminae with their heavily sclerotised and sculptured carapace-like metasomas, were actually closely related. Molecular analyses have now confirmed this (Jasso-Martínez *et al.*, 2021) and they are now treated as synonyms. In addition, the subfamily includes several small groups whose relationships have long been uncertain, such as the Chremylini and Cedriini (Fig. 9.10 b) which were included in Pambolinae by van Achterberg (1995b). Both *Cedria* and *Chremylus* occur in the region and are characterised by having 12 or fewer flagellomeres. Other genera included in the Chremylini by van Achterberg (1995b) are unrelated.

Cedriini

A poorly known group whose systematic placement and composition have fluctuated considerably; they now appear to be basal members of the Hormiinae (Jasso-Martínez *et al.*, 2021). From what little is known based on observations of a single species from India, *Cedria paradoxa*, their biology is unusual in that the females show parental care, remaining with their broods (Beeson and Chatterjee, 1935; Chu, 1935; Mathur, 1959). Quicke *et al.* (2017b) described a new species of *Cedria* (Fig. 9.10 b) from Thailand.

HORMIUS GROUP (HORMIINI). *Hormius* (Fig. 9.10 a) and *Parahormius* are common and a few other genera may occur, but the group is seriously understudied. Biology is only known for extralimital species which are gregarious ectoparasitoids of weakly concealed caterpillars such as Tortricidae (Shaw and Huddleston, 1991; Sharkey *et al.*, 2021 a). Interestingly, at least some species simply lay their egg(s) in the vicinity of the host and final host location is done by the 1st instar parasitoid larva.

LYSITERMINI. Small, quite distinctive small cyclostomes with a carapace-like metasoma formed of three tergites. A key to the genera was provided by Belokobylskij *et al.* (2007). The tribe comprises two genus groups, the *Lysitermus* group and the *Aulosaphes* group, which differ in the positions of fore wing veins 2cu-a and m-cu relative to the parts of 2CU.

The Lysitermini, represented in the region by the common genus *Acanthormius* (Fig. 9.10 d, e), *Aulosaphes* and *Aulosaphoides*. *Acanthormius* and *Aulosaphes* species, appear to be specialist parasitoids of bag-worm moth caterpillars (Psychidae) (van Achterberg, 1995b; Gupta and Quicke, 2018), including of some pest species such as *Metisa plana* (Fuat *et al.*, 2022). A few species of *Acanthormius* are brachypterous. Van Achterberg (1995b) provided keys to the species of *Acanthormius*, *Aulosaphes* and *Aulosaphoides*, although it is clear that there are many undescribed species in S.E. Asia.

PENTATERMINI. The sole genus *Pentatermus* (Fig. 9.10 c) is common in Malaise trap samples and at light traps. It has five visible tergites and is superficially rather like a small rogadine.

Pambolinae

A small subfamily with numerous undescribed S.E. Asian species, but which are easy to recognise to subfamily because of their propodeal apophyses (spines), especially in combination with the form of the metasoma which has the first and usually the second at least partially sculptured but the posterior tergites entirely or largely smooth and very shiny (Fig. 9.11 a, b). Although properly cyclostome in terms of the hypoclypeus being turned inwards exposing the labrum, the latter is not concave and shiny, but rather flat and sculptured (Fig. 9.11 c). Somewhat unfortunately, the type species of *Pambolus* is apterous and the winged species have often been placed in the genus *Phaenodus*, especially by Russian workers, although most researchers now consider *Phaenodus* just as a subgenus of *Pambolus*. Only the winged species occur in the region. Apparently a revision of Oriental species is in preparation. They have been reared from beetle infested wood.

Rhysipolinae

A small subfamily of small wasps, but not infrequently collected. The biology of tropical species is hardly known, but the temperate species of *Rhysipolis* (Fig. 9.12), which is the commonest genus in S.E. Asia, are very interesting because they show the unusual biological combination of being koinobiont ectoparasitoids of caterpillars in weakly concealed situations (Shaw, 1983). The wasp's sting causes temporary host paralysis, allowing the wasp time to glue a large yolky egg onto one of the host's intersegmental membranes. Following recovery from paralysis the caterpillar resumes feeding, but the initial envenomation causes it to cease normal moults and instead causes the host to construct its cocoon and enter a premature prepupal state, at which time the *Rhysipolis* larva commences feeding properly.

Belokobylskij and Con (1988) recorded the subfamily from S.E. Asia (Vietnam) for the first time with the description of a new species, *R. parnarae* reared from the common straight swift butterfly, *Parnara guttata* (Hesperiidae); van Achterberg and Lau (2022) presented some very nice photographs and biological observation for *R. taiwanicus*, which occurs in China, Taiwan and Vietnam. Long and Belokobylskij (2003) recorded four species of *Rhysipolis* from Vietnam, most of which are also present in China and included in the key to Oriental and East Palaearctic species

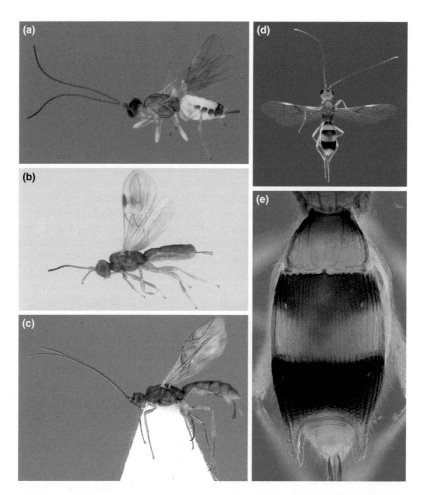

Fig. 9.10. A diversity of Thai Hormiinae: **(a)** *Hormius* sp. (Hormiini), showing weakly sclerotised central regions of metasomal tergites 2–5, but note that they are not always so conspicuous; **(b)** *Cedria* sp. (Cedriini) with fewer than 13 flagellomeres and fore wing vein rs-m absent; **(c)** *Pentatermus* nr. *striatus* (Pentatermini), showing five fully sclerotised and sculptured metasomal tergites; **(d, e)** *Acanthormius* sp. (Lysitermini), showing three-segmented metasomal carapace, and posterolateral projections, although they are not always so conspicuous.

by Zhang *et al.* (2016). They are collected fairly frequently at light traps in Thailand.

Rhyssalinae

Phylogenies concur that the small subfamily Rhyssalinae is the most basal of the traditional cyclostome subfamilies. Five tribes are recognised, of which Acrisidini, Histeromerini and Rhyssalini occur in S.E. Asia. These can be separated using the key by Quicke *et al.* (2020b).

They are generally uncommon and poorly represented in the tropics. In S.E. Asia they are represented by *Rhyssalus*, *Oncophanes* and *Tobiason* (Rhyssalinae), *Histeromerus* (Histeromerini) (Fig. 9.13 a) and undescribed Acrisidini (Fig. 9.13 b). They are poorly known biologically, although some details of the biology of an *Histeromerus* species were provided by Shaw (1995a). These are gregarious idiobiont ectoparasitoids of concealed beetle larvae. The biology of an extralimital *Histeromerus* was described in some detail by Shaw (1995a) and plausible host records include Buprestidae, Cerambycidae, Lucanidae and Lyctidae.

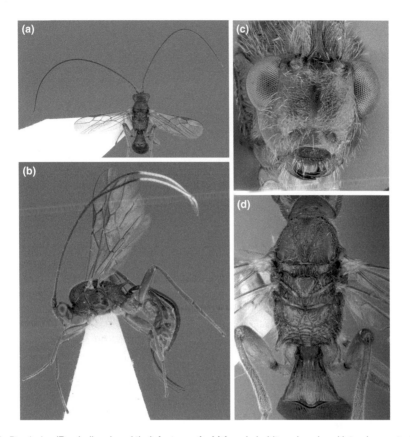

Fig. 9.11. *Pambolus* (Pambolinae) and their features: **(a, b)** female habitus, dorsal and lateral aspects; **(c)** face, note the exposed sculptured and flat labrum; **(d)** mesosoma dorsal view, note the propodeal apophyses.

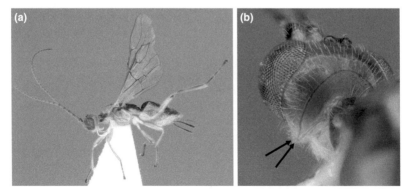

Fig. 9.12. *Rhysipolis* sp. (Rhysipolinae) and features: **(a)** female habitus, lateral view; **(b)** posteroventral view of head showing how occipital (genal) and hypostomal carinae run closely parallel but reach mandible separately.

Rogadinae

This is one of the larger subfamilies, and species of several genera are quite common at light traps but others are diurnal. In terms of species diversity and abundance they are dominated by the cosmopolitan genus *Aleiodes*. The state of this group's taxonomy is exemplified by Butcher *et al.*'s (2012) study of Thai *Aleiodes* in which combining traditional

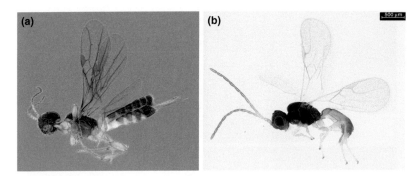

Fig. 9.13. Exemplar Rhyssalinae from Thailand: **(a)** *Histeromerus orientalis* (Histeromerini); **(b)** unidentified acrisidine, possibly *Acrisis* sp.

morphology with evidence from DNA barcoding led to the recognition and description of 179 species new to science. There can be little doubt that applying the same approach anywhere in the tropics would have similar results.

They are of particular biological and evolutionary interest because they all pupate within the mummified remains of their host caterpillars, and that means that if the host mummy and emerged parasitoids are kept together, as they should be, they provide an absolutely accurate host–parasitoid interaction record. Whilst in the better-known temperate region it may be possible for experienced entomologists to identify the host caterpillar remains visually, in the tropics this is hardly ever going to be possible, but DNA barcoding can be extremely valuable here because it will almost certainly be possible to sequence host DNA from the mummy and thus provide an identification (Quicke *et al.*, 2012).

Six tribes are recognised: Aleiodini (Fig. 9.14 f, g), Betylobraconini, Clinocentrini (Fig. 9.14 b, c), Facitorini *stat. rev.* (Fig. 9.14 a), Rogadini (Fig. 9.14 e), Yeliconini (Fig. 9.14 d) and the extralimital Stiropiini. Relationships between the tribes and genera were investigated by Zaldivar-Riverón *et al.* (2009) and more extensively by Quicke *et al.* (2021). Members of the Clinocentrini have relatively longer ovipositors than most other genera in the region, and are represented by the cosmopolitan *Clinocentrus*, and the rather aberrant *Tebennotoma* (Fig. 9.14 c). Hosts of the former are semi-concealed caterpillars typically inhabiting silk webs; hosts of the latter are unknown. Although Facitorina were recovered as monophyletic in Quicke *et al.*'s (2021) four-gene investigation, they were recovered as a grade leading to Aleiodini by both Jasso-Martínez *et al.*'s (2021) phylogenomics and Jasso-Martínez *et al.*'s (2022a) mitogenomics studies.

There is no key to the genera of Rogadinae specifically for S.E. Asia, but most genera are included in Chen and He's (1997) key to Chinese species. The obvious exceptions are *Cornutorogas* and the extremely rare genus *Spinariella*, which can be identified using van Achterberg's (2007) revision of the Spinariina (Rogadini). Quicke *et al.* (1997) revised the species of *Yelicones*, including the East Palaearctic species; Quicke and Butcher (2011b) described two new genera from Thailand; Oanh *et al.* (2021) described six new species of *Colastomion* from Vietnam. Ranjith *et al.* (2022b) recorded the clinocentrine genus *Kerevata* from India and S.E. Asia for the first time. Long *et al.* (2023) provided an up-to-date checklist of *Aleiodes* and the closely related genus *Heterogamus* from Vietnam, including many new records, and additionally commented on three species that exert some control on agricultural pests.

Most host records are from extralimital taxa. There is some degree of host group specialisation at genus and tribe level; for example, *Macrostomion* appear to be specialist gregarious parasitoids of sphingids (Shaw, 2002), *Colastomion* on choreutids (Sakagami *et al.*, 2020), Old World *Triraphis*

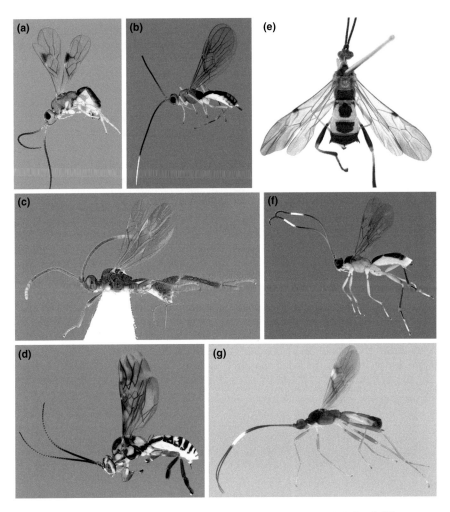

Fig. 9.14. Exemplar Rogadinae from Thailand: **(a)** *Conobregma* sp. (Facitorini); **(b, c)** *Clinocentrus* and *Tebennotoma* spp., respectively (Clinocentrini); **(d)** *Yelicones* sp. (Yeliconini); **(e)** *Spinaria* sp. (Rogadini); **(f, g)** *Aleiodes concoronarius* and *Heterogamus* sp., respectively (Aleiodini).

on Limacodidae and Zygaenidae (Quicke *et al.*, 2021), *Yelicones* are only known from Pyralidae (Quicke *et al.*, 2018), etc.

Alysioid Subcomplex

Four subfamilies are included: the Opiinae and Alysiinae, which are entirely parasitoids of Diptera and have long been recognised as closely related; the Telengaiinae, the name now used for the Gnamptodontinae, which are morphologically similar to the first two; and the Exothecine. Together these seem to form a sister group to the Braconinae.

Alysiines, telengaiines and opiines share possession of a fairly distinct type of male tergal exocrine gland, called Hagen's glands (Buckingham and Sharkey, 1988).

The first and only study specifically to address the phylogenetic relationships between the Opiinae and Alysiinae was by Gimeno *et al.* (1997) but the results were inconclusive.

Alysiinae

Highly distinctive wasps that are most diverse in temperate regions, although their tropical diversity has hardly been explored. Many are rather small-bodied, making morphological study difficult. They are unique among Old World braconids in having exodont mandibles (Figs 9.4 a, b, 9.15 a, b). Exodont is defined as the tips of the mandibles not overlapping when closed, which is not difficult to see, but fortunately these wasps often die with their mandibles splayed apart. The other thing is that the mandibles have more than two teeth, and the relative sizes and widths of these, and other mandibular features, play a large role in their identification.

Two tribes are recognised and are generally easily distinguished: the Alysiinae, which have the fore wing 2nd submarginal cell closed; and the Dacnusini, which lack vein rs-m and in these vein 3RS has a fairly distinctive curve towards the wing margin (Fig. 9.15 d). Dacnusines are parasitoids of phytophagous Diptera such as leaf-miners and gall-formers. Developmental biology of a widespread *Dacnusa* species was described by Croft and Copland (1994).

Because of their hosts, many are associated with moist habitats; alysiines are most abundant and diverse in forested and high-altitude locations. Wharton (1984) reviewed what was known at the time about the biology of the Alysiini and not a great deal has changed since then.

There is no key to the S.E. Asian genera but the key to Chinese genera and subgenera by Zhu *et al.* (2017) should work for most specimens. Zhang *et al.* (2020) provide a key to the distinctive *Bobekia* group of genera. A number of small genera and subgenera with S.E. Asian species have recently been revised, several using a combination of morphology and DNA barcoding: *Hylcalosia* (Yaakop *et al.*, 2009; Yao *et al.*, 2019a), *Neurolarthra* (Yao *et al.*, 2018a), *Separatatus* (Yao *et al.*, 2018b), *Senwot* (Yao *et al.*, 2019b), *Orthostigma*

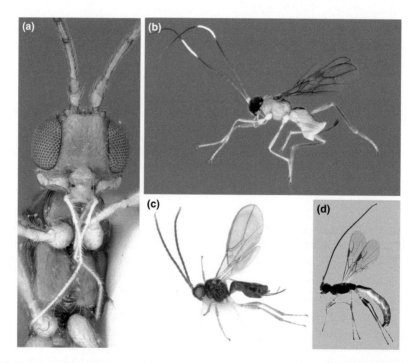

Fig. 9.15. Representative Alysiinae and characters: **(a)** ventral view of head and mesosoma of gen. sp. indet. showing exodont mandibles and absence of prepectal carina; **(b)** gen. sp. (Alysiini); **(c)** gen. sp. (Alysiini); **(d)** gen. sp. (Dacnusini), note absence of fore wing vein rs-m and straight then strongly curved vein 3RS.

(*Patrisaspilota*) (Peris-Felipo *et al.*, 2019), *Anamalysia* (Yao *et al.*, 2022a) and *Phaenospila* (Yao *et al.*, 2022b). In S.E. Asia, the dacnusines we most commonly encounter belong to the *Coelinius* group, including *Coeliniaspis*, which was originally described from Cambodia, and a key to the genera of this group was provided by Zheng *et al.* (2017).

Exothecinae

Historically these small wasps were generally placed in a very broad concept of the Rogadinae (e.g. Shaw and Huddleston, 1991) or of Hormiinae (Wharton, 1993) but were elevated to subfamily level by Quicke and van Achterberg (1990). They are not common in S.E. Asia and only five genera are recorded from the region: *Colastes*, *Colastinus*, *Orientocolastes*, *Shawiana* and *Vietcolastes*. Singh *et al.* (2022) described a new genus from India and provided a generic key for the region that includes also *Pseudophanomeris*, but does not include *Shawiana*. Exothecines are idiobiont ectoparasitoids of leaf-mining or gall-forming holometabolous larvae but detailed biological observations are only available for a few Holarctic genera.

Telengaiinae (=Gnamptodontinae)

The Gnamptodontinae was erected as a subfamily separate from the Opiinae by van Achterberg (1983 b). The Telengaiinae was erected for a single species from Turkmenia by Tobias (1962). Molecular data (Jasso-Martínez *et al.*, 2022a) show that they are probably derived within the former Gnamptodontinae, and since the name Telengaiinae has priority, that is now the correct name for the subfamily. This relationship was recognised by Tian *et al.* (2020), who gave the dates of all relevant publications, but then still referred to them as Gnamptodontinae

Telengaiines are small wasps, body length less than 2 mm with short ovipositors (Fig. 9.16 a, b), and have the anterior margin of the 2nd metasomal tergite differentiated as a sometimes smooth, raised area (Fig. 9.16 c). Some species of both groups resemble Braconinae in that they possess a pair of anterolateral diverging grooves on the 3rd metasomal tergite demarking triangular areas.

Based on temperate rearing records, telengaiines are specialist parasitoids of neptulid leaf-mining caterpillars. It is not known for certain, but it is strongly suspected that they are koinobionts (Quicke, 2015).

They are represented in S.E. Asia by *Gnamptodon*, *Pseudognamptodon* and *Tamdaona*. The highly aberrant genus *Tamdaona* was originally placed in the Exothecinae, but van Achterberg (1995b) and Ranjith *et al.* (2017) considered it to belong to the Rogadinae. Tian *et al.* (2020) formally transferred it to the 'Gnamptodontini' and placed it in a separate tribe, Tamdaonini. Its inclusion in Telengaiinae is now well confirmed by molecular data (Jasso-Martínez *et al.*, 2022a). Tian *et al.* (2020) provided a key to world genera and keys to world species of *Pseudognamptodon* and *Tamdaona*.

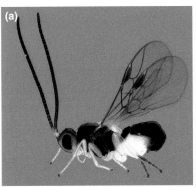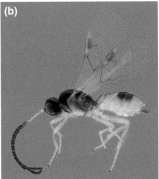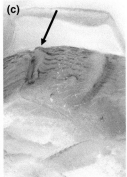

Fig. 9.16. Exemplar Thai Telengaiinae and character: **(a, b)** *Gnamptodon* sp.; **(c)** *Pseudognamptodon* sp., 2nd metasomal segment showing transverse basal raised area bordered posteriorly by a groove (arrow).

Opiinae

Most opiines are non-cyclostome (a secondary character state reversal) but a few have a well-developed hypopclypeal depression. They have mostly been recognised in keys by the occipital carina being present laterally but absent dorsally (Fig. 9.17 a, b), but there are a few exceptions, some completely lacking the carina and a few species having it weakly continuous mediodorsally (Fig. 9.17 c). They also lack a prepectal carina but both these character systems are homoplastic within the family.

Earlier systematic works often lumped a number of genera belonging to other subfamilies within the Opiinae, or they were all included within a broad concept of Rogadinnae. Wharton (1988) made a start on redefining it as a monophyletic group and defining some generic groupings, however, many additional genera have been described since then. Worldwide they are dominated by the huge, possibly polyphyletic, genus *Opius* (Fig. 9.18 e), which are predominantly small-bodied wasps. In the tropics there are more somewhat larger species because of greater abundance of year-round fruit in which fruit flies (Tephritidae) develop, and these are the hosts of many opiines. Because of this, the Opiinae is one of the most economically important braconid subfamilies, since members of several genera are specialist parasitoids of fruit fly pests of commercial crops. The main genera of tephritid parasitoids were reviewed by Wharton (1997), who reassigned several species (see Table 12.4 in Quicke, 2015). The most important ones of these are *Diachasmimorpha*, especially *D. longicaudata* (Fig. 9.18 a), *Fopius*, *Psyttalia* (Fig. 9.18 b) and *Utetes* together with the extralimital *Doryctobracon*. There have been no thorough molecular phylogenetic analyses of the subfamily. The early one by Gimeno et al. (1997) provided some support that the larger-bodied tephritid parasitoids probably comprise a monophyletic group.

Developmental stages have been described for several, predominantly economically important, species (e.g. Rocha et al., 2004). Mass rearing techniques have been developed for some opiine parasitoids of fruit flies (Bautista et al., 1999; González et al., 2007).

These are all rather uniform morphologically and there are few species level identification keys. Further, it is likely that some of the recognised species are actually cryptic species complexes. For example, using allozyme data, Kitthawee et al. (1999) found an absence of heterozygotes in samples of *D. longicaudata* from a range of Thai localities and this was followed by Kitthawee (2013) for *D. longicaudata* in Thailand which ITS2 data show is probably a complex of three separate species. Such studies could have important implications for biocontrol since '*D. longicaudata*' is an important parasitoid of *Batrocera correcta* and *B. dorsalis*, but perhaps

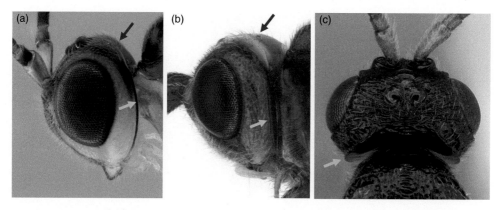

Fig. 9.17. Showing the dorsally incomplete (red arrows) but laterally present (green arrows) occipital carina in three opiine genera.

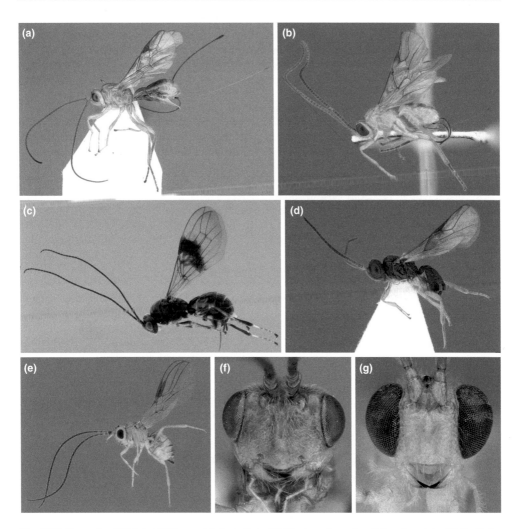

Fig. 9.18. Exemplar S.E. Asian Opiinae and characters: **(a)** *Diachasmimorpha* aff. *longicaudata*; **(b)** *Psyttalia* sp.; **(c)** gen. sp., nr *Biosteres*; **(d)** gen. sp. nr *Coleopius*, note carapace-like metasoma; **(e)** *Opius* sp.; **(f)** *Psyttalia* sp., face, note the non-cyclostome condition; **(g)** gen. sp., with partly developed hypoclypeal depression and largely exposed labrum.

these hosts are preferred by different cryptic parasitoid species. Finally, Kitthawee and Dujardin (2016) showed that, knowing which species each specimen belonged to, it then became possible to find discriminatory morphology. Economically important opiine fruitfly parasitoids were surveyed in Thailand and Malaysia by Chinajariyawong *et al.* (2000).

There is no key to the S.E. Asian genera but the one to the central Chinese (Hunan) fauna (Li *et al.*, 2013) includes most of them. The revision of *Psyttalia* for northern parts of China and the East Palaearctic by Wu *et al.* (2016) may also be useful, although obviously it will not include many species from S.E. Asia. Tan *et al.* (2016) described a new genus, *Carinopius*, from China and Vietnam.

Non-Cyclostomes

The non-cyclostomes comprise 23 subfamilies worldwide with only Khoikhoinae (S. Africa),

Mendeselinae (New World) and Microtypinae (Holarctic) not known from the region. Most of the subfamilies form a well-supported clade divided into three large groupings termed informally helconoids, microgastroids and sigalphoids, the first of these being split further by Jasso-Martínez et al. (2022b) into separate helconoid and euphoroid clusters.

The remaining subfamily Meteorideinae has a single genus in the region, *Meteoridea*, whose placement is far from certain. Sharanowski et al. (2011) recovered it as the sister to the microgastroids+sigalphoids but with little support, whereas the phylogenomics paper of Jasso-Martínez et al. (2022b) recovered it as sister to the whole non-cyclostome lineage plus the highly aberrant strictly Australian Trachypetinae (or Trachypetidae).

The Sigalphoid Complex

A close relationship between the Agathidinae and Sigalphinae was strongly suspected based on morphology, in particular most species of both have a well-developed hind wing vein 2-CU (see Fig. 9.20 b). Earlier, Quicke and van Achterberg (1990) had recovered them together on the basis of morphological phylogenetic analyses, although they also recovered them as closely related to the extralimital Trachypetinae (including Cercobarconinae), which now appears to be erroneous. The Neotropical genus *Pselaphanus*, previously placed in its own subfamily, is now regarded as a member of the Sigalphinae.

Both subfamilies are koinobiont endoparasitoids of Lepidoptera caterpillars, and have similar modified polypodiform 1st instar larvae (Shaw and Quicke, 2000).

Agathidinae

Agathidinae is a large non-cyclostome group with 1061 species worldwide and 238 in the Oriental Region (Yu et al., 2016). However, Sharkey et al. (2006) estimated that there are about 1000–2000 species of Agathidinae still waiting to be discovered. The biology of most members of the Agathidinae is unknown, but they are normally koinobiont endoparasitoids of lepidopteran larvae; they can be nocturnal or diurnal; almost all are solitary but gregarious development has been reported for a Neotropical species (Sarmiento et al., 2004). Those that attack exposed hosts have short ovipositors (Fig. 9.19 a, d) and those attacking concealed hosts such as caterpillars inhabiting leaf rolls or stems have longer ones (Fig. 9.19 b, c). Any larval instar may be attacked, depending on the species (Sharkey et al., 2009). The larger species are often colourful with partly or entirely darkened wings (Fig. 9.19 a, c, d) (Chen and van Achterberg, 2019). A few species can be used as natural enemies in biological control programmes. Agathidinae comprises at least 50 genera.

Characteristics of the subfamily: occipital carina absent, fore wing M+CU not tubular at the basal 1/3 or more (see Fig. 9.5), fore wing 3RSb terminating approximately half way between apex of pterostigma and wing tip, hind wing vein CUb usually indicated (van Achterberg and Long, 2010). Claws may be simple, bifid or with a pointed basal lobe, which makes for a good character to start identification keys.

The biology has been described for a few extralimital taxa (e.g. Odebiyi and Oatman, 1972). The first instar larvae are polypodiform (Aoyama and Ohshima, 2019).

Van Achterberg and Long (2010) revised Agathidinae from Vietnam; 17 genera were recognised, with three new erected genera (*Coronagathis*, *Gyragathis* and *Zelodia*). A total of 65 species have been discovered, with 12 new records and 42 species that are new to science. Sharkey and Clutts (2011) revised Agathidinae from Thailand and provided a key to the oriental genera. Sharkey and Chapman (2017a) described a further ten new S.E. Asian genera, and consequently provided an updated key to the genera of Agathidini to accommodate them. Together these will probably permit identification of virtually all S.E. Asian genera. The specimens on which that work was based were collected from Malaise traps during 2006–2009 from the TIGER project. To the genera recorded definitely from S.E. Asia by these workers, we can add *Aneurobracon* as

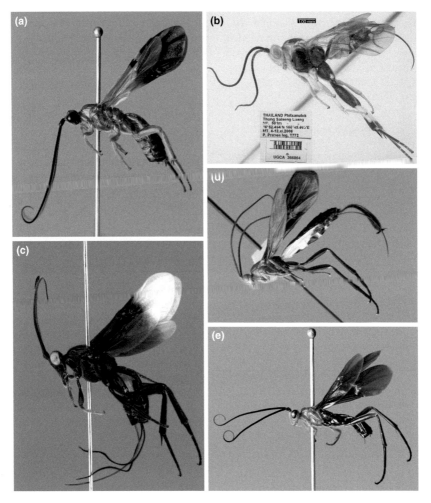

Fig. 9.19. Exemplar Agathidinae: **(a)** *Troticus* sp., Singapore; **(b)** *Cremnoptoides yui*; **(c)** *Biroia fuscicornis*, Sakaerat Biosphere Research Station, Thailand; **(d)** *Braunsia sumatrana*, a species with flattened ovipositor sheaths; **(e)** *Euagathis* sp. (Disophryni), Singapore. (Source: **(b)** and **(d)** photographs by Michael Sharkey (Hymenoptera Institute, California) reproduced with permission.)

was predicted. They noted that the principally cool temperate genus *Agathis* had not been recorded from the region but suggested that it was likely to occur at high altitudes.

Sharkey and Chapman (2017b) presented a molecular phylogeny of the subfamily based on combined mitochondrial gene COI and the nuclear ribosomal gene 28S sequence data, and recognised seven tribes: Agathidini, Agathirsini (extralimital), Cremnoptini, Disophryni, Earinini, Lytopylini and Mesocoelini.

Van Achterberg and Chen (2002) revised the *Euagathis* species of China and northern Vietnam, van Achterberg (2004) revised the species from Wallacia (i.e. Sulawesi and Papua New Guinea), and the Thai species were revised by van Achterberg *et al.* (2014). Sharkey and Stoelb (2012a, b) revised the Thai species of *Therophilus* and *Zelodia*, respectively. Sharkey and Stoelb (2013) revised *Agathacrista*. Very surprisingly, Long and van Achterberg (2016) reported the discovery of a new species of *Plesiocoelus* from Vietnam, a genus that was previously known from only a single Neotropical species. This illustrates the need to consult more than just regional keys to make identifications of strange taxa. Thai *Zosteragathis* species were revised by Sharkey and Chapman (2018).

Sigalphinae

These are poorly known, rather rare wasps. Little is known of their biology, other than that they are koinobiont endoparasitoids of Lepidoptera. The tip of the female ovipositor is correspondingly very fine and lacks a pre-apical notch.

The only species for which early stages and oviposition behaviour are known is an extralimital species of *Acampsis* (Shaw and Quicke, 2000). Oviposition was into 1st to 3rd instar host caterpillars, and the egg was placed within a host ventral nerve cord ganglion, a presumed strategy to avoid host haemocytes. As with Agathidinae, its putative sister group, the first instar larva is polypodiform. A preliminary molecular phylogeny was published by Quicke *et al.* (2008).

The world genera were revised by Sharkey and Janzen (1995) but for S.E. Asia this was superseded by van Achterberg (1995a) in which several new species from the region were described. Only *Sigalphus* (Fig. 9.20) has been recorded from S.E. Asia to date. Sharkey and Janzen (1995) noted that, to their surprise, the mature larvae of one of their Costa Rican species had a very low success rate at constructing cocoons. Van Achterberg (2014) described a new species of *Sigalphus* from southern Vietnam and provided a key to the world species.

The Helconoid Complex

Almost all molecular phylogenetic analyses recover a monophyletic group of subfamilies including the traditional subfamilies Helconinae s.l., Macrocentrinae, Acampsohelconinae, Blacinae, Charmontinae, Homolobinae, Xiphozelinae and Amicrocentrinae (extralimital).

Acampsohelconinae

This subfamily actually takes its name from an extinct group based on the fossil genus *Acampsohelcon*. There are three extant genera: *Afrocampsis* from Africa, *Canalicephalus* from S.E. Asia (Fig. 9.21) and *Urosigalphus* from the New World. Initially, *Afrocampsis* was placed in the Sigalphinae because of the combination of a carapace and possession of hind wing vein CUb, but molecular analyses, prompted by Michael Sharkey, which included representatives of all extant genera, recovered it among the Helconinae s.l. (Quicke *et al.*, 2002).

The 15 species of *Canalicephalus* were revised by van Achterberg (2002).

Helconoid subcomplex

BRACHISTINAE. This subfamily was for a long time regarded as a tribe of Helconinae, along with the apparently closely related Diospilini.

Fig. 9.20. *Sigalphus ?gyrodontus* from Sakaerat Research Station, Central Thailand.

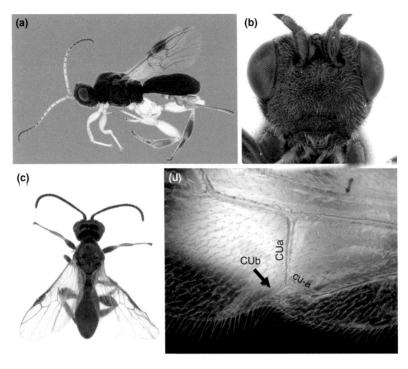

Fig. 9.21. *Canalicephalus* spp. from Thailand: **(a)** habitus, lateral; **(b)** face; **(c)** habitus dorsal; **(d)** base of hind wing showing presence of vein CUb.

It is a large subfamily of small to large-sized parasitoid wasps with about 794 described species worldwide (Sharanowski et al., 2011; Yu et al., 2016). The subfamily is divided into five tribes as follows: Blacini, Brachistini, Brulleiini, Diospilini and tentatively the Tainitermini, which were originally placed in the Euphorinae (Chen and van Achterberg, 2019; Sharanowski et al., 2011). Blacines were also for a long while regarded as a separate subfamily and were revised as such on a world basis by van Achterberg (1988). However, molecular analyses place them within the Brachistinae (Sharanowski et al., 2011). Some are difficult to separate from Euphorinae.

Despite their abundance, very little has been published on brachistine biology. Obrtel (1960) described the developmental stages of an extralimital *Triaspis* sp. Brachistinae parasitise Coleoptera larvae in the families Brentidae (Apioninae), Anobiidae, Cerambycidae, Chrysomelidae, Curculionidae, and Mordellidae, including very important agricultural pests (Belokobylskij et al., 2004a). Two new species of Brachistinae had been reared from seeds of Dipterocarpaceae (mainly *Shorea* species) in West Malaysia. These are probably parasitoids of *Alcidodes dipterocarpi* and/or *Alcidodes humeralis*, common weevils in *Shorea* seeds (van Achterberg et al., 2009).

Identification of Brachistinae can be difficult. Some have only one submarginal cell in the fore wing (Fig. 9.22 b, c), others two (Fig. 9.22 d); the metasoma is generally short and broad (Shaw and Huddleston, 1991). Some genera (such as *Triaspis* and *Schizoprymnus*) have the first three tergites forming a fuse carapace (either with two transverse furrows in the former genus or none in the later) concealed beneath the rest of metasomal tergites; and the ovipositor is distinctly long. Both resemble chelonines but differ in wing venation and ovipositor length (Shaw and Huddleston, 1991). *Foerteria* and *Polydegmon* also have the three-segmented carapace, but with a flexible articulation between the first and the second

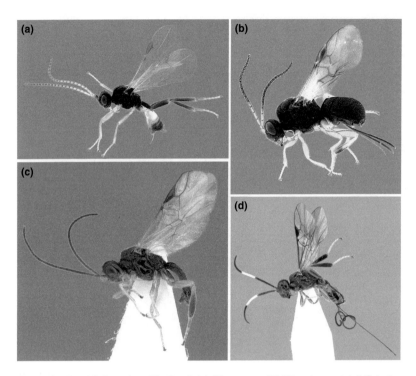

Fig. 9.22. Exemplar Brachistinae from Thailand: **(a)** *Blacus* sp.; **(b)** *Triaspis* sp.; **(c)** ? *Polydegma* sp.; **(d)** nr *Atree* but claws with pointed basal lobe.

tergites (van Achterberg, 1990). Ranjith *et al.* (2022c) described a new genus of Diospilini from India and provided a key to the Indo-Malayan genera of that tribe.

Van Achterberg (1983c) revised the tribe Brulleiini, but since then numerous new species have been described. Members of this tribe are rather larger bodied than most other brachistines, with body lengths c. 12–16 mm. Long *et al.* (2018) recorded *Brulleia* from Vietnam for the first time and described three new species. They revised the genus from the Oriental Region and provided a key to the 21 species known. The second genus, *Parabrulleia* has also been recorded from Vietnam (Long and van Achterberg, 2014).

HELCONINAE. Rather uncommon medium to moderately large wasps, many associated with xylophagous hosts. Only three genera are recorded from the region, viz. *Helcon*, *Ussurohelcon* and *Wroughtonia* (Fig. 9.23), the latter probably being the commonest. Long *et al.* (2020) revised the 17 *Wroughtonia* species known from Vietnam including 12 newly described ones, and one new subgenus.

Macrocentroid subcomplex

CHARMONTINAE. Charmontinae is a small subfamily, erected by Quicke and van Achterberg (1990) for species that were previously classified in the Homolobinae (van Achterberg, 1979a). Currently, members of this subfamily comprise only two genera, *Charmon* and *Charmontia*, the latter only occurring in Chile. *Charmon* is known from eight species distributed mainly across the Holarctic and Meso-America, but with a single species described from the Oriental Region (Pakistan), the Indian Ocean (Reunion Island) and Papua New Guinea (Rousse, 2013; Sabahatullah *et al.*, 2014; Loncle *et al.*, 2022).

Charmontines are koinobiont endoparasitoids of various weakly concealed Lepidoptera larva (Shaw and Huddleston, 1991; Yu *et al.*, 2016), and *Charmon extensor* alone is reported to parasitise members of

Fig. 9.23. *Wroughtonia* sp. female from Thailand: **(a)** habitus; **(b)** detail of hind femur showing ventral tooth diagnostic of the genus; **(c)** detail of fore wing venation.

12 lepidopteran families. Unique morphological characters of the subfamily are usually given as a row of pits along the lower margin of the clypeus, second submarginal cell in fore wing absent, hind wing anal cross-vein present, and antennae and ovipositor very long with the terminal flagellomere strongly acuminate (Mason, 1974; van Achterberg, 1993a). Some members of the Charmontinae share some characters with the Macrocentrinae, for example, both have longitudinally ridged ovipositors (Quicke and van Achterberg, 1990). *Charmon* belongs to the macrocentroid subcomplex and appears to the sister group to the Macrocentrinae (Sharanowski *et al.*, 2011) and as sister group to the Macrocentrinae (Jasso-Martínez *et al.*, 2022a, b).

HOMOBOLINAE. A small subfamily of medium-sized, koinobiont endoparasitoids of Lepidoptera caterpillars. S.E. Asian species have typical ophionoid facies associated with their predominantly nocturnal habits and have short ovipositors. The key character is the well-developed antescutal depression forming crescent-triangular gap between the pronotum and the mesoscutum (Fig. 9.24 c–e). Their general appearance and wing venation are very similar to short-ovipositored macrocentrine, *Austrozele* (Fig. 9.25 c).

MACROCENTRINAE. These medium-sized (approximately 4–12 mm body length) koinobiont endoparasitoids of Lepidoptera caterpillars are common. They have an almost unique identifying feature: a group of pegs on the mid- and usually the hind trochantellus (Fig. 9.25 a) and sometimes on the base of the adjacent femur (Fig. 9.25 a). Similar pegs are found on the uncommon orgiline genus, *Orgilonia*, and in one exceptionally rare braconine from Papua New Guinea. Some are nocturnal and frequently come to light traps. There are three genera in S.E. Asia: *Macrocentrus* (Fig. 9.25 b) which is by far the commonest, *Aulacocentrum* and *Austrozele* (Fig. 9.25 c). Members of the first two have long ovipositors, the latter have short ovipositors. Many temperate *Macrocentrus* species attack Tortricidae larvae often living in stems, shoots and leaf-rolls. Macrocentrines all appear to be polyembryonic, even the solitary species (Krugner *et al.* 2005; Lu *et al.*, 2007). Avoidance of encapsulation by the host, at least in the important extalimital species *Macrocentrus cingulum*, appears to be passive (Hu *et al.*, 2003).

Developmental stages of *Aulacocentrum philippinensis* were described (as *Macrocentrus*) by Kuwana *et al.* (1940). Barrion (2003) recorded *Aulacocentrum philippinense* from three crambid pest moths in the Philippines.

The genera may be identified using van Achterberg (1993b), although there may be more undescribed genera in the region. *Aulacocentrum* was revised by He and van Achterberg (1994).

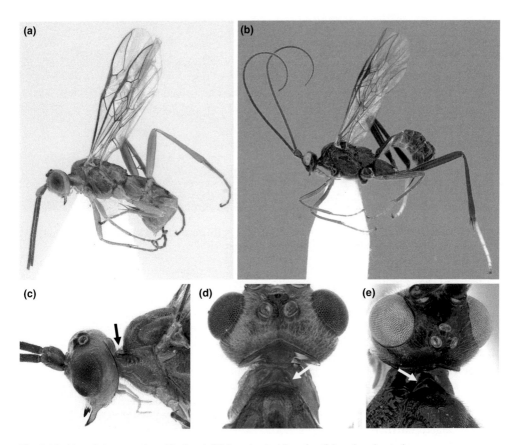

Fig. 9.24. *Homolobus* spp. from Thailand: **(b)** females habitus; **(c–d)** head and anterior mesosoma antero-lateral and dorsal views with arrows pointing to antescutal depression and its margins which are easier to see in the photograph of the less common largely black species.

ORGILINAE. The S.E. fauna is dominated by two genera: *Stantonia* (Fig. 9.26 a), which are often quite brightly coloured and frequent flowers; and *Orgilus* (Fig. 9.26 b), which are generally smaller and less conspicuous. These genera belong to the tribes Mimagathidini and Orgilini, respectively, the former being distinguished by having a relatively short hind wing vein M+CU (Fig. 9.26 a). In addition, two other smaller genera, *Eleonoria* and *Orgilonia*, occur in the region.

A key to the world genera of Orgilinae was provided by van Achterberg (1987) which also included a key to the Indo-Australian species of *Stantonia*. Braet et al. (2000) described a new S.E. Asian genus, *Eleonoria*, which possesses trochantellar pegs as in the Macrocentrinae, but can be distinguished from that subfamily by the absence of fore wing vein rs-m. Braet et al. (2000) provided keys to the species of both *Eleonoria* and *Orgilonia*, another genus that is present in S.E. Asia. Braet and Quicke. (2003) presented a morphological phylogenetic analysis of the Mimagathidinae and a key to the 57 species of *Stantonia* known at that time.

Oatman et al. (1969) described the biology of an extralimital *Orgilus* species. Until recently, they were thought to be exclusively solitary koinobiont endoparasitoids of 'microlepidoptera' caterpillars, living in weakly concealed situations, especially ones whose early instars are leaf-miners or tunnellers (Shaw and Huddleston, 1991). However, this was based entirely on evidence from temperate species, and caterpillar rearing in Costa

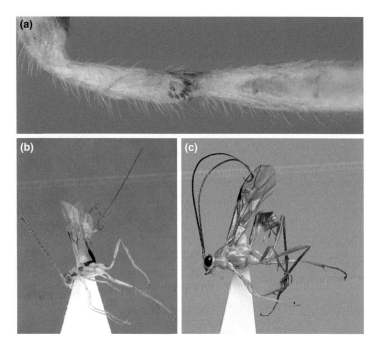

Fig. 9.25. Exemplar Thai macrocentrines and subfamily features: **(a)** *Macrocentrus* sp., basal part of hind leg showing cluster of pegs on the trochantellus and two on the adjacent part of the rest of the femur; **(b)** *Macrocentrus* sp., female; **(c)** *Austrozele* sp., female.

Rica has revealed gregarious species of *Orgilus* (Sharkey *et al.*, 2021a).

XIPHOZELINAE. This small subfamily of large and conspicuous, mainly nocturnal braconids was erected by van Achterberg (1979b) for a species that had previously been placed in the Macrocentrinae. Two genera are known: *Xiphozele* from the South East Palaearctic and from India through to S.E. Asia and Indonesia; and *Distilirella* from Papua New Guinea and Taiwan. Specimens are rare in collections. One species of *Xiphozele* from South India was reported as having been reared from the 'pupa' (probably cocoon) of *Ophiusa simillima* (Lepidoptera: Noctuidae). Van Achterberg (2008) described a new species of *Xiphozele* from Vietnam and provided a key to the four known species.

The Euphoroid Complex

Two subfamilies only are included in this group.: the very common and highly speciose Euphorinae (including Meteorinae) and the rare Cenocoeliinae. Both are cosmopolitan.

Cenocoeliinae

Cenocoeliines are easily recognised from other braconids by the very high insertion of the metasoma on the propodeum, reminiscent of Evanioidea. The genera were revised by van Achterberg (1994a), who additionally included *Ussurohelcon* because its metasoma is inserted a short distance above the hind coxae, but these have now been demonstrated to belong to the Helconinae.

Euphorinae (including Meteorinae)

The Euphorinae and Meteorinae were for a long while regarded as separate but closely related subfamilies, but now Meteorini are generally regarded as just a euphorine tribe since molecular studies consistently recover them together (Stigenberg *et al.*, 2015) and often nested within the latter (Sharanaowski *et al.*, 2011; Jasso-Martínez *et al.*, 2022a, b).

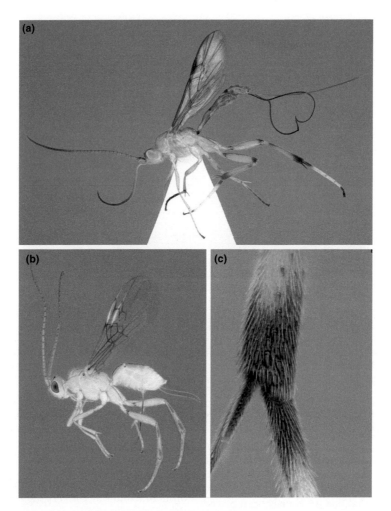

Fig. 9.26. Exemplar Thai orgilines and subfamily features: **(a)** *Stantonia* sp., showing very short hind wing subbasal cell; **(b)** *Orgilus* sp., showing longer hind wing subbasal cell; **(c)** *Stantonia* sp., outer apex of hind tibia showing cluster of pegs.

The former Euphorinae includes numerous tribes and genera, some with highly divergent morphologies, whereas Meteorinae comprises only two genera and is dominated by the highly speciose genus *Meteorus*. So diverse is the morphology that several groups were previously described as separate subfamilies and these were often treated as belonging to quite different groups. These include the extralimital Neoneurini and Planitorini (=Mannokeraiini) and the S.E. Asian and African Ecnomiini.

Euphorinae phylogeny has been explored in several morphological (e.g. Shaw, 1988) and molecular (e.g. Li *et al.*, 2003a; Stigenberg *et al.*, 2015) studies. The last of these recognised 14 tribes: Centistini, Cosmophorini, Dinocampini, Ecnomiini, Euphorini, Helorimorphini, Meteorini, Myiocephalini, Neoneurini, Perilitini, Planitorini, Pygostolini, Syntretini and Townesilitini. With the exceptions of Neoneurini and Planitorini, probably all of these occur in the region.

Meteorini are koinobiont endoparasitoids predominantly of Lepidoptera caterpillars, although a few extralimital species specialise on Coleoptera. This tribe is represented in the region by the very common genus *Meteorus* (Fig. 9.27 a, b). *Meteorus* larvae typically emerge from near one of the abdominal spiracles of their host before it is fully grown, and then drop down on a silken thread at the end of which they spin their cocoon and pupate (Shaw and Huddleston, 1991). Some species are gregarious and form a mass of cocoons from a single silk thread. Shirai and Maetô (2009) demonstrated that the suspended cocoons provide a defence against predation by ants. Some *Meteorus* species are highly polyphagous, *M. pulchricornis* being known to parasitise lepidopteran hosts in 15 families and 10 superfamilies (Maetô, 2018).

Whilst Meteorini have standard koinobiont endoparasitoid biology, the other euphorines are endoparasitoids of nymphal and also of adult insects belonging to several orders inclduing some hemimetabola. The latter strategy is hard to define as either idio- or koinobiont because although the host continues to remain active for a while after parasitism it soon stops feeding and so as a food resource for the developing parasitoid they are nearly limited to what was available when the host was attacked, a strategy termed imagobiont by Shaw (2004). Attacking such diverse and often hard-bodied hosts has led to an enormous morphological diversity between genera and this makes keying them really quite tricky; for example, in van Achterberg's (1993a) subfamily key they appear in five separate couplets.

Euphorine development involves teratocytes, cells which disassociate from the embryonic membranes and grow massively and help with host regulation and provide a food source for the developing larva (Barratt and Sutherland, 2001; Bell *et al.*, 2000, 2004; Quicke, 2015). It is now known that several euphorines have associated viruses which are injected into their hosts during oviposition, and in some cases may be involved with host behavioural manipulation (Dheilly *et al.*, 2015).

Dinocampus coccinellae is a very well-known parasitoid of ladybird beetles (Coccinellidae) attacking many host species, and is virtually cosmopolitan (Ceryngier *et al.*, 2023). In S.E. Asia it has thus far only been recorded from Vietnam, but seems highly likely to be more widespread in the region.

Streblocera (Streblocerini) has perhaps received an unduly large amount of taxonomic interest because its members are so very easily recognised by their presumably raptorial antennae (Fig. 9.27 c). We write 'presumably' because, despite its commonness, no one has ever reared a *Streblocera* or seen one attacking a host. What has happened is that quite a few papers have now been published describing various and often numerous new species of the genus, particularly from China, Taiwan, South Korea and India, as always some with better descriptions than others. The differentiating characters are often slight and sometimes ambiguously described, making it very difficult, if not impossible, to decide whether specimens from S.E. Asia are new or just a range extension of species from nearby countries.

The genera occurring in the region can probably mostly be identified using the key to Chinese genera by Chen and van Achterberg (1997). There are no specific keys for the great majority of genera occurring in S.E. Asia, and those that exist mostly deal with highly distinctive genera. *Cosmophorus* (Fig. 9.27 e) was revised by van Achterberg and Quicke (2000) although we have seen several additional undescribed species from Thailand; most tropical species are known from only one or two specimens, and therefore the total number of species in the genus is likely to be far greater than the number described. *Ecnomius* was revised by van Achterberg (1995) and more recently by Braet (2014).

The Microgastroid Complex

This group has a core of six subfamilies (Cardiochilinae, Cheloninae, Khoikhoinae (extralimital), Mendeselinae (extralimital), Microgastrinae and Miracinae) with the Proteropinae and Ichneutinae probably forming the sister group to these. Of these, it is known that all members of the Cardiochilinae, Cheloninae, Microgastrinae and Miracinae have associated polydnaviruses (see p. 17). These originated from insect nudiviruses approximately 100 million years ago (Whitfield, 1997, 2002; Murphy *et al.*, 2008; Bézier *et al.*, 2009, 2013; Whitfield and O'Connor,

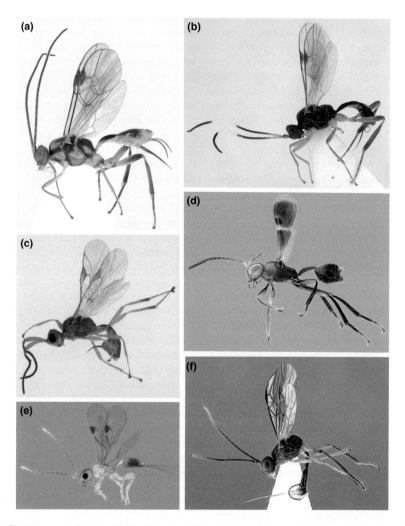

Fig. 9.27. Showing a small range of the diversity of S.E. Asian euphorines: **(a, b)** two species of *Meteorus*; **(c)** *Streblocera* sp., note the (presumed) raptorial antennae; **(d)** *Leiophron* sp.; **(e)** *Cosmophorus* sp.; **(f)** gen. sp. indet.

2012), but part of the ancestral viral genome became integrated into the ancestral wasp's germline DNA, and so they co-evolved. The polydnaviruses are replicated in huge numbers in the calyx glands at the distal end of the wasp's lateral oviducts and co-injected into the host at oviposition, as particles containing numerous and diverse circular viral DNA molecules. These particles enter cells of various host tissues, where they are transcribed into proteins that cause a range of changes in host cells, including the death of some host tissues.

In addition to having polydnaviruses, members of at least the Microgastrinae and Cardiochilinae produce teratocytes which also have effects on host caterpillar physiology (Pennacchio *et al.*, 1992; Schepers *et al.*, 1998; Quicke, 2015).

Cardiochilinae

A medium-sized subfamily of medium-sized wasps, which are usually fairly easy to recognise by their general appearance. Many species have the apex of the fore wing somewhat darkened (Fig. 9.28 a–d), but so too do some members of other groups.

Until at least 1979, Cardiochilinae was treated as a tribe in Microgastrinae (Wharton et al., 1997). They are a small cosmopolitan non-cyclostome subfamily; morphological characters look similar to those of microgastrines, with fore wing vein rs-m, but they have vein 3RS strongly arched at its base (Fig. 9.28 c, d) and they have a Y-shaped suture on the first metasomal tergite. There are approximately 18 genera and 240 described species worldwide (Yu et al., 2016; Kang et al., 2022). The third largest subfamily in microgastroid complex, members of this subfamily are cosmopolitan, highly diverse in the tropic and arid regions (Dangerfield et al., 1999).

Cardiochilines are solitary koinobiont endoparasitoids of caterpillars, especially Noctuinae and Pyralidae, including important agricultural pests such as the tobacco bud-worm, *Heliothis virescens* (Fabricius, 1777), and cotton bollworm, *Heliocoverpa armigera* (Huddleston and Walker 1988), *Chloridea virescens* (Noctuidae), *Cnaphalocrocis medinalis* (Crambidae), *Diaphania hyalinata* (Crambidae) and *Heliothis armigera* (Noctuidae) (Kang et al., 2022). *Cardiochiles philippensis* is reported as an endoparasitoid reared from larvae of the rice leaf-folder, *Cnaphalocrocis medinalis* (Lepidoptera: Pyralidae) (Long and Belokobylskij 2003). Emergence of cardiochilines from the host may be either while they are still on the host plant or once they have spun a cocoon and become prepupal. Larval developmental stages have been described for a few species including one from India (Singh and Baldev Parshad, 1970).

The North American species *Toxoneuron nigriceps* (previously often caslled *Cardiochiles nigriceps*) is an important biological control agent against tobacco budworm, *Heliothis virescens*, and it is one of the most intensively investigated braconids, which says something about the money involved with this crop. It has also been important in the study of bracoviruses, and of teratocytes (Pennacchio et al., 1991, 1992). Cônsoli and Vinson (2004) detailed its postembryonic development, and there are many papers on how it regulates host development (e.g. Li et al., 2003b). The final (3rd) instar parasitoid larva egresses from the host and reinserts

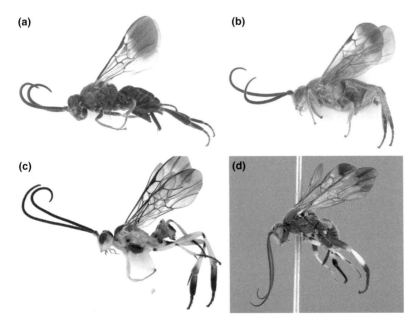

Fig. 9.28. A selection of common cardiochilines from Thailand: **(a)** *Cardiochiles* sp.; **(b)** *Schoenlandella* sp.; **(c)** *Hartemita* sp., note the expanded and flattened hind basitarsus; **(d)** *Austrocardiochiles* sp.

its head to continue feeding from an external position (see Fig. 2.1) (Kuriachan et al., 2011), a strategy employed by a number of non-cyclostome braconids.

Cardiochiline genera were revised by Dangerfield et al. (1999). About two decades later, Kang et al. (2020) erected *Orientocardiochiles* for some species from Malaysia and Vietnam, bringing the total of reported genera from the Oriental region to nine. Long and van Achterberg (2011) described two new species of the distinctive genus *Hartemita* from Vietnam.

Cheloninae (including Adeliinae)

Easily recognised largely on the basis of their metasomal carapace (Fig. 9.29), but it must be remembered that at least some members of a few other groups possess a carapace too, for example Acampsolconinae (see Fig. 9.21 a), Brachistinae (see Fig. 9.22 b) and Microgastrinae (see Fig. 9.31 d). Therefore, until you are familiar with the group, you need to check some other characters. The easiest is wing venation; with the exception of the Adeliini, which are small and rare, chelonines have a closed 2nd submarginal cell and a postpectal carina (= posterior transverse mesopleural carina) but this appears to evolve as a co-adapted feature in some other groups with a metasomal carapace.

For a long time the Adeliinae (which lack a metasomal carapace, are smaller than most chelonines and have reduced venation) were treated as a separate subfamily. However, following earlier suggestions based on morphological and molecular analyses, they were formally included in the Cheloninae as a separate tribe, Adeliini, by Kittel et al. (2016).

Chelonines are exclusively egg-larval koinobiont endoparasitoids of Lepidoptera. They inject the host egg with polydnaviruses along with their own egg. All temperate species studied to date are solitary parasitoids, but recently one Costa Rican species of *Chelonus* has been found to be gregarious (Sharkey et al., 2021a). The biologies of several extralimital species of *Chelonus* have been investigated intensively. Chelonine development has been fairly extensively studied largely in parallel with investigation of their polydnaviruses (e.g. Hafez et al., 1980; Kawakami, 1985; Grossniklaus-Bürgin et al., 1994; Hentz et al., 1997). *Chelonus formosanus* is a widespread parasitoid of *Spodoptera frugiperda* (Noctuidae) and a potential method for its mass rearing was described by Gupta et al. (2020).

The only specific phylogenetic study concentrating on the subfamily is that of Kittell et al. (2016) based on three gene fragments and morphology. Their results corroborated monophyly of the Chelonini and Odontosphaeropygini, but not of Phanerotomini or Pseudophanerotomini, and Adeliini were place as sister group to the other representatives exclusive of the Chelonini.

Zettel (1990) provided a key to genera, and Kittel and Austin (2014) keyed the Australian genera. There are no relevant keys to the species of any of the larger genera. The uncommon genus *Megascogaster* was revised by Kittel (2014). Species-level identification for S.E. Asia except for a few small genera or subgenera is effectively impossible. Quicke and Butcher (2018) recorded the Afrotropical genus *Odontosphaeropyx* from S.E. Asia for the first time with description of a new species from Thailand, and shortly thereafter Long et al. (2019) described another new species from Vietnam. Ranjith and Priyadarsanan (2023) provided a key to the seven subgenera of *Chelonus*.

Dirrhopinae

A monotypic, very small, principally Holarctic, and very rare subfamily, known from the region based on a single species described from Vietnam and Australia (Belokobylskij et al., 2003), though probably it occurs everywhere. Ranjith et al. (2021) provided a key to the six known species. They are small wasps with body lengths less than 2 mm, and the only host record is for a north American species which was reared from a nepticulid moth.

Ichneutinae

A rather small group with those genera parasitising sawfly larvae being more common in the temperate regions, and the smaller-bodied genera that attack leaf-mining caterpillars.

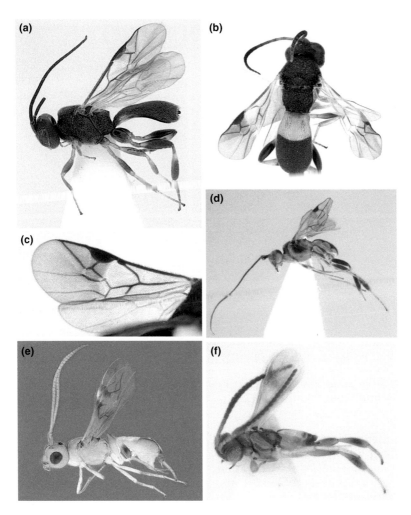

Fig. 9.29. Selected Thai Cheloninae: **(a–c)** *Chelonus* spp., note absence of fore wing vein (RS+M)a; **(d)** *Phanerotomella* sp.; **(e)** *Phanerotoma* sp., note presence of two transverse sutures on metasomal carapace; **(f)** *Adelius* sp., note secondary loss of metasomal carapace but swollen hind tibiae typical of subfamily.

In the past these Lepidoptera parasitoid genera (*Oligoneurus* (Fig. 9.30) and *Paroligoneurus*) were sometimes treated as a tribe, the Muesebeckiini, but the morphological phylogenetic analyses of Sharkey and Wharton (1994) showed that this rendered the 'Ichneutini' paraphyletic. Although the genus *Proterops* was included in the subfamily for a long time, it has recently been elevated to subfamily status (Chen and van Achterberg, 2019), and indeed it lacks the strong anterior bend of vein 1M that is a synapomorphy of the remaining genera.

Sharkey and Wharton (1994) provided a key to the world genera. *Oligoneurus* is the only genus recorded from S.E. Asia. Species are rather seldom collected, but are undoubtedly present nearly everywhere, yet only two species have been recorded, both from Vietnam (Long and Belokobylskij, 2003).

Microgastrinae

Without doubt the largest and commonest subfamily globally (Rodriguez *et al.*, 2013), and probably in S.E. Asia too. It has long

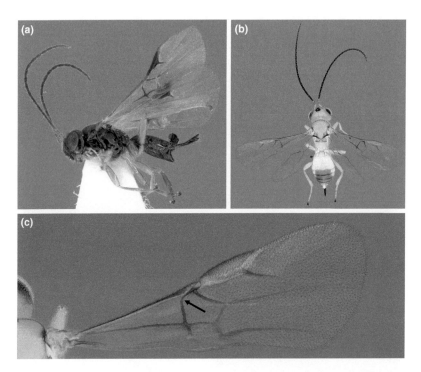

Fig. 9.30. *Oligoneurus* spp. from Thailand: **(a)** habitus, lateral; **(b)** habitus, dorsal; **(c)** fore wing showing strong anterior bend in vein 1M.

been recognised as a natural group although up until about 1979 the Dirrhopinae and Miracinae were also included as tribes within it (Nixon, 1965). As their name suggests, they have relatively small gasters (metasomas) and, given their general appearance and characteristic wing venation, they are seldom mistaken for any other group.

Whitfield *et al.* (2018) reviewed much of what is known about their biology, diversity and evolution. Earlier concepts of tribes have now been abandoned since molecular phylogenetic studies based on increasing numbers of Sanger-sequenced genes have failed to yield robust hypotheses of higher-level relationships. This strongly suggests that the subfamily underwent a very rapid radiation early in its evolution. There is still much uncertainty about the most basal genera, although the common genus *Microplitis* (Fig. 9.31 a) appears to be one of them.

Most species attack hosts in their early instars, sometimes just after they emerge from their eggs. A few are egg-larval parasitoids, sometimes facultatively (Ruberson and Whitfield, 1996). However, completion of their development after moulting to the 1st instar is delayed until the host caterpillar has nearly completed its development. The ancestral biology is probably solitary parasitism but gregariousness appears to have evolved on numerous occasions. Some gregarious species have broods within a single large host of a 100 or so young, but others may have only a handful. Emergence of gregarious broods from the host is highly synchronised. When it comes to species attacking much larger hosts, the microgastrine larvae have first to construct an internal supporting structure against which they can push in order to egress (Nakamatsu *et al.*, 2006).

Mason (1981) started formal major changes in the generic classification of the subfamily by splitting the huge genus *Apanteles* into a number of monophyletic groups. It had previously been recognised that *Apanteles sensu lato* was a polyphyletic mess based on the homoplastic loss of fore wing vein rs-m,

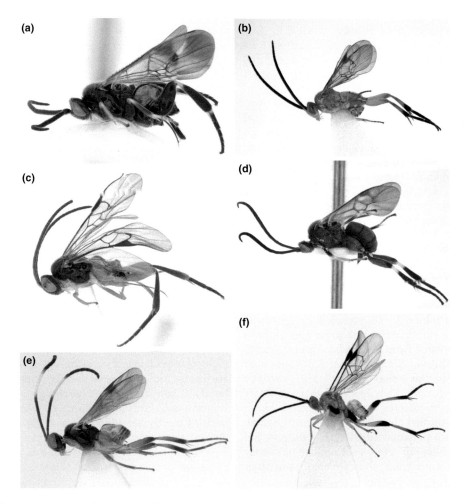

Fig. 9.31. Assorted Microgastrinae from Thailand: **(a)** *Microplitis* sp.; **(b)** *Jimwhitfieldius* sp.; **(c)** *Agupta* sp., note the relatively long ovipositor; **(d)** *Fornicia* sp., showing metasomal carapace; **(e)** *Diolcogaster* sp., showing short ovipositor; **(f)** *Wilkinsonellus* sp. Note that **(a–c)** possess a closed areolet (2nd submarginal cell).

thus increasing the number of genera from 19 to 50. Although a lot of workers at that time did not approve of many of Mason's concepts (see Shaw and Huddleston, 1991) most of Mason's genera are still recognised. One issue was that Mason, working in Canada, based his conclusions largely on the Holarctic and New World faunas. *Apanteles* still exists as a valid and rather large genus, but many species were transferred to *Cotesia*. These include a lot of fairly well-known species such as the large white butterfly (*Pieris brassicae*) parasitoid *Cotesia glomerata*.

Since Mason (1981), quite a few more genera have been described, e.g. 17 by Fernandez-Triana and Boudreault (2018). World species were catalogued by Fernandez-Triana *et al.* (2020) which included 318 new generic combinations. There is still no key to the world genera or to those of S.E. Asia. Fernández-Triana and Goulet (2009) provided a world key to species of the small genus *Philoplitis* which includes two S.E. Asian species, similarly for *Paraplitis* (Fernández-Triana *et al.*, 2013). Fernández-Triana *et al.* (2014) recorded the previously Australian endemic genus *Miropotes* from S.E. Asia (Thailand and Vietnam) for the first time with description of a new species, and Fernández-Triana *et al.* (2016) provided a key to the world species of *Exoryza* which includes one that occurs in S.E. Asia. Fernández-Triana and Boudreault (2018) described a number of new genera with S.E. Asian representatives, viz. *Agupta, Austinicotesia, Billmasonius, Jimwhitfieldius, Markshawius, Tobleronius, Ungunicus, Ypsiloigaster* and *Zachterbergius*. DNA barcoding will probably necessarily have to play a large role in sorting out the species of the larger genera in the tropics (Smith *et al.*, 2012).

Miracinae

This is a cosmopolitan subfamily of small wasps that are koinobiont endoparasitoids of leaf-mining Lepidoptera belonging mainly to the Nepticulidae but some species have been reared from Heliozelidae and Lyonetiidae (Quicke, 2015; Yu *et al.*, 2016), but host records for the tropics are virtually non-existent. They are probably the sister group to Microgastrinae. Despite being little-studied they have been confirmed to produce polydnaviruses (Wharton and Sittertz-Bhatkar, 2002). Miracinae are very easy to recognise; they are small, less than 2 mm, fore wing vein r-rs is either incredibly short (Fig. 9.32 a) or veins 2RS and 3RS originate from more or less the same point on the pterostigma, and the sclerotised parts of the 1st and anterior of the 2nd are very narrow and surrounded by pleated membrane.

Two genera are recognised: *Centistidea* and *Mirax*, the former having a midlongitudinal propodeal carina and at least moderately well-developed notauli. See Ranjith *et al.* (2019) for a key to the known species of *Centistidea*.

Proteropinae

A small subfamily comprising the widespread genus *Proterops* plus the Neotropical genera *Hebichneutes* and *Masonbeckia* (van Achterberg and Desmier de Chenon, 2009). *Proterops* was for a long time classified within the Ichneutinae because of a combination of general habitus appearance and parasitism of sawflies; however, based upon molecular evidence, they represent a separate lineage and are treated as a separate subfamily (Chen and van Achterberg, 2019). Proteropines are koinobiont endoparasitoids of sawfly larvae, predominantly Argidae but also sometimes Tenthredinidae.

Two species are known from S.E. Asia, *Proterops borneoensis* and *P. fumosus* from

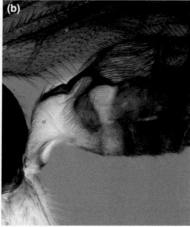

Fig. 9.32. *Centistidea* sp. from Thailand: **(a)** showing characteristic fore wing venation; **(b)** weakly sclerotised and pleated parts of anterior metasomal tergites and location of 1st spiracle in the latero-tergal region rather than in heavily sclerotised notum.

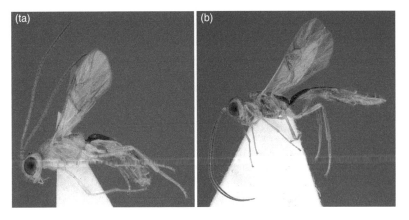

Fig. 9.33. Female and male of *Meteoridea* sp. collected at light trap.

Indonesia (Borneo and Sumatra) and Vietnam, respectively. Van Achterberg and Desmier de Chenon (2009) provided a key to the Old World species of *Proterops* and recorded *P. borneoensis* as a parasitoid of the argid sawfly, *Cibdela janthina*, in China.

Unplaced

Meteorideinae

This is a peculiar small group with only two genera worldwide: *Meteoridea* (Fig. 9.33), which is common in S.E. Asia and which often come to light traps; and *Pronkia*, which is restricted to New Zealand. Molecular data suggest that they may be very basal among the non-cyclostomes (Jasso Martínez *et al.*, 2022b), or at least that they are not closely associated with any of the subfamily groups dealt with above.

Only a few *Meteoridea* species are recognised, but DNA barcoding of non-reared material suggests that there may be rather a lot of undescribed species. One named species, *M. hutsoni*, is a common and probably economically important parasitoid of the black-headed caterpillar, *Opisina arenosella*, a serious pest of coconuts. Its biology as a larval-pupal parasitoid was described by Ghosh and Abdurahiman (1984).

A key to the nine species known at the time was provided by Penteado-Dias (1996) but that does not include the seven species described from China by He *et al.* (2000).

References

Aoyama, H. and Ohshima, I. (2019) Observation of *Aneurobracon philippinensis* (Hymenoptera: Braconidae) immatures shows how koinobiont offspring flexibly adjust their development to host growth. *Annals of the Entomological Society of America* 112, 490–496.

Banks, J.C. and Whitfield, J.B. (2006) Dissecting the ancient rapid radiation of microgastrine wasp genera using additional nuclear genes. *Molecular Phylogenetics and Evolution* 41, 690–703.

Barratt, B.I.P. and Sutherland, M. (2001) Development of teratocytes associated with *Microctonus aethiopoides* Loan (Hymenoptera: Braconidae) in natural and novel host species. *Journal of Insect Physiology* 47, 257–262.

Barrion, A.T. (2003) New hosts for the parasitoid, *Aulacocentrum philippinense* (Ashmead) (Hymenoptera: Braconidae) in Luzon Island, Philippines. *Philippine Entomologist* 17, 90–93.

Bautista, R.C., Mochizuki, N., Spencer, J.R., Harris, E.J. and Ichimura, D.M. (1999) Mass-rearing of the tephritid fruit fly parasitoid *Fopius arisanus* (Hymenoptera: Braconidae). *Biological Control* 15, 137–144.

Beeson, C.F.C. and Chatterjee, S.N. (1935) On the biology of the Braconidae (Hymenoptera). *Indian Forest Records* 1, 105–138.

Bell, H.A., Marris, G.C., Bell, J. and Edwards, J.P. (2000) The biology of *Meteorus gyrator* (Hymenoptera: Braconidae), a solitary endoparasitoid of the tomato moth, *Lacanobia oleracea* (Lepidoptera: Noctuidae). *Bulletin of Entomological Research* 90, 299–308.

Bell, H.A., Kirkbride-Smith, A.E., Marris, G.C. and Edwards, J.P. (2004) Teratocytes of the solitary endoparasitoid *Meteorus gyrator* (Hymenoptera: Braconidae): morphology, numbers and possible functions. *Physiological Entomology* 29, 335–343.

Belokobylskij S.A. (1993) East Asiatic species of the genus *Neurocrassus* (Hymenoptera: Braconidae). *Zoosystematica Rossica* 2, 161–172.

Belokobylskij, S.A. (2002) Two new Oriental genera of Doryctinae (Hymenoptera, Braconidae) from termite nests. *Journal of Natural History* 36, 953–962.

Belokobylskij, S.A. and Con, Q.V. (1988) Discovery of the genus *Rhysipolis* Förster (Hymenoptera, Braconidae) in the Indomalayan Region and description of a new species from Vietnam. *Entomologicheskoye Obozreniye* 67, 162–165.

Belokobylskij, S. and Quicke, D.L.J. (1999) A new genus and two new species of brachypterous Lysiterminae (Hymenoptera: Braconidae). *Journal of Hymenoptera Research* 8, 120–125.

Belokobylskij, S.A. and Zaldivar-Riverón, A. (2021) Reclassification of the doryctine tribe Rhaconotini (Hymenoptera, Braconidae). *European Journal of Taxonomy* 741, 1–168.

Belokobylskij S.A., Iqbal, M. and Austin, A.D. (2003) First record of the subfamily Dirrhopinae (Hymenoptera: Braconidae) from the Australian region, with a discussion of relationships and biology. *Australian Journal of Entomology* 42, 260–265.

Belokobylskij, S.A., Güçlü, C. and Özbek, H. (2004a) A new species of the genus *Schizoprymnus* Förster (Hymenoptera, Braconidae, Brachistinae) from Turkey. *Zoosystematica Rossica* 12, 245–248.

Belokobylskij S.A., Iqbal M. and Austin A.D. (2004b) Systematics, distribution and diversity of the Australasian doryctine wasps (Hymenoptera, Braconidae, Doryctinae). *Records of the South Australian Museum Monograph Series* 8, 1–150.

Belokobylskij, S., Zaldivar-Riveron, A., Leon-Regagnon, V. and Quicke, D.L.J. (2007) A new genus of Lysitermini (Hymenoptera: Braconidae: Lysiterminae) from Madagascar and its taxonomic placement based on 28S rDNA sequence data. *Zootaxa* 1461, 25–37.

Belshaw, R. and Quicke, D.L.J. (1997) A molecular phylogeny of the Aphidiinae (Hymenoptera: Braconidae). *Molecular Phylogenetics and Evolution* 7, 281–293.

Belshaw, R., Fitton, M., Herniou, E., Gimeno, C. and Quicke, D.L.J. (1998) A phylogenetic reconstruction of the Ichneumonoidea (Hymenoptera) based on the D2 variable region of 28S ribosomal RNA. *Systematic Entomology* 23, 109–123.

Belshaw, R., Dowton, M., Quicke, D.L.J. and Austin, A.D. (2000) Estimating ancestral geographical distributions: a Gondwanan origin for aphid parasitoids? *Proceedings of the Royal Society of London B* 267, 491–496.

Bézier, A., Annaheim, M., Herbinière, J., Wetterwald, C., Gyapay, G., Bernard-Samain, S., Wincker, P., Roditi, I., Heller, M., Belghazi, M., Pfister-Wilhem, R., Periquet, G., Dupuy, C., Huguet, E., Volkoff, A.-N., Lanzrein, B. and Drezen, J.-M. (2009) Polydnaviruses of braconid wasps derive from an ancestral nudivirus. *Science* 323, 926–930.

Bézier A., Louis, F., Jancek, S., Periquet, G., Thézé, J., Gyapay, G., Musset, K., Lesobre, J., Lenoble, P., Dupuy, C., Gundersen-Rindal, D., Herniou, E.A. and Drezen, J.-M. (2013) Functional endogenous viral elements in the genome of the parasitoid wasp *Cotesia congregata*: insights into the evolutionary dynamics of bracoviruses. *Philosophical Transactions of the Royal Society of London B: Biological Sciences* 368(1626), 20130047.

Braet, Y. (2014) One new Afrotropical species of the genus *Ecnomios* Mason, 1979 (Hymenoptera: Braconidae, Ecnomiinae). *Bulletin de la Société royale belge d'Entomologie* 150, 163–167.

Braet, Y. and Quicke, D.L.J. (2003) A phylogenetic analysis of the Mimagathidini with revisionary notes on the genus *Stantonia* Ashmead, 1904 (Hymenoptera: Braconidae: Orgilinae). *Journal of Natural History* 38, 1489–1589.

Braet, Y., van Achterberg, C. and Chen, X.-X. (2000) Notes on the tribe Mimagathidini Enderlein, with the description of a new genus (Hymenoptera: Braconidae: Orgilinae). *Zoologische Mededelingen, Leiden* 73, 465–486.

Buckingham, G.R. and Sharkey, M.J. (1988) Abdominal exocrine glands in Braconidae (Hymenoptera). In: Gupta, V.K. (ed.) *Advances in Parasitic Hymenoptera Research*. E.J. Brill, Leiden, pp. 199–242.

Butcher, B.A. and Quicke, D.L.J. (2010) Revision of the Indo-Australian braconine wasp genus *Ischnobracon* Baltazar (Hymenoptera: Braconidae) with description of six new species from Thailand, Laos and Sri Lanka. *Journal of Natural History* 44, 2187–2212.

Butcher, B.A., Smith, M.A., Sharkey, M.J. and Quicke, D.L.J. (2012) A turbo-taxonomic study of Thai *Aleiodes* (*Aleiodes*) and *Aleiodes* (*Arcaleiodes*) (Hymenoptera: Braconidae: Rogadinae) based largely on COI bar-coded specimens, with rapid descriptions of 179 new species. *Zootaxa* 3457, 1–232.

Ceryngier, P., Franz, K.W. and Romanowsky, J. (2023) Distribution, host range and host preferences of *Dinocampus coccinellae* (Hymenoptera: Braconidae): a worldwide database. *European Journal of Entomology* 120, 26–34.

Chansri, K., Quicke, D.L.J. and Butcher, B.A. (2022) Four new species of *Physaraia* (Hymenoptera: Braconidae: Braconinae) from Thailand. *Zootaxa* 5182, 479–488.

Chen, X. and He, J. (1997) Revision of the subfamily Rogadinae (Hymenoptera: Braconidae) from China. *Zoologische Verhandelingen, Leiden* 308, 1–187.

Chen, X. and van Achterberg, C. (1997) Revision of the subfamily Euphorinae (excluding the tribe Meteorini Cresson) (Hymenoptera: Braconidae) from China. *Zoologische Verhandelingen, Leiden* 313, 1–217.

Chen, X.X. and van Achterberg, C (2019) Systematics, phylogeny, and evolution of braconid wasps: 30 years of progress. *Annual Review of Entomology* 64, 335–358.

Chinajariyawong, A., Clarke, A.R., Jirasurat, M., Kritsaneepiboon, S., Lahey, H.A., Vijaysegaran, S. and Walter, G.H. (2000) Survey of opiine parasitoids of fruit flies (Diptera: Tephritidae) in Thailand and Malaysia. *Raffles Bulletin of Zoology* 48, 71–101.

Chishti, M.J.K. and Quicke, D.L.J. (1994) Revision of the Indo-Australian parasitic wasp genus *Macrobracon* with the description of two new species (Hymenoptera: Braconidae). *Zoological Journal of the Linnean Society* 111, 265–296.

Chishti, M.J.K. and Quicke, D.L.J. (1996) A revision of the Indo-Australian species of *Stenobracon* (Hymenoptera: Braconidae) parasitoids of lepidopterous stem-borers of graminaceous crops. *Bulletin of Entomological Research* 86, 227–245.

Chomphukhiao, N., Takano, S., Takasu, K. and Uraichuen, S. (2018) Existence of two strains of *Habrobracon hebetor* (Hymenoptera: Braconidae): a complex in Thailand and Japan. *Applied Entomology and Zoology* 53, 373–380.

Chu, J. (1935) Notes on the biology of *Cedria paradoxa* Wilkinson, a hymenopterous parasite of the mulberry pyralid (*Mararonia pyloalis* Walker). *Chekiang Province Bureau of Entomology Yearbook* 1933, 193–202.

Cônsoli F.L. and Vinson, S.B. (2004) Host regulation and the embryonic development of the endoparasitoid *Toxoneuron nigriceps* (Hymenoptera: Braconidae). *Comparative Biochemistry & Physiology B* 137, 463–73.

Croft, P. and Copland, J.W. (1994) Larval morphology and development of the parasitoid *Dacnusa sibirica* (Hym., Braconidae) in the leafminer host *Chromatomyia syngenesiae*. *Entomophaga* 39, 85–93.

Dangerfield, P.C., Austin A.D. and Whitfield J.B. (1999) Systematics of the world genera of Cardiochilinae (Hymenoptera: Braconidae). *Invertebrate Systematics* 13, 917–976.

Dheilly N.M., Maure, F., Ravalle, M., Galinie, R., Doyon, J., Duval, D., Leger, L., Volkoff, A.-N., Missé, D., Nidelet, S., Demolombe, V., Brodeur, J., Gourbal, B., Thomas, F. and Mitta, G. (2015) Who is the puppet master? Replication of a parasitic wasp-associated virus correlates with host behaviour manipulation. *Proceedings of the Royal Society, Series B* 28220142773.

Dowton, M. and Austin, A.D. (1998) Phylogenetic relationships among the microgastroid wasps (Hymenoptera: Braconidae): combined analysis of 16S and 28S rDNA genes and morphological data. *Molecular Phylogenetics and Evolution* 10, 354–366.

Dowton, M., Austin, A.D. and Antolin, M.F. (1998) Evolutionary relationships among the Braconidae (Hymenoptera: Ichneumonoidea) inferred from partial 16S rDNA gene sequences. *Insect Molecular Biology* 7, 129–150.

El-Heneidy, A. H. and Quicke, D.L.J. (1991) The Indo-Australian species of the braconine wasp genus *Zaglyptogastra* Ashmead. *Journal of Natural History* 25, 183–201.

Fernández-Triana, J. and Boudreault, C. (2018) Seventeen new genera of microgastrine parasitoid wasps (Hymenoptera, Braconidae) from tropical areas of the world. *Journal of Hymenoptera Research* 64, 25–140.

Fernández-Triana, J. and Goulet, H. (2009) A review of the genus *Philoplitis* Nixon (Hymenoptera: Braconidae: Microgastrinae). *ZooKeys* 20, 285–298.

Fernández-Triana, J., Ward, D. F., Cardinal, S., and van Achterberg, C. (2013). A review of *Paroplitis* (Braconidae, Microgastrinae), and description of a new genus from New Zealand, Shireplitis, with convergent morphological traits. *Zootaxa* 3722, 549–568.

Fernández-Triana. J., van Achterberg, K.[C.]. and Whitfield, J.B. (2014) Australasian endemic no more: four new species of *Miropotes* Nixon (Hymenoptera, Braconidae, Microgastrinae), with the first record from the Oriental region. *Tijdschrift voor Entomologie* 157, 59–77.

Fernández-Triana, J.L., Whitfield, J.B., Smith, M.A., Dapkey, T., Hallwachs, W. and Janzen, D.H. (2016) Review of the world species of *Exoryza* (Hymenoptera, Braconidae, Microgastrinae), with description of five new species. *Deutsche Entomologische Zeitschrift* 63, 195–210.

Fernández-Triana, J., Shaw, M.R., Boudreault, C., Beaudin, M. and Broad, G.R. (2020) Annotated and illustrated world checklist of Microgastrinae parasitoid wasps (Hymenoptera, Braconidae). *ZooKeys* 920, 1–1089.

Flores, S., Nassar, J.M. and Quicke, D.L.J. (2005) Reproductive phenology and pre-dispersal seed predation in *Protium tovarense* (Burseraceae), with description of the first known phytophagous "*Bracon*" species (Hymenoptera: Braconidae: Braconinae). *Journal of Natural History* 39, 3663–3685.

Fuat, S., Adam, N.A., Hazmi, I.R. and Yaakop, S. (2022) Interactions between *Metisa plana*, its hyperparasitoids and primary parasitoids from good agriculture practices (GAP) and non-gap oil palm plantations. *Community Ecology* 23, 429–438.

Gadallah, N.S., Gharari, H. and Shaw, S.R. (eds) (2022a) *Braconidae of the Middle East (Hymenoptera): Taxonomy, Distribution, Biology, and Biocontrol Benefits of Parasitoid Wasps*. Academic Press (Elsevier Inc.), London, 596 pp.

Gadallah, N.S., Ghahari, H., Shaw, S.R. and Quicke, D.L.J. (2022b) Introduction to the Braconidae of the Middle East. In: Gadallah, N.S., Gharari, H. and Shaw, S.R. (eds) *Braconidae of the Middle East (Hymenoptera): Taxonomy, Distribution, Biology, and Biocontrol Benefits of Parasitoid Wasps*. Academic Press (Elsevier Inc.), London, pp. 2–32.

Gadallah, N.S., Kavallieratos, N.G., Gharari, H. and Shaw, S.R. (2022c) Subfamily Aphidiinae Haliday, 1833. In: Gadallah, N.S., Gharari, H. and Shaw, S.R. (eds) *Braconidae of the Middle East (Hymenoptera): Taxonomy, Distribution, Biology, and Biocontrol Benefits of Parasitoid Wasps*. Academic Press (Elsevier Inc.), London, pp. 92–155.

Ghosh, S.M. and Abdurahiman, U.C. (1984) Bioethology of *Meteoridea hutsoni* (Nixon) (Hymenoptera: Braconidae), a parasite of *Opisina arenosella* Walker, the black headed caterpillar pest of coconut. *Entomon* 9, 31–34.

Gimeno, C., Belshaw, R. and Quicke, D.L.J. (1997) Phylogenetic relationships of the Opiinae/Alysiinae (Hymenoptera: Braconidae) and the utility of cytochrome b, 16S and 28S D2 rDNA. *Insect Molecular Biology* 6, 273–284.

González, P.I., Montoya, P., Perez-Lachaud, G., Cancino, J. and Liedo, P. (2007) Superparasitism in mass reared *Diachasmimorpha longicaudata* (Ashmead) (Hymenoptera: Braconidae), a parasitoid of fruit flies (Diptera: Tephritidae). *Biological Control* 40, 320–326.

Grossniklaus-Bürgin, C., Wyler, T., Pfister-Wilhelm, R. and Lanzrein, B. (1994) Biology and morphology of the parasitoid *Chelonus inanitus* (Braconidae, Hymenoptera) and effects on the development of its host *Spodoptera littoralis* (Noctuidae, Lepidoptera). *Invertebrate Reproduction & Development* 25, 143–158.

Gupta, A. and Quicke, D.L.J. (2018) A new species of *Acanthormius* (Braconidae: Lysiterminae) reared as a gregarious parasitoid of psychid caterpillar (Lepidoptera: Psychidae) from India. *Zootaxa* 4388, 425–430.

Gupta, A., van Achterberg, C. and Chitrala, M. (2016) A new species of *Crinibracon* Quicke (Hymenoptera: Braconidae) parasitic on pupae of *Hasora chromus* (Cramer) (Lepidoptera: Hesperiidae) from India. *Zootaxa* 4158, 281–291.

Gupta, A., Lalitha, Y., Varshney, R., Shylesha, A.N. and van Achterberg, C. (2020) *Chelonus formosanus* Sonan (Hymenoptera: Braconidae) an egg-larval parasitoid of the invasive pest *Spodoptera frugiperda* (J. E. Smith) (Lepidoptera: Noctuidae) amenable to laboratory mass production in India. *Journal of Entomology and Zoology Studies* 8, 1521–1524.

Hafez, M., Tawfik, M.F.S. and Ibrahim, A.A. (1980) The immature stages of *Chelonus inanitus* (L.), a parasite of the cotton leafworm, *Spodoptera littoralis* (BOISD.), in Egypt (Hym., Braconidae). *Deutsche Entomologische Zeitschrift* 27, 29–38.

He, J.H. and van Achterberg, C. (1994) A revision of the genus *Aulacocentrum* Brues (Hymenoptera: Braconidae: Macrocentrinae) from China. *Zoologische Mededelingen, Leiden* 68, 159–171.

He, J.H., Chen, X.X. and Ma, Y. (2000) *Hymenoptera Braconidae., Fauna Sinica. Insecta Vol.* 18. Science Press, Beijing, 757 pp.

Hentz, M., Ellsworth, P, and Naranjo, S. (1997) Biology and morphology of *Chelonus* sp. nr. *curvimaculatus* (Hymenoptera: Braconidae) as a parasitoid of *Pectinophora gossypiella* (Lepidoptera: Gelechiidae). *Annals of the Entomological Society of America* 90, 631–639.

Hu, J., Zhu, X.-X. and Fu, W.-J. (2003) Passive evasion of encapsulation in *Macrocentrus cingulum* Brischke (Hymenoptera: Braconidae), a polyembryonic parasitoid of *Ostrinia furnacalis* Guenee (Lepidoptera: Pyralidae). *Journal of Insect Physiology* 49, 367–375.

Huddleston, T. and Walker, A.K. (1988) *Cardiochiles* (Hymenoptera: Braconidae), a parasitoid of lepidopterous larvae, in the Sahel of Africa, with a review of the biology and host relationships of the genus. *Bulletin of Entomological Research* 78, 435–461.

Jasso-Martínez, J.M., Quicke, D.L.J., Belokobylskij, S.A., Meza-Lázaro, R.N. and Zaldívar-Riverón A. (2021) Phylogenomics of the lepidopteran endoparasitoid wasp subfamily Rogadinae (Hymenoptera: Braconidae) and related subfamilies. *Systematic Entomology* 46, 83–95.

Jasso-Martínez, J.M., Quicke, D.L.J., Belokobylskij, S.A., Santos, B.F., Fernández-Triana, J.L., Kula, R.R. and Zaldívar-Riverón, A. (2022a) Mitochondrial phylogenomics and mitogenome organization in the parasitoid wasp family Braconidae (Hymenoptera: Ichneumonoidea). *BMC Ecology and Evolution* 22, 46.

Jasso-Martínez, J.M., Santos, B.F., Zaldívar-Riverón, A., Fernández-Triana, J., Sharanowski, B.J., Richter, R., Dettman, J.R., Blaimer, B.B., Brady, S.G. and Kula, R.R. (2022b) Phylogenomics of braconid wasps (Hymenoptera, Braconidae) sheds light on classification and the evolution of parasitoid life history traits. *Molecular Phylogenetics and Evolution* 173, e107452. doi: 10.1016/j.ympev.2022.107452

Kang, I., Long, K.D., Sharkey, M.J., Whitfield, J.B. and Lord, N.P. (2020) *Orientocardiochiles*, a new genus of Cardiochilinae (Hymenoptera, Braconidae), with descriptions of two new species from Malaysia and Vietnam. *ZooKeys* 971, 1–15.

Kang, I., Whitfield, J.B., Owens, B.E. and Chen, J. (2022) Resurrection of *Neocardiochiles* Szépligeti 1908 (Hymenoptera, Braconidae, Cardiochilinae) with descriptions of five new species from the Neotropical region. *Journal of Hymenoptera Research* 91, 41–68.

Kawakami, T. (1985) Development of the immature stages of *Ascogaster reticulatus* Watanabe (Hymenoptera: Braconidae), an egg-larval parasitoid of the smaller teat tortrix moth, *Adoxophyes* sp. (Lepidoptera: Tortricidae). *Applied Entomology and Zoology* 20, 380–386.

Kittel, R.N. (2014) Revision of *Megascogaster* (Hymenoptera, Braconidae, Cheloninae), with a new species from Sulawesi, Indonesia. *Zootaxa* 3860, 371–378.

Kittel, R.N. and Austin, A.D. (2014) Synopsis of Australian chelonine wasps (Hymenoptera: Braconidae: Cheloninae) with description of two new genera. *Austral Entomology* 53, 183–202.

Kittel, R.N. and Maeto, K. (2019) Revalidation of *Habrobracon brevicornis* stat. rest. (Hymenoptera: Braconidae) based on the CO1, 16S, and 28S gene fragments. *Journal of Economic Entomology* 112, 906–911.

Kittel, R.N., Austin, A.D. and Klopfstein, S. (2016) Molecular and morphological phylogenetics of chelonine parasitoid wasps (Hymenoptera: Braconidae), with a critical assessment of divergence time estimations. *Molecular Phylogenetics and Evolution* 101, 224–241.

Kitthawee, S. (2013) ITS2 sequence variations among members of *Diachasmimorpha longicaudata* complex (Hymenoptera: Braconidae) in Thailand. *Journal of Asia-Pacific Entomology* 16, 173–179.

Kitthawee, S. and Dujardin, J.-P. (2016) The *Diachasmimorpha longicaudata* complex in Thailand discriminated by its wing venation: reference-based morphometric identification. *Zoomorphology* 135, 323–332.

Kitthawee, S., Julsilikul, D., Sharpe, R.G. and Baimai, V. (1999) Protein polymorphism in natural populations of *Diachasmimorpha longicaudata* (Hymenoptera: Braconidae) in Thailand. *Genetica* 105, 125–131.

Krugner, R., Daane, K.M., Lawson, A.B. and Yokota, G.Y. (2005) Biology of *Macrocentrus iridescens* (Hymenoptera: Braconidae): a parasitoid of the oblique banded leafroller (Lepidoptera: Tortricidae). *Environmental Entomology* 34, 336–343.

Kuriachan, I., Henderson, R., Laca, R. and Vinson, S.B. (2011) Post-egression host tissue feeding is another strategy of host regulation by the koinobiont wasp, *Toxoneuron nigriceps*. *Journal of Insect Science* 11:3. Available at insectscience.org/11.3

Kuwana, Z., Ishii, G. and Kurosawa, T. (1940) Studies on the Hymenopterous parasites of *Margaronia pyloalis* Walk. (Lepidoptera). II. *Macrocentrus philippinensis* Ashm. (Braconidae). *Reports of the Sericultural Experimental Station, Tokyo* 10, 1–26.

Li, X.-Y., van Achterberg, C. and Tan, J.-C. (2013) Revision of the subfamily Opiinae (Hymenoptera, Braconidae) from Hunan (China), including thirty-six new species and two new genera. *ZooKeys* 268, 1–186.

Li, Y., van Achterberg, C. and Chen, X.-X. (2017) Review of the genus *Craspedolcus* Enderlein sensu lato in China, with the description of a new genus and four new species (Hymenoptera, Braconidae, Braconinae). *Zookeys* 647, 37–65.

Li, Y., He, J.H. and Chen, X.X. (2020) The subgenera *Glabrobracon* Fahringer, *Lucobracon* Fahringer and *Uncobracon* Papp of the genus *Bracon* Fabricius (Hymenoptera, Braconidae, Braconinae) in China, with the description of eleven new species. *Deutsche Entomologische Zeitschrift* 67, 209–253.

Li, Y., van Achterberg, C. and Chen, X.-X. (2021) A new genus and eight newly recorded genera of Braconinae Nees (Hymenoptera, Braconidae) from China, with descriptions of fourteen new species. *ZooKeys* 1038, 105–178.

Loncle, M.K., Quicke, D.L.J., Sureerat, D. and Butcher, B.A. (2022) The first record of *Charmon* Haliday, 1833 (Braconidae: Charmontinae) from Southeast Asia with description of a new species from Thailand. *Zootaxa* 5213, 93–100.

Long, K.D. and Belokobylskij, S.A. (2003) A preliminary list of the Braconidae (Hymenoptera) of Vietnam. *Russian Entomological Journal* 12, 385–398.

Long K.D. and Belokobylskij S.A. (2011) Vietnamese species of the genus *Spathius* Nees (Hymenoptera: Braconidae: Doryctinae) with reduced first radiomedial vein of the forewing. *American Museum Novitates* 3721, 1–26.

Long, K.D. and van Achterberg, C. (2011) Two new species of the genus *Hartemita* Cameron, 1910 (Hymenoptera: Braconidae: Cardiochilinae) from Vietnam. *Tijdschrift voor Entomologie* 154, 223–228.

Long, K.D. and van Achterberg, C. (2014) An additional list with new records of braconid wasps of the family Braconidae (Hymenoptera) from Vietnam. *Tap Chi Sinh Hoc* 36, 397–415.

Long, K.D. and van Achterberg, C. (2016) A new species if the genus *Plesiocoelus* van Achterberg (Hymenoptera: Braconidae: Agathidinae) from Vietnam. *Tap Chi Sinh Hoc* 28, 304–309.

Long, K.D., Hoa, D.H. and Quynh Nga, C.H. (2018) New record of the genus *Brulleia* Szépligeti, 1904 (Hymenoptera: Braconidae: Brachistinae), with description of three new species from Vietnam. *Academia Journal of Biology* 40, 1–12.

Long, K.D., Dzuong, D.V. and Hoa, D.T. (2019) New records of rare genera of the subfamily Cheloninae (Hymenoptera: Braconidae), with description of two new species from Vietnam. *Academia Journal of Biology* 41, 1–9.

Long, K.D., van Achterberg, C., Carpenter, J.M. and Oanh, N.T. (2020) Review of the genus *Wroughtonia* Cameron, 1899 (Hymenoptera, Braconidae, Helconinae), with the description of 12 new species from Vietnam. *American Museum Novitates* 3953, 54 pp.

Long, K.D., Mai, P.Q., Nhi, P.T., Hiep, N.D., Duong, T.D. and Hoa, D.T. (2023) An updated checklist of *Aleiodes* and *Heterogamus* species (Hymenoptera, Braconidae, Rogadinae) in Vietnam, with new records. *Academia Journal of Biology* 45, 1–9.

Lu, J.F., Feng, C.J., Hu, J. and Fu, W.J. (2007) Extraembryonic membrane of the polyembryonic parasitoid *Macrocentrus cingulum* Brischke (Hym., Braconidae) is essential for evasion of encapsulation. *Journal of Applied Entomology* 131, 472–477.

Ma, W.-J., Kuijper, B., de Boer, J.G., van de Zande, L., Beukeboom, L.W., Wertheim, B. et al. (2013) Absence of complementary sex determination in the parasitoid wasp genus *Asobara* (Hymenoptera: Braconidae). *PLoS ONE* 8, e60459.

Maetô, K. (2018) Polyphagous koinobiosis: the biology and biocontrol potential of a braconid endoparasitoid of exophytic caterpillars. *Applied Entomology and Zoology* 53, 433–446.

Mai, P.Q., Long, K.D., Hiep, N.D., Hoa, D.T. and Duong, T.D. (2023) First record of the genus *Testudobracon* Quicke, 1986 (Hymenoptera: Braconidae: Braconinae) in Vietnam, with description of four new species. *Zootaxa* 5244, 485–500.

Martens, A.P., Buffington, M.L., Quicke, D.L.J., Raweearamwong, M., Butcher, B.A. and Johnson, P.J. (2021) *Ishtarella thailandica* Martens, new genus, new species (Hymenoptera: Braconidae: Aphidiinae) of aphid parasitoid from Thailand, with a country checklist of Aphidiinae. *Insecta Mundi* 0904, 1–6.

Mason, W.R.M. (1974) A generic synopsis of Brachisitini (Hymenoptera: Braconidae) and recognition of the name *Charmon* Haliday. *Proceedings of the Entomological Society of Washington* 76, 235–246.

Mason, W.R.M. (1981) The polyphyletic nature of *Apanteles* Foerster (Hymenoptera: Braconidae): a phylogeny and reclassification of Microgastrinae. *Memoirs of the Entomological Society of Canada* 115, 1–147.

Mathur, R.N. (1959) Breeding technique and utilisation of *Cedria paradoxa* Wilk. (Hym., Braconidae) for biological control of teak defoliator. *Entomologist's Monthly Magazine* 95, 248–250.

Matsuo, K., Uechi, N., Tokuda, M., Maeto, K., and Yukawa, J. (2016) Host range of braconid species (Hymenoptera: Braconidae) that attack Asphondyliini (Diptera: Cecidomyiidae) in Japan. *Entomological Science* 19, 3–8.

Murphy, N., Banks, J., Whitfield, J.B. and Austin, A. (2008) Phylogeny of the parasitic microgastroid subfamilies (Hymenoptera: Braconidae) based on sequence data from seven genes, with an improved time estimate of the origin of the lineage. *Molecular Phylogenetics and Evolution* 47, 378–395.

Nakamatsu, Y., Harvey, J.A. and Tanaka, T. (2006) The mechanism of the emergence of *Cotesia kariyai* (Hymenopotera: Braconidae) larvae from the host. *European Journal of Entomology* 103, 355–360.

Nixon, G.E.J. (1965) A reclassification of the tribe Microgasterini (Hymenoptera: Braconidae). *Bulletin of the British Museum (Natural History): Entomology* 2, 1–284.

Oanh, N.T., Long, K.D., Nhi, P.T. and Hoa, D.T. (2021) Six new braconid wasps of the genus *Colastomion* Baker, 1917 (Hymenoptera: Braconidae: Rogadinae) from Vietnam. *Zootaxa* 5040, 215–237.

Oanh, N.T., Long, K.D. and Nhhiep, H.T. (2023) Three new species of the genus *Spinadesha* Quicke (Hymenoptera: Braconidae: Braconinae) from Vietnam. *Zootaxa* 5258, 573–582.

Oatman, E.R., Platner, G.R. and Greany, P.D. (1969) The biology of *Orgilus lepidus* (Hymenoptera: Braconidae), a primary parasite of the potato tuberworm. *Annals of the Entomological Society of America* 62, 1407–1414.

Obrtel, R. (1960) Premature stages of *Triaspis caudatus* (Nees) (Hymenoptera: Braconidae). *Folia Zoologica* 9, 35–38.

Odebiyi, J.A. and Oatman, E.R. (1972) Biology of *Agathis gibbosa* (Hymenoptera: Braconidae), a primary parasite of the potato tuberworm. *Annals of the Entomological Society of America* 65, 1104–1014.

Pennacchio, F, and Digilio, M.C. (1990) Morphology and development of larval instars of *Aphidius ervi* Haliday (Hymenoptera, Braconidae, Aphidiinae). *Bolletino del Laboratorio di Entomologia Agrarian 'Filippo Silvestri'* 46, 163–174.

Pennacchio, F., Vinson, S.B. and Tremblay, E. (1991) Effects of *Cardiochiles nigriceps* Viereck (Hymenoptera: Braconidae) teratocytes on physiology of its host *Heliothis virescens* (F.) (Lepidoptera; Noctuidae). *Redia, Geornale de Zoology* 124, 433–438.

Pennacchio, F., Vinson, S.B. and Tremblay, E. (1992) Host regulation effects on *Heliothis virescens* (F.) larvae induced by teratocytes of *Cardiochiles nigriceps* Viereck (Lepidoptera: Noctuidae – Hymenoptera: Braconidae). *Archives of Insect Biochemistry and Physiology* 19, 177–192.

Penteado-Dias, A.M. (1996) New Neotropical species of the genus *Meteoridea* Ashmead (Hymenoptera: Braconidae: Meteorideinae). *Zoologische Mededelingen, Leiden* 70, 197–205,

Perioto, N.W., Lara, R.I.R., Ferrerira, C.S., Fernandes, D.R.R., Pedroso, E.D.C., Volpe, H.X.L., Nais, J., Correa, L.R.B. and Viel, S.R. (2011) A new phytophagous *Bracon* Fabricius (Hymenoptera: Braconidae) associated with *Protium ovatum* (Burseraceae) fruits from Brazilian savannah. *Zootaxa* 3000, 59–65.

Peris-Felipo, F. J., Stigenberg, J., Quicke, D.L.J. and Belokobylskij, S.A. (2019) Revision of the Oriental subgenus *Patrisaspilota* Fischer, 1995 (Hymenoptera: Braconidae: Alysiinae: *Orthostigma* Ratzeburg, 1844) with description of a new species from Papua New Guinea. *Zootaxa* 4629, 365–378.

Quicke, D.L.J. (1984) Evidence for the role of white-tipped ovipositor sheath in the Braconinae (Hymenoptera: Braconidae). *Proceedings and Transactions of the British Entomology and Natural History Society* 17, 71–79.

Quicke, D.L.J. (1987) The Old World genera of braconine wasps (Hymenoptera: Braconidae). *Journal of Natural History* 21, 43–157.

Quicke, D.L.J. (1988) Four new genera of the *Plesiobracon* Cameron group (Insecta, Hymenoptera, Braconinae). *Zoologica Scripta* 17, 411–418.

Quicke, D.L.J. (1989) The Indo-Australian and E. Palaearctic braconine genus Euurobracon (Hymenoptera: Braconidae: Braconinae). *Journal of Natural History* 23, 775–802.

Quicke, D.L.J. (1989a) Three new genera of Braconini from Australasia and Malaysia (Insecta, Hymenoptera, Braconidae). *Zoologica Scripta* 18, 295–302.

Quicke, D.L.J. (1989b) Two new genera and species of Braconinae (Insecta, Hymenoptera, Braconidae) from Brunei. *Zoologica Scripta* 18, 139–143.

Quicke, D.L.J. (1991) Ovipositor mechanics of the braconine wasp genus *Zaglyptogastra* and the ichneumonid genus *Pristomerus*. *Journal of Natural History* 25, 971–977.

Quicke, D.L.J. (2015) *The Braconid and Ichneumonid Parasitic Wasps: Biology, Systematics, Evolution and Ecology*. Wiley Blackwell, Oxford, 688 pp.

Quicke, D.L.J. and Belshaw, R. (1999) Incongruence between morphological data sets: an example from the evolution of endoparasitism among parasitic wasps (Hymenoptera: Braconidae). *Systematic Biology* 48, 436–454.

Quicke, D.L.J. and Butcher, B.A. (2011a) Corrigendum to revision of the genus *Ischnobracon* Baltazar (Hymenoptera: Braconidae: Braconinae) by Areekul Butcher & Quicke (2010). *Journal of Natural History* 45, 39–40.

Quicke, D.L.J. and Butcher, B.A. (2011b) Two new genera of Rogadinae (Insecta: Hymenoptera: Braconidae) from Thailand. *Journal of Hymenoptera Research* 23, 23–34.

Quicke, D.L.J. and Butcher, B.A. (2018) First record of *Odontosphaeropyx* Cameron, 1910 from the Oriental Region with description of a new species from Thailand (Hymenoptera, Braconidae, Cheloninae). *ZooKeys* 809, 41–47. doi: 10.3897/zookeys.809.30742

Quicke, D.L.J. and Ingram, S.N. (1993) Braconine wasps of Australia. *Memoirs of the Queensland Museum* 33, 299–336.

Quicke, D.L.J. and Laurenne, N.M. (2005) Notes on host searching by the parasitic wasp *Zaglyptogastra* Ashmead (Hymenoptera: Braconidae: Braconinae) in Kibale Forest, Uganda. *Journal of Hymenoptera Research* 14, 177–181.

Quicke, D.L.J. and Stanton, M.L. (2005) *Trigastrotheca laikipiensis* n. sp. (Hymenoptera: Braconidae): A new species of brood parasitic wasp that attacks foundress queens of three coexisting acacia-ant species in Kenya. *Journal of Hymenoptera Research* 14, 182–190.

Quicke, D.L.J. and van Achterberg, C. (1990) Phylogeny of the subfamilies of the family Braconidae (Hymenoptera: Ichneumonidae). *Zoologische Verhandelingen* 258, 1–95.

Quicke, D.L.J. and Walker, C. (1991) A new Indo-Australian genus of Braconini (Insecta, Hymenoptera, Braconidae). *Zoologica Scripta* 20, 419–424.

Quicke, D.L.J., Ficken, L., Fitton, M.G. (1992) New diagnostic ovipositor characters for doryctine wasps (Hymenoptera, Braconidae). *Journal of Natural History* 26, 1035–1046.

Quicke, D.L.J., Chen. J., Chishti, M.J.K. and Kruft, R.A. (1997) A revision of *Yelicones* (Hymenoptera: Braconidae: Rogadinae) from the East Palaearctic and Oriental regions with descriptions of 4 new species. *Journal of Natural History* 31, 779–797.

Quicke, D.L.J., Manzari, S. and van Achterberg, C. (2002) The systematic placement of *Afrocampsis* van Achterberg & Quicke (Hymenoptera: Braconidae): molecular and morphological evidence indicate that it belongs to Helconinae s.l. not Sigalphinae. *Zoologische Mededelingen, Leiden* 76, 443–450.

Quicke, D.L.J., Sharkey, M.J., Laurenne, N.M. and Dowling, A. (2008) A preliminary molecular phylogeny of the Sigalphinae (Hymenoptera: Braconidae), including *Pselaphanus* Szépligeti, based on 28S rDNA, with descriptions of new Afrotropical and Madagascan *Minanga* and *Malasigalphus* species. *Journal of Natural History* 42, 2703–2719.

Quicke, D.L.J., Smith, M. A., van Achterberg, C., Miller, S. E. and Hrcek, J. (2012) A new genus and three new species of parasitoid wasp from Papua New Guinea and redescription of *Trigonophatnus* Cameron (Hymenoptera, Braconidae, Rogadinae). *Journal of Natural History* 46, 1369–1385.

Quicke, D.L.J., Hogan J.E., Bennett A.M.R., Broad, G.R. and Butcher, B.A. (2017a) Partial revision of the Indo-Australian braconine wasp genus *Gammabracon* Quicke (Hymenoptera: Braconidae) with descriptions of new species from Indonesia (Moluccas), Malaysia, Philippines and Thailand. *Journal of Natural History* 51, 1249–1294. doi: 10.1080/0022933.2017.1324055

Quicke, D.L.J., Belokobylskij, S.A., Raweearamwong, M. and Butcher, B.A. (2017b) A new species of *Cedria* Wilkinson (Hymenoptera: Braconidae: Lysiterminae) from Thailand. *Zootaxa* 4365, 395–400.

Quicke, D.L.J., Butcher, B.A., Ranjith, A.P. and Belokobylskij, S.A. (2017c) Revision of the non-Afrotropical species of *Trigastrotheca* Cameron (Hymenoptera: Braconidae: Braconinae) with descriptions of four new species. *Zootaxa* 4242, 95–110.

Quicke D.L.J., Kuslitzky, W. and Butcher, B.A. (2018) First host record for Old World *Yelicones* (Hymenoptera: Braconidae: Rogadinae) adds to evidence that they are strictly parasitoids of Pyralidae (Lepidoptera). *Israel Journal of Entomology* 48, 33–40.

Quicke, D.L.J., Austin, A.D., Fagan-Jeffries, E., Hebert, P.D.N. and Butcher, B.A. (2019a) Molecular phylogeny places the enigmatic subfamily Masoninae within the Ichneumonidae, not the Braconidae. *Zoologica Scripta* 49, 64–71

Quicke, D.L.J., Chaul, J.C.M. and Butcher, B.A. (2019b) First South American record of the rare ichneumonoid subfamily Masoninae van Achterberg (Hymenoptera: Ichneumonoidea: Ichneumonidae) with description of a new species from Brazil. *Zootaxa* 4664, 587–593.

Quicke, D.L.J., Austin, A.D., Fagan-Jeffries, E.P., Hebert, P.D.N. and Butcher, B.A. (2020a) Recognition of the Trachypetidae stat.n. as a new extant family of Ichneumonoidea (Hymenoptera), based on molecular and morphological evidence. *Systematic Entomology* 45, 771–782.

Quicke, D.L.J., Belokobylskij, S.A., Braet, Y., van Achterberg, C., Hebert, P.D.N., Prosser, S.W.J., Austin, A.D., Fagan-Jeffries, E.P., Ward, D.F., Shaw, M.R. and Butcher, B.A. (2020b) Phylogenetic reassignment of basal cyclostome braconid parasitoid wasps (Hymenoptera) with description of a new, enigmatic Afrotropical tribe with a highly anomalous 28S D2 secondary structure. *Zoological Journal of the Linnean Society* 190, 1002–1019.

Quicke, D.L.J., Fagan-Jeffries, E.P., Jasso-Martínez, J.M., Zaldívar-Riverón, A., Shaw, M.R., Janzen, D.H., Hallwachs, W., Smith, M.A., Hebert, P.D.N., Hrcek, J., Miller, S., Sharkey, M.J., Shaw, S.R. and Butcher, B.A. (2021) A molecular phylogeny of the parasitoid wasp subfamily Rogadinae (Ichneumonoidea: Braconidae) with descriptions of three new genera. *Systematic Entomology* 46, 1019–1044.

Quicke, D.L.J., Gafar, D., Watanabe, K. and Butcher, B.A. (2022) A new species of the long-tailed wasp genus *Euurobracon* Ashmead (Hymenoptera, Braconidae, Braconinae) from Java, Indonesia, is described and the type species redescribed. *ZooKeys* 1116, 71–83.

Quicke, D.L., Jasso-Martínez, J.M., Ranjith, A.P., Sharkey, M.J., Manjunath, R., Naik, S., Herbert, P.D.N., Priyadarsanan, D.R., Thurman, J. and Butcher, B.A. (2023a) Phylogeny of the Braconinae (Hymenoptera: Braconidae): A new tribal order!. *Systematic Entomology*, 1–26. https://doi.org/10.1111/syen.12608

Quicke, D.L., Ranjith, A.P., Priyadarsanan, D.R., Nasser, M., Hebert, P.D., and Butcher, B.A. (2023b) Two new genera and one new species of the tribe Adeshini (Hymenoptera, Braconidae, Braconinae) from India and South Africa. *ZooKeys* 1166, 235–259.

Ranjith, A.P. and Priyadarsanan, D.H. (2023) New subgeneric reports of the genus *Chelonus* (Hymenoptera: Braconidae) from India and Sri Lanka with description of nine species. *Zootaxa* 5278(3), 461–492.

Ranjith, A.P., Nasser, M., Rajmohana, K. and Quicke, D.L.J. (2016a) A new genus of Braconinae (Hymenoptera: Braconidae) from India with remarkable head ornamentation. *Zootaxa* 4061, 173–180.

Ranjith, A.P., Quicke, D.L.J., Saleem, U.K.A., Butcher, B.A., Zaldívar-Riverón, A. and Nasser, M. (2016b) Entomophytophagy ('sequential predatory, then phytophagous behaviour') in an Indian braconid 'parasitoid' wasp (Hymenoptera): specialized larval morphology, biology and description of a new species. *PLoS One* 11, e0156997.

Ranjith, A.P., van Achterberg, C., Priyadarsanan, D.R., Kim, I.-K., Keloth, R., Mukundan, S. and Nasser, M. (2019) First Indian record of *Centistidea* Rohwer (Hymenoptera: Braconidae, Miracinae) with description of eight new species. *Insect Systematics & Evolution* 50, 407–444.

Ranjith, A.P., Samartsev, K.G. and Nasser, M. (2021) Discovery of the braconid subfamily Dirrhopinae van Achterberg (Hymenoptera: Ichneumonoidea) from the Indian subcontinent with the description of a new species from south India. *Zootaxa* 4908, 251–262

Ranjith, A.P., Quicke, D.L.J., Manjusha, K., Butcher, B.A. and Nasser, M. (2022a) Hunting parasitoid wasp: first report of mite predation by a 'parasitoid wasp'. *Scientific Reports* 12, 1747.

Ranjith, A.P., Quicke, D.L.J., Belokobylskij, S.A. and Priyadarsanan, D.R. (2022b) *Kerevata* Belokobylskij (Hymenoptera: Braconidae: Rogadinae) is no longer a Papua New Guinean endemic with descriptions of three new species from the Indomalayan Region. *Zootaxa* 5091, 341–356.

Ranjith, A.P., van Achterberg, C. and Priyadarsanan, D.R. (2022c) *Atree*, a remarkable new genus of the subfamily Brachistinae (Hymenoptera: Braconidae) and the first report of the tribe Diospilini from India. *Zootaxa* 5105, 571–580.

Raychaudhuri, D. (1990) *Aphidiids (Hymenoptera) of Northeast India*. Indira Publishing House, Oak Park, Michigan. 155 pp.

Richards, O.W. (1956) An interpretation of the ventral region of the hymenopterous thorax. *Proceedings of the Royal Entomological Society of London. Series A, General Entomology* 31, 99–104.

Richards, O.W. (1977) *Hymenoptera, Introduction and Key to Families. Handbooks for the Identification of British Insects*, Vol. VI, Part 1, 2nd edn. Royal Entomological Society of London, London, 110 pp.

Rocha, K.L., Mangine, T., Harris, E.J. and Lawrence, P.O. (2004) Immature stages of *Fopius arisanus* (Hymenoptera: Braconidae) in *Bactrocera dorsalis* (Diptera: Tephritidae). *Florida Entomologist* 87, 164–168.

Rodriguez, J.J., Fernández-Triana, J.L., Smith, M.A., Janzen, D.H., Hallwachs, W., Erwin, T.L. and Whitfield, J.B. (2013) Extrapolations from field studies and known faunas converge on dramatically increased estimates of global microgastrine parasitoid wasp species richness (Hymenoptera: Braconidae). *Insect Conservation and Diversity* 6, 530–536.

Rousse, P. (2013) *Charmon ramagei* sp. nov., a new Charmontinae (Hymenoptera: Braconidae) from Reunion, with a synopsis of world species. *Zootaxa* 3626, 583–588.

Ruberson, J.R. and Whitfield, J.B. (1996) Facultative egg-larval parasitism of the beet armyworm, *Spodoptera exigua* (Lepidoptera: Noctuidae) by *Cotesia marginiventris* (Hymenoptera: Braconidae). *Florida Entomologist* 79, 296–302.

Sabahatullah, M., Mashwani, M.A., Tahira, Q.A. and Inayatullah, M. (2014) New record of subfamily Charmontinae (Braconidae: Hymenoptera) in Pakistan with the description of a new species. *Pakistan Journal of Agricultural Research* 27, 296–302.

Sakagami, K., Shimizu, S., Fujie, S. and Maeto, K. (2020) Revisiting the host use and phylogeny of *Colastomion* Baker (Hymenoptera, Braconidae, Rogadinae), with a new host record from Japan. *Journal of Hymenoptera Research* 77, 175–186.

Samartsev, K. and Ku, D.-S. (2020) New species of the genera *Bracon* Fabricius and *Syntomernus* Enderlein (Hymenoptera, Braconidae, Braconinae) from South Korea. *ZooKeys* 999, 1–47.

Sanchis, A., Latorre, A., Gonzalez-Candelas, F. and Michelena, J.M. (2000) An 18S rDNA-based molecular phylogeny of Aphidiinae (Hymenoptera: Braconidae). *Molecular Phylogenetics and Evolution* 14, 180–194.

Sarmiento, C.E., Sharkey, M.J. and Janzen, D.H. (2004) The first gregarious species of the Agathidinae (Hymenoptera: Braconidae). *Journal of Hymenoptera Research* 13, 295–301.

Schepers, E.J., Dahlman, D.L. and Zhang, D. (1998) *Microplitis croceipes* teratocytes: in vitro culture and biological activity of teratocyte secreted protein. *Journal of Insect Physiology* 44, 767–777.

Sharanowski, B.J., Dowling, A.P.G. and Sharkey, M.J. (2011) Molecular phylogenetics of Braconidae (Hymenoptera: Ichneumonoidea), based on multiple nuclear genes, and implications for classification. *Systematic Entomology* 36, 549–572.

Sharanowski, B.J., Ridenbaugh, R.D., Piekarski, P.K., Broad, G.R., Burke, G.R., Deans, A.D., Lemmon, A.R., Moriarty Lemmon, E.C., Diehl, G.J., Whitfield, J.B. and Hines, H.M. (2021) Phylogenomics of Ichneumonoidea (Hymenoptera) and implications for evolution of mode of parasitism and viral endogenization. *Molecular Phylogenetics and Evolution* 156, 107023

Sharkey, M.J. (1993) Family Braconidae. In: Goulet, H. and Huber, J. (eds) *Hymenoptera of the World: An Identification Guide to Families (Vol. VII)*. Research Branch, Agriculture Canada, Ottawa, pp. 362–395.

Sharkey, M.J. and Chapman, R. (2017a) Ten new genera of Agathidini (Hymenoptera, Braconidae, Agathidinae) from Southeast Asia. *ZooKeys* 660, 107–150.

Sharkey, M.J. and Chapman, R. (2017b) Phylogeny of the Agathidinae (Hymenoptera: Braconidae) with a revised tribal classification and the description of a new genus. *Proceedings of the Entomological Society of Washington* 119, 823–842.

Sharkey, M.J. and Chapman, E.G. (2018) Revision of *Zosteragathis* Sharkey of Thailand (Hymenoptera, Braconidae, Agathidinae, Agathidini). *Deutsches Entomologische Zeitschrift* 65, 225–253.

Sharkey, M.J. and Clutts, S.A. (2011) A revision of Thai Agathidinae (Hymenoptera, Braconidae), with descriptions of six new species. *Journal of Hymenoptera Research* 22, 69–132.

Sharkey, M.J. and Janzen, D.H. (1995) Review of the world species of *Sigalphus* (Hymenoptera: Braconidae: Sigalphinae) and biology of *Sigalphus romeroi*, new species. *Journal of Hymenoptera Research* 4, 99–109.

Sharkey, M.J. and Stoelb, S.A.C. (2012a) Revision of *Therophilus* s.s. (Hymenoptera, Braconidae, Agathidinae) from Thailand. *Journal of Hymenoptera Research* 27, 1–36.

Sharkey, M.J. and Stoelb, S.A.C. (2012b) Revision of *Zelodia* (Hymenoptera, Braconidae, Agathidinae) from Thailand. *Journal of Hymenoptera Research* 26, 31–71.

Sharkey, M.J. and Stoelb, S.A.C. (2013) Revision of *Agathacrista* new genus (Hymenoptera, Braconidae, Agathidinae, Agathidini). *Journal of Hymenoptera Research* 33, 99–112.

Sharkey M.J. and Wharton, R.A. (1994) A revision of the genera of the world Ichneutinae (Hymenoptera: Braconidae). *Journal of Natural History* 28, 873–912.

Sharkey, M.J. and Wharton, R.A. (1997) Morphology and terminology. In: Wharton, R.A., Marsh, P.M. and Sharkey, M.J. (eds) Identification Manual to the New World Genera of Braconidae, Special Publication of the International Society of Hymenopterists, Vol. 1 International Society of Hymenopterists, Washington, DC, pp. 19–37.

Sharkey, M.J., Laurenne, N.M., Quicke, D.L.J., Sharanowshi, B. and Murray, D. (2006) Revision of the Agathidinae (Hymenoptera: Braconidae) with comparisons of static and dynamic alignments. *Cladistics* 22, 546–567.

Sharkey, M.J., Yu, D.S., van Noort, S., Seltmann, K. and Penev, L. (2009) Revision of the Oriental genera of Agathidinae (Hymenoptera, Braconidae) with an emphasis on Thailand including interactive keys to genera published in three different formats. *ZooKeys* 21, 19–54.

Sharkey, M.J., Janzen, D.H., Hallwachs, W., Chapman, E.G., Smith, M.A., Dapkey, T., Brown, A., Ratnasigham, S., Naik, S., Manjunat, R., Perez, K., Milton, M., Hebert, P.D.N., Shaw, S.R., Kittel, R.N., Solis, A., Metz, M., Goldstein, P.Z., Brown, J.W., Quicke, D.L.J., van Achterberg, C., Brown B.V. and Burns, J.M. (2021a) Minimalist revision and description of 411 new species in 11 subfamilies of Costa Rican braconid parasitic wasps, including host records. *Zootaxa* 1013, 1–665.

Sharkey, M., Athey, K.J., Fernández-Triana, J.L., Penteado-Dias, A.M., Monckton, S.K. and Quicke, D.L.J. (2023) Key to the New World subfamilies of the family Braconidae (Hymenoptera: Ichneumonoidea). *Canadian Journal of Arthropod Identification* No. 49, 43 pp.

Shaw, M.R. (1983) On[e] evolution of endoparasitism; the biology of some genera of Rogadinae (Braconidae). *Contributions of the American Entomological Institute* 20, 307–328.

Shaw, M.R. (1995a) Observations on the adult behaviour and biology of *Histeromerus mystacinus* Wesmael (Hymenoptera: Braconidae). *The Entomologist* 114, 1–13.

Shaw, M.R. (2002) A new species of *Macrostomion* Szépligeti (Hymenoptera: Braconidae: Rogadinae) from Papua New Guinea, with notes on the biology of the genus. *Zoologische Mededelingen, Leiden* 76, 133–140.

Shaw, M.R. and Huddleston, T. (1991) Classification and biology of braconid wasps (Hymenoptera: Braconidae). *Handbooks for the Identification of British Insects* 7, pp. 1–126.

Shaw, M.R. and Quicke, D.L.J. (2000) The biology and early stages of *Acampsis alternipes* (Nees), with comments on the relationships of the Sigalphinae (Hymenoptera: Braconidae). *Journal of Natural History* 34, 611–628.

Shaw, S.R. (1988) Euphorine phylogeny: the evolution of diversity in host-utilization by parasitoid wasps (Hymenoptera: Braconidae). *Ecological Entomology* 13, 323–335.

Shaw, S.R. (1995b) Braconidae. In: Hanson, P.E. and Gauld, I.D. (eds) *The Hymenoptera of Costa Rica*. Oxford University Press, New York, and The Natural History Museum, London, pp. 431–463.

Shaw, S.R. (2004) Essay on the evolution of adult-parasitism in the subfamily Euphorinae (Hymenoptera: Braconidae). *Proceedings of the Russian Entomological Society, St. Petersburg* 75, 1–15. [English with Russian abstract]

Shi, M. and Chen, X.X. (2005) Molecular phylogeny of the Aphidiinae (Hymenoptera: Braconidae) based on DNA sequences of 16S rRNA, 103 iDNA and AI Pase 6 genes. *European Journal of Entomology* 102, 133–138.

Shirai, S. and Maetô, K. (2009) Suspending cocoons to evade ant predation in *Meteorus pulchricornis*, a braconid parasitoid of exposed-living lepidopteran larvae. *Entomological Science* 12, 107–109.

Singh, L.R.K., van Achterberg, C. and Sheela, S. (2022) Studies on the subfamily Exothecinae (Hymenoptera: Braconidae) with the description of a new genus and a new species from India. *Zootaxa* 5133, 40–52.

Singh, R.P. and Baldev Parshad (1970) Biological notes on *Cardiochiles hymeniae* Max Fischer and Baldev Parshad (Braconidae: Hymenoptera). *Indian Journal of Entomology* 32, 127–129.

Smith, M.A., Fernandez-Triana, J.J., Eveleigh, E., Gomez, J., Guclu, C., Hallwachs, W., Hebert, P.D.N., Hrcek, J., Huber, J.T., Janzen, D., Mason, P.G., Miller, S., Quicke, D.L.J., Rodriguez, J.J., Rougerie, R., Shaw, M.R., Várkonyi, G., Ward, D.F., Whitfield, J.B. and Zaldivar-Riverón, A. (2012) DNA barcoding and the taxonomy of Microgastrinae wasps (Hymenoptera, Braconidae): impacts after 8 years and nearly 20,000 sequences. *Molecular Ecology Resources* 13, 168–176.

Smith, P.T., Kambhampati, S., Völkl, W. and Mackauer, M. (1999) A phylogeny of aphid parasitoids (Hymenoptera: Braconidae: Aphidiinae) inferred from mitochondrial NADH 1 dehydrogenase gene sequence. *Molecular Phylogenetics and Evolution* 11, 236–245.

Starý, P. and Ghosh, A.K. (1983) *Aphid parasitoids of India and adjacent countries (Hymenoptera: Aphidiidae)*. Technical Monograph No. 7. Zoological Survey of India, Calcutta, India, 73 pp. + 230 figures.

Starý, P. and Schlinger, E.I. (1967) Far East Asian Aphid Host–Parasite List. In: *A Revision of the Far East Asian Aphidiidae (Hymenoptera)*. Springer, Dordrecht, pp. 129–132.

Starý, P., Sharkey, M. and Hutacharern, C. (2008) Aphid parasitoids sampled by Malaise traps in the National parks of Thailand (Hymenoptera, Braconidae, Aphidiinae). *Thai Journal of Agricultural Science* 4, 37–43.

Starý, P., Rakhshani, E., Tomanovic, Z., Kavallieratos, N.G. and Sharkey, M.J. (2010a) Aphid parasitoids (Hymenoptera, Braconidae, Aphidiinae) from Thailand. *Zootaxa* 2498, 47–52.

Starý, P., Rakhshani, E., Havelka, J., Tomanovic, Z., Kavallieratos, N.G. and Sharkey, M.J. (2010b) Review and key to the world parasitoids (Hymenoptera: Braconidae: Aphidiinae) of Greenideinae aphids (Hemiptera: Aphididae), including notes on invasive pest species. *Annals of the Entomological Society of America* 103, 307–321.

Stigenberg, J., Boring, C.A. and Ronquist, F. (2015) Phylogeny of the parasitic wasp subfamily Euphorinae (Braconidae) and evolution of its host preferences. *Systematic Entomology* 40, 570–591.

Szépligeti, G.V. (1904) Hymenoptera, Fam. Braconidae. In: Wytsman, P. (ed., 1902–32) *Genera Insectorum* 22, 1–253 + 32 figures.

Tan J.-L., Tan Q.-Q., van Achterberg, C. and Chen X.-X. (2016) A new genus *Carinopius* gen. n. of the subfamily Opiinae (Hymenoptera, Braconidae) from China and Vietnam, with description of a new species. *Zootaxa* 4061, 569–574.

Tian, X.-X., van Achterberg, C., Wu, J.-X. and Tan, J.-L. (2020) New Gnamptodontinae (Hymenoptera: Braconidae) from China and Vietnam, with two genera new for China and seven new species. *Zootaxa* 4778, 471–508.

Tobias, V.I. (1962) A new subfamily of braconids (Hymenoptera, Braconidae) from Central Asia. *Trudy Zoologicheskogo Instituta, Akademiya Nauk SSSR* 30, 268–270.

van Achterberg, C. (1974) The features of the petiolar segment in some Braconidae. *Entomologische Berichten, Amsterdam* 34, 21–23.

van Achterberg, C. (1976a) A revision of the tribus Blacini (Hymenoptera, Braconidae, Helconinae). *Tijdschrift voor Entomologie* 118, 159–322.

van Achterberg, C. (1976b) A preliminary key to the subfamilies of the Braconidae (Hymenoptera). *Tijdschrift voor Entomologie* 119, 33–78.

van Achterberg, C. (1979a) A revision of the subfamily Zelinae auct. (Hymenoptera, Braconidae). *Tijdschrift voor Entomologie* 122, 241–479.

van Achterberg, C. (1979b) A revision of the new subfamily Xiphozelinae (Hymenoptera: Braconidae). *Tijdschrift voor Entomologie* 122, 29–46.

van Achterberg, C. (1983a) Revisionary notes on the Palaearctic genera and species of the tribe Exothecini Foerster (Hymenoptera: Braconidae). *Zoologische Mededelingen, Leiden* 57, 339–355.

van Achterberg, C. (1983b) Revisionary notes on the subfamily Gnamptodontinae, with description of eleven new species (Hymenoptera, Braconidae). *Tijdschrift voor Entomologie* 126, 25–57.

van Achterberg, C. (1983c) A revision of the new tribe Brulleiini (Hymenoptera: Braconidae). *Contributions of the American Entomological Institute* 20, 281–306.

van Achterberg, C. (1987) Revisionary notes on the subfamily Orgilinae (Hymenoptera, Braconidae). *Zoologisches Verhandelingen, Leiden* 242, 1–111.

van Achterberg, C. (1988) Revision of the subfamily Blacinae Foerster (Hymenoptera, Braconidae). *Zoologisches Verhandelingen, Leiden* 249, 1–324.

van Achterberg, C. (1990) Revision of the genera *Foersteria* Szépligeti and *Polydegmon* Foerster (Hymenoptera: Braconidae) with the description of a new genus. *Zoologische Verhandelingen Leiden* 257, 1–32.

van Achterberg, C (1991) Revision of the genus *Trispinaria* Quicke (Hymenoptera: Braconidae). *Zoologische Verhandelingen, Leiden* 65, 181–198.

van Achterberg, C. (1993a) Illustrated key to the subfamilies of the Braconidae (Hymenoptera: Ichneumonoidea). *Zoologische Verhandelingen* 283, 1–189.

van Achterberg, C. (1993b) Revision of the subfamily Macrocentrinae Foerster (Hymenoptera: Braconidae) from the Palaearctic region. *Zoologische Verhandelingen, Leiden* 286, 1–110.

van Achterberg, C. (1994a) New morphological terms. *Ichnews* 14, 5. Available at https://naturalhistory.si.edu/sites/default/files/media/file/ichnews14-1994.pdf (accessed 26 January, 2023)

van Achterberg, C. (1994b) Generic revision of the subfamily Cenocoeliinae Szépligeti (Hymenoptera: Braconidae). *Zoologisches Verhandelingen, Leiden* 292, 1–52.

van Achterberg, C. (1995a) New taxa of the subfamilies Betylobraconinae, Cenocoeliinae, Ecnomiinae, Homolobinae, and Sigalphinae (Hymenoptera: Braconidae) from East Indonesia. *Zoologische Mededelingen, Leiden* 69, 307–328.

van Achterberg C. (1995b) Generic revision of the subfamily Betylobraconinae (Hymenoptera: Braconidae) and other groups with modified fore tarsus. *Zoologisches Verhandelingen, Leiden* 298, 1–242.

van Achterberg, C. (2002) Revision of the genus *Canalicephalus* Gibson and the recognition of the Acampsohelconinae (Hymenoptera: Braconidae) as extant. *Zoologische Mededelingen* 76, 17-24, 347–370.

van Achterberg, C. (2004) Revision of the *Euagathis* species (Hymenoptera: Braconidae: Agathidinae) from Wallacea and Papua. *Zoologische Mededelingen, Leiden* 78, 27.viii.2004: 1–76.

van Achterberg, C. (2007) Revision of the genus *Spinaria* Brullé (Hymenoptera: Braconidae: Rogadinae), with keys to genera and species of the subtribe Spinariina van Achterberg. *Zoologische Mededelingen, Leiden* 81, 11–83.

van Achterberg, C. (2008) A new species of *Xiphozele* Cameron (Hymenoptera: Braconidae) from South Vietnam. *Zoologische Mededelingen, Leiden* 82, 1–8.

van Achterberg, C. (2014) *Sigalphus anjae* spec. nov. (Hymenoptera: Braconidae: Sigalphinae) from southern Vietnam. *Zoologische Mededelingen, Leiden* 88, 9–17.

van Achterberg, C. and Austin, A.D. (1992) Revision of the genera of the subfamily Sigalphinae (Hymenoptera: Braconidae), including a revision of the Australian species. *Zoologische Verhandelingen, Leiden* 280, 1–44.

van Achterberg, C. and Chen, X. (2002) Revision of the *Euagathis* species (Hymenoptera: Braconidae: Agathidinae) from China and northern Vietnam. *Zoologische Mededelingen, Leiden* 76, 309–346.

van Achterberg, C. and Desmier de Chenon, R. (2009) The first report of the biology of *Proterops borneoensis* Szépligeti (Hymenoptera: Braconidae: Ichneutinae), with the description of a new species from China. *Journal of Natural History* 43, 619–633

van Achterberg, C. and Lau, C.S.K. (2022) Biological notes on *Rhysipolis taiwanicus* Belokobylskij (Hymenoptera, Braconidae, Rhysipolinae). *Journal of Hymenoptera Research* 93, 81–87.

van Achterberg, K.[C.] and Long, K.D. (2010) Revision of the Agathidinae (Hymenoptera, Braconidae) of Vietnam, with the description of forty-two new species and three new genera. *ZooKeys* 54, 1–184.

van Achterberg, C. and Quicke, D.L.J. (1992) Phylogeny of the subfamilies of the family Braconidae: a reassessment assessed. *Cladistics* 8, 237–264.

van Achterberg, C. and Quicke, D.L.J. (2000) The palaeotropical species of the tribe Cosmophorini (Hymenoptera: Braconidae: Euphorinae) with descriptions of twenty-two new species. *Zoologische Mededelingen, Leiden* 74, 283–338.

van Achterberg, C. and Weiblen, G.D. (2000) *Ficobracon brusi* gen. nov. & spec. nov. (Hymenoptera: Braconidae), a parasitoid reared from figs in Papua New Guinea. *Zoologische Mededelingen, Leiden* 74, 51–55.

van Achterberg, C., Hosaka, K.T., Ng, Y.F. and Ghani, B.A. (2009) The braconid parasitoids (Hymenoptera: Braconidae) associated with seeds of Dipterocarpaceae in Malaysia, *Journal of Natural History* 43, 635–686.

van Achterberg, C., Sharkey, M.J. and Chapman, E.G. (2014) Revision of the genus *Euagathis* Szépligeti (Hymenoptera, Braconidae, Agathidinae) from Thailand, with description of three new species. *Journal of Hymenoptera Research* 36, 1–25.

Wharton, R.A. (1984) Biology of the Alysiini (Hymenoptera: Braconidae), parasitoids of cyclorrhaphous Diptera. *Technical Monographs* 11, 1–39. Texas Agricultural Experiment Station, College Station, Texas.

Wharton, R.A. (1988) Classification of the braconid subfamily Opiinae (Hymenoptera). *The Canadian Entomologist* 120, 333–360.

Wharton, R.A. (1993) Bionomics of the Braconidae. *Annual Review of Entomology* 38, 121–143.

Wharton, R.A. (1993) Review of Hormiini (Hymenoptera; Braconidae) with a description of new taxa. *Journal of Natural History* 27, 107–171.

Wharton, R.A. (1997) Generic relationships of opiine Braconidae (Hymenoptera) parasitic on fruit-infesting Tephritidae (Diptera). *Contributions of the American Entomological Institute* 30, 1–53.

Wharton, R.A. and Sitterz-Bhatkar, H. (2002) Polydnaviruses in the genus *Mirax* Haliday (Hymenoptera: Braconidae). *Journal of Hymenoptera Research* 11, 358–364.

Wharton, R.A., Shaw, S.R., Sharkey, M.J., Wahl, D.B., Woolley, J.B., Whitfield, J.B., Marsh, P.M. and Johnson, J.W. (1992) Phylogeny of the subfamilies of the family Braconidae (Hymenoptera: Ichneumonoidea): a reassessment. *Cladistics* 8, 199–235.

Wharton, R.A., Marsh P.M. and Sharkey, M.J. (eds) (1997) *Manual of the New World Genera of the Family Braconidae (Hymenoptera)*. International Society of Hymenopterists, Washington, DC.

Whitfield, J.B. (1997) Molecular and morphological data suggest a common origin for the polydnaviruses among braconid wasps. *Naturwissenschaften* 34, 502–507.

Whitfield, J.B. (2002) Estimating the age of the polydnavirus/braconid wasp symbiosis. *Proceedings of the National Academy of Science, USA* 99, 7508–7513.

Whitfield, J.B. and O'Connor, J.M. (2012) Molecular systematics of wasp and polydnavirus genomes and their coevolution. In: Beckage, N.E. and Drezen, J.-M. (eds) *Parasitoid Viruses: Symbionts and Pathogens*. Elsevier, London, pp. 89–97.

Whitfield, J.B. and Wagner, D.L. (1991) Annotated key to the genera of Braconidae (Hymenoptera) attacking leafmining Lepidoptera in the Holarctic Region. *Journal of Natural History* 25, 733–754.

Whitfield, J.B. and Wharton, R.A. (1997) Hormiinae. In: Wharton, R.A., Marsh P.M. and Sharkey, M.J. (eds) *Manual of the New World Genera of the Family Braconidae (Hymenoptera)*. International Society of Hymenopterists, Washington, DC, pp. 285–301.

Whitfield, J.B., Austin, A.D. and Fernandez-Triana, J.L. (2018) Systematics, biology, and evolution of microgastrine parasitoid wasps. *Annual Review of Entomology* 63, 389–406.

Wu, Q., Achterberg, C. van, Tan, J.L. and Chen, X.X. (2016) Review of the East Palaearctic and North Oriental *Psyttalia* Walker, with the description of three new species (Hymenoptera, Braconidae, Opiinae). *ZooKeys* 629, 103–151.

Yaakop, S., van Achterberg, C. and Ghani, I.B.A. (2009) *Heratemis* Walker (Hymenoptera: Braconidae: Alysiinae: Alysiini): revision and reconstruction of the phylogeny combining molecular data and morphology. *Tijdschrift voor Entomologie* 152, 3–64.

Yao, J., Achterberg, C.V., Sharkey, M.J., Chapman, E.G. and Chen, J. (2018a) Two species and a genus new for Thailand, with description of a new species of *Neurolarthra* Fischer (Hymenoptera: Braconidae: Alysiinae). *Zootaxa* 4438, 551–560.

Yao, J., van Achterberg, C., Sharkey, M.J. and Chen, J. (2018b) *Separatatus* Chen Wu (Hymenoptera: Braconidae: Alysiinae) newly recorded from Thailand, with description of one new species and one new combination. *Zootaxa* 4433, 187–194.

Yao, J., van Achterberg, C., Sharkey, M.J., Chapman, E.G. and Chen, J. (2019a) *Senwot* Wharton, a new genus for the Oriental region with description of two new species from Thailand (Hymenoptera: Braconidae: Alysiinae). *Journal of Asia-Pacific Entomology* 22, 103–109.

Yao, J., van Achterberg, C., Sharkey, M.J., Chapman, E.G. and Chen, J. (2019b) Five species and a genus new for Thailand, with description of five new species of *Hylcalosia* Fischer (Hymenoptera: Braconidae: Alysiinae). *Insect Systematics and Evolution*. doi.org/10.1163/1876312X-00002303

Yao, J., van Achterberg, C., Yaakop, S., Long, K.D., Sharkey, M.J. and Chapman, E.G. (2022a) A new genus *Anamalysia* van Achterberg (Hymenoptera, Braconidae, Alysiinae), six new species, and two new combinations from India, Indonesia, Malaysia, Singapore, Thailand, and Vietnam. *ZooKeys* 1126, 131–154.

Yao, J., van Achterberg, C., Sharkey, M.J., Chapman, E.G., Fang, S. Aizezi, A. and Li, J. (2022b) *Phaenospila* gen. nov. (Hymenoptera: Braconidae: Alysiinae) and three new species from Thailand. *Zootaxa* 5195, 468–484.

Yu, D.S., van Achterberg, C. and Horstmann, K. (2005) *World Ichneumonoidea 2004. Taxonomy, biology, morphology and distribution*. CD/DVD. Taxapad, Vancouver.

Yu, D.S.K., van Achterberg, C. and Horstmann, K. (2012) *World Ichneumonoidea 2011. Taxonomy, Biology, Morphology and Distribution*. Taxapad, Interactive Catalogue database on flash-drive, Ottawa, Canada. www.taxapad.com

Yu, D.S.K., van Achterberg, C. and Horstmann K. (2016) *Taxapad 2016, Ichneumonoidea 2015*. Taxapad database on flash-drive, Ottawa, Canada.

Zaldivar-Riverón, A., Mori, M. and Quicke, D.L.J. (2006) Systematics of the cyclostome subfamilies of braconid parasitic wasps (Hymenoptera: Ichneumonoidea): a simultaneous molecular and morphological Bayesian approach. *Molecular Phylogenetics and Evolution* 38, 130–145.

Zaldivar-Riverón, A., Belokobylskij, S. A., León-Regagnon, V., Briceno, R. and Quicke, D.L.J. (2008) Molecular phylogeny and historical biogeography of the cosmopolitan parasitic wasp subfamily Doryctinae (Hymenoptera: Braconidae). *Invertebrate Systematics* 22, 345–363.

Zaldivar-Riverón, A., Shaw, M.R., Saez, A.G., Mori, M., Belokobylskij, S.A., Shaw, S.R. and Quicke, D.L.J. (2009) Evolution of the parasitic wasp subfamily Rogadinae (Braconidae): phylogeny and evolution of lepidopteran host ranges and mummy characteristics. *BMC Evolutionary Biology* 8, 329.

Zettel, H. (1990) Eine Revision der Gattungen der Cheloninae (Hymenoptera, Braconidae) mit Beschreibungen neuer Gattungen und Arten. *Annales Naturhistorisches Museum Wien* 91, 147–196.

Zhang, R.-N., van Achterberg, C., Tian, X.-X. and Tan, J.-L. (2020) Review of the *Bobekia*-group (Braconidae, Alysiinae, Alysiini), with description of a new genus and a new subgenus. *ZooKeys* 926, 25–51.

Zhang, Y., Xiong, Z. C., van Achterberg, K. and Li, T. (2016) A key to the East Palaearctic and Oriental species of the genus *Rhysipolis* Foerster, and the first host records of *Rhysipolis longicaudatus* Belokobylskij (Hymenoptera: Braconidae: Rhysipolinae). *Biodiversity Data Journal* 4.

Zheng, M.-L., Chen, J.-H. and van Achterberg, C. (2017) First report of the genus *Coeliniaspis* Fischer (Hymenoptera, Braconidae, Alysiinae) from China and Russia. *Journal of Hymenoptera Research* 57, 135–142.

Zhu, J.-C., van Achterberg, C. and Chen, X.-X. (2017) An illustrated key to the genera and subgenera of the Alysiini (Hymenoptera, Braconidae, Alysiinae), with three genera new for China. *ZooKeys* 722, 37–79.

Žikíc, V., Lazarevíc, M. and Milŏsevíc, D. (2017) Host range patterning of parasitoid wasps Aphidiinae (Hymenoptera: Braconidae). *Zoologischer Anzeiger* 268, 75–83.

10

Ichneumonidae – Darwin Wasps

Abstract

The abundant and diverse family Ichneumonidae is described and various specific morphological terms used in relation to the family are illustrated. A simplified identification key to most of the subfamilies occurring in the Region is presented. World and regional generic and specific diversities of subfamilies are tabulated. Each major group and subfamily are then discussed individually, including what is known of their biology. Representative species and identification characters are illustrated photographically. References are provided to appropriate identification works and to recent taxonomic description relevant to S.E. Asia.

This is a huge family of moderately small to very large species, and very common. Worldwide there are some 25,000 described species classified into 37 subfamilies and a total of more than 1450 genera. Their species richness and biologies are summarised in Table 10.1. Although there are of course a very large number of tropical species, they are arguably a group that does not increase in species diversity towards the equator as much as many other insect groups (see Chapter 4, *Anomalous Diversity*, and Quicke, 2012, for discussion).

To the beginner this is probably one of the most daunting groups of parasitoids. Relative to many of the 'microhymenopteran' groups, their taxonomy is actually quite well-known, especially for the larger-bodied species. However, it is widely appreciated that, compared with say the Braconidae, recognition of them to subfamily using keys is challenging. Although males of some subfamilies are distinctive, we strongly urge beginners to stick with only trying to identify females (Gauld, 1984) which generally have clearer characters. Also, most keys rely on ovipositor features at some point.

In order to promote study of and taxonomic work on the family, a meeting was held in Basel (Switzerland) in 2019, and the attendees decided that a common name for the ichneumonids might be helpful (Klopfstein et al., 2019a). Thus many papers on the group now include that name 'Darwin wasp' in their titles, as we do here. This refers to Charles Darwin's comment in a letter to the famous American botanist, Asa Gray: '*I cannot persuade myself that a beneficent and omnipotent God would have designedly created the Ichneumonidae with the express intention of their feeding within the living bodies of caterpillars …*' (Darwin, 1860). We actually suspect that Darwin was probably also thinking of braconid parasitoids.

Table 10.1. Summary of currently recognised subfamilies of Ichneumonidae recorded from S.E. Asia, their described generic and species richness, biology, and host (prey) groups. (Source: data updated from various sources, notably Yu *et al.*, 2016 and Broad *et al.*, 2018, and more recent descriptions of S.E. Asian species.)

Subfamily	World genera	World species	S.E. Asia genera	Described S.E. Asia species	Life history strategy	Host order(s)
Acaenitinae	29	281	3	33	Koinobiont	C, (L*)
Agriotypinae	1	16	1	2	Idiobiont	T
Anomaloninae	45	750	11	44	Koinobiont	C, L
Ateleutinae	2	46	1	1	Idiobiont	L
Banchinae	64	1760	16	80	Koinobiont	L, C
Brachycyrtinae	1	12	1	2	?	N
Campopleginae	66	2100	18	95	Koinobiont	L, [C]
Cremastinae	35	790	5	27	Koinobiont	L
Cryptinae	275	3100	87	380	Idiobiont, occasionally kleptoparasitoid predator, egg predator	L, C, H
Ctenopelmatinae	107	1350	3	5; several more present but no published records	Koinobiont	H
Cylloceriinae	4	38	1	1	Presumed koinobiont endoparasitism	D
Diplazontinae	22	355	8	14		D
Eucerotinae	1	41	1	present but no published records	Koinobiont hyperparasitoid	
Ichneumoninae	440	4100	160	512	Idiobionts although some technically koinobionts	L
Labeninae	12	150	1	6		C, H
Lycorininae	1	34	1	1	Koinobiont	L
Mesochorinae	28	921	3	49	Koinobiont hyperparasitoid	L via H or D
Metopiinae	27	840	7	10	Koinobiont	L
Microleptinae	1	14	1	1	Endoparasitoid but otherwise uncertain	D
Nesomesochorinae	3	1	48	8	Unknown, presumably koinobiont	L
Ophioninae	28	>1000	7	246	Koinobiont	L
Orthocentrinae	30	470	2	3	Koinobiont	D
Oxytorinae	1	23	1	2	Unknown	unknown
Phygadeuontinae	59	392	25	49	Idiobiont (incl. as pseudo-hyperparasitoid), egg predator,	L, C, D
Pimplinae	79	33	1700	378	Idiobiont, koinobiont on spiders, egg predator	A, L, C
Poemeniinae	10	80	4	22	Idiobiont	C, H
Rhyssinae	8	250	6	70	Idiobiont	C, H
Sisyrostolinae	6	46	1	5	Presumed koinobiont	C
Stilbopinae	3	27	1	2	Koinobiont	L
Tersilochinae	24	450	6	18	Koinobiont	C

Continued

Table 10.1. Continued.

Subfamily	World genera	World species	S.E. Asia genera	Described S.E. Asia species	Life history strategy	Host order(s)
Tryphoninae	54	1250	8	42	Koinobiont ectoparasitoid	H, L
Xoridinae	4	210	3	29	Idiobiont	C, (H)

Abbreviations: A, Arachnida; C, Coleoptera; D, Diptera; H, Hymenoptera; K, kleptoparasitoids; L, Lepidoptera; N, Neuroptera; T, Trichoptera. Host orders in [] are uncommon associations; * indicates very uncertain records but possible.

Subfamily identification keys that should work for most females found in the region started with Townes et al. (1961) with a key specifically for those of the Indo-Australia Region. This is of course rather out of date as several subfamilies have been split since then, but if the reader is aware of the changes it could be quite useful.

Baltazar (1964) created a key to 215 ichneumonid genera known at that time from the Philippines and, although out of date, it may be of some help for difficult groups. Still the most general and important identification works are those of Henry Townes (1969, 1970a, b, 1971). Wahl's web-based 'Key to the Subfamilies of North and Central American Ichneumonidae' (with ten downloadable .pdf files of the sub-keys), and his key in *Hymenoptera of the World* (Wahl, 1993a) are both very good, though they do not include recent recognition of Phygadeuontinae and Ateleutinae as separate from Cryptinae. Based on the British and to a lesser degree the West Palaearctic fauna, the key in Broad et al. (2018) is also very useful but it does not include several subfamilies from outside Europe, and as regards the S.E. Asian fauna the Nesomesochorinae, Labeninae and Sisyrostolinae are missing. Therefore, we recommend using both, if the specimen at hand keys to Cryptinae in Wahl.

The formal phylogenetic history of the family really started with combined morphological and molecular analyses of Quicke et al. (2009). Certainly, more famous ichneumonid experts had earlier constructed and analysed morphological matrices, but never published the results, probably largely because they suggested relationships that they felt were simply wrong, based upon their long study of the group. This is not surprising, because the family is rife with morphological homoplasy (Gauld and Mound, 1982).

Quicke et al. (2000, 2009) extended the idea of informal groupings of subfamilies and recognised three large groups called the pimpliformes, ophioniformes and ichneumoniformes, which had been proposed previously, and added three other smaller ones: labeniformes, orthopelmatiformes and xoridiformes.

Beginners will often get confused because some of the most important identification literature by the world-famous ichneumonidologist Henry Townes and co-authors used an idiosyncratic way of naming higher taxononomic groups instead of following the rules and recommendations of the International Commission on Zoological Nomenclature (ICZN). Townes based his choices on which included genus was the one first described in a tribe or subfamily, whereas the ICZN rule is that the family-level group's name should be based on that name which was first used for a family-level group. In many other instances this would not be of great importance, but Henry Townes's work on Ichneumonidae is so fundamental, and his works still widely used, that it is. Table 10.2 gives the correspondences between the ICZN and Townes's systems. In addition, Townes referred to the genus *Pimpla* as *Coccygomimus*, *Ephialtes* as *Pimpla*, and *Apechthis* as *Ephialtes*. Fitton and Gauld (1976) and Fitton et al. (1988) discussed the history of these name choices.

Townes's concept of Orthocentrinae was more restricted than at present, with

Table 10.2. Correspondence between Henry Townes's family group names in Ichneumonidae and the correct names in accordance with ICZN.

Name used by Townes	Name according to ICZN rules	Notes
Anomalinae	Anomaloninae	
Ephialtinae	Pimplinae	Townes (1969) included Diacritinae, Poemeniinae and Rhyssinae as tribes
Ephialtini	Pimplini	
Gelinae	Cryptinae *s.l.*	Includes Ateleutinae and Phygadeuontinae
Joppini	Ichneumonini	
Labiinae	Labeninae	
Microleptinae	Microleptinae	Townes (1969) included the Oxytorinae, Cylloceriinae and *Helictes* group of Orthocentrinae here
Pimplini	Ephiatini	
Porizontinae	Campopleginae	
Scolobatinae	Ctenopelmatinae	

only what is now called the *Orthocentrus* group in his Orthocentrinae and most other genera placed in his greatly expanded Microleptinae. Nowadays, Microleptinae is restricted to *Microleptes*, but Townes also included in his concept, *Oxytorus* (Oxytorinae), the two genera of Cylloceriinae and the South American genus *Tatogaster*, which is now placed in the Tatogastrinae. Whilst Cylloceriinae are part of the 'pimpliformes' group of subfamilies, *Tatogaster* and *Oxytorus* are both more closely related to various parts of the Ctenopelmatinae, i.e. basal ophioniformes (Bennett *et al.*, 2019). This just serves to highlight how difficult ichneumonid morphological systematics can be.

Several other possible confusions surround the names used for Orthocentrinae. Perkins (1960) termed them Plectiscinae, whereas Gauld (1991) used the names Helictinae and Orthocentrinae, the former for the Microleptinae *sensu* Townes (but excluding Cylloceriinae, Microleptinae, Oxytorinae and Tatogastrinae), and the latter for the *Orthocentrus* group of genera. *Oxytorus*, mentioned above, and the name Oxytorinae was used by van Rossem (1990, and previous) for what Gauld called Helictinae.

Pham and Long (2016) provided a catalogue of the Ichneumonidae known at that time from Vietnam, and Yu *et al.* (2016) catalogued the world species.

Morphological Terminology

Tyloids

The antennal flagellum of males of several ichneumonid subfamilies bears distinctive, raised structures called tyloids (Fig. 10.1), whose form and distribution can be of considerable taxonomic help. These have been shown to be secretory (release-and-spread) structures (Bin *et al.*, 1999; Bordera and Hernández-Rodríguez, 2003) which are rubbed or held against receptor regions of the antennae of conspecific females. The presence or absence of tyloids is a useful subfamily-level character for males, and their arrangement can be important for generic and specific identification. In the Diplazoninae, it has been shown that the distribution of tyloids is on those male flagellomeres that wrap around and contact the female antenna in courtship (Steiner *et al.*, 2010) – the male flagellum may wrap around the female's up to three times. Tyloid morphology of the Ichneumoninae and Cryptinae was investigated by Gokhman and Krutov (1996).

Epomia

The epomia is a vertical or diagonal ridge or carina that partly separates the anterior

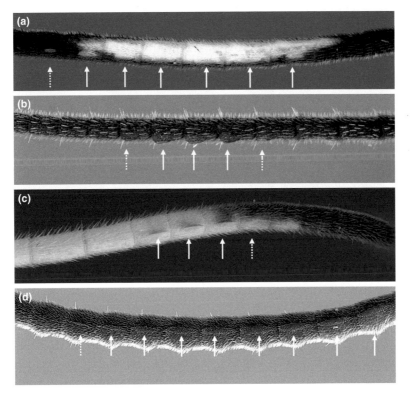

Fig. 10.1. Antennal segments of male ichneumonids with tyloids: **(a)** Pimplinae; **(b, c)** Cryptinae; **(d)** Ichneumoninae. Solid arrows indicate flagellomeres with full-sized tyloids, dashed arrows those with weakly developed ones.

(medial) from the lateral part of the pronotum (Fig. 10.2). Epomiae are present in Anomaloninae, Campopleginae, Cremastinae, Cryptinae, Ichneumoninae, Metopiinae, Mesochorinae, Nesomesochoinae, Orthocentrinae, Pimplinae, Rhyssinae and Xoridinae; absent in Ateleutinae, Diplazontimae, Ophioninae, Tersilochinae, and variable in most other groups. They are also present in many other families.

Epicnemial Carina

Homologous to what is called the prepectal carina in Braconidae, the epicnemial carina, when present, runs just posterior to the anterior margins of the mesopleuron and mesosternum (see Fig. 6.5). Its condition and shape are often used in keys.

Sternaulus

Many ichneumonids and braconids have a more or less longitudinal depression on the side of the mesopleuron/mesosternum. This depression may be short or long, and is often more coarsely sculptured than the surrounding area. These grooves do not, however, appear to be homologous. The groove in ichneumonids is located lower down the side of the mesosoma and is called a sternaulus, whereas the groove in braconids is situated slightly higher and is called the mesopleural sulcus (van Achterberg, 1988). In one opiine braconid genus, *Sternaulopius*, both grooves are present (see Fig. 9.2) which enabled Wharton (2006) to study the internal musculature arrangements of both. He showed that the true sternaulus as found in Ichneumonidae is defined by an internal ridge that

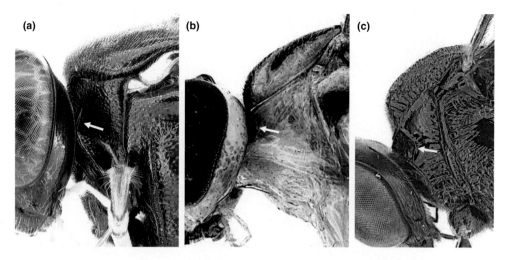

Fig. 10.2. Photographs of the pronotum in various ichneumonids showing epomia (arrows): **(a, b)** Ephialtini species (Pimplinae); **(c)** *Anomalon* sp. (Anomaloninae).

supports the origin of the mesopleural-basalare muscle, and that this is not the case with the normal braconid precoxal sulcus.

Episternal Scrobe

A small pit approximately halfway up and slightly anterior to the mesopleural suture which borders the posterior margin of the mesopleuron (=mesepisternum, =mesepimeron) (a little left of black arrow head, see Fig. 10.40 b). A useful landmark in Pimplinae.

Postpectal Carina

Easiest viewed ventrally, a sort of posterior analogue of the epicnemial (= prepectal) carina, which when present runs close to the posterior margin of the mesosternum just anterior to the middle coxal cavities. It is called the posterior transverse carina of mesosternum by many ichneumonoidologists, and may be complete (e.g. Cremastinae), partially present but interrupted medially or in front of the coxal cavities, or absent (e.g. Banchinae).

Propodeal Carinae

In the Ichneumonidae the presence and absence of various propodeal carinae is of great importance for the recognition of many groups, and these are illustrated in Fig. 10.3. Unfortunately, Henry Townes employed a different terminology to that widely used today and the equivalents are also given in that figure. Where the lateral longitudinal and posterior transverse carinae meet, there is sometimes a tooth/spine-like projection called an apophysis; these are particularly frequent among the Cryptinae, Ichneumoninae and Phygadeuontinae.

Wings

Beginners, and even those with some expertise, will have to get to grips with the terminology of wing veins. We follow Sharkey and Wharton's (1997) interpretation of homologies and nomenclature as throughout this book, and we illustrate that in Fig. 10.4. However, various ichneumonid workers have employed different systems. Because of the importance of many works using these alternative systems, especially those of Henry Townes, we provide terminological comparisons in Tables 10.3 and 10.4.

Areolet

The small fore wing cell arising posteriorly from r-rs (the vein arising posteriorly from the 'middle' of the pterostigma) in ichneumonids,

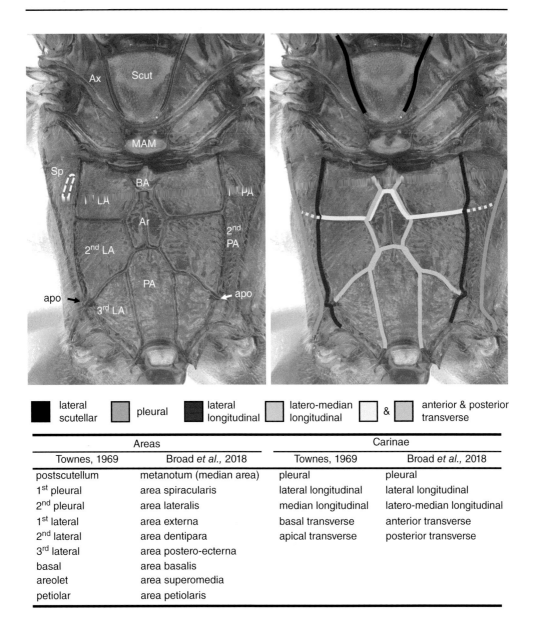

Fig. 10.3. The names of the various carinae and the areas they define.

Areas		Carinae	
Townes, 1969	Broad et al., 2018	Townes, 1969	Broad et al., 2018
postscutellum	metanotum (median area)	pleural	pleural
1st pleural	area spiracularis	lateral longitudinal	lateral longitudinal
2nd pleural	area lateralis	median longitudinal	latero-median longitudinal
1st lateral	area externa	basal transverse	anterior transverse
2nd lateral	area dentipara	apical transverse	posterior transverse
3rd lateral	area postero-ecterna		
basal	area basalis		
areolet	area superomedia		
petiolar	area petiolaris		

when present or indicated, is termed the areolet, and its form is very important to ichneumonid identification. Although many braconids have only a smallish or small rhombic or triangular cell there, many cyclostomes (e.g., Braconinae, Mesostoinae, Rhyssalinae) together with the almost certainly basal Chilean genus Apozyx (see Jasso-Martínez et al., 2022), have a very elongate cell. In these, there has been no debate that the basal-most of the veins forming this cell is part of RS (2RS in present terminology), and therefore the more distal, irrespective of how far distally it is located, is an rs-m cross vein and this is usually referred to as rs-m (or r-m). Because the corresponding

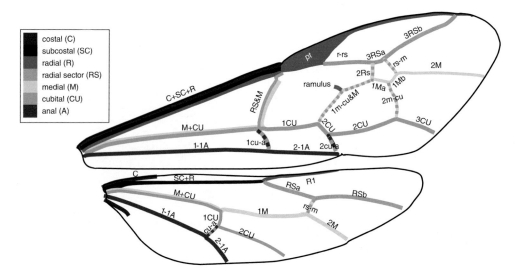

Fig. 10.4. Stylised ichneumonid wing venation using terminology of Sharkey and Wharton (1997).

cell in Ichneumonidae is small its basal-most transverse vein is located more distally in comparison with that of the 2nd submarginal cell of many braconids, there has been considerable debate about whether the veins bordering it are homologous between the two families (Sharkey and Wahl, 1992).

In the interpretation used here, vein 2RS was retained and one rs-m cross vein lost (Fig. 10.4). The alternative view held by many ichneumonidologists (based considerably on the position of the most basal transverse vein) is that either the braconid cell is bordered distally by a homologue of 2rs-m, and that in ichneumonids the transverse veins are 2rs-m and 3rs-m (vein 1rs-m having been lost in all non-xyeleid Hymenoptera, see Vein Naming Systems, Chapter 6, this volume), and that vein SR1 has also lost (see Fitton et al., 1988; Broad et al., 2018). This is clearly not a parsimonious explanation. Here we have adopted the system that the basal areolet cross vein is 2RS because that only involves the evolutionary loss of one cross vein. The alternative interpretations are shown in Fig. 10.5.

It seems relevant also, that the extinct Tanychorinae, which everyone seems happy to include as a basal Ichneumonidae from the Cretaceous period approximately 125 mya, has a relatively huge areolet compared with those of all extant members of the family. In Tanychora, the configuration of its retained fore wing vein (RS+M)a and the basal vein of the 'areolet' strongly suggest that the latter is 2RS (see Fig. 13.2 in Quicke, 2015).

Wing bullae

Hymenoptera wings are not static flat plates but fold and flex dynamically during flight. This is possible because the wings have flexion lines, and where they pass across a wing vein the latter is especially weakened, non-tubular, for a short length. These breaks are called bulli or fenestrae. In the Ichneumonidae, a particularly important character is whether the flexion line that passes through fore wing vein 2m-cu divides into two wing membrane folds before reaching 2m-cu or after if it splits basally, there are two bulli (Fig. 10.6 b). If it splits after (distally), there

Table 10.3. Fore wing vein nomenclature equivalents in various systems.

Townes (1969)	Broad et al. (2018)	Huber and Sharkey (1993); Quicke (2015) and current work
radius (1st abscissa)	2r&RS	r-rs
radius (2nd abscissa, if areolet pentagonal)	RS	3RSa
radius (3rd abscissa)	RS	3RSb
1st recurrent	1m-cu&M (posterior part)	M+1m-cu
cubitus	1m-cu&M (anterior part)	M+1m-cu
medius	M₁CU	M₁CU
discoideus (1st abscissa)	CU (1st abscissa)	1CU
discoideus (2nd abscissa)	CU (2nd abscissa)	2CU
discoideus (3rd abscissa)	2cu-a	2cu-a
2nd recurrent	2m-cu	2m-cu
submedius	AA	1-1A
brachius	AA	2-1A
1st intercubitus	2rs-m	2RS
2nd intercubitus (transverse part)	3rs-m	rs-m
2nd intercubitus (distal longitudinal part)	M	2M

Table 10.4. Hind wing vein nomenclature equivalents in various systems.

Townes (1969)	Broad et al. (2018)	Bennet et al. (2019)	Sharkey and Wharton (1997); Quicke (2015) and current work
subcostellar	SC+R	R, R1	
radiella	RS	Rs	RSa, RSb
intercubitella	rs-m	1rs-m	rs-m
mediella	M+CU	M+CU	M+CU
cubitella	M	M	1M, 2M
1st abscissa of nervellus	CU	1/Cu	CU1a
discoidella	CU	2/Cu	CU1b
2nd abscissa of nervellus	cu-a	cu-a	cu-a
submediella followed by brachiella	AA	1A	1-A, 2-A

is a single bulla in the vein (Fig. 10.6 a). This is a very important character in ichneumonid subfamily keys.

Glymmae

The side of the 1st metasomal tergite in ichneumonids may have a distinct pit or longitudinal groove or a pit in a groove; this is called a glymma. Its position, depth and shape are frequently used in keys. They are present, for example, in Cremastinae, Metopiinae and Mesochorinae.

Gastrocoeli and Thyridea

In Ichneumoninae in particular, there is a pair of fairly coarsely sculptured depressions antero-laterally on the 2nd metasomal tergite just posterior the anterior margin of the tergite (Fig. 10.7). Slightly posterior to these is a pair of small, more finely sculptured, very weakly depressed zones called thyridea.

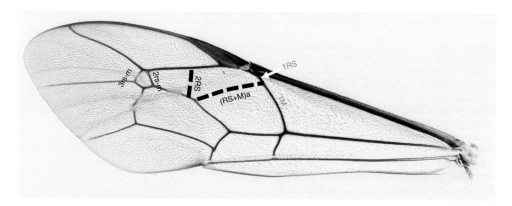

Fig. 10.5. Explanation of the wing vein names used by many workers, in which it is assumed that the areolet is bordered basally and distally by two rs-m veins and that 2RS has been lost.

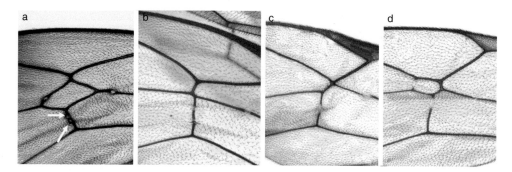

Fig. 10.6. Two states of the bullae in the left fore wing vein 2m-cu: (a, b) two bulli (arrows in (a)) because flexion line bifurcates distal to the vein (Metopius and Acaenitinae, respectively); (c, d) single bulla because flexion line bifurcates basal to the vein (a perilissine Ctenopelmatinae and Ateleute, respectively).

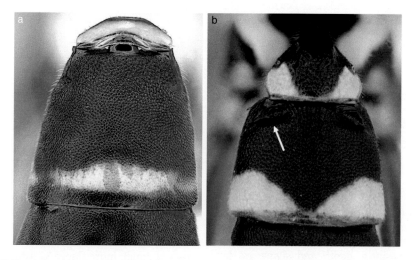

Fig. 10.7. Examples of 2nd metasomal tergite without (a) and with well-developed (b, arrow) thyridium–gastrocoelus complex. The finely sculptured oval area next to the arrow head in (b) is the thyridium, immediately anterior to it is (in this case crenulate) depression.

Highly Simplified Key to Females (and a Few Males) of Relatively Common or Distinctive S.E. Asian Subfamilies

N.B. Does not include Labeninae, Microleptinae, Sisyrostolinae, Stilbopinae or *Phrudus* group of Tersilochinae.

1	Wings totally absent (Fig. 10.17 a).................................... some **Phygadeuontinae**	
-	Wings present... 2	
2(1)	Flagellum extremely strongly expanded and flattened medially (Fig. 10.11 b); metasoma sessile (somewhat uncommon)... male **Eucerotinae**	
-	Flagellum usually unmodified, occasionally slightly and swollen flattened medially; metasoma variable 3	
3(2)	Parameres of male genitalia needle-like, long and thin (Fig. 10.25 f, g) .... male **Mesochorinae** (and a very few Cryptinae)	
-	Parameres not needle-like... 4	
4(3)	Fore wing with glabrous area basal to pterostigma, almost always with this containing one or two brown thickened scleromes (Fig. 10.28 d); usually with ophionoid facies (pale brown, long antennae, very large ocelli) (common, especially at light at night); with a spurious vein running close to but separated from postero-distal part of fore wing (Fig. 10.28 d, arrow) .... most **Ophioninae**	
-	Fore wing without glabrous area basal to pterostigma and without any scleromes...... 5	
5(4)	Fore wing with only one transverse vein between RS and M, i.e. areolet open **AND** the transverse vein intercepting M markedly distal to 2m-cu, not in-line with it (relatively uncommon, a few species key here); general habitus variable 6	
-	Fore wing with two transverse veins between RS and M **OR** with a single vein originating from RS but which then branches so that two parts of it intercept M, i.e. areolet closed **OR IF** with single cross vein then this intercepts M in-line with or basal to 2m-cu (only very occasionally, slightly distal to it) 8	
6(5)	Fore wing with well pigmented, brown spurious vein running close to and parallel to the postero-distal wing margin beyond 2cu-a (common, especially at light at night); (a few uncommon spp.) (Fig. 10.28 d, arrow)............................**Ophioninae**	
-	Fore wing without and pigmented spurious vein close to distal wing margin beyond 2cu-a.. 7	
7(6)	Propodeum uniformly areolate-rugose without any carinae (Fig. 10.18 d); hypopygium normal, short, apically truncate (Fig. 10.18 b); (moderately common).............. ....some **Anomaloninae**	
-	Propodeum generally smooth with distinct carinae; hypopygium very large, elongate, extending beyond apex of tergites, acutely pointed (Fig. 10.34 e)............................ ....some **Acaenitinae**	
8(5)	Mesoscutum with strong transverse ridges (Fig. 10.47 b); large wasps, females with long ovipositors and body length often greater than 15 mm (Fig. 10.46), ovipositor always longer than body (often found searching for hosts on dead or dying trees or logs)..**Rhyssinae**	
-	Mesoscutum smooth, punctate, or otherwise sculptured but never with strong transverse carination 9	
9(8)	Temples with distinct zone of scabriculous or dentate sculpture (Fig. 10.45 b); prepectal (=epicnemial) carina absent; (uncommon; often found searching for hosts on dead or dying trees or logs).. **Poemeniinae**	
-	Temple sculpture essentially the same as rest of top and side of head; prepectal carina usually at least partly developed 10	

10(9)	Claws large, with a modified, apically at least slightly spatulate bristle that touches the tip of the claw (Fig. 10.40 f–h); spiracles of 1st metasomal tergite at or before middle of tergum; body length usually > 1 cm) (common)............some **Pimplinae**
-	Claws without a modified, apically spatulate seta covering the tip; body length variable.. 11
11(10)	Female antenna strongly curved or acutely angled before apex (Fig. 10.48 a, c) **AND** with 1 to a few specialised, thickened setae projecting ventrally from the angulation; (rather uncommon, generally found searching for hosts on dead or dying trees or logs without bark)..**Xoridinae**
-	Antenna **NEVER** with group of specialised, thickened peg-like setae with blunt tips preapically.. 12
12(11)	Face with a large shield-shaped area demarked laterally by carina (Fig. 10.26 a), extending from mouth to antennal sockets (and projecting between them as a point); mid tibia with 1 spur; (large robust wasps, body length > 15 mm) ....**Metopiinae** (*Metopius*)
-	Face and clypeus may be confluent, bulging and projected as a point between antennal sockets but without delineated shield-shaped area; mid tibia usually with two spurs... 13
13(12)	Claws with large pointed basal lobe (Fig. 10.40 i); (common)........some **Pimplinae**
-	Claws with rounded basal lobe, sometimes with pecten (comb), rarely bifurcate 14
14(13)	With egg (or eggs) suspended from ovipositor, attached by anchor which will be hidden inside the egg-canal (Fig. 10.32 d, e) ....some **Tryphoninae**
-	Without egg (or eggs) suspended from ovipositor... 15
15(14)	Spiracle of 1st metasomal tergite clearly posterior to the middle of the segment (usually in the posterior third) (Figs 10.10 a, b, 10.20 e) 16
-	Spiracle of 1st metasomal tergite more or less at the middle or clearly anterior to the middle of the segment (Figs 10.19 d, 10.26 g) 24
16(15)	Mesosoma short, anteriorly hunched, almost round in profile (Fig. 10.10 a, b) **AND** fore wing vein 1cu-a postfurcal by more than its own length; (uncommon) ....**Brachycyrtinae**
-	Mesosoma almost always markedly longer than high; fore wing vein 1cu-a interstitial or only marginally postfurcal 17
17(16)	Fore wing with only one transverse vein between RS and M, i.e., areolet open **AND** veins around the remaining vein and 1M thickened (Fig. 10.31 d); clypeal margin with fringe of strong setae; hind wing vein M+CU largely unpigmented (spectral) and strongly arched (Fig. 10.31 e) ....**Tersilochinae**
-	Fore wing areolet veins not especially thickened, other characters different......... 18
18(17)	Propodeum uniformly with reticulate or areolate sculpture and without any of the standard pattern of carinae **AND** fore wing with only one transverse vein between RS and M, i.e. areolet open; propodeum often produced posteriorly into a neck (Fig. 10.18 a, b, d) ....some **Anomaloninae**
-	Propodeum with at least some of the standard pattern of carinae (see Fig. 10.3); areolet present or absent 19
19(18)	Areolet large and rhombic with sides of more or less equal length (Fig. 10.25 b, d) **AND** 1st metasomal tergite with deep glymmae that are separated medially by only a thin, transparent piece of cuticle **AND** face and clypeus forming a weakly convex surface (Fig. 10.25 a, b); ovipositor without a pre-apical dorsal notch or protruding nodus; ovipositor sheaths not flexible (Fig. 10.25 a, b); claws pectinate (Fig. 10.25 e); (very common) ....some **Mesochorinae**

–	Areolet not large and rhonbic **OR** if intermediate **THEN** other characters differ 20
20(19)	Eyes strongly converging ventrally, almost touching (Fig. 10.27 b); face and clypeus separated by distinct groove, the clypeus bulging in lateral view (Fig. 10.27 b); labial palp very short, with three segments (Fig. 10.27 d) but this can be hard so see; (very uncommon, small, 3–8 mm long) .... **Nesomesochorinae**
–	Eyes not strongly converging ventrally, well separated above mandibles; clypeus variable; labial palp much longer, with four segments (common) 21
21(20)	Ovipositor with pre-apical dorsal notch; (Figs 10.21 c, e, 10.27 e) claws usually pectinate (see Fig. 10.25 e); fore wing vein m-cu with only 1 bulla; areolet absent **OR** if present petiolate, rarely quadrate not pentagonal; metasoma often laterally compressed (Figs 10.20 c-e, 10.21 a, e, 10.29) 22
–	Ovipositor without pre-apical dorsal notch, sometimes with protruding nodus; claws without pecten; fore wing vein m-cu with 1or 2 bullae; areolet present, usually pentagonal **OR** if fore wing vein rs-m absent, pentagonal shape of areolet still apparent by shape of veins; metasoma dorsoventrally compressed (Figs 10.15, 10.16, 10.17) .... **Ichneumoniformes** (see p. 165)
22(21)	Maxillary palp very long, reaching well beyound fore coxa, reaching or exceeding mid coxa (Fig. 10.29); (very rare, far north only) .... **Oxytorinae**
–	Maxillary palp much shorter, only reaching or exceeding mid coxa 23
23(22)	Face and clypeus not, or hardly, differentiated by a groove, face densely short silvery (or golden) setose (Fig. 10.20 a, b); hind tibial spurs arising from same membranous area as hind basitarsus; (exceedingly common) .... **Campopleginae**
–	Face and clypeus clearly separate by a groove, face usually less conspicuously setose (Fig. 10.21 b, e); hind tibial spurs arising from separate membranous sockets and separated from where the basitarsus inserts by zone of thick cuticle (Fig. 10.21 d), consequently origin of spurs distinctly more basal (Fig. 10.21 f) relative to alternative option; (occasional) .... **Cremastinae**
24(15)	Apex of fore tibia with a tooth (Fig. 10.22 b), although this is sometimes difficult to see and may require the basitarsus to be manipulated out of the way and viewing from the end (Fig. 10.22 c); mandibles large, hardly narrowing distally, lower tooth usually at least as long as upper tooth; (uncommon, mostly high altitudes) .... most **Ctenopelmatinae**
–	Apex of fore tibia without a tooth-like process 25
25(24)	Ventral membrane of metasoma with numerous conspicuous short dark setae arising from small dark circles (Fig. 10.37 c, d); scapus sometimes cylindrical approximately 2 × longer than wide (Fig. 10.37 a) .... most **Orthocentrinae**
–	Ventral membrane of metasoma without conspicuous short dark setae arising from dark circles; scapus considerably less than 2 × longer than wide 26
26(25)	Metasomal tergites 2–4 with triangular areas defined by posteriorly diverging grooves arising near the mid-anterior margin (Figs 10.19 e, 10.24 b) 27
–	Metasomal tergites 2-4 without triangular areas defined by posteriorly diverging grooves 28
27(26)	Triangular areas on tergites 2–4 bordered posteriorly by a subposterior groove (Fig. 10.24 b); (uncommon) .... **Lycorininae**
–	Triangular areas on tergites 2–4 not bordered posteriorly by a groove (Fig. 10.19e); (common) .... **Banchinae: Glyptini**

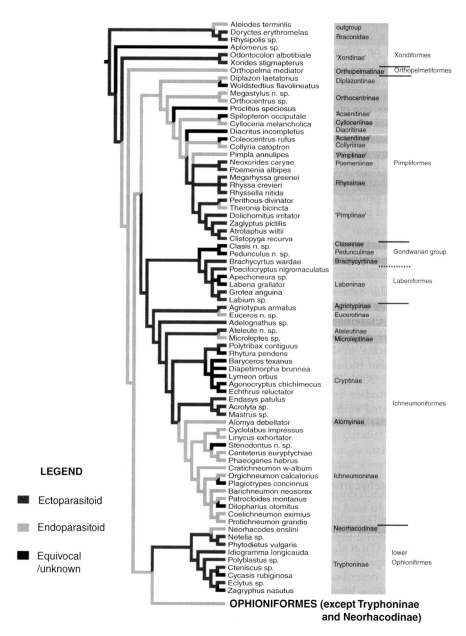

Fig. 10.8. Phylogenetic relationships of ichneumonid wasps (part 1). The topology was inferred from total evidence analysis by Bennett *et al.* (2019). Branches are colour coded according to mode of parasitism. (Continued in Fig. 10.9.)

he investigated laid about 3000 eggs each over a two- to three-week adult lifespan. The female *Euceros* lays several hundred of her small stalked eggs per day on vegetation and these darken and hatch after a few days; the planidium usually remains on the egg pedicel awaiting a caterpillar or sawfly larva to come near and it attaches to this. What happens next depends on whether or not the secondary host (caterpillar) has already been

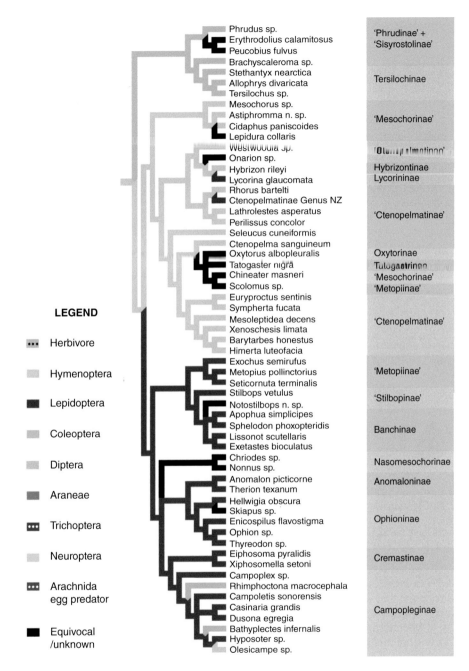

Fig. 10.9. Phylogenetic relationships of ichneumonid wasps (part 2). The topology was inferred from total evidence analysis by Bennett et al. (2019). Branches are colour coded according to host group; all are endoparasitoids. (Continued from Fig. 10.8.)

parasitised by a primary parasitoid, which will become the primary host of the *Euceros*. If so, the *Euceros* planidium burrows into the caterpillar and then for a short period into its primary parasitoid host, then exiting from it to feed externally on it.

Fig. 10.10. *Brachycyrtus* sp., female from Thailand: **(a)** habitus, note the slightly clubbed antennae; **(b)** detail of mesosoma, note how nearly round in profile it is and the strong anterior overhang; **(c)** fore wing detail, note the long abscissa of the medial vein bordering the areolet basally.

Males of *Euceros* are very easy to recognise, having the middle part of the flagellum extremely swollen and flattened (Fig. 10.11 a, b) whereas the equivalent part in females is normal and filiform (Fig. 10.11 c). The females have the anterior of the pronotum, just behind its transverse groove, modified into a bilobed flange or process (Fig. 10.11 d, e). The ovipositor and sheaths are hardly visible.

Gauld and Wahl (2002) stated that the Eucerotinae:

"... has a strikingly anomalous distribution as it is amphitropical (i.e. present in the north and south temperate areas but almost absent from the tropics). Although isolated eucerotines occur in tropical Australia and New Guinea, and in Madagascar (Barron, 1976, 1978), the group is absent from tropical Asia, tropical America and mainland Africa."

Euceros gilvus, which occurs in Papua New Guinea as well as Australia, is the only described species in the general Region, but we frequently collect specimens in Thailand (Fig. 10.11).

The Ichneumoniformes

The Agriotypinae, Alomyinae, Ateleutinae, Cryptinae, Ichneumoninae, Microleptinae and Phygadeuontinae constitute the ichneumoniformes that are present in S.E. Asia. Adelognathinae are absent from the region. Until recently, Alomyinae were known from China and Japan, but they have now been discovered in Laos (Riedel, 2019). It is currently treated as the most basal tribe of Ichneumoninae, but it is quite aberrant and does not fit comfortably in that group, therefore we treat it here as a separate subfamily.

Separation of Ichneumoninae from traditional 'Cryptinae' is often somewhat of a problem for beginners so we present here a short key and figures (Fig. 10.12).

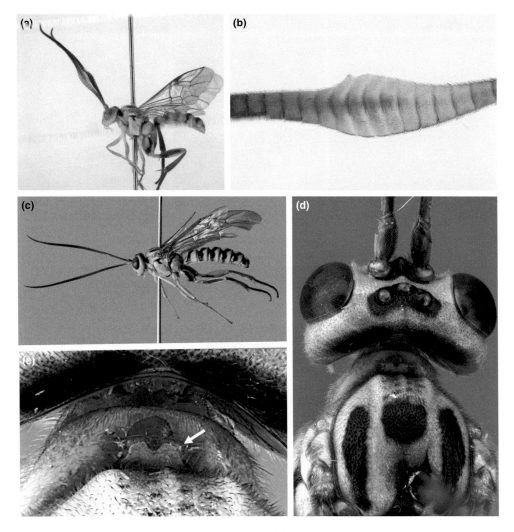

Fig. 10.11. *Euceros* sp. from Thailand: **(a)** male habitus; **(b)** detail of swollen part of male flagellum; **(c)** female habitus; **(d)** dorsal view of female anterior mesosoma and head; **(e)** detail of female anterior pronotum showing bilobed flange (arrow).

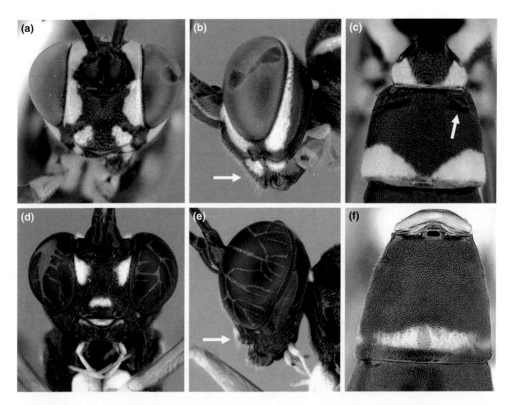

Fig. 10.12. Features of differentiating ichneumonines: **(a–c)** from cryptines+phygadeuontines **(d–f)**. **(a)** showing wide flat clypeus; **(b)** showing flat profile of clypeus (arrow) more or less in line with face; **(c)** showing deeply impressed thyridea (arrow); **(d)** note the shiny, narrow, reflexed lower border of clypeus; **(e)** showing the bulging clypeal profile (arrow); **(f)** showing absence of deep anterolateral impressions of metasomal tergite 2.

Key to Separate Female Ateleutinae, Cryptinae, Ichneumoninae and Phygadeuontinae

1	Ovipositor usually not protruding (Fig. 10.16) **OR** protruding by no more than 0.3 length of metasoma; ovipositor sheaths short, stiff and straight; clypeus wide, almost flat (observe in side view) more or less continuous in profile with face, not protruding and apical margin truncate (Fig. 10.12 a, b); metasomal tergite 2 with impressed thyridea (Fig. 10.7 b, 10.12 c); sternaulus absent or short, reaching only to mid-length of mesopleuron (90%) ....**Ichneumoninae**
-	Ovipositor usually at least 0.3 × as long as metasoma (95%) (Figs 10.14, 10.15 a, b, f); ovipositor sheaths flexible; clypeus usually relatively narrower and rather bulging (observe in side view) (Fig. 10.12 c, d); and with apical margin turned inwards; metasomal tergite 2 without impressed thyridea (Fig. 10.7 a, 10.12 f); sternaulus usually reaching close to posterior margin of mesopleuron.....................................2
2(1)	Fore wing vein 2m-cu with two bulli (Fig. 10.17 d).....................**Phygadeuontinae**
-	Fore wing vein 2m-cu with one bulla (Fig. 10.13 b)..3
3(2)	Areolet wider than high (Fig. 10.13 b) **AND** with vein 2M distinctly shorter than 3M (Fig. 10.13 d); vein 3rs-m absent or spectral (Fig. 10.4 d), never tubular; small wasps, fore wing 2.2–7.5 mm..**Ateleutinae**

-	Areolet wider than high **AND** with vein 2M distinctly shorter than 3M; vein 3rs-m usually present and at least partially tubular (rarely absent); body size variable.......... 4
4(3)	Sternaulus posteriorly ending or fading out pointing above the ventral corner of mesopleuron (Fig. 10.17 b-d); small wasps, fore wing 2.2-7.5 mm .... **Phygadeuontinae**
-	Sternaulus posteriorly ending or fading out pointing below the ventral corner of mesopleuron; size variable..**Cryptinae**

Agriotypinae

This extremely uncommon, principally (north) temperate group are specialised idiobiont ectoparasitoids of caddisfly (Trichoptera) larvae in their cases under water in fast-flowing streams. There is a single genus, *Agriotypus*, with 16 described species (Bennett, 2001). Because of their specialised biology they are seldom encountered by hymenopterists unless they go specifically in search of them.

They display several adaptations to their amphibious life, as larvae, pupae and adults. Indeed, unlike other Ichneumonidae, they have their metasomal sternites equally sclerotized as their tergites, which, among other things, led early workers to place them in their own family, Agriotypidae. As recently as 50 years ago, Mason (1971) still regarded them as a separate family and suggested that they should be classified in the Proctotrupoidea (based on fusion of the first metasomal tergite and sternite and the completely strongly sclerotised remaining metasomal sterna). Despite their relative scarcity in collections, their biology has actually received considerable attention because it is so unusual for the family (Elliot, 1982; Quicke, 2015).

They are easily recognised by the presence of a long, upcurved spine originating posteriorly from the scutellum in combination with the heavily sclerotized metasomal tergites. Beginners sometimes see the scutellar protruberances of the banchine genus *Banchus* or the tryphonine *Atopotropos* and think they may have an *Agriotypus*, but these (very few) other ichneumonids have obvious unevenly and far less sclerotised metasomal sternites, often with pale membranous cuticle visible.

The Agriotypinae have only recently been discovered in S.E. Asia when Bennett (2001) described *Agriotypus chaoi* and *A. masneri*, both from Vietnam. Mason (1971) had earlier described a species from India.

Alomyinae

This small subfamily has long been recognised as quite different from Ichneumoninae (Perkins, 1960; Broad et al., 2018). In molecular studies it has been recovered in various positions in the ichneumoniformes (Laurenne et al., 2006; Quicke et al., 2009). The most recent study using UCE data (Santos et al., 2021) placed it as sister group to remaining Ichneumoninae and consequently the two were synonymised yet again. However, contrary to Wahl and Mason (1995), they are not synonymous with the ichneumonine tribe Phaeogenini, the similarities between the two groups apparently being due in part to symplesiomorphies and in part to convergences. Alomyines differ from other Ichneumoniformes in lacking a differentiated fore trochantellus.

The subfamily comprises two genera, the W. Palaearctic *Alomya*, and the East Palaearctic and S.E. Asian *Megalomya*. Both are parasitoids of ghost moths (Lepidoptera, Hepialidae). Riedel (2019) described four species of *Megalomya* from Laos.

Ateleutinae

Ateleutinae and Phygadeuontinae were raised to subfamily status from within Cryptinae by Santos (2017). With a single S.E. Asian genus, *Ateleute*, this is a rather easily recognised subfamily. The areolet has a rather distinctive shape with 2m-cu intercepting M well before its middle (Fig. 10.13 b). The hind tibia is usually somewhat robust with numerous distinct short spines, and

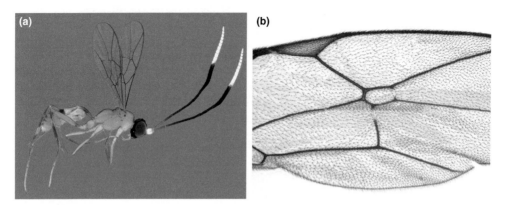

Fig. 10.13. A species of *Ateleute*, probably undescribed, from Thailand: **(a)** habitus, note robust hind tibia; **(b)** distal part of fore wing, note the shape of the areolet.

the spiracles of the first metasomal tergite are situated close to the middle (unlike most Phygadeuontinae and Cryptinae, in which they are far behind the middle).

Ateleutine biology is hardly known; one paper (Momoi *et al.*, 1965) records a Japanese species as an idiobiont ectoparasitoid of a bagworm moth (Psychidae).

We have collected many individuals in Malaise traps operating at a few sites in North and Central Thailand and these comprise at least eight species, all, or nearly all, undescribed.

Cryptinae

This is by far the most abundant and diverse ichneumonid subfamily even after the recent removal to separate subfamilies of the Ateleutinae and Phygadeuontinae (q.v.). It takes a great deal of practice to identify them, even in the temperate region. Worldwide there are 408 genera and more than 4500 species (Santos, 2017).

The only comprehensive, but dated, key to the genera of the whole group is Townes (1970b) in which the three subfamilies were collectively referred to as Gelinae. Townes's work first keyed them to tribes and then provided generic keys for each (Cryptini [as Mesostenini], and Echthrini) and the subtribes. This has advantages and disadvantages. However, the tribes that Townes recognised have been shown to be largely phylogenetically meaningless, based on molecular studies (Quicke *et al.*, 2009; Santos, 2017), although some subtribes are probably largely composed of closely related genera. As commented on by Broad *et al.* (2018), Townes's presentation of keys to (largely non-monophyletic) subtribes and then to genera has made cryptine generic identification probably more difficult than necessary.

Cryptine phylogeny was reassessed by Laurenne *et al.* (2006) based on the D2-D3 expansion region of the nuclear 28S rDNA gene, and by Santos (2017) employing data from seven molecular loci plus 109 morphological characters. The two both obtained broadly similar results. Based on Santos, the Cryptinae comprises two tribes, Aptesini and Cryptini. The Aptesini are largely what Townes referred to as the Echthrini, but the nominal genus *Echthrus* of that tribe actually belongs to the Cryptini and is not close to *Aptesis*. To avoid confusion with the use of names of subtribes which he and Laurenne *et al.* had shown mostly to be polyphyletic, he proposed 11 genus groups (labelled clades A–K) based respectively around *Gabunia*, *Ischnus*, *Xylophrurus*, *Cryptus*, *Mesostenus*, *Ceratomansa*, *Glodianus*, *Lymeon*, *Osphrynchotus*, *Trachysphyrus* and *Agrothereutes*. Groups represented in S.E. Asia are as follows.

- The *Gabunia* group is largely the same as Townes's Gabuniina and the wasps are parasitoids of wood-borers and also of twig-nesting aculeates (Barthélémy and Broad, 2012).

- The *Cryptus* group is a largely Old World group and morphologically heterogeneous. Townes considered *Cryptus* an invalid name, ignoring the ICZN, and used the name *Itamnoplex*, which he put in the Iscnina.
- The *Mesostenus* group is also mainly Old World is very diverse and includes several species-rich genera in the region, e.g. *Friona*, *Goryphus*, *Gotra* and *Isotima*.
- The *Ceratomansa* group is principally Australian but *Lorio* and *Wuda* occur on New Guinea.
- The *Osphrynchotus* group is principally New World and Afrotropical (*Osphrynchotus*) but the largely Palaearctic genus *Acroricnus* occurs in Thailand and Malaysia.
- The *Agrothereutes* group largely corresponds to Townes's Agrothereutina and is principally Holarctic with a few genera extending into S.E. Asia, for example *Agrothereutes* itself, *Thrybius* and *Etha*. However, *Amauromorpha* was recovered in the *Mesostenus* group.

A considerable number of genera have the apical flagellomere modified to form hammers that are used for host location in a way similar to the echolocation of bats and dolphins (see Pimplina, this Chapter below, also *Echolocation* in Chapter 3 and Chapter 8, this volume). Since the echoes are transmitted through a solid substrate, this is technically referred to as vibrational sounding. The apparent phylogenetic distribution of these hammers suggests that this mode of host detection has evolved independently on several occasions (Laurenne *et al.*, 2009). The Gabuniina are generally large wasps that use this method of host location (Quicke *et al.*, 2003) and includes the impressive S.E. Asian genus *Dinocryptus* which reaches up to 25 mm body length (Fig. 10.14). In addition to most of the traditional Gabuniina, the *Gabunia* group recovered by Santos (2017) included *Echthrus* from Townes's nominal Echthrini. Interestingly, this extralimital genus also uses vibrational sounding.

The Sphecophagina of Townes, represented in the region by *Arthula* (Fig. 10.15 d), were recovered as a distinct clade within the *Cryptus* group by Santos (2017). They are atypical in that the female ovipositor is very short, as in Ichneumoninae, making difficulties for diagnosing the subfamily. Based on available host records, the Sphecophagina are idiobiont ectoparasitoids of vespine larvae within the wasps' nests, and their larvae feed on pharate adults (Makino, 1983; Sayama and Konichi, 1996; Berry *et al.*, 1997; Ubaidillah *et al.*, 2009). The biology of the European *Sphecophaga vesparum* has been intensively investigated (Donovan, 1991) and it, and also *S. orientalis*, have been introduced into New Zealand to control the invasive *Vespula germanica* and *V. vulgaris* (Donovan and Read 1987; Donovan *et al.*, 2002).

S. Gupta and Gupta (1983) revised the Gabuniini of most of the region, but excluding southern Indonesia and East Timor. Oh *et al.* (2010) provided a key to the genera of Sphecophagina, but unfortunately the genus *Lentocerus* from China, which they included due to its original placement, is actually a synonym of *Euceros* (Eucerotinae) (Broad, 2021).

The single species of *Amauromorpha* from the region is a parasitoid of rice stem-borers (Gupta, 1999). Yoshida and Matsumoto (2015) revised, nicely illustrated and included life history

Fig. 10.14. Specimen of the largest-bodied cryptine, *Dinocryptus niger* from Thailand.

information for the genus *Chlorocryptus* which includes a single species in the region (Fig. 10.15 e).

Ichneumoninae

In the north temperate region ichneumonines are one of the commonest and most frequently collected groups of ichneumonids. However, whilst numerous and diverse in the tropics, they appear relatively less frequently in sweep and Malaise samples in the tropics. Norhafiza and Idris (2013) recorded only 38 species from Malaysia (and Singapore) compared with 383 species in 95 genera from the British Isles (Broad *et al.*, 2018).

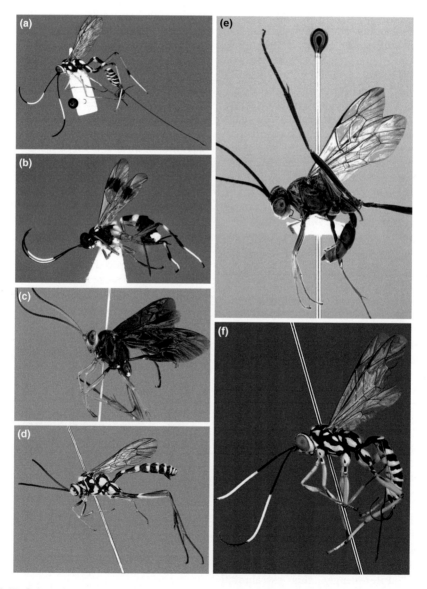

Fig. 10.15. Selected representatives of Cryptinae: **(a)** *Stenarella insidiator*, female; **(b)** cf. *Friona*, female; **(c)** *Mansa* sp., male; **(d)** *Arthula* sp., female, note very short ovipositor in this genus; **(e)** *Chlorocryptus purpuratus*, male; **(f)** *Arhytis* sp., female.

Ichneumonines are closely related to cryptines s.l. and to the extralimital Adelognathinae. Their general habitus is very similar to that of cryptines and distinguishing males can require some experience. Unlike most cryptines, most ichneumonines have a hardly exserted ovipositor, so any member of the Ichneumoniformes with a long (half metasoma length or more) ovipositor will not be an ichneumonine.

Quite a lot of older books and papers refer to two subgroups of ichneumonines, the 'stenopneusticae' and 'eulopneusticae' referring to the shape of the propodeal spiracle – elongate and slit-shaped versus round, respectively. In effect the latter group comprised the Phaeogenini and Platylabini (called Pristicerotini by some), both of which comprise small-bodied taxa. Another distinction can be made between ambipygous ones, i.e. those that have a very short ovipositor and blunt hypopygium and attack exposed caterpillars (koinobionts), compared with ichneumonines with more exserted ovipositors and pointed hypopygiums which attack already pupated hosts (idiobionts) (Hinz, 1983). However, as recent molecular work has shown (Santos et al., 2021), the groupings suggested by this are not monophyletic and parasitism of pupae and associated short ovipositors appear to have evolved on multiple occasions.

Molecular studies have generally failed to recover the larger tribes except for Phaeogenini and Platylabini, which were found to be largely monophyletic. Recently, analyses of UCE data (Santos et al., 2021) supported a revised classification with just seven tribes, viz. Alomyini (treated here as a separated subfamily), Phaeogenini, Notosemini, Eurylabini, Platylabini, Ichneumonini and a new monogeneric tribe, Abzariini. In addition to Abrarini, the Eurylabini and Notosemini are also monogeneric groups, the vast majority of genera (333) belonging to the Ichneumonini and around 40 to each of the Platylabini and Phaeogenini.

Tereshkin (2009) provided an illustrated key to the tribes of Ichneumoninae as well as to the genera of the cosmopolitan Platylabini. He recognised 17 tribes, of which 13 occur in S.E. Asia (Compsophorini, Eurylabini, Geoedartiini, Heresiarchini, Ichneumonini, Ischnojoppini, Joppocryptini, Listrodomini, Oedicephalini, Phaeogenini, Platylabini, Protichneumonini and Trogini). However, as Gauld (1984) noted, the limits of most of these, especially the larger ones, are vague. Sime and Wahl (2002) in a morphological phylogenetic analysis provided evidence that the Heresiarchini was paraphyletic with respect to the Trogini and so synonymised them, but this was not followed by Tereshkin, who did not cite Sime and Wahl's paper.

Sime and Wahl (2002) provided a key to genera of the *Callajoppa* group (formerly Trogini, see above) which includes a number of S.E. Asian genera, viz. *Cobunus, Dimaetha, Facydes, Holcojoppa, Neofacydes, Queequeg* and *Yeppoona* (New Guinea). Riedel (2011) started the first serious work on S.E. Asian ichneumonines with a treatment if the tribes Clypeodromini, Listrodromini, Goedartini, Compsophorini and Platylabini in which he described 15 new species and one new genus, and provided keys to the species of *Clypeodromus* and Oriental species of *Listrodomus*. The new taxa were mostly based on material from Laos and Vietnam. Riedel (2013) described 23 species from S.E. Asia as well as providing a summary of the known species, plus a key to the species of *Heresiarches*; this was followed by Riedel (2017a) in which mainly the fauna of Laos was treated as well as descriptions of six more new species plus the first descriptions of some males. Riedel (2017b) reviewed the S.E. Asian Platylabini, Eurylabini and Oedicephalini. Sheng et al. (2016) described three new species of *Notosemus* (Fig. 10.16 a) from China and provided a key to the world species. Riedel (2023a, b) provides keys to the Oriental species of *Anisobas, Listrodomus* and *Ichneumon*, and Riedel (2023c) describes four new genera and 31 species of Ichneumoninae from Indonesia and Sabah.

Microleptinae

Although the subfamily name has been used more extensively in numerous works, it is nowadays applied in a restricted sense including only the genus *Microleptes*. They have been reported recently from India and

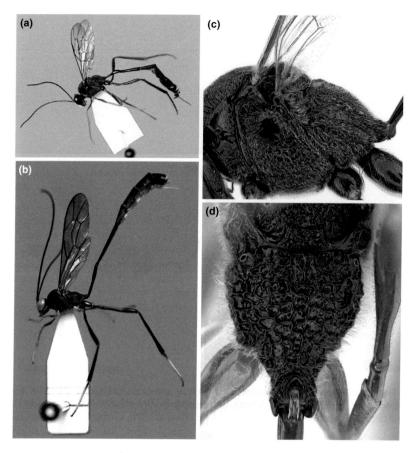

Fig. 10.18. Exemplar Anomalinae from Thailand, and characters: **(a)** *Anomalon* sp. (Anomalonini), habitus; **(b)** *Perisphincter* sp., (Gravenhorstiini), habitus; **(c)** *Anomalon* sp., propodeum; **(d)** *Clatha* sp. propodeum showing uniform foveate-reticulate sculpture and absence of carinae.

least some members of which are koinobiont endoparasitoids of concealed beetle larvae (Tenebrionidae), although the literature also contains unsubstantiated records of lepidopteran hosts. In contrast, all members of the Gravenhorstiini are parasitoids of Lepidoptera, and collectively they attack members of numerous families. However, relatively few biological details are known, and essentially nothing for any S.E. Asian or tropical species in general. What is known was summarised by Broad *et al.* (2018) and includes observations that female anomalonines are only interested in moving host caterpillars and they oviposit very rapidly without any associated temporary host paralysis.

The only available generic-level identification key for S.E. Asia is that of Townes (1971), which is understandably somewhat out of date. Even using genus keys to the European fauna is tricky, and there is much need for an updated revision.

Banchinae

Banchines are classified into four tribes, viz. Atropini (=Lissonotini), Banchini, Glyptini and the rare Townesionini. They are medium- to large-sized ichneumonids and frequently collected. Banchines are now know to have polydnaviruses.

Key to female tribes of Banchinae relevant to S.E. Asia

1		Metasoma with only four visible tergites, all heavily sculptured.......... **Sachtlebenia**
-		Metasoma with six visible, smooth or less strongly sculptured tergites.................. 2
2(1)		Tergites 2–4 with a pair of posteriorly diverging grooves.........................**Glyptini**
-		Tergites without grooves...3
3(2)		Hind wing vein 1-CU (almost always) longer than cu-a (assuming they are distinguished by the presence of at least a trace of 2-CU); ovipositor at least as long as metasoma; areolet sometimes open..**Atrophini**
-		Hind wing vein 1-CU (almost always) much shorter than cu-a (assuming they are distinguished by the presence of at least a trace of 2-CU); ovipositor shorter than half length of metasoma; areolet always closed .... **Banchini (*Exetastes*)**

Atrophini (= Lissonotini)

These are generally the most commonly collected banchine genera, and in the region are frequent in Malaise trap samples. The following genera occur in S.E. Asia: *Alloplasta*, *Amphirhachis*, *Cryptopimpla*, *Glyptopimpla* (treated as synonym of *Telutaea* by Townes), *Himertosoma* (Philippines only), *Leptobatopsis*, *Lissonota*, *Stictolissonota*, *Syzeuctus* (Fig. 10.19 a) and *Tossinola* (Philippines only). However, nearly all the records of described species (Yu et al., 2016) are from Myanmar. Of these, *Leptobatopsis* and *Syzeuctus* are probably the commonest. *Cryptopimpla* is principally a temperate genus with very few tropical records but it has been recorded from mountains of Sulawesi (Takasuka et al., 2011a).

The World species of *Amphirhachis* were revised by Watanabe (2017).

Banchini

Two genera of Banchini are present in S.E. Asia, *Banchus* and *Exetastes* (Fig. 10.19 c). Described species of both genera are only recorded from Myanmar but Fitton (1987) showed a distribution map of *Banchus* that includes Indonesia and the Philippines. Both range from approximately 5 to 15 mm in body length. *Exetastes* are not uncommon in Thai Malaise trap samples. They have a rather large, rhombic areolet. Townes (1970b) noted that they tend to inhabit more open grassy or shrubby areas, and that at least in the Holarctic some species frequently visit flowers.

Banchini have short ovipositors and parasitise exposed or very weakly concealed hosts.

Glyptini

Only two genera have been recorded in the region: *Apophua* (Fig. 10.19 b) and *Teleutaea*. The commonest genus in S.E. Asia is probably *Teleutaea*, which is widespread but most species-rich in the East Palaearctic with a few species in the region. Vas (2017) provided a key to the Oriental (including S.E. Asian) species of this genus. Extralimital species of *Apophua* have been investigated from the point of view of biological control of tortricids.

Townesionini

Kasparyan (1993) erected a new subfamily, the Townesioninae, for his new and highly aberrant East Palaearctic genus *Sachtlebenia* which is closely related and includes a few S.E. Asian species. Gauld and Wahl (2000) synonymised the subfamily with the Banchinae, placing it as a member of tribe Glyptini. In the combined molecular and morphological analyses of Quicke et al. (2009) the genus *Townesion* was instead generally recovered as the sister group to the remaining Banchinae, and therefore we treat it as a separate tribe. The species of *Sachtlebenia* were revised by Broad (2014).

Campopleginae (= Porizontinae)

One of the commonest, most diverse, yet most difficult subfamilies. Recognising a campoplegine is fairly easy: the face and clypeus are not or guardedly differentiated and covered with dense silvery (or yellowish)

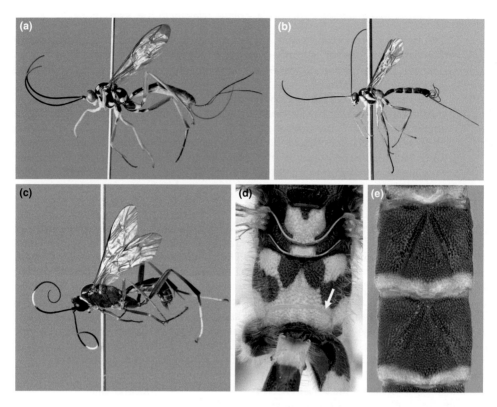

Fig. 10.19. Exemplar Banchinae from Thailand: **(a)** *Syzeuctus* sp (Atrophini); **(b)** *Apophua* sp. (Glyptini); **(c)** *Exetastes* sp. (Banchini); **(d)** *Syzeuctus* sp., posterior mesosoma showing posterior transverse propodeal carina (arrow) and anteriorly located 1st metasomal spiracles; **(e)** *Apophua* sp., metasomal tergites 2 and 3 showing posteriorly diverging oblique grooves characteristic of Glyptini.

setosity (Fig. 10.20 a, b); the 1st metasomal tergite is long and petiolate with spiracles far behind the middle (Fig. 10.20 d, e); if an areolet is present then it is usually petiolate or 2RS and rs-m originate anteriorly touching one another; and the ovipositor has a pre-apical dorsal notch. Identification to genus is not that easy. Townes (1970b) included the Nesomesochorinae (as Nonnini) and *Hellwigia* and *Skiapus*, which are now both included in the Ophioninae (Quicke *et al.*, 2005). The remaining true campoplegines he treated as two tribes, Campoplegini and Porizontini, which he distinguished on the basis of features of the first tergite, but these have now been abandoned, and instead Wahl (1991) recognised a small number of genus groups. However, preliminary molecular analyses have not supported these and the phylogenetics of the subfamily is greatly in need of work.

Campoplegines are koinobiont endoparasitoids, mostly of Lepidoptera caterpillars, although some species are known to use larvae of tenthredinoid symphytans, Raphidioptera, Trichoptera and chrysomelid, curculionid and cerambycid beetles as hosts. Numerous species are economically important, with some being used in biocontrol programmes, and, given their relative abundance, are undoubtedly important regulators of the populations of their hosts in natural ecosystems. Most species are larval parasitoids but a few are egg-larval (Shimazaki *et al.*, 2011). The fully grown campoplegine larva may pupate inside the mummified remains of the host caterpillar, or underneath it, or make a silken thread and spin its cocoon suspended from adjacent vegetation.

A few genera and species have been studied intensively because they are one of the few groups of ichneumonoids that produce polydnaviruses, which play a vital role in protecting the campoplegine egg and larva from the host's immune response, and

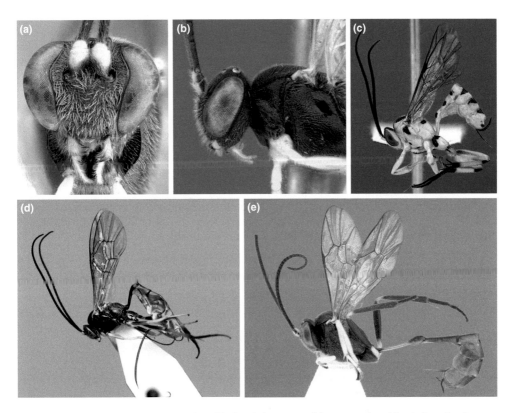

Fig. 10.20. Exemplar Campopleginae from Thailand characters: **(a)** gen. sp., head front view showing dense silvery setosity and hardly any differentiation between face and clypeus; **(b)** head and anterior mesosoma showing facial profile; **(c)** *Xanthocampoplex* sp.; **(d)** *Eriborus* sp.; **(e)** *Charops* sp.

also have regulatory roles on host development. Polydnaviruses are also produced by at least some banchine ichneumonids, and by all members of the microgastroid lineage of Braconidae (q.v.). These appear to have different evolutionary origins, although there are similarities between the two groups of ichnoviruses.

The genus *Diadegma* includes several species that are important in the biological control of the diamondback moth, *Plutella xylostella*, a pest of various Brassicaceae worldwide, and some have been introduced into S.E. Asia (Verkerk and Wright, 1997; Aziz et al., 2000; Talekar, 2004). *Campoletis chlorideae* is an important parasitoid of the chick pea pod borer, *Helicoverpa armigera* (Lepidoptera: Noctuidae) in Myanmar (Yin et al., 2005).

Very little taxonomic work has been carried out on the S.E. Asian fauna. Kusigemati (1990) described three new species (in three genera) from Thailand.

Cremastinae

This is a rather small subfamily with relatively few genera and species in cooler regions. Genera-wise, it is most diverse in the New World and species-rich in the Mediterranean climates and Australia. Although they are seldom common, one or two frequently turn up in Malaise trap samples. Superficially they resemble Campopleginae and Nesomesochorinae with rather long petiolate 1st metasomal tergite. They are recognisable by the presence of a solid cuticle bridge that separates the membranous insertions of the tibial spurs from that of the basitarsus of the hind leg. This may be difficult to see in some specimens, especially if the basitarsus is bent downwards relative to the tibia. This character also means that in lateral view the spur insertions are distinctly more basal than the apex of the tibia. Care needs to be taken by beginners for this but with practice it becomes easier to see whether the bridge is present. Otherwise, they superficially resemble campoplegines and nesomesochorines. In addition to the hind tibial character, cremastines differ from campoplegines in that the clypeus is well differentiated from the face.

The world fauna comprises more than 25 genera, but only four genera are recorded from S.E. Asia: *Mecotes*, *Pristomerus*, *Temelucha* and *Trathala*. *Mecotes* is known only from a single species from Indonesia (Townes, 1971).

Some genera, including all members of the cosmopolitan genus *Pristomerus*, have the apex of the ovipositor formed into a distinct wavy profile (Fig. 10.21 e), which is part of an ovipositor steering mechanism (Quicke, 1991). Many, but not all, *Pristomerus* also have a tooth on the ventral surface of the hind femur (Fig. 10.21 f).

Cremastines are koinobiont endoparasitoids, predominantly of weakly concealed Lepidoptera larvae such as leaf-rollers, buds, galls and some that bore into fruit.

Ctenopelmatinae

Medium-sized to large parasitoids of sawfly larvae that are generally thought to be uncommon in the tropics (Gauld *et al.*, 1997).

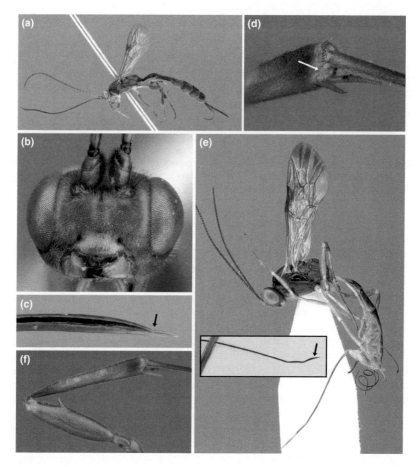

Fig. 10.21. Exemplar Cremastinae and characters: **(a–c)** rather aberrant Thai *Temelucha* sp., habitus, face and ovipositor apex (arrow showing pre-apical dorsal notch), respectively; **(d–f)** *Pristomerus* sp. from New Guinea, apex of hind tibia (arrow indicating sclerotized bridge separating spurs from basitarsus), habitus and insert showing wavy ovipositor (arrow indicating where small pre-apical notch is located), and hind leg showing femoral spine and somewhat basally offset origin of tibial spurs.

The S.E. Asian ctenopelmatines are certainly not well known and, although they are largely restricted to more northern mountainous regions where their hosts are most abundant (e.g. Reshchikov et al., 2017), there are numerous undescribed species.

Although they are not the only group of ichneumonids to possess a tooth on the disto-lateral margin of the fore tibia, this is the main practical way of separating ctenopelmatines from tryphonines (Fig. 10.22). However, this tooth is often rather hard to discern until one has had some practice; it helps to bend the basitarsus at an angle and to move the specimen about under the microscope to view the apex of the fore tibia from a range of angles. The tooth is actually a slightly projecting point, sometimes darkened, where the apical tibial margin forms an angle (Fig. 10.22 c). In addition, the mandibles are conspicuously large.

The subfamily is classified into eight: Ctenopelmatini, Euryproctini, Mesoleiini, Olethrodotini, Perilissini, Pionini, Scolobatini and Westwoodiini. Monophyly of the subfamily is not certain, although it seems agreed that they are closely related to the Tryphoninae. In the combined 28S rDNA and morphological analyses of Quicke et al. (2009) and Bennett et al. (2019), it was never recovered as monophyletic.

Nearly everything known about ctenopelmatine biology concerns European or North American species. Some have been quite well studied because their sawfly hosts are pests (e.g. the European pine sawfly, *Neodiprion sertifer*). Whereas most koinobiont parasitoids are pro-ovigenic, or nearly so, i.e. their full egg complement is mature soon after eclosion, all ctenopelmatines appear to be synovigenic (Quednau and Guevremont, 1975; Cummins et al., 2011). This might have something to do with hosts being rather dispersed and hard to locate.

Pionines are mostly egg-larval parasitoids and their ovipositors are correspondingly slender with a sharp tip and no pre-apical dorsal notch, although some *Rhorus* species are known to attack early larval host instars.

The total number of ctenopelmatines recorded from S.E. Asia is small. Reshchikov and van Achterberg (2014) described a new species of the rare perilissine genus *Metopheltes* from Vietnam and illustrated all species of this predominantly east Palaearctic genus. Reshchikov et al. (2014) reviewed the Asian species of *Neurogenia*, a genus with highly distinctive wing venation, and provided a key to species, of which two are known from Vietnam. Reshchikov et al. (2017) recorded two new species of *Rhorus*

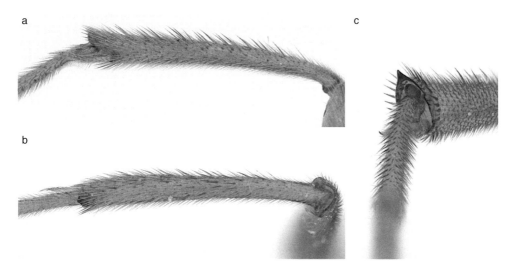

Fig. 10.22. Three aspects of a (particularly visible) fore tibial spur in a *Lathrolestes* species.

(Fig. 10.23 a) from Thailand. Reshchikov and van Achterberg (2018) described a new genus and species of Perilissini, *Gilen* (Fig. 10.23 b), from Laos, Thailand and Vietnam which has a large facial protuberance vaguely and fancifully reminiscent of a unicorn's horn. Another new genus, *Rhytidaphora* belonging to the Euryproctini, was recently described from northern Thailand (Reshchikov *et al.*, 2022).

Lycorininae

This is a small and generally uncommon but cosmopolitan subfamily comprising only the genus *Lycorina*. They are easily recognised by the presence of a distinctive, anteriorly pointed, triangular area on tergites 2–4 (Fig. 10.24). Beginners often confuse various Pimplinae and Banchinae for lycorines because both have demarcated triangular areas on their tergites. However, the shape of these is quite different. Those of *Lycorina* are pointed anteriorly and the point does not reach the base of the tergite. The egg possesses an anchor similar to that of tryphonines and Coronado-Rivera et al. (2004) showed that it was a koinobiont and that it had an external feeding phase. A more detailed account was published the same year (Shaw, 2014), in which it was shown that the female *Lycorina* oviposits inside the host caterpillars anus, which together with the egg morphology led Shaw to postulate that the parasitoid egg's anchor is probably inserted through the wall of the rectum but that the egg's body and earlier instars are technically external within the rectum and so develop as ectoparasitoids throughout their development.

The systematic position of *Lycorina* has changed over the years. Historically it was regarded as a pimpline, probably because of the metasomal triangles. Townes and Townes (1951) moved it to the Banchinae, then it was put in the tribe Glyptini, then Townes and Townes (1966) moved it specifically to the Glyptini, but since Townes (1970b) it has been regarded as constituting a separate subfamily. At least one species oviposits into its caterpillar host through its anus (Shaw, 2004).

Lycorina is cosmopolitan, and the nominal subgenus is widespread through the Palaearctic and Indo-Australian region where it is moderately diverse but with only a handful of the species being described. *L. borneoensis* was known from Malaysia since 1966 (Momoi, 1966) and a second S.E. Asian species was described from Indonesia recently by Shimizu and Ogawa (2018), who provided a key (largely based on colour pattern) to the Oriental species.

Mesochorinae

The Mesochorinae is a very common subfamily virtually everywhere. An updated key to the world genera was provided by Araujo *et al.* (2018), and Wahl (1993b) provided a morphological phylogenetic analysis. It is dominated by the genus *Mesochorus*, which are typically small to medium-sized wasps. In S.E. Asia the only other genera are *Astiphromma* and *Cidaphus*. Two names commonly encountered in

Fig. 10.23. Members of two genera of Ctenopelatinae: **(a)** *Rhorus* sp. (Pionini); **(b)** *Gilen orientalis* (Perilissinae), note the facial protuberance characteristic of the genus.

Fig. 10.24. *Lycorina* sp. male from Thailand: **(a)** habitus; **(b)** metasoma dorsal view showing anteriorly pointing triangular areas.

the literature, *Plectochorus* and *Stictopisthus*, are now regarded as synonyms of *Mesochorus*.

Despite their abundance and importance, there are few detailed studies on their biology. Day (2002) found that the North American *M. curvulus*, which is a hyperparasitoid of mirid plant bugs via member of the euphorine braconid genus *Peristenus*, appeared to have little effect on the population of its host. Life history and developmental details are provided for some other extralimital species by Blunck (1944) and Yeargan and Braman (1989). The sharp, thin ovipositor appears to be used to locate and oviposit within its primary parasitoid host.

Although beginners sometimes get confused with the standard identification key of them having a large rhombic areolet, once you have seen one or two they are unmistakeable (Fig. 10.25 a, b, d). The areolet is really relatively large and often broad diamond-shaped (Fig. 10.25 d) and similarly large diamond-shaped ones are uncommon (for example, *Metopius*) and not found in combination with a narrow petiolate 1st tergite. In addition, the claws are strongly pectinate (Fig. 10.25 e). The face is wide, somewhat protruding, and the clypeus only weakly differentiated. Males are also easy to recognise by their elongate, needle-like parameres (gonosquama) (Fig. 10.25 f, g), a condition that occurs only rarely in a few cryptines and also in the ctenopelmatine genus *Neurogenia* which occurs in the region. *Cidaphus* species are relatively large (8.5–14.0 mm). *Mesochorus* species have a transverse carina running across the top of the face just below the margin of the antennal sockets (Fig. 10.25 c).

The small number of species that have been placed in *Plectochorus*, which is predominantly East and S.E. Asian, were revised by Lee (1992). Prior to this, identification of the S.E. Asian *Mesochorus* was essentially a non-starter, but Riedel (2023) provided a key to the species, including 39 newly described ones. In addition, he described a new genus, *Orientochorus*, based on a species from Vietnam, and also a new species of *Astiphromma*.

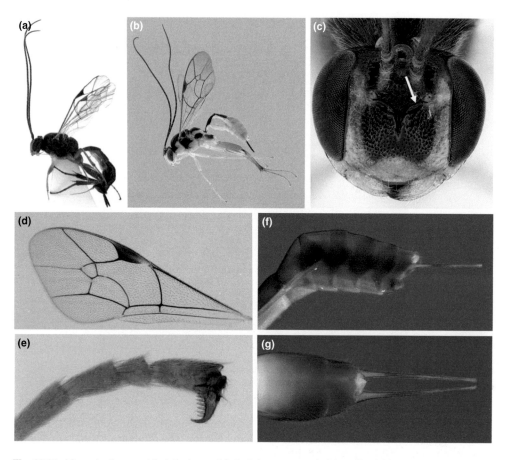

Fig. 10.25. Mesochorines and their features: **(a)** *Astiphromma* sp., habitus; **(b, c)** *Mesochorus* sp., habitus and face showing oblique transverse carina just below antennal sockets; **(d)** *Mesochorus* sp., fore wing showing very large areolet with second abscissa of vein M almost as long as 2m-cu; **(e)** *Mesochorus* sp., claw with pecten; **(f, g)** *Mesochorus* sp., male, showing very narrow and elongate parameres.

Metopiinae

Rather commonly collected in the tropics, most of these medium-sized to large wasps have a rather characteristic habitus and face. The key features are that the metasoma is sessile (broad at the base), the face and clypeus forming a bulge with an undivided surface (no groove separating clypeus and face), usually robust legs with swollen femora, and very short ovipositors (Fig. 10.26 d, e). The only other group one might confuse them with is the generally far smaller *Orthocentrus* group of the Orthocentrinae. *Metopius* itself is one of the easiest of all ichneumonid genera to recognise because their face has a large shield-shaped area bordered by a carina (Fig. 10.26 a).

Seven genera are known from S.E. Asia: *Acerataspis* (Fig. 10.26 c, f, g), *Colpotrochia*, *Drepanoctonus*, *Exochus* (Fig. 10.26 b, e), *Hypsicera*, *Metopius* (Fig. 10.26 a, d), *Seticornuta* and *Triclistus* (Yu et al., 2016; Zheng et al., 2021). These can be recognised using the key in Townes (1971).

Very few details of metopiine biology are known. Metopiines are solitary koinobiont larval-pupal parasitoids of Lepidoptera (Fitton, 1984; Quicke, 2015). They have short ovipositors and mostly attack exposed hosts or weakly concealed ones such as leaf-rollers. In the latter case the female accesses the host

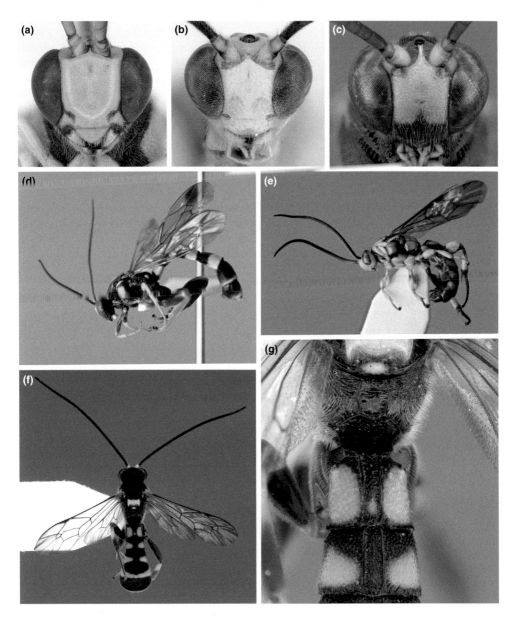

Fig. 10.26. Thai metopiines and features: **(a)** *Metopius* face showing margined shield shape; **(b)** *Exochus* face, showing inter-antennal process; **(c)** *Aceratapsis* face; **(d)** *Metopius*; **(e)** *Exochus*, note bulging face; **(f, g)** *Aceratapsis*, showing submedial carinae on tergites 1 and 2 that are present in approximately half of S.E. Asian genera.

by entering the roll. Development is completed within the host pupa, where the parasitoid larva spins a flimsy cocoon. The emerged metopiine cuts off the anterior of the host pupa to escape.

Triclistus spp. are parasitoids of smaller lepidopterans, particularly the family Tortricidae (Fitton, 1984). The S.E. Asian *T. aitkini* has been recorded as a parasitoid of the crambid rice pest *Cnaphalocrocis medinalis*.

Drepanoctonus spp. are only reliably known to attack Drepanidae. Aeschlimann (1975) provided some more detailed biological information for an extralimital *Triclistus* species. Mohamed *et al.* (2014) provided a key to separate the two species of *Triclistus* known from Malaysia.

Nesomesochorinae

A small subfamily of medium-sized but generally uncommon wasps. Three genera of Nesomesochorinae (*Nonnus* from North and South America, and the Old World *Chriodes*

and *Klutiana*) were long classified in the Campopleginae, which they strongly resemble, especially the shape of the 1st tergite (Fig. 10.27 a, c). A fourth genus, *Bina*, has recently been described from Peru (Shimizu and Alvarado, 2020). However, in an early molecular study of ichneumonid relationships, Belshaw and Quicke (2002) recovered *Nonnus* and *Chriodes* as the sister group of the Anomaloninae, rather removed from the campoplegines. Subsequent work (Quicke *et al.*, 2005) led to *Chriodes* and the closely related *Klutiana* being placed in the Nesomesochorinae; the name derives from *Nesomesochorus*, which is a junior subjective synonym of *Chriodes*.

Fig. 10.27. Species of *Klutiana* from Thailand: **(a)** lateral habitus; **(b)** face, showing strongly converging eyes; **(c)** dorsal mesosoma and anterior metasoma; **(d)** head and anterior mesosoma showing very short palps (arrow); **(e)** apical third of ovipositor showing pre-apical dorsal notch.

The most obvious feature of the Old World Nesomesochorinae is that the eyes are large and converge strongly ventrally (Fig. 10.27 b), indeed almost touching. Although a few taxa in other subfamilies have similar eyes, nesomesochorines also have extremely well-developed (almost complete) propodeal carination (Fig. 10.27 c), very short palps with a reduced number of labial palp segments (fewer than four) (Fig. 10.27 d), ovipositor with a pre-apical dorsal notch (Fig. 10.27 e), and strongly pectinate claws (but the pecten not reaching to the apex).

The host association published by Conlong (1994) for an Afrotropical species of *Chriodes* is based on a misidentification of a campoplegine (Broad et al., 2018), and therefore nothing is as yet known about nesomesochorine biology.

Gupta and Maueswari (1974) revised the Oriental species of *Chriodes*, and Gupta (1979) those of *Klutiana*. *Chriodes* may be separated from *Klutiana* by the latter lacking hind wing vein CU1b (discoidella sensu Townes) and lacking lateral carinae of the scutellum (Fig. 10.27 c).

Ophioninae

A predominantly crepuscular and nocturnal group of medium-sized to large wasps, most species displaying the eponymous 'ophionoid facies' of yellow-brown coloration, long antennae, large eyes and large ocelli (Fig. 10.28 b, c). There are more than 1000 described species worldwide, in 31 genera, although in general, two genera, *Enicospilus* and *Ophion*, predominate. In the Old World tropics, the most frequently encountered (at light traps) are members of the genus *Enicospilus*. Although the presence of scleromes in a glabrous area of the discosubmarginal cell is often used as a diagnostic character for *Enicospilus*, it is not reliable. Instead, *Ophion* are separable from *Enicospilus* because they possess a transparent membranous flange behind the setal comb on the fore tibial spur

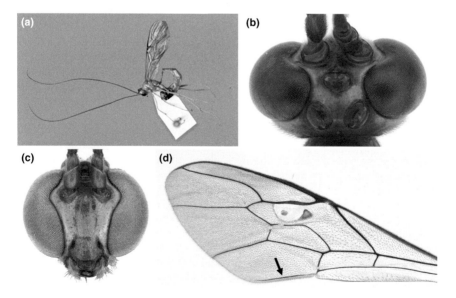

Fig. 10.28. *Enicospilus* spp.: **(a)** female habitus showing (very) long antennae and pale brown-yellow coloration; **(b)** head dorsal view showing large ocelli; **(c)** face showing large eyes that are strongly emarginate opposite antennal sockets; **(d)** fore wing, showing the far distal RS vein relative to 2m-cu, a glabrous area in the discocubital cell, in this species also associated with well-developed sclauromes that are important for species identification. Note also the adventitious fore wing vein (arrow) that is diagnostic of the family.

(antenna cleaner) whereas it is lacking in *Enicospilus* species.

Ophionines are solitary koinobiont endoparasitoids mostly of exposed lepidopterous larvae, especially larger species of the families Noctuidae, Geometridae and Lymantriidae. The ophionine larva usually completes development in the host prepupa from which it exits to spin a robust cocoon. In at least some species the cocoon has a paler coloured central band.

Historically two tribes, Enicospilini and Ophionini were recognised (Townes, 1971). Gauld (1985) conducted a large morphological phylogenetic analysis and Rousse et al. (2016) presented the most detailed molecular phylogenetic analysis of the subfamily. Gauld had basically abandoned formal tribes because evidence suggested that the traditional Ophionini were paraphyletic. However, most genera could be placed into his *Ophion* genus group or *Enicospilus* genus group. Rousse similarly recovered these two groups of genera, and additionally a third, the *Thyreodon* group. Although many genera were unplaced, they reinstated Enicospilini and Ophionini and added Thyreodonini.

Enicospilus is a very large genus worldwide, and indeed comparatively well-studied, but species identification is a long process, and available keys rely heavily on ratios of wing vein lengths (Gauld and Mitchell, 1981). Gauld and Mitchell (1981) revised the S.E. Asian (Indo-Papuan) species of the subfamily. Recent molecular work (Shimizu et al., 2020) indicated the existence of numerous synonymies as well as cryptic species. Several species appear to be widespread, extending for example from India, Japan and China all the way to Australia and the Philippines (Shimizu et al., 2020). Shimizu and Konishi (2018) recorded 14 species from Laos and provided photographs of whole insects and diagnostic wing features for eight of them, but this is likely far fewer than the total number. Yusof and Ghani (2009) described a new species from Malaysia. Other genera that occur in the region are *Dicamptus, Dictyonotus, Leptophion, Ophion* (only two species) and *Stauropoctonus* (only two species).

Townes (1971) provided keys to the world genera. S.E. Asian *Stauropoctonus* comprise the widespread *S. torresi* and *S. townesorum*, which is known only from the Philippines. They may be recognised using the key to world species by Lee and Kim (2002). Pham et al. (2021a) recorded three species of *Leptophion* from Vietnam for the first time. Shimizu and Konishi (2016) recorded *Dictyonotus* and Pham et al. (2020) recorded *Dicamptus* from Vietnam for the first time.

Oxytorinae

This monotypic subfamily with only 25 described species was until very recently known only from the New World and Palaearctic Region (Broad et al., 2018). They have long been suspected as being related to the probably para-/polyphyletic Ctenopelmatinae assemblage, and phylogenetic analyses tend to confirm this (Quicke et al., 2009). Note that the name was used by Gauld (1984) for the *Helictes* group of Orthocentrinae, which are not related to *Oxytorus*.

Oxytorus (Fig. 10.29) are not easy to recognise, especially males, which may explain the dearth of records from S.E. Asia. There is no single diagnostic character (or even simple combination of a few characters) that are diagnostic for the subfamily. The main features are: long antennae, clypeus large and separated from face by a groove, large mandibles (as in Ctenopelmatinae), with lower tooth slightly shorter than upper, elongate maxillary palpi, reaching at least to the middle of the mesosternum, no notauli, sternaulus weak, fore wing vein 2m-cu with only one bulla, no fore tibial tooth, metasomal segment 1 elongate and narrow, with prominent longitudinal carinae and with spiracle at or beyond middle, ovipositor sheath quite rigid, short and rather leaf-like in profile, and ovipositor with pre-apical dorsal notch. The only autapomorphy might be that the antennae have a central cluster of small placoid sensilla on their distal segments but this requires a very good light microscope, or SEM to see clearly (Gauld, 2000).

Nothing is known about their biology, but ovipositor morphology (very short with pre-apical dorsal notch) strongly suggests

Fig. 10.29. Holotype female of *Oxytorus rufopropodealis*. (Source: from Riedel *et al.*, 2021, reproduced under terms Creative Commons Attribution License CC BY 4.0.)

they are endoparasitoids of some easily accessible host; the elongate metasoma might suggest that they use this to obtain access to the host.

Riedel *et al.* (2021) described two species from Vietnam, viz. *Oxytorus carinatus* from the central part of the country at an altitude > 1200 m, and *O. rufopropodealis* (Fig. 10.29) from the northern part at altitudes of 300–600 m.

Sisyrostolinae (= Brachyscleromatinae)

For a long while, members of this subfamily were included in the Phrudinae (now included in the Tersilochinae), but Townes *et al.* (1961) placed them in their own subfamily, Brachyscleromatinae, because of the interrupted posterior mesosternal carina and pendent laterotergites (epipleura). but in 1969, Townes placed them back in the Phrudinae. In the combined molecular and morphological phylogeny of Quicke *et al.* (2009) they were recovered separate from Phrudinae *s.s.* and the Brachyscleromatinae was resurrected. The correct name for this subfamily is now considered to be Sisyrostolinae (Bennett *et al.*, 2013).

In S.E. Asia these small to medium-sized ichneumonids are rarely collected; some Afrotropical species are rather large. Of the six known genera, only *Brachyscleroma* occurs in S.E. Asia (Fig. 10.30 a) (Gupta, 1994). Biological data (for *Brachyscleroma* and the whole subfamily) are only available for one species, *B. apoderi*, which was reared from the attelabid beetle genus *Apoderus* (Cushman, 1940). Five species, including *B. apoderi*, are known from Indonesia and Malaysia (Yu *et al.*, 2016). Gupta (1994) followed by He *et al.* (2000) provided keys to the known species of *Brachyscleroma* of the world. Sheng and Sun (2011) described a new genus, *Laxiareola* (Fig. 10.30 b), from China and provided a key to world genera.

Stilbopinae

This is an almost entirely Palaearctic group comprising the single genus *Stilbops*. Only two species have thus far been recorded from S.E. Asia, both from northern Indo-China: *S. acicularis* from Myanmar and *S. gorokhovi* from Vietnam.

Tersilochinae

This is a particularly poorly known subfamily in general and particularly from the tropics.

Most of the S.E. Asian tersilochines in the original sense are easy to recognise because

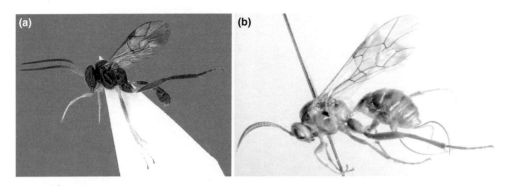

Fig. 10.30. Representatives of two Oriental genera of Sisyrostolinae: **(a)** *Brachyscleroma* sp., from Thailand; **(b)** *Laxiareola ochracea* from China. (Source: reproduced from Sheng and Sun (2011) under terms of Creative Commons Generic Licence CC-BY-2.5.)

the fore wing veins 'around' the tiny open areolet are thickened (Fig. 10.31 c, d) and the ovipositor has a preapical dorsal notch. Quicke *et al.* (2009) synonymised the Phrudinae *s.s.* and the extralimital Neorhacodinae with Tersilochinae. However, whilst current views agree with the former, the Neorhacodinae have been re-elevated to subfamily status (see Broad *et al.*, 2018). The traditional phrudines do not share the wing venation modification.

They are mostly koinobiont endoparasitoids of Coleoptera larvae. In Europe, two *Tersilochus* species play a role in the control of a stem-boring beetle larvae (Chrysomelidae and Curculionidae) in oilseed rape, *Brassica napi* (Barari *et al.*, 2005).

A great deal of what we know about the taxonomy of S.E. Asian tersilochines is due to the efforts of Andrey Khalaim of the Russian Academy of Sciences, St Petersburg, mainly concerning the fauna of Vietnam. Khalaim (2011) surveyed the 60 species of tersilochines in eight known from S.E. Asian and some adjacent regions, and provided a key to the known genera and species. In that work, he described one new genus and 21 new species all from S.E. Asia in the sense of this book (principally from Brunei, Vietnam and Laos). *Diaparsis* was found to be by far the commonest genus in S.E. Asia. He provided a key to the eight genera known from the region plus two others that seem likely to occur. The work also included the description of one new genus, *Slonopotamus*, from India and Laos. More recently, Khalaim (2017a) added *Barycnemis* to the S.E. Asia tally (Vietnam) with one new record and one new species, and Khalaim (2017b) described a new species of *Allophrys* from Indonesia. Yue *et al.* (2017) described another new species of *Allophrys* from the Philippines. Khalaim (2018a, 2019) described five new species of *Probles* from Vietnam but did not provide a key. Khalaim (2018b) revised the genera *Allophrys* and *Aneuclis* from Vietnam with keys to the species. Khalaim and Villemant (2019) described several new species of *Allophrys* from Papua New Guinea and include a key to all known Papuan species, which includes two known from elsewhere in S.E. Asia.

Tryphoninae

A medium-sized subfamily of predominantly Holarctic, medium-sized wasps. The world genera were revised by Bennett (2015). This impressive work includes a key to tribes, and for each, their genera. It also includes an overall single key to world genera, and also individual keys to the genera of different zoogeographical regions, including one for the Australasian region and one for the Oriental region.

Most genera of Tryphoninae are koinobiont ectoparasitoids (an uncommon combination) of sawfly larvae, and as these hosts are principally north temperate, that is where the bulk of the subfamily's diversity is to be found. Undoubtedly, some sawfly parasitising species

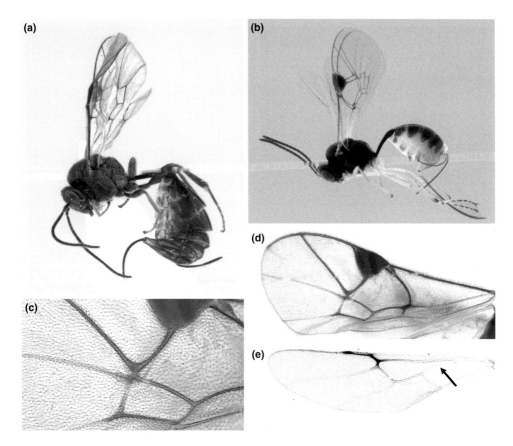

Fig. 10.31. Various exemplar Thai Tersilochinae and features: in **(b–d)**, note fore wing showing wide pterostigma and detail of thickened veins around areolet; in **(e)** note hind wing, showing strongly curved and spectral vein M+CU.

will be found at high altitudes in S.E. Asia where their potential hosts are quite abundant and diverse. However, three tryphonine tribes, the Oedemopsini, Phytodietini (Fig. 10.33) and the rare Sphinctini, are parasitoids of lepidopteran caterpillars and some of these are commonly collected in lowland tropics. Some of the most frequent hymenopteran visitors to UV light traps are members of the phytodietine genus *Netelia*, which are easy to confuse with Ophioninae from a distance. Females of both can deliver a sharp sting but the pain subsides fairly quickly.

Oedemopsini

As regards the Oedemopsini, three genera (*Acaenitellus*, *Atopotrophos* (Fig. 10.32 c) and *Kristotomus* (Fig. 10.32 d, e), collectively represented by six species, are recorded from S.E. Asia. In addition, Bennett (2006) described *Scudderopsis* from Papua New Guinea. Unlike most other tryphonines (excluding Phytodietina), oedemopsines are parasitoids of weakly concealed Lepidoptera caterpillars, notably of the Choreutidae, Coleophoridae, Gelechiidae, Heliodinidae, Oecophoridae, Tortricidae and Ypsolophidae. *Acaenitellus* had originally been placed in the Orthocentrinae (as Microleptinae) by Townes (1970b), but Gupta (1988) found that the body of the egg travelled externally down the ovipositor and it was accordingly transferred to the Tryphoninae. This goes to show just how difficult the correct placement of some ichneumonids can be, even to world experts.

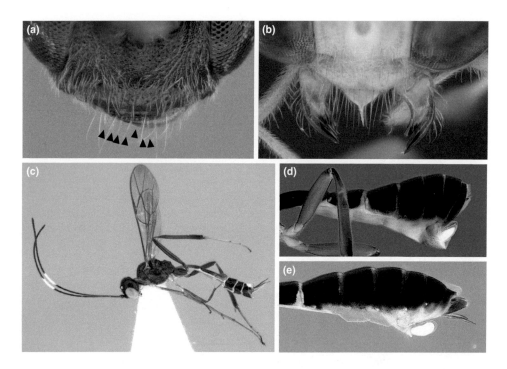

Fig. 10.32. Tryphoninae exemplar and features: **(a, b)** gen. sp. and *Netelia* sp., lower part of face showing fringe of evenly spaced bristles or their sockets (arrow heads) along lower clypeal margin; **(c)** *Atopotrophos* sp.; **(d, e)** *Kristotomus* species showing egg suspended below ovipositor, note the thin stalk originating distally and curving over the egg, which then enters the oviposotor egg canal where the egg's anchor is located.

Bennett (2006) provided a key to the world genera of Oedomopsini, and this was followed by a revised version in Bennett (2015).

Phytodietini

The phytodietine genus *Netelia* (Figs 10.32 b, 10.33 a) is cosmopolitan and comprises medium-sized to large-bodied species, most of which are crepuscular/nocturnal and usually display an ophionoid facies. The taxonomy of tropical *Netelia* species has hardly begun, but progress is being made in countries such as Japan (Konishi *et al.*, 2014, 2022; Pham *et al.*, 2021). Only 15 species of *Netelia* and seven species of *Phytodietus* have thus far been recorded from S.E. Asia (Yu *et al.*, 2016), but this is a gross underestimate; for example, the genus was only first recorded from Thailand in 2016 (Kostro-Ambroziak and Reshchikov, 2016). Both genera have well-developed pectinate claws (Fig. 10.33 e). Biology has only been described for extralimital species, e.g. Simmonds (1947).

Sphinctini

In addition to Oedemopsini, one species of Sphinctini, *Sphinctus chinensis*, has been recorded from Myanmar. The hosts of *Sphinctus* are limacodid moth caterpillars and the biology of a European species, *S. serotinus*, has been described in detail by Shaw and Voogd (2016). *S. chinensis* has also been reared in China, we believe from a limacodid.

Tryphonini

This is an almost entirely temperate group, although we have recorded *Monoblastus* (det. Alexey Reshchikov) from Thailand.

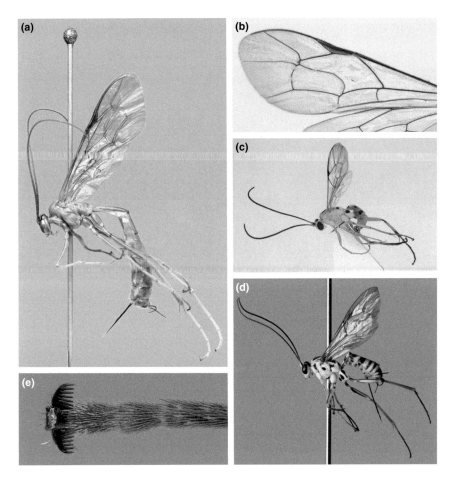

Fig. 10.33. Exemplars of the two genera of Phytodietini: **(a, b)** *Netelia* sp., habitus and fore wing detail showing strong basal bend in vein 3RS; **(c, d)** females of two Thai *Phytodietus* species; **(e)** detail of claw of a *Phytodietus* showing very well-developed pecten.

The Pimpliformes

This grouping has been well-supported by molecular studies. It comprises the Acaenitinae, Cylloceriinae, Collyriinae (extralimital), Diacritinae, Diplazontinae, Orthocentrinae, Pimplinae, Poemeniinae and Rhyssinae. Many of these are medium-sized to large wasps and are frequently prominent in collections. Until fairly recently (e.g. Fitton *et al.*, 1988) the Diacritinae, Poemeniinae and Rhyssinae were treated as tribes within the Pimplinae. See Quicke *et al.* (2009) for discussion on inclusion of Collyriinae in the Pimpliformes. The Rhyssinae were elevated to subfamily status by Gauld (1991).

With the exception of orthocentrines, their generally large body size has made them especially appealing for ecological and evolutionary investigation. The phylogeny of this group has been investigated directly or indirectly more than any other (Wahl and Gauld, 1998; Klopfstein *et al.*, 2019b).

Acaenitinae

These are generally rather uncommon, medium-sized to large ichneumonids whose females are easily recognised by their very large pointed hypopygium (Fig. 10.34 e) and their front and middle claws have an accessory tooth. They also often have a peribasal projection from the first metasomal sternite which is sharply pointed (Fig. 10.34 d), and

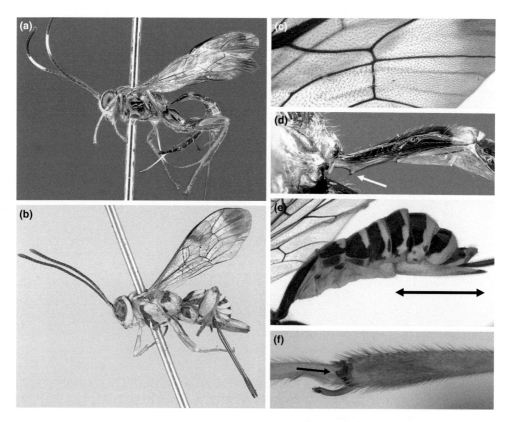

Fig. 10.34. Acaenitines and selected characters: **(a)** *Yezoceryx* sp., habitus; **(b)** *Y. shengi*, habitus; **(c)** *Jezarotes* ?*yanensis*, part of fore wing showing distal (postfurcal) location of 2-RS relative to 2m-cu; **(d)** *Yezoceryx* sp., tergite and sternite 1 showing strong sternal projection (arrow); **(e, f)** *J.* ?*yanensis*, in **(e)** metasoma showing extremely large and acute hypopygium, and **(f)** fore tibia showing tooth (arrow).

whose characters are important for generic identification (Townes, 1971). The apex of the fore tibia is modified and superficially tooth-like (Fig. 10.34 f) and at least the fore and mid legs have ancillary adpressed claws adjacent to the main claws. Spiracles of 1st tergite usually just in front of the middle. The world fauna comprises 29 genera but until 2017 only three (*Phalgea*, *Siphimedia* and *Yezoceryx*) had been recorded from S.E. Asia. However, several papers since then have added a number of species of *Ishigakia*, *Jezarotes* and *Spilopteron*.

Although there are numerous records from xylophagous Coleoptera, relatively little detailed biology is known. Other host associations should be considered dubious until supported by carefully isolated rearings (Broad et al., 2018). Based upon their close relationship to the Pimplinae, some workers had tacitly assumed that they would be idiobiont ectoparasitoids, even though it was understood that other members of the 'Pimpliformes' were koinobiont endoparasitoids (e.g. Diplazontinae, Orthocentrinae). The European species *Acaenitus dubitator* is a koinobiont endoparasitoid of an endophytic weevil (Curculionidae), oviposition being into a 1st and 2nd instar host larva (Shaw and Wahl, 1989).

Townes (1971) provided a key to the genera, although Sheng and Sun (2014) described a new one from China that could well also occur in S.E. Asia. *Yesoceryx* is by far the largest genus with more than 70 described species, with most being from China and other parts of the Oriental Region. Ito and

Maeto (2015) revised *Jezarotes* and described a new species from Laos. Pham and van Achterberg (2015) described new *Phalgea* species from Vietnam. Pham *et al.* (2017) recorded this genus from Vietnam for the first time, described nine new species and provided a key to the Vietnamese species. Pham *et al.* (2018a) described two new Vietnamese species of *Ishigakia*, and Pham and Long (2019) described another new *Ishigakia*, provided an updated key to world species of the genus, described another Vietnamese *Yeszoceryx* species and recorded *Siphimedia nigriscuta* from Vietnam for the first time. Pham *et al.* (2019) recorded seven species of *Spilopteron* from Vietnam, including five which they described as new.

Cylloceriinae

A small subfamily known almost entirely from the Holarctic and northern South America, together with one species from southern China. A new species from northern Myanmar and northern Thailand was recently described by Liu and Reshchikov (2019). The subfamily comprises just two genera, the most widespread and largest being *Cylloceria*, a relatively easy-to-recognise genus because of the exceedingly long first flagellomere (Fig. 10.35) in combination with an ovipositor that has a pre-apical dorsal notch. Although a few cryptines also have very elongate flagellomeres, their ovipositors usually have a protuding pre-apical dorsal nodus and never a notch. Male *Cylloceria* have a near-semicircular notch at the apex of the 3rd flagellomere and a second notch at the base of the next flagellomere; these form a structure that is presumed to 'grasp' the female antenna during courtship, but this has never been observed.

Biology is only known for the west Palaearctic species *Cylloceria melancholica*, which is a parasitoid of *Tipula irrorata* (Diptera: Tipulidae) (Broad *et al.*, 2018), and the wasps are usually associated with damp habitats.

Diplazontinae

These typically smallish wasps are obligate parasitoids of Syrphidae (hoverfly larvae)

Fig. 10.35. An extralimital species of *Cylloceria* from China. (Source: photograph by and © Alexey Reshchikov, reproduced with permission.)

and almost always of syrphids that feed upon aphids. Their distribution is therefore limited to where their principally temperate hosts occur, and in S.E. Asia that nearly always means at higher elevations. Thus far the subfamily has hardly been investigated in the region and to date only *Diplazon laetatorius* (possibly the commonest ichneumonid in the world), *D. orientalis*, *D. visayensis*, *Homotropus lucidus*, *Promethes philippinensis* and *Sussaba sugiharai* have been recorded from the region (Klopfstein *et al.*, 2011; Yu *et al.*, 2016), but we have collected several additional species and those of S.E. Asia really need revising.

They are fairly distinctive in appearance, the mesosoma–metasoma junction is broad (metasoma is sessile), the ovipositor is very short, and they have distinctive mandibles with the upper tooth partly divided by a notch (Fig. 10.36 d).

Klopfstein (2014) provided a key to the 17 genera of Diplazontinae occurring in the West Palaearctic which should work for all S.E. Asian taxa. Many diplazontines harbour *Wolbachia* bacterial endosymbionts and these can cause highly female-biased sex ratios and also can badly distort DNA barcoding as a means of species delimitation and recognition (Klopfstein *et al.*, 2016).

Fig. 10.36. Thai Diplazontinae and features: **(a)** *Diplazon* sp., female habitus; **(b, c)** *Diplazon* sp., female habitus (missing antennae) and anterior metasoma, dorsal view showing sessile 1st tergite, respectively; **(d)** gen. sp. indet., head showing the notched upper mandibular tooth

Orthocentrinae (includes Helictinae, Microleptinae and Oxytorinae of Authors)

Probably the taxonomically most neglected ichneumonid subfamily in the world. They are all rather small (typically 3–5 mm), but morphologically heterogeneous. The world fauna comprises more than 500 described species. The tropical fauna has been almost entirely neglected until recently. It comprises two genus groups: the *Orthocentrus* group and the rest (Fig. 10.37). Several workers have treated these as separate subfamilies, and included them separately in identification keys (e.g. Townes, 1971; Gauld, 1984). Molecular, morphological and combined phylogenetic studies have consistently recovered them together (Wahl and Gauld, 1998; Quicke *et al.*, 2009; Klopfstein *et al.*, 2019b). The *Orthocentrus* group is distinguished in large part by having a rather large cylindrical scape (Fig. 10.37 a), the face and clypeus forming an evenly convex surface without a distinct separating groove, and the mandible short, strongly tapered and sometimes unidentate.

Most orthocentrines for which reliable host records are available are koinobiont endoparasitoids of Diptera, especially those belonging to the nematocerans and mostly members of the family Mycetophilidae (Humala, 2003). Of course, not many people rear wild nematocerans or can easily identify most of them, so host records are very sparse, with none for S.E. Asia. Correspondingly, they tend to be collected near the ground in damp places, or associated with quite rotten wood.

The taxonomic literature on S.E. Asian species is incredibly sparse. Baltazar (1964) recorded the genera *Aperileptus*, *Eusterinx*, *Gnathochorisis*, *Hemiphanes*, *Ischyracis*, *Megastylus*, *Orthocentrus*, *Pantisarthrus*, *Plectiscidea*, *Plectiscus*, *Proclitus* and *Symplecis* from the Philippines; however, these are not listed in the Taxapad online database (Yu *et al.*, 2016) because the records are not based upon described

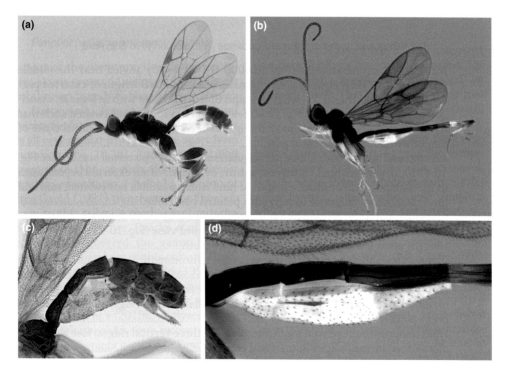

Fig. 10.37. Exemplar orthocentrines and characters: **(a)** *Orthocentrus* sp., note the large scapus; **(b)** *Stenomacrus* sp.; **(c, d)** metasomas of *Orthocentrus* and *Stenomacrus*, respectively, showing fairly characteristic contrasting dark setae and surrounding cuticle of the ventral membranous regions.

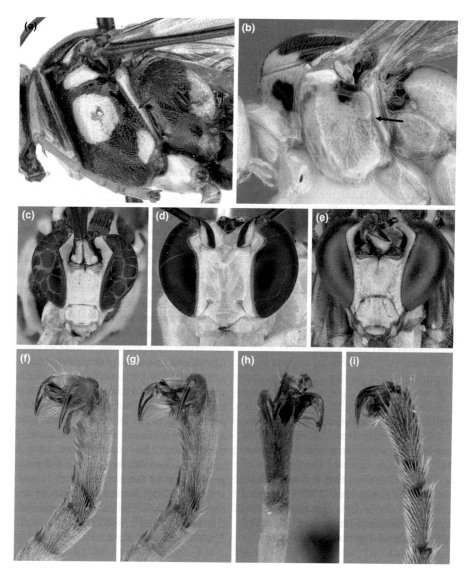

Fig. 10.40. Features of some non-ephialtine Pimplinae: (a) *Echthromorpha* sp., mesosoma showing straight mesoplural furrow (white arrow); (b) *Xanthopimpla* sp., showing weakly angled mesoplural furrow (black arrow); (c, d, e) faces of, respectively, *Theronia* sp., *Xanthopimpla* sp. showing transverse clypeal groove, and *Echthromorpha* sp.; (f, g) *Theronia* sp., apex of hind tarsus, showing the strongly spatulate specialised seta; (h) *Xanthopimpla* sp. claw, showing weakly spatulate modified bristle; (i) *Flavipimpla* sp. claw showing large pointed basal lobe.

et al. (2009); Takasuka and Matsumoto (2011). Gauld and Dubois (2006) carried out a phylogenetic analysis of the group and presented a good key to genera that is much updated since Townes (1969). *Zatypota albicoxa* (Fig. 10.42) was recorded from Indonesia by Takasuka *et al.* (2011b) and its oviposition behaviour has been described by Takasuka *et al.* (2009) and Takasuka and Matsumoto (2011) based on a different but related host. The species was subsequently also recorded from Vietnam (Takasuka and Watanabe, 2012).

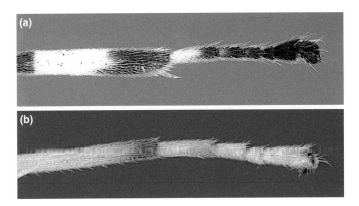

Fig. 10.41. Comparison of the relative width of the telotarsus to the basitarsus in a typical member of the *Polysphincta* group and a fairly closely related non-member: **(a)** *Zatypota* sp.; **(b)** *Zaglyptus* sp.

Fig. 10.42. Larva of *Zatypota albicoxa* attached to and feeding upon the postsoma of a female *Parasteatoda merapiensis* host spider in Java. (Source: photograph by and © Keizo Takasuka (Keio University), reproduced with permission.)

Fig. 10.43. Apical part of ovipositor of *Zaglyptus* sp., a larval parasitoid/predator of spider adults and their eggs within their nests.

Matsumoto (2016) presented a molecular phylogenetic analysis of the group and recognised three subgroups based around the genera *Acrodactyla*, *Polysphincta* and *Schizopyga*. Members of the first two subgroups are associated with spiders that construct aerial webs, the Acrodactyla subgroup mostly on Tetragnathidae and Linyphiidae, and the *Polysphincta* subgroup on Araneidae and some Theridiidae. In contrast, the *Schizopyga* subgroup are specialised parasitoids of spiders that make egg-laying chambers or funnel webs, such as Agelenidae and Clubionidae.

Closely related to the *Polysphincta* group are some other genera, such as *Tromatobia* (*T. hutachararerni* is recorded from Thailand), and *Zaglyptus*, species which are egg predators within spider egg sacs, and it seems likely that this provided the route by which pimplines evolved to be able to attack sub-adult and adult spiders themselves. Old observations of extralimital taxa (Nielsen, 1935) indicated that the female *Zaglyptus* first stings to death the adult spider that is guarding its own egg-sac, into which the parasitoid lays a few eggs. Her larvae then feed on the eggs and dead adult spider. This may explain why this genus has the proximal lower ovipositor serration elongated and outwardly curved (Fig. 10.43).

Other ephialtine genera in S.E. Asia include *Acropimpla* (Varga and Reshchikov, 2015b), *Camptotypus*, *Chablisea*, *Flavopimpla*, *Leptopimpla*, *Liotryphon*, *Lissopimpla*, *Sericopimpla*, *Tromatobia*, and *Zaglyptus* (Idris and Rizki, 2005; Pham et al., 2013b), *Pimplaetus* and, less commonly collected, *Dolichomitus* (Fig. 10.44).

Gupta and Tikar (1976) provided keys to the Oriental Pimplinae (as Pimplini) known at the time. Rizki and Idris (2006) described a new Malaysian species of *Camptotypus*, and Rizki and Idris (2010) described a new Malaysian species of *Zaglyptus*; both papers provided

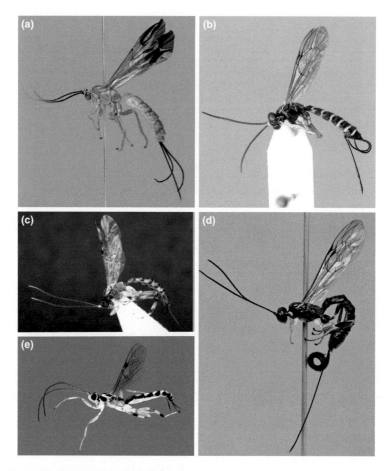

Fig. 10.44. Exemplar Ephialtini species: **(a)** *Camptotypus* sp.; **(b)** *Sericopimpla* sp.; **(c)** *Zaglyptus* sp.; **(d)** *Flavopimpla* sp.; **(e)** *Zatypota yambar*, note strongly protruding face of this species.

keys to the Malay species of the respective genera. Similarly, Pham *et al.* (2013b) described a new species of *Zaglyptus* from Vietnam and provided a key to the three species known from there. Varga and Reshchikov (2015b) revised *Acropimpla* for the northern part of S.E. Asia; nothing is known of their hosts in the region. Pham *et al.* (2011d) described two new species of *Chablisea* from Vietnam and provided a key to known species.

Some species of *Acropimpla* and *Camptotypus* are important parasitoids of the teak defoliator *Hyblaea puera* (Gupta and Tikar, 1976). *Camptotypus* and members of the *Sericopimpla* group attack cocooned Lepidoptera and spider egg sacs (Idris and Rizki, 2005).

Poemeniinae

A small subfamily of moderately sized to large ichneumonids, associated with hosts in dead or dying wood, therefore most often collected in wooded or forested areas. Four genera are recorded from S.E. Asia: *Cnastis* (one species from India to Sulawesi), *Deuteroxorides* (one species from Myanmar), *Eugalta* (Fig. 10.46; 12 species widespread) and *Poemenia* (one species from Myanmar). Gupta (1980) revised the Oriental species (as the Poemeniini of the Pimplinae *s.l.*). He also recognised *Pseudeugalta* as separate from *Eugalta* but this has not been followed by many subsequent workers. They are superficially similar to some xoridines with their elongate cylindrical mesosomas, general form and typical habitat (Fig. 10.45). They differ most obviously in the absence of a prepectal carina (Fig. 10.45 c) and, in S.E. Asian genera, unidentate mandibles (Fig. 10.45 d). Gupta (1985) made a number of nomenclatural changes based on the Townes's system of name priority and referred to the group (as a tribe of Pimplinae) under the name Neoxoridini.

Host records for poemeniines are rather scarce, and based entirely on temperate representatives. From what is known, these are idiobiont ectoparasitoids of wood-boring insect larvae (Townes, 1969). Species of the extralimital genus *Pseudorhyssa*, if indeed it belongs to the Poeminiinae (see Klopfstein et al., 2019b) are cleptoparasitoids of rhyssines (Spradbery, 1968) who locate the oviposition hole 'drilled' by a rhyssine and then insert their own slender ovipositor down it to lay a large egg close to that of the rhyssine. Her egg hatches quickly, and the first instar larva seeks out and kills the rhyssine egg/larva and then feeds on the paralysed wood-wasp larva.

Several genera have some sort of coarse sculpture on the top of the head (Fig. 10.45 b) and this has been hypothesised to help move debris from blocked tunnels inside the log and also in strengthening the head capsule. Extralimital species of *Podoschistus* and *Neoxorides* have been recorded as parasitoids of wood-boring beetles of the families Cerambycidae, Buprestidae and sometimes Curculionidae (Townes and Townes, 1960). However, extralimital *Poemenia* species are mainly parasitoids of Crabronidae and Megachilidae nesting in beetle borings in wood or galls (Carlson, 1979; Fitton et al., 1988) and they correspondingly lack the coarse sculpture of the temples and chisel-like mandibles that are associated with being parasitoids of wood-borers and are present in the other genera.

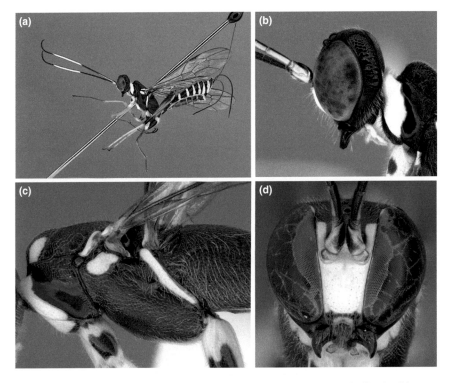

Fig. 10.45. *Eugalta* sp. from Thailand: **(a)** habitus; **(b)** detail of head showing scale-like denticles on temples; **(c)** mesosoma, lateral view, showing absence of epicnemial carina; **(d)** face showing heavily sclerotized, unidentate mandibles.

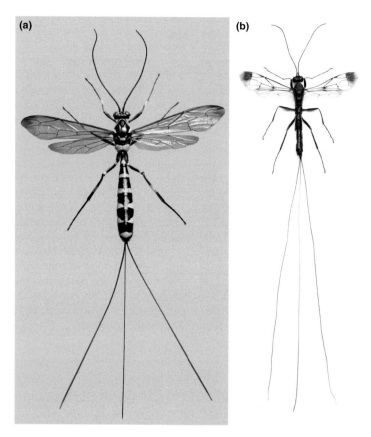

Fig. 10.46. Two large S.E. Asian rhyssines: **(a)** *Cyrtorhyssa moellerii* from Thailand, body length 38 mm and ovipositor length 60 mm; **(b)** *Myllenyxis* sp. from Indonesia (Java), body length 22 mm and ovipositor length 65 mm. (Source: (b) photograph by and © Edy Bhaskhara, reproduced with permission.)

Rhyssinae

Rhyssines are medium-sized to very large wasps with long ovipositors. They are most species-rich in the lowland forests of S.E. Asia (Kamath and Gupta, 1972). Worldwide, the subfamily includes 234 described species in eight genera, and the S.E. Asian fauna includes *Cyrtorhyssa* (Fig. 10.46 a), *Epirhyssa*, (Fig. 10.47 b, c) *Megarhyssa*, *Myllenyxis* (Fig. 10.46 b), *Sychnostigma* (Fig. 10.47 a) and *Triancyra*. *Epirhyssa* is widespread including the Neotropical Region, Africa and Japan, with species recorded from Malaysia, Myanmar, Philippines, Singapore and Vietnam. *Megarhyssa* has only recently been recorded from the region (Pham, 2018b).

They are one of the most easily recognised subfamilies in S.E. Asia because of their large size, very long ovipositors (Figs 10.46, 10.47 a) and the transverse mesoscutal ridges (Fig. 10.47 b), although an extralimital pimpline genus *Pseudorhyssa* also shares this feature as do (to a lesser extent) some Xoridinae (Quicke *et al.*, 2009; Klopfstein *et al.*, 2019b). The transverse mesoscutal ridges are presumed to aid egress from their pupation site and also occur in Ibaliidae and Aulacidae. The mandibles are very robust with the upper tooth chisel-shaped (Wahl and Gauld, 1998) (Fig. 10.47 c). *Myllenyxis* species have the apical part of the ovipositor strongly laterally compressed and sinuate in profile, which is likely an adaptation for

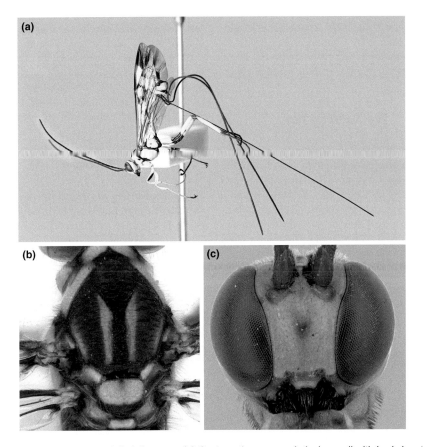

Fig. 10.47. Thai Rhyssinae and their features: **(a)** *Sychnostigma* sp., relatively small with body length only about 8 mm; **(b)** *Epirhyssa* sp., mesoscutum showing sharp transverse ridges; **(c)** *Epirhyssa* sp., face showing robust, very hardened, sclerotised (black) mandibles.

being able to steer the tip inside pre-existing fissures or tunnels (see Quicke, 1991).

Rhyssines are idiobiont ectoparasitoids of wood-boring insect larvae. In the temperate region, hosts are larvae of siricoid woodwasps (Siricidae and Xiphydriidae). However, wood-boring sawflies are very uncommon in the tropics and for a long while it has been suspected that in these regions, rhyssines attack wood-boring beetle larvae. Sheng and Sun (2010) reported several rhyssine species and genera (including *Rhyssa persuasoria*) as having been reared from various cerambycid beetles, which rather goes against all the vast number of N. American and W. Palaearctic rearings that are 100% from woodwasps. Recently, another species has been reared from a beetle larva in Thailand with full biological details (Chansri et al., 2023).

They are protandrous and males, which are often far smaller, locate tree trunks/logs where females are soon going to emerge from. Often several males aggregate at these hotspots (Eggleton, 1991) and in some genera (e.g. *Megarhyssa*) males have very slender metasomas enabling them to mate with females that have not yet managed to chew their way out of the wood.

The genera of Rhyssinae (as Ephialtinae: Rhyssini) can be identified using the key in Townes (1969). The S.E. Asian genera and species were revised by Kamath and Gupta (1972). Pham et al. (2018b) revised *Triancyra* for Vietnam. Very few molecular data are available for tropical rhyssines so the only limited molecular phylogenetic hypothesis about generic relationships and, is by Chansri *et al.* (2023). As some genera are not identifiable by any single character, they might not all be monophyletic.

The Xoridiformes

That the Xoridinae, the only subfamily placed in the Xoridiformes, was probably basal among extant Ichneumonidae was proposed by Quicke *et al.* (2000) based on a combination of morphological and biological data, and they proposed the informal group name Xoridiformes. They were used for rooting the family in the analyses of Quicke *et al.* (2009) and Klopfstein *et al.* (2018). Bennett *et al.* (2019) in their combined morphological and molecular analyses also found that the Xoridinae, or at least a subset of xoridine taxa, formed the sister group to the remaining Ichneumonidae.

Xoridinae

Female xoridines are typically rather large wasps, although males are often much smaller, as is the case of many idiobionts. They are all parasitoids of wood-boring hosts,

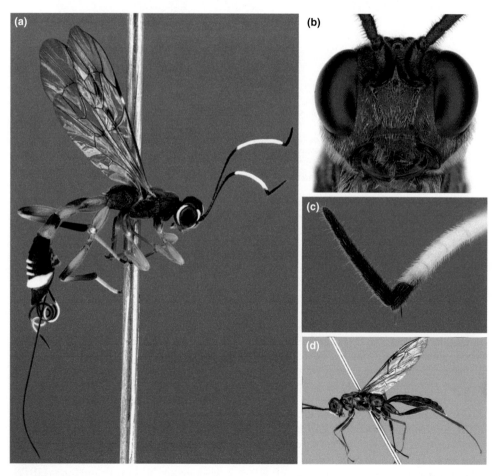

Fig. 10.48. Selected xoridines from Thailand: **(a)** *Xorides* (*Epixorides*) sp., habitus; **(b)** *Xorides* (*Cyanoxorides*) sp., face showing unidentate, very robust mandibles; **(c)** *Xorides* (*Epixorides*) sp., apex of female antenna showing angulation and specialised ventral hammer pegs; **(d)** *Xorides* (*Cyanoxorides*) sp., male, lateral habitus.

mainly Coleoptera larvae. In S.E. Asia, only three genera are known, and in terms of species richness these are dominated by *Xorides*, which in turn is divided into ten subgenera (Townes, 1969). Thirteen species of *Xorides* are known from S.E. Asia, plus one species of *Odontocolon* from Myanmar (Chandra, 1978) and two of *Aplomerus* from Vietnam and Thailand (Varga *et al.*, 2014).

Although Quicke *et al.* (2009) and Bennett *et al.* (2019) did not always recover the Xoridinae as monophyletic, it seems likely that they are. Certainly most, but not all, analyses to date have recovered them as sister group to the remainder of the extant Ichneumonidae (Quicke *et al.*, 1999; Klopfstein *et al.*, 2019b; Sharanowski *et al.*, 2021). Most recently, Zheng *et al.* (2022) using a mitogenomics approach recovered the brachycyrtiformes as most basal with the Xoridinae, then being sister group to the remainder.

Female *Xorides* are characterised by having a few stiff, blunt solid spines arising from a bend near the tip of the antenna, usually at an angulation (Fig. 10.48 c). Members of the subfamily have a fairly distinctive body shape similar to that of poemeniines, with an ovipositor usually a little longer than the body. The areolet is open and vein 2-RS (as interpreted here; = 2rs-m of Broad *et al.*, 2018) is very short. The upper mandible tooth is broad and chisel-like (Fig. 10.48 b) and this is used for chewing the emerging wasp's exit hole from the wood substrate of its host. *Xorides* biology has been described for a number of extralimital species (Sharifi and Javadi, 1971),

Species of *X.* (*Cyanoxorides*) and some members of other subgenera are metallic blue metallic or green or have metallic markings (Fig. 10.48 d).

Note

[1] Klopfstein *et al.* (2019b) noted that the clypeus of *Echthromorpha* is not divided by a transverse groove *contra* Townes (1969), although it superficially appears to be so.

References

Aeschlimann, J.-P. (1975) Biologie, comportement et lacher experimental de *Triclistus pygmaeus* Cresson (Hym., Ichn.). *Mittleilungen der Schweizerischen Anstalt für das forstliche Versuchswesen* 48, 165–171.

Akhtar, M.S., Kumawat, M.M. and Ramamurthy, V.V. (2010) An annotated checklist of *Xanthopimpla* Saussure (Hymenoptera: Ichneumonidae). *Oriental Insects* 44, 243–269. doi: 10.1080/00305316.2010.10417618

Amanda, T.P.O., Yaakop, S. and Idris, A.B. (2011) A catalogue of the genus-group *Theronia* (Hymenoptera: Ichneumonidae: Pimplinae) from Sundaland. *Serranga* 16, 75–89.

Araujo, R.O., Vivallo, R. and Santos, B.F. (2018) Ichneumonid wasps of the subfamily Mesochorinae: new replacement names, combinations and an updated key to the World genera (Hymenoptera: Ichneumonidae). *Zootaxa* 4521, 52–60.

Aziz, A., Fitton, M.G. and Quicke, D.L.J. (2000) Key to the *Diadegma* species (Hymenoptera: Ichneumonidae) parasitising diamondback moth, *Plutella xylostella*, with the description of a new species. *Bulletin of Entomological Research* 90, 375–389.

Baltazar, C.R. (1964) The genera of parasitic Hymenoptera in the Philippines, Part 2. *Pacific Insects* 6, 15–67.

Barari, H., Ferguson, A.W., Piper, R.W., Smith, E., Quicke, D.L.J. and Williams, I.H. (2005) The separation of two hymenopteran parasitoids, *Tersilocus obscurator* and *Tersilochus microgaster* (Ichneumonidae), of stem-mining pests of winter oilseed rape using DNA, morphometric and ecological data. *Bulletin of Entomological Research* 95, 299–307.

Barron, J.R. (1976) Systematics of Nearctic *Euceros* (Hymenoptera: Ichneumonidae: Eucerotinae). *Naturaliste Canadien* 103, 285–375.

Barron, J.R. (1978) Systematics of world Eucerotinae (Hymenoptera: Ichneumonidae). Part 2. Non-Nearctic species. *Naturaliste Canadien* 105, 327–374.

Barthélémy, C. and Broad, G.R. (2012) A new species of *Hadrocryptus* (Hymenoptera, Ichneumonidae, Cryptinae), with the first account of the biology for the genus. *Journal of Hymenoptera Research* 24, 47–58. doi: 10.3897/JHR.24.1888

Belshaw, R. and Quicke, D.L.J. (2002) Robustness of ancestral state estimates: evolution of life history strategy in ichneumonoid parasitoids. *Systematic Biology* 51, 450–477.

Bennett, A.M.R. (2001) Phylogeny of Agriotypinae (Hymenoptera: Ichneumonidae), with comments on the subfamily relationships of the basal Ichneumonidae. *Systematic Entomology* 26, 329–356.

Bennett, A.M.R. (2006) A new genus of Oedemopsini (Hymenoptera: Ichneumonidae: Tryphoninae) with a key to the genera of the tribe. *Canadian Entomologist* 138, 464–472.

Bennett, A.M.R. (2015) Revision of the world genera of Tryphoninae (Hymenoptera: Ichneumonidae). *Memoirs of the American Entomological Institute* 86, 1–387.

Bennett, A.M.R., Sääksjärvi, I.E. and Broad, G.R. (2013) Revision of the New World species of *Erythrodolius* (Hymenoptera: Ichneumonidae: Sisyrostolinae), with a key to the world species. *Zootaxa* 3702, 425–436.

Bennett, A.M.R., Cardinal, S., Gauld, I.D. and Wahl, D.B. (2019) Phylogeny of the subfamilies of Ichneumonidae (Hymenoptera). *Journal of Hymenoptera Research* 71, 1–156.

Berry, J.A., Harris, R.J., Read, P.E.C. and Donovan, B.J. (1997) Morphological and colour differences between subspecies of *Sphecophaga vesparum* (Hymenoptera: Ichneumonidae). *New Zealand Journal of Zoology* 24, 35–46.

Bin, F., Wäckers, F., Romani, R. and Isidoro, N. (1999) Tyloids in *Pimpla turionellae* (L.) are release structures of male antennal glands involved in courtship behaviour (Hymenoptera: Ichneumonidae). *International Journal of Insect Morphology and Embryology* 28, 61–68. doi: org/10.1016/S0020-7322(99)00015-X

Blunck, H. (1944) Zur Kenntnis der Hyperparasiten von *Pierris brassicae* L. *Zeitschrift für Angewandte Entomologie* 30, 418–491.

Bordera, S. and Hernández-Rodríguez, E. (2003) Description of two new species of *Enclisis* (Hymenoptera: Ichneumonidae) and support for the secretory role of tyloids in ichneumonoid males. *European Journal of Entomology* 100, 401–409.

Broad, G.R. (2014) A revision of *Sachtlebenia* Townes, with notes on the species of *Townesion* Kasparyan (Hymenoptera: Ichneumonidae: Banchinae). *Proceedings of the Russian Entomological Society* 85, 63–76.

Broad, G.R. (2021) Taxonomic changes in Ichneumonoidea (Hymenoptera), and notes on certain type specimens. *Zootaxa* 4941, 511–541.

Broad, G.R. and Quicke, D.L.J. (2000) The adaptive significance of host location by vibrational sounding in parasitoid wasps. *Proceedings of the Royal Society B: Biological Sciences* 267, 2403–2409.

Broad, G.R., Laurenne, N.M. and Quicke, D.L.J. (2004) The genus *Nipponaetes* Uchida (Hymenoptera: Ichneumonidae: Cryptinae) in Costa Rica, with a reassessment of the generic limits. *European Journal of Entomology* 101, 651–655.

Broad, G.R., Shaw, M.R., Fitton, M.G. (2018) *The Ichneumonid Wasps of Britain and Ireland (Hymenoptera: Ichneumonidae)*. Royal Entomological Society, Telford, UK.

Carlson, R.W. (1979) The Ichneumonidae. In: Krombein, K.V., Hurd, J.P.D., Smith, D.R. and Burks, B.D. (eds) *Catalog of Hymenoptera in America north of Mexico*. Smithsonian Institution Press, Washington, DC, pp. 315–739.

Chandra, G. (1978) Oriental Species of *Ischnoceros* and *Odontocolon* (Hymenoptera: Ichneumonidae). *Oriental Insects* 12, 319–325. doi: 10.1080/00305316.1978.10432092

Chansri, K., Somsiri, K., Quicke, D.L.J. and Butcher, B.A. (2023) First confirmed parasitism of pleasing fungus beetles (Coleoptera, Erotylidae) by a tropical rhyssine ichneumonid, and first record for *Cyrtorhyssa moellerii* Bingham (Hymenoptera, Ichneumonidae) from Thailand. Journal of Hymenoptera Research...

Conlong, D.E. (1994) A review and perspectives for the biological control of the African sugarcane stalkborer *Eldana saccharina* Walker (Lepidoptera: Pyralidae). *Agriculture, Ecosystems & Environment* 48, 9–17.

Coronado-Rivera, J., Gonzalez-Herrera, A., Gauld, I.D. and Hanson, P. (2004) The enigmatic biology of the ichneumonid subfamily Lycorininae. *Journal of Hymenoptera Research* 13, 223–227.

Cummins, H.M., Wharton, R.A. and Colvin, A.M. (2011) Eggs and egg loads of field-collected Ctenoplematinae (Hymenoptera: Ichneumonidae): evidence for phylogenetic constraints and life-history trade-offs. *Annals of the Entomological Society of America* 104, 465–475.

Cushman, R.A. (1940) New genera and species of ichneumon-flies with taxonomic notes. *Proceedings of the United States National Museum* 88, 355–372.

Darwin, C. (1860) Letter no. 2814, to Asa Gray. Darwin Correspondence Project. Available at https://www.darwinproject.ac.uk/letter/DCP-LETT-2814.xml

David, A.D. and Idris, A.B. (2016) Morphological phylogenetic analysis of genus *Xanthopimpla* Saussure 1892 (Hymenoptera: Ichneumonidae: Pimplinae) from Malaysia. *Serangga* 21, 1–19.

Day, W.H. (2002) Biology, host preferences, and abundance of *Mesochorus curvulus* (Hymenoptera: Ichneumonidae), a hyperparasite of *Peristenus* spp. (Hymenoptera: Braconidae) parasitizing plant bugs (Miridae: Hemiptera) in alfalfa-grass forage crops. *Annals of the Entomological Society of America* 95, 218–222.

Donovan, B.J. (1991) Life cycle of *Sphecophaga vesparum* (Curtis) (Hymenoptera: Ichneumonidae), a parasitoid of some vespid wasps. *New Zealand Journal of Zoology* 18, 181–192. doi: 10.1080/03014223.1991.10757965

Donovan, B.J., and Read, P.E.C. (1987) Attempted biological control of social wasps, *Vespula* spp. (Hymenoptera: Vespidae) with *Sphecophaga vesparum* (Curtis) (Hymenoptera: Ichneumonidae) in New Zealand. *New Zealand Journal of Zoology* 14, 329–335.

Donovan, B.J., Havron, A., Leathwick, D.M. and Ishay, J.S. (2002) Release of *Sphecophaga orientalis* Donovan (Hymenoptera: Ichneumonidae: Cryptinae) in New Zealand as a possible 'new association' biocontrol agent for the adventive social wasps *Vespula germanica* (F.) and *Vespula vulgaris* (L.) (Hymenoptera: Vespidae: Vespinae). *New Zealand Entomologist* 25, 17–25.

Eggleton, P. (1991) Patterns in male mating strategies of the Rhyssini: a holophyletic group of parasitoid wasps (Hymenoptera: Ichneumonidae). *Animal Behaviour* 41, 829–838.

Elliott, J.M. (1982) The life cycle and spatial distribution of the aquatic parasitoid *Agriotypus armatus* (Hymenoptera: Agriotypidae) and its caddis host, *Silo pallipes* (Trichoptera: Goeridae). *Journal of Animal Ecology* 51, 923–941.

Fischer, S., Samietz, J. and Dorn, S. (2003). Efficiency of vibrational sounding in parasitoid host location depends on substrate density. *Journal of Comparative Physiology* A 189, 723–730.

Fitton, M.G. (1984) Subfamily Metopiinae. In: Gauld, I.D. (ed.) *An Introduction to the Ichneumonidae of Australia*. British Museum (Natural History), London, UK, pp. 353–370.

Fitton, M.G. (1987) A review of the *Banchus*-group of ichneumon-flies, with a revision of the Australian genus *Philogalleria* (Hymenoptera: Ichneumonidae). *Systematic Entomology* 12, 33–45. doi: 10.1111/j.1365-3113.1987.tb00545.x

Fitton, M.G. and Gauld, I.D. (1976) The family group names of the Ichneumonidae (excluding Ichneumoninae) (Hymenoptera). *Systematic Entomology* 1, 247–258.

Fitton, M.G., Shaw, M.R. and Gauld, I.D. (1988) Pimpline ichneumon-flies. In: Barnard, P.C. and Askew, R.R. (eds) Hymenoptera, Ichneumonidae (Pimplinae). *RES Handbooks for the Identification of British Insects*, Vol. 7, Pt 1. British Museum (Natural History), London, 112 pp.

Gauld, I.D. (1984) *An Introduction to the Ichneumonidae of Australia, with a Contribution on the Metopiinae by M. G. Fitton*. British Museum (Natural History), London, 413 pp.

Gauld, I.D. (1985) The phylogeny, classification and evolution of parasitic wasps of the subfamily Ophioninae (Ichneumonidae). *Bulletin of the British Museum (Natural History), Entomology Series* 51, 61–185.

Gauld, I.D. (1991) The Ichneumonidae of Costa Rica, 1. *Memoirs of the American Entomological Institute* 47, 1–589.

Gauld, I.D. (1997) The Ichneumonidae of Costa Rica, 2. *Memoirs of the American Entomological Institute* 57, 1–485.

Gauld, I.D. (2000) The Ichneumonidae of Costa Rica, 3. Introduction and keys to species of the subfamilies: Brachycyrtinae, Cremastinae, Labeninae, and Oxytorinae, and with an appendix on the Anomaloninae; with contributions to the Brachycyrtinae by Sondra Ward, and the Oxytorinae by Victor Mallet. *Memoirs of the American Entomological Institute* 63, 1–453.

Gauld, I.D. and Dubois, J. (2006) Phylogeny of the *Polysphincta* group of genera (Hymenoptera: Ichneumonidae; Pimplinae): a taxonomic revision of spider ectoparasitoids. *Systematic Entomology* 31, 529–564. doi: 10.1111/j.1365-3113.2006.00334.x

Gauld, I.D. and Mitchell, P.A. (1981) *The Taxonomy, Distribution and Host Preferences of Indo-Papuan Parasitic Wasps of the Subfamily Ophioninae*. CABI, Slough, UK.

Gauld, I.D. and Mound, L.A. (1982) Homoplasy and the delineation of holophyletic genera in some insect groups. *Systematic Entomology* 7, 73–86.

Gauld, I.D. and Wahl, D.B. (2000) The Townesioninae: a distinct subfamily of Ichneumonidae (Hymenoptera) or a clade of the Banchinae? *Transactions of the American Entomological Society* 126, 279–292.

Gauld, I.D. and Wahl, D.B. (2002) The Eucerotinae: a Gondwanan origin for a cosmopolitan group of Ichneumonidae? *Journal of Natural History* 36, 2229–2248.

Gauld, I.D., Wahl, D., Bradshaw, K., Hanson, P. and Ward, S. (1997) The Ichneumonidae of Costa Rica, 2. Introduction and keys to species of the smaller subfamilies, Anomaloninae, Ctenopelmatinae, Diplazontinae, Lycorininae, Phrudinae, Tryphoninae (excluding Netelia) and Xoridinae, with an appendices on the Rhyssinae. *Memoirs of the American Entomological Institute* 57, 1–485.

Gauld, I.D., Wahl, D.B. and Broad, G.R. (2002) The suprageneric groups of the Pimplinae (Hymenoptera: Ichneumonidae): a cladistic re-evaluation and evolutionary biological study. *Zoological Journal of the Linnean Society* 136, 421–485. doi: 10.1046/j.1096-3642.2002.00031.x

Gokhman, V.E. and Krutov, V.V. (1996) On external structure of male antennae in the subfamily Ichneumoninae (Hymenoptera Ichneumonidae) and related groups. *Zoologichesky Zhurnal* 75, 1182–1194.

Gupta, A., Sampathkumar, M., Mohan, M., Shylesha, A.N., Venkatesan, T., Shashank, P.R., Dhanyakumar, O., Ramkumar, P., Sakthivel, N. and Geetha, B. (2021) Assessing adverse impact of the native biological control disruptors in the colonies of the recent invasive pest *Phenacoccus manihoti* Matile-Ferrero (Hemiptera: Pseudococcidae) in India. *Global Ecology and Conservation* 32, e01878. doi: 10.1016/j.gecco.2021.e01878

Gupta, S. and Gupta, V. (1983) *Ichneumonologia orientalis. Part IX. The tribe Gabuniini (Hymenoptera: Ichneumonidae)*. Association for the Study of Oriental Insects, Gainesville, Florida, 313 pp.

Gupta, V.K. (1962) Taxonomy, zoogeography, and evolution of Indo-Australian *Theronia* (Hymenoptera: Ichneumonidae). *Pacific Insects Monograph* 4, 1–142.

Gupta, V.K. (1979) Oriental species of *Klutiana* (Hymenoptera: Ichneumonidae). *Oriental Insects* 13, 323–339. doi: 10.1080/00305316.1979.10433626

Gupta, V.K. (1980) A revision of the tribe Poemeniini in the oriental region (Hymenoptera: Ichneumonidae). *Oriental Insects* 14, 73–130. doi:10.1080/00305316.1980.10434585

Gupta, V.K. (1985) a review of the Neoxoridini of the world (Hymenoptera: Ichneumonidae: Pimplinae). *Oriental Insects* 19, 323–329.

Gupta, V.K. (1988) Relationships of the genera of the Tryphonine tribe Oedemopsini and a revision of *Acaenitellus* Morley. In: Gupta, V.K. (ed.) *Advances in Parasitic Hymenoptera Research*. E.J. Brill, Leiden/New York, pp. 243–258.

Gupta, V.K. (1994) A review of the genus *Brachyscleroma* with descriptions of new species from Africa and the Orient (Hymenoptera: Ichneumonidae: Phrudinae). *Oriental Insects* 28, 353–382.

Gupta, V.K. (1999) A review of the mesostenine genus *Amauromorpha* (Hymenopera: Ichneumonidae), *Oriental Insects* 33, 267–277. doi: 10.1080/00305316.1999.10433794

Gupta, V.K. and Mauesuwari, S. (1974) The Oriental species of *Chriodes* (Hymenoptera: Ichneumonidae). *Oriental Insects* 8, 199–218.

Gupta, V.K. and Tikar, D.T. (1976) *Ichneumonologia orientalis or a monographic study of Ichneumonidae of the Oriental Region. Part 1. The tribe Pimplini (Hymenoptera: Pimplinae)*. Oriental Insects Mononograph 1. Association for the Study of Oriental Insects, Delhi, 313 pp.

He, J., Chen, X. and Ma, Y. (2000) Revision of the genus *Grachyscleroma* [*Brachyscleroma*] Cushman (Hymenoptera: Ichneumonidae) from China with a key to known species of the world. In: Zhang, Y. (ed.) *Systematic and Faunistic Research on Chinese Insects. Proceedings of the 5th National Congress of Insect Taxonomy*. China Agriculture Press, Beijing, pp. 235–245.

He, J., Chen, X. and Ma, Y. (2001) A new species of the genus Rothneya Cameron (Hymenoptera: Ichneumonidae) from Xi Zang, China. *Entomologia Sinica* 8, 111–114.

Hinz, R. (1983) The biology of the European species of the genus *Ichneumon* and related species (Hym., Ichneumonidae). *Contributions of the American Entomological Institute* 20, 151–152.

Huber, J.T. and Sharkey, M.J. (1993) Structure. In: Goulet, H. and Huber, T.J. (eds) *Hymenoptera of the World: An Identification Guide to Families*. Research Branch Agriculture Canada Publication 1894/E, Ottawa, pp. 13–59.

Humala, A.E. (2003) *The ichneumonid wasps in the fauna of Russia and adjacent countries: Subfamilies Microleptinae and Oxytorinae (Hymenoptera: Ichneumonidae)*. Nauka, Moscow, 176 pp. [in Russian]

Humala, A.E. (2016) A new species of the genus *Eusterinx* Förster, 1869 (Hymenoptera: Ichneumonidae: Orthocentrinae) from Malaysia. *Euroasian Entomological Journal* 15, 171–174.

Humala, A.E. (2021) First records of the genus *Gnathochorisis* Förster (Hymenoptera, Ichneumonidae, Orthocentrinae) in the Oriental region. In: Proshchalykin, M.Yu. and Gokhman, V.E. (eds) *Hymenoptera Studies Through Space and Time: A Collection of Papers Dedicated to the 75th Anniversary of Arkady S. Lelej. Journal of Hymenoptera Research* 84, 103–113. doi: 10.3897/jhr.84.68700

Idris, A.B. (1999) Catalogue of Pimplinae (Hymenoptera: Ichneumonidae) from Peninsular Malaysia. *The Pan-Pacific Entomologist* 75, 73–81.

Idris, A.B. and Rizki, A. (2005) Notes on the tribe Ephialtini (Hymenoptera: Ichneumonidae: Pimplinae) of Malaysia. *Serranga* 10, 111–126.

Ito, M. and Maeto, K. (2015) Revision of the genus *Jezarotes* Uchida (Hymenoptera: Ichneumonidae: Acaenitinae), with the description of a new species from Laos. *Zootaxa* 3946, 416–426.

Jasso-Martínez, J.M., Santos, B.F., Zaldívar-Riverón, A., Fernandez-Triana, J., Sharanowski, B.J., Richter, R., Dettman, J.R., Blaimer, B.B., Brady, S.G. and Kula, R.R. (2022) Phylogenomics of braconid wasps (Hymenoptera, Braconidae) sheds light on classification and the evolution of parasitoid life history traits. *Molecular Phylogenetics and Evolution* 173, e107452. doi: 10.1016/j.ympev.2022.107452

Kamath, M.K. and Gupta, V.K. (1972) Ichneumonologia Orientalis, Part II. The Tribe Rhyssini (Hymenoptera: Ichneumonidae). *Oriental Insects Monographs* 2, 1–300.

Kasparyan, D.R. (1993) Townesioninae, a new ichneumonid subfamily from the Eastern Palearctic (Hymenoptera: Ichneumonidae). *Zoosystematica Rossica* 2, 155–159.

Khalaim, A.I. (2011) Tersilochinae of South, Southeast and East Asia, excluding Mongolia and Japan (Hymenoptera: Ichneumonidae). *Zoosystematica Rossica* 20, 96–148.

Khalaim, A.I. (2017a) Tersilochinae (Hymenoptera: Ichneumonidae) of Vietnam, Part 2: genus *Barycnemis* Förster, 1869. *Proceedings of the Zoological Institute RAS* 321, 371–376.

Khalaim, A.I. (2017b) A new species of *Allophrys* Förster, 1869 (Hymenoptera: Ichneumonidae: Tersilochinae) with large propodeal spiracles from Indonesia. *Proceedings of the Zoological Institute RAS* 321, 365–370.

Khalaim, A.I. (2018a) A new remarkable species of *Probles* with clavate antennae from Vietnam (Hymenoptera: Ichneumonidae: Tersilochinae). *Zoosystematica Rossica* 27, 234–238.

Khalaim, A.I. (2018b) The genera *Allophrys* Förster and *Aneuclis* Förster (Hymenoptera: Ichneumonidae: Tersilochinae) of Vietnam. *Zootaxa* 4378, 414–428.

Khalaim, A.P. (2019) Four new species of the genus *Probles* Förster (Hymenoptera: Ichneumonidae: Tersilochinae) from Vietnam. *Zoosystematica Rossica* 28, 120–131.

Khalaim, A.I. and Villemant, C. (2019) Tersilochinae (Hymenoptera: Ichneumonidae) of Papua New Guinea: genera *Allophrys* Förster and *Probles* Förster. *Zootaxa* 4544, 235–250.

Klopfstein, S. (2014) Revision of the Western Palaearctic Diplazontinae (Hymenoptera, Ichneumonidae). *Zootaxa* 3801, 1–143. doi: 10.11646/zootaxa.3801.1.1

Klopfstein, S., Quicke, D.L.J., Kropf, C. and Frick, H. (2011) Molecular and morphological phylogeny of Diplazontinae (Hymenoptera, Ichneumonidae). *Zoologica Scripta* 40, 379–402.

Klopfstein, S., Kropf, C. and Baur, H. (2016) *Wolbachia* endosymbionts distort DNA barcoding in the parasitoid wasp genus *Diplazon* (Hymenoptera: Ichneumonidae). *Zoological Journal of the Linnean Society* 177, 541–557. doi: 10.1111/zoj.12380

Klopfstein, S., van Der Schyff, G.M., Tierney, S. and Austin, A.A. (2018) *Wolbachia* infections in Australian ichneumonid parasitoid wasps (Hymenoptera: Ichneumonidae): evidence for adherence to the global equilibrium hypothesis. *Biological Journal of the Linnean Society* 123, 518–534.

Klopfstein, S., Santos, B.F., Shaw, M.R., Alvarado, M., Bennett, A.M.R., Pos, D.D., Giannotta, M., Florez, A.F.H., Karlsson, D., Khalaim, A.I., Lima, A.R., Mikó, I., Sääksjärvi, I.E., Shimizu, S., Spasojevic, T., van Noort, S., Vilhelmsen, L. and Broad, G.R. (2019a) Darwin wasps: a new name heralds renewed efforts to unravel the evolutionary history of Ichneumonidae. *Entomological Communications* 1, ec01006.

Klopfstein, S., Langille, B., Spasojevic, T., Broad, G.R., Cooper, S.J., Austin, A.D. and Niehuis, O. (2019b) Hybrid capture data unravel a rapid radiation of pimpliform parasitoid wasps (Hymenoptera: Ichneumonidae: Pimpliformes). *Systematic Entomology* 44, 361–383. doi: 10.1111/syen.12333 44, 361–383

Konishi, K., Chen, H.P. and Pham, N.T. (2022) A taxonomic review of the genus *Netelia*, subgenus Monomacrodon (Hymenoptera: Ichneumonidae: Tryphoninae), with description of a new species. *Raffles Bulletin of Zoology* 70, 376–384.

Konishi, K. (2014) A revision of the subgenus *Bessobates* of the genus *Netelia* from Japan (Hymenoptera, Ichneumonidae, Tryphoninae). *Zootaxa* 3755(4), 301–346.

Kostro-Ambroziak, A. and Reshchikov, A. (2016) First report of the genus *Phytodietus* Gravenhorst, 1829 (Hymenoptera: Ichneumonidae: Tryphoninae) from Thailand. *Biodiversity Data Journal* 2016, e8027. doi: 10.3897/BDJ.4.e8027

Kusigemati, K. (1990) Some Banchinae, Porizontinae and Metopiinae of south east Asia, with descriptions of three new species (Hymenoptera, Ichneumonidae). 昆蟲, 58(2), 397–404.

Laurenne, N. and Quicke, D.L.J. (2009) Antennal hammers: echos of sensilla past. In Pontarotti, P. (ed.) *Evolutionary Biology: Concepts, Molecular and Morphological Evolution*. Proceedings of the 13th Evolutionary Biology Meeting, Marseille, pp. 271–282.

Laurenne, N.M., Broad, G.R. and Quicke, D.L.J. (2006) Direct optimization and multiple alignment of 28S D2-3 rDNA sequences: problems with indels on the way to a molecular phylogeny of the cryptine ichneumon wasps (Insecta: Hymenoptera). *Cladistics* 22, 442–473.

Laurenne, N.M., Karatolos, N. and Quicke, D.L.J. (2009) Hammering homoplasy: multiple gains and losses of vibrational sounding in cryptine wasps (Insecta: Hymenoptera: Ichneumonidae). *Biological Journal of the Linnean Society* 96, 82–102.

Lee, J.-W. (1992) A revision of the genus *Plectochorus* (Hymenoptera, Ichneumonidae: Mesochorinae). *Oriental Insects* 26, 241–263.

Lee, J.-W. and Kim, K.B. (2002) Taxonomical review of the subfamily Ophioninae (Hymenoptera: Ichneumonidae) I. Genus *Stauropoctonus* Brauns. *Korean Journal of Entomology* 32, 81–86. doi: 10.1111/j.1748-5967.2002.tb00015.x

Liu, J.-X. and Reshchikov, A. (2019) A new species of *Cylloceria* Schiødte, 1838 (Hymenoptera, Ichneumonidae) from the Oriental Region. *Zootaxa* 4609, 139–148. doi: 10.11646/zootaxa.4609.1.6

Makino, S.I. (1983) Biology of *Latibulus argiolus* (Hymenoptera, Ichneumonidae), a parasitoid of the paper wasp *Polistes biglumis* (Hymenoptera, Vespidae). *Kontyu* 51, 426–434.

Mason, W.R.M. (1971) An Indian *Agriotypus* (Hymenoptera: Agriotypidae). *The Canadian Entomologist* 103, 1521–1524.

Matsumoto, R. (2016) Molecular phylogeny and systematics of the *Polysphincta* group of genera (Hymenoptera, Ichneumonidae, Pimplinae). *Systematic Entomology* 41, 854–864. doi: 10.1111/syen.12196

Matsumoto, R. and Konishi, K. (2007) Life histories of two ichneumonid parasitoids of *Cyclosa octotuberculata* (Araneae): *Reclinervellus tuberculatus* (Uchida) and its new sympatric congener (Hymenoptera: Ichneumonidae: Pimplinae). *Entomological Science* 10, 267–278.

Mohamed, N., Yaakop, S.B. and Ghani, I.D. (2014) A taxonomic review of the genus *Triclistus* Foerster, 1868 (Hymenoptera: Ichneumonidae: Metopiinae) from Malaysia. *AIP Conference Proceedings* 1571, 311–316.

Momoi, S. (1966) Some new Ichneumonidae (Hymenoptera) from New Guinea and adjacent areas. *Pacific Insects* 8, 152–164.

Momoi, S., Kishitani, Y. and Iwata, K. (1965) Studies on an ichneumonid parasite (Hymenoptera) of *Clania miniscura* (Lepidoptera). *Science Reports of the Hyogo University of Agriculture. Series: Plant Protection* 7, 25–31.

Moore, S.D. and Kfir, R. (1996) Biological studies of *Xanthopimpla stemmator* (Thunberg) (Hymenoptera: Ichneumonidae), a parasitoid of lepidopteran stem borers. *African Entomology* 4, 131–136.

Nielsen, E. (1935) A third supplementary note upon the life histories of polysphinctas (Hym. Ichneum.). *Entomologiske Meddelelser* 19, 191–215.

Norhafiza, A.F. and Idris, A.B. (2013) Current status of subfamily Ichneumoninae (Hymenoptera: Ichneumonidae) from Malaysia and Singapore. Proceedings of the Universiti Kebangsaan Malaysia, Faculty of Science and Technology Postgraduate Colloquium – Selangor, Malaysia (3–4 July 2013), AIP, Malaysia, pp. 308–310. doi: 10.1063/1.4858674

Oh, S., Lee, J., and Choi, W. (2010) A new species of *Arthula* (Hymenoptera: Ichneumonidae: Cryptinae) from Korea, and its intraspecific seasonal variation. *The Canadian Entomologist* 142, 200–211. doi: 10.4039/n10-007

Otten, H., Wäckers, F.L., Battini, M. and Dorn, S. (2000) Efficiency of vibrational sounding in the parasitoids *Pimpla turionellae* is affected by female size. *Animal Behaviour* 61, 671–677.

Otten, H., Wäckers, F.L., Isidoro, N., Romani, R. and Dorn, S. (2002) The subgenual organ in *Pimpla turionellae* L. (Hymenoptera, Ichneumonidae): ultrastructure and behavioral evidence for its involvement in vibrational sounding. *Redia* 85, 61–76.

Perkins, J.F. (1960) Hymenoptera. Ichneumonoidea. Ichneumonidae, subfamilies Ichneumoninae II, Alomyinae, Agriotypinae and Lycorininae. *Handbooks for the Identification of British Insects* 7(2aii), 117–213.

Pham, N.T. (2013) *Taxonomy and distributional pattern of Pimplinae (Hymenoptera: Ichneumonidae) from Vietnam*. Unpublished doctoral dissertation, Rheinische Friedrich-Wilhelms-Universität Bonn, Bonn, Germany. https://nbn-resolving.org/urn:nbn:de:hbz:5n-32063 (accessed 23 August 2022)

Pham, N.T. and Long, K.D. (2016) A checklist of the family Ichneumonidae (Hymenoptera: Ichneumonoidea) from Vietnam. *Academia Journal of Biology* 38, 411–441. doi: 10.15625/0866-7160/v38n4.8883

Pham, N.T. and Long, K.D. (2019) Contribution to the knowledge of Acaenitinae (Hymenoptera: Ichneumonidae) from Vietnam, with descriptions of two new species. *Journal of Asia-Pacific Entomology* 22, 820–825. doi: 10.1016/j.aspen.2019.06.009

Pham, N.T. and van Achterberg, C. (2015) A review of the genus *Phalgea* Cameron (Hymenoptera: Ichneumonidae: Acaenitinae) with description of a new species from Vietnam. *Zootaxa* 3947, 146–150. doi: 10.11646/zootaxa.3947.1.11

Pham, N.T., Broad, G.R. and Lampe, K.-H. (2010) Descriptions of two new species of *Augerella* Gupta (Hymenoptera: Ichneumonidae: Pimplinae) and the first record of *A. orientalis* (Gupta) from Vietnam. *Zootaxa* 2654, 17–29.

Pham, N.T., Broad, G.R., Matsumoto, R. and Wägele, W.J. (2011a) Revision of the genus *Xanthopimpla* Saussure (Hymenoptera: Ichneumonidae: Pimplinae) in Vietnam, with descriptions of fourteen new species. *Zootaxa* 3056, 1–67.

Pham N.T., Broad, G.R and Wägele W.J. (2011b) The genus *Acropimpla* Townes (Hymenoptera: Ichneumonidae: Pimplinae) in Vietnam, with descriptions of three new species. *Zootaxa* 2921, 1–12.

Pham, N.T., Broad, G.R. and Lampe, K.-H. (2011c) Descriptions of two new species of *Augerella* Gupta (Hymenoptera: Ichneumonidae: Pimplinae) from Vietnam. *Zootaxa* 2745, 68.

Pham, N.T., Broad, G.R., Matsumoto, R. and Wägele W.J. (2011d) Two new species of the genus *Chablisea* Gauld et Dubois (Hymenoptera: Ichneumonidae: Pimplinae) from Vietnam. *Biologia* 66, 1134–1139. doi: 10.2478/s11756-011-0117-z

Pham, N.T., Broad, G.R., Matsumoto, R., and Wägele W.J. (2012) First record of the genus *Brachyzapus* Gauld and Dubois (Hymenoptera: Ichneumonidae: Pimplinae) from Vietnam, with descriptions of six new species. *Journal of Natural History* 46, 1639–1631. doi: 10.1080/00222933.2012.679640

Pham, N.T., Broad, G.R. and Wägele W.J. (2013a) A review of the *Theronia* genus-group (Hymenoptera: Ichneumonidae: Pimplinae) from Vietnam with descriptions of five new species. *Journal of Natural History* 47(23–24), 1501–1538. doi: 10.1080/00222933.2012.763105

Pham, N.T., Broad, G.R. and Wägele W.J. (2013b) Review of the genus *Flavopimpla* (Hymenoptera: Ichneumonidae: Pimplinae) from Vietnam with descriptions of two new species. *Biologia* 68, 720–726. 10.2478/s11756-013-0206-2

Pham, N.T., Broad, G.R., Dang, H.T. and Böhme, W. (2013c) A review of the genus *Pimpla* Fabricius, 1804 (Hymenoptera: Ichneumonidae: Pimplinae) from Vietnam with descriptions of two new species. *Organisms Diversity & Evolution* 13, 397–407.

Pham, N.T., Broad, G.R., Matsumoto, R. and van Achterberg, C. (2017) First record of the genus *Yezoceryx* Uchida (Ichneumonidae: Acaenitinae) from Vietnam, with descriptions of nine new species. *Zootaxa* 4311, 345–372. doi:10.11646/zootaxa.4311.3.2

Pham, N.T., Ito, M., Matsumoto, R. and van Achterberg, C. (2018a) Two new species of the genus *Ishigakia* (Hymenoptera: Ichneumonidae, Acaenitinae) from Vietnam based on morphological and molecular evidence. *Zootaxa* 4442, 539–550. doi: 10.11646/zootaxa.4442.4.3

Pham, N.T., Broad, G.R., Zhu, C.-D. and van Achterberg, C. (2018b) A review of the genus *Triancyra* Baltazar (Ichneumonidae: Rhyssinae) from Vietnam, with descriptions of three new species. *Zootaxa* 4377, 565–574.

Pham, N.T., Broad, G.R., Matsumoto, R. and van Achterberg, C. (2019) First records of the genus *Spilopteron* Townes (Hymenoptera: Ichneumonidae: Acaenitinae) from Vietnam, with descriptions of five new species. *Zootaxa* 4590, 153–165. doi: 10.11646/zootaxa.4590.1.6.

Pham, N.T., Matsumoto, R. and Shimizu, S. (2020) *Dicamptus* Szépligeti (Hymenoptera: Ichneumonidae: Ophioninae) from Vietnam. *Zootaxa* 4830, 371–382. doi: 10.11646/zootaxa.4830.2.8

Pham, N.T., Long, K.D., Thu, C.T.K., Nga, C.T.Q., Tru, H.V., Phu, P.V. and Hanh, L.M. (2021a) First records of the genus *Leptophion* Cameron (Hymenoptera: Ichneumonidae: Ophioninae) from Vietnam. *Academia Journal of Biology* 43(1). doi: 10.15625/2615-9023/15716

Pham, N.T., Chen, H.P. and Konishi, K. (2021b) Notes on the genus *Netelia*, subgenus *Parabates* (Ichneumonidae, Tryphoninae), with description of a new species from Oriental Region. *Zootaxa* 4974(3), 577584.

Pham, N.T., Matsumoto, R. and Broad, G.R (2022) First record of the genus *Stauropoctonus* Brauns (Ichneumonidae: Ophioninae) from Vietnam, with description of a new species. *Zootaxa* 5155, 142–150. doi: 10.11646/zootaxa.5155.1.8

Pham, N.T. (2018) A review of the genus *pham and Ionyssa* (Hymenoptera: Ichneumonidae: Rhyssinae) from Vietnam, with three new country records. *Journal of Vietnamese Environment* 9, 22–25.

Quednau, F.W. and Guévremont, H. (1975) Observations on mating and oviposition behaviour of *Priopoda nigricollis* (Hymenoptera: Ichneumonidae), a parasite of the birch leaf-miner, *Fenusa pusilla* (Hymenoptera: Tenthredinidae). *The Canadian Entomologist* 107, 1199–204.

Quicke, D.L.J. (1991) Ovipositor steering mechanics of the braconine wasp genus *Zaglyptogastra* and the ichneumonid genus *Pristomerus*. *Journal of Natural History* 25, 971–977.

Quicke, D.L.J. (2012) We know too little about parasitoid wasp distributions to draw any conclusions about latitudinal trends in species richness, body size and biology. *PLoS ONE* 7(2), e32101.

Quicke, D.L.J. (2015) *The Braconid and Ichneumonid Parasitic Wasps: Biology, Systematics, Evolution and Ecology.* Wiley Blackwell, Oxford, 688 pp.

Quicke, D.L.J., Fitton, M.G., Tunstead, J., Ingram, S.N. and Gaitens, P.V. (1994) Ovipositor structure and relationships within the Hymenoptera, with special reference to the Ichneumonoidea. *Journal of Natural History* 28, 635–682.

Quicke, D.L.J., Lopez-Vaamonde, C. and Belshaw, R. (1999) The basal Ichneumonidae (Insecta: Hymenoptera): 28S rDNA considerations of the Brachycyrtinae, Labeninae, Paxylommatinae and Xoridinae. *Zoologica Scripta* 28, 203–210.

Quicke, D.L.J., Fitton, M. G., Notton, D.G. Belshaw, R., Broad, G.R. and Dolphin, K. (2000) Phylogeny of the Ichneumonidae (Hymenoptera): a simultaneous molecular and morphological analysis. In: Austin, A.D. (ed.) *Hymenoptera: Evolution, Biodiversity and Biological Control.* CSIRO, Canberra, pp. 74–83.

Quicke, D.L.J., Laurenne, N.M., Broad, G.R. and Barclay, M. (2003) Host location behaviour and a new host record for *Gabunia* aff. *togoensis* Krieger (Hymenoptera: Ichneumonidae: Cryptinae) in Kibale Forest National Park, West Uganda. *African Entomology* 11, 308–310.

Quicke, D.L.J., Fitton, M.G., Broad, G.R., Crocker, B., Laurenne, N.M. and Miah, I.M. (2005) The parasitic wasp genera *Skiapus*, *Hellwigia*, *Nonnus*, *Chriodes* and *Klutiana* (Hymenoptera, Ichneumonidae): recognition of the Nesomesochorinae stat. rev. and Nonninae stat. nov. and transfer of *Skiapus* and *Hellwigia* to the Ophioninae. *Journal of Natural History* 39, 2559–2578.

Quicke, D.L.J., Laurenne, N.M., Fitton, M. G. and Broad, G. R. (2009) A thousand and one wasps: a 28S rDNA and morphological phylogeny of the Ichneumonidae (Insecta: Hymenoptera) with an investigation into alignment parameter space and elision. *Journal of Natural History* 43, 1305–1421.

Quicke, D.L.J., Austin, A.D., Fagan-Jeffries, E., Hebert, P.D.N., Butcher, B.A. (2019) Molecular phylogeny places the enigmatic subfamily Masoninae within the Ichneumonidae, not the Braconidae. *Zoologica Scripta* 49, 64–71.

Reshchikov, A. and van Achterberg, C. (2014) Review of the genus *Metopheltes* Uchida, 1932 (Hymenoptera, Ichneumonidae) with description of a new species from Vietnam. *Biodiversity Data Journal* 2, e1061. doi: 10.3897/BDJ.2.e1061

Reshchikov, A. and van Achterberg, C. (2018) The Unicorn exists! A remarkable new genus and species of Perilissini (Hymenoptera: Ichneumonidae) from South East Asia. *Acta Entomologica Musei Nationalis Pragae* 58, 523–529. doi: 10.2478/aemnp-2018-0041

Reshchikov, A., Kumar, G. and van Achterberg, C. (2014) Review of Asiatic *Neurogenia* Roman, 1910 (Hymenoptera, Ichneumonidae) with description of three new species. *Tijdschrift voor Entomologie* 157, 123–135.

Reshchikov, A., Choi, J.-K., Xu, Z.-F. and Pang, H. (2017) Two new species of the genus *Rhorus* Förster, 1869 from Thailand (Hymenoptera, Ichneumonidae). *Journal of Hymenoptera Research* 54, 79–92. doi: 10.3897/ jhr.54.11662

Reshchikov, A., Santos, B. F., Liu, J.-X. and Barthélémy, C. (2019) Review of *Palpostilpnus* Aubert (Hymenoptera, Ichneumonidae, Phygadeuontinae), with the description of ten new species. *European Journal of Taxonomy* 582, 1–63. doi: 10.5852/ejt.2019.582

Reshchikov, A., Quicke, D.L.J. and Butcher, B.A. (2022) A remarkable new genus and species of Euryproctini (Hymenoptera: Ichneumonidae, Ctenopelmatinae) from Thailand. *European Journal of Entomology* 834, 102–116. doi: 10.5852/ejt.2022.834.1903

Riedel, M. (2011) Contribution to the Ichneumoninae (Hymenoptera, Ichneumonidae) of Southeastern Asia: 1. Tribes Clypeodromini, Listrodromini, Goedartini, Compsophorini, and Platylabini. *Linzer Biologische Beiträge* 43, 1549–1572.

Riedel, M. (2013) Contribution to the Ichneumoninae (Hymenoptera, Ichneumonidae) of Southeastern Asia: 2. Tribe Heresiarchini. *Linzer Biologische Beiträge* 45, 2025–2076.

Riedel, M. (2017a) Contribution to the Ichneumoninae (Hymenoptera, Ichneumonidae) of Southeastern Asia: 3. Heresiarchini. *Linzer Biologische Beiträge* 49, 895–917.

Riedel, M. (2017b) Contribution to the Ichneumoninae (Hymenoptera, Ichneumonidae) of Southeastern Asia: 4. Platylabini, Eurylabini, and Oedicephalini. *Linzer Biologische Beiträge* 49, 1275–1307.

Riedel, M. (2019) Four new species of the genus *Megalomya* Uchida (Hymenoptera, Ichneumonidae, Alomyinae) from Laos. *Linzer Biologische Beiträge* 51, 179–187.

Riedel, M. (2023a) New contribution to the Oriental species of *Anisobas* Wesmael and *Listrodromus* Wesmael (Hymenoptera, Ichneumonidae, Ichneumoninae). *Linzer Biologische Beiträge* 54(2), 625–639.

Riedel, M. (2023b) Five new Oriental species of the genus *Ichneumon* Linnaeus (Hymenoptera, Ichneumonidae, Ichneumoninae). *Linzer Biologische Beiträge* 55, 47–60.

Riedel, M. (2023c) Contribution to the knowledge of the Ichneumoninae (Hymenoptera, Ichneumonidae) from maritime Southeast Asia. *Zootaxa* 5363(1), 1–94.

Riedel, M. (2023d) Contribution to the taxonomy of the Southeast Asian Mesochorinae (Hymenoptera, Ichneumonidae). *Zootaxa* 5245, 1–072.

Riedel, M., Vu Van. L. and Schmidt, S. (2021) First record of the subfamily Oxytorinae (Insecta, Hymenoptera, Ichneumonidae) from the Oriental Region, with descriptions of two new species from Vietnam. *Biodiversity Data Journal* 9, e69867. doi: 10.3897/BDJ.9.e69867

Rizki, A. and Idris, A.B. (2006) A new species of *Camptotypus* Kriechbaumer (Hymenoptera: Ichneumonidae: Pimplinae) from Malaysia. *Serangga* 11, 107–115.

Rizki, A. and Idris, A.B. (2010) One new species and one new record of the genus *Zaglyptus* Foerster (Hymenoptera: Ichneumonidae: Pimplinae) from Malaysia. *Serangga* 15, 9–24.

Romani, R., Ruschioni, R., Riolo, P. and Isidoro, N. (2018) Transmission and scanning electron microscopic observations on antennal apical pegs in the wasp species Pimplinae (Insecta, Hymenoptera). *Micron* 107, 72–78.

Rousse, P., Quicke, D.L.J., Matthee, C.A., Lefeuvre, P. and van Noort S. (2016) A molecular and morphological reassessment of the phylogeny of the subfamily Ophioninae (Hymenoptera: Ichneumonidae). *Zoological Journal of the Linnean Society* 178, 128–148.

Santos, B.F. (2017) Phylogeny and reclassification of Cryptini (Hymenoptera, Ichneumonidae, Cryptinae), with implications for ichneumonid higher level classification. *Systematic Entomology* 42, 650–676. https://doi.org/10.1111/syen.12238

Santos, B.F., Wahl, D.B., Rousse, P., Bennett, A.M.R., Kula, R. and Brady, S.G. (2021) Phylogenomics of Ichneumoninae (Hymenoptera, Ichneumonidae) reveals pervasive morphological convergence and the shortcomings of previous classifications. *Systematic Entomology* 46, 704–724. doi: 10.1111/syen.12484

Sayama, K. and Konishi, K. (1996) Biological notes on *Sphecophaga vesparum* (Curtis) (Hymenoptera: Ichneumonidae) from Japan. *Japanese Journal of Entomology* 64, 889–890.

Sharanowski, B., Ridenbauch, R.D., Piekarski, P.K., Broad, G., Burke, G.R., Deans, A., Lemmon, A.R., Lemmon, E.C.M., Diehl, G., Whitfield, J.B. and Hines, H.M. (2021) Phylogenomics of Ichneumonoidea (Hymenoptera) and implications for evolution of mode of parasitism and viral endogenization. *Molecular Phylogenetics and Evolution* 156, 107023. doi: 10.1016/j.ympev.2020.107023

Sharifi, S. and Javadi, I. (1971) Biology of *Xorides corcyrensis* Kriech. (Hymenoptera: Ichneumonidae), a parasite of the Rosaceae branch borer *Osphranteria coerulescens* Redt. (Coleoptera: Cerambycidae). *Zeitschrift für Angewandte Entomologie* 68, 25–31.

Sharkey, M.J. and Wahl, D.B. (1992) Cladistics of the Ichneumonoidea (Hymenoptera). *Journal of Hymenoptera Research* 1, 15–24.

Sharkey, M.J. and Wharton, R.A. (1997) Morphology and terminology. In: Wharton, R.A., Marsh, P.M. and Sharkey, M.J. (eds) *Manual of the New World genera of the family Braconidae (Hymenoptera)*. Special Publication of the International Society of Hymenopterists No. 1, Washington, DC, pp. 19–37.

Shaw, M.R. (2004). Notes on the biology of *Lycorina triangulifera* Holmgren (Hymenoptera: Ichnemonidae: Lycorininae). *Journal of Hymenoptera Research* 13, 302–308.

Shaw, M.R. (2009) Notes on the host-feeding and hyperparasitic behaviours of *Itoplectis* species (Hymenoptera: Ichneumonidae, Pimplinae). *Entomologist's Gazette* 60, 113–116.

Shaw, M.R. (2014) Illustrated notes on the biology of two European species of *Euceros* Gravenhorst (Hymenoptera: Ichneumonidae: Eucerotinae). *Proceedings of the Russian Entomological Society* 85, 122–132.

Shaw, M.R. and Askew, R.R. (1976) Ichneumonoidea (Hymenoptera) parasitic upon leaf-mining insects of the orders Lepidoptera, Hymenoptera and Coleoptera. *Ecological Entomology* 1(2), 127–133.

Shaw, M.R. and Borisova, N.V. (2018) A novel host of *Itoplectis viduata* (Gravenhorst) (Hymenoptera: Ichneumonidae, Pimplinae), with some wider rearing records. *Entomologist's Gazette* 69, 25–27.

Shaw, M.R. and Voogd, J. (2016) Illustrated notes on the biology of *Sphinctus serotinus* Gravenhorst (Hymenoptera, Tryphoninae, Sphinctini). *Journal of Hymenoptera Research* 49, 81–93.

Shaw, M.R. and Wahl, D.B. (1989) The biology, egg and larvae of *Acaenitus dubitator* (Panzer) (Hymenoptera, Ichneumonidae: Acaenitinae). *Systematic Entomology* 14(1), 117–125.

Sheng, M.L. and Sun, S.P. (2010) *Parasitic Ichneumonids on Woodborers in China (Hymenoptera: Ichneumonidae)*. Science Press, Beijing, 338 pp. [in Chinese with English summary]

Sheng, M.L. and Sun, S.P. (2011) A new genus and species of Brachyscleromatinae (Hymenoptera: Ichneumonidae) from China, *Laxiareola ochracea*. *Journal of Insect Science* 11, 27. Available at: insectscience.org/11.27

Sheng, M.-L. and Sun, S.-P. (2014) *Combivena* gen. n. (Hymenoptera: Ichneumonidae: Acaenitinae) from China. *Journal of Insect Science* 14, 158. doi: 10.1093/jisesa/ieu020

Sheng, M.L., Sun, S.P., Zhang, Y. and Liang, Y.-P. (2016) Three new species of the genus *Notosemus* Förster, 1869 (Hymenoptera, Ichneumonidae, Ichneumoninae) from China, with a key to the world species. *European Journal of Taxonomy* 209, 1–19. doi: 10.5852/ejt.2016.209

Shimazaki, M., Watanabe, K. and Shimazaki, Y. (2011) A record of the koinobiont endoparasitoid wasp, *Melalophacharops everese* (Hymenoptera, Ichneumonidae), attacking eggs of a lycaenid butterfly, *Acytolepis puspa*. *Lepidoptera Science* 62, 151–155.

Shimizu, S. and Alvarado, M. (2020) A new genus and two new species of the subfamily Nesomesochorinae Ashmead (Insecta: Hymenoptera: Ichneumonidae). *Neotropical Entomology* 49, 704–712. doi: 10.1007/s13744-020-00778-7

Shimizu, S. and Konishi, K. (2016) New record of the genus *Dictyonotus* Kriechbaumer, 1894 (Hymenoptera: Ichneumonidae: Ophioninae) from Vietnam. *Japanese Journal of Systematic Entomology* 22, 159–160.

Shimizu, S. and Konishi, K. (2018) A preliminary checklist of the Laotian species of the genus *Enicospilus* Stephens, 1835 (Hymenoptera: Ichneumonidae: Ophioninae), with 11 new species records from Laos. *Japanese Journal of Systematic Entomology* 24, 153–162.

Shimizu, S. and Ogawa, R. (2018) Discovery of the subfamily Lycorininae Cushman & Rohwer, 1920 (Hymenoptera: Ichneumonidae) from Indonesia, based on *Lycorina longicauda* Shimizu, sp. nov., with a key to the Oriental *Lycorina* species. *Austral Entomology* 58, 148–155. doi: 10.1111/aen.12347

Shimizu, S., Broad, G.R. and Maeto, K. (2020) Integrative taxonomy and analysis of species richness patterns of nocturnal Darwin wasps of the genus *Enicospilus* Stephens (Hymenoptera, Ichneumonidae, Ophioninae) in Japan. *ZooKeys* 990, 1–144. doi: 10.3897/zookeys.990.55542

Sime, K.R. and Wahl, D.B. (2002) The cladistics and biology of the *Callajoppa* genus-group (Hymenoptera: Ichneumonidae, Ichneumoninae). *Zoological Journal of the Linnean Society* 134, 1–56. doi: 10.1046/j.1096-3642.2002.00006.x

Simmonds, F.J. (1947) The biology of *Phytodietus pulcherrimus* (Cress.) (Ichneumonidae, Tryphoninae) parasitic of Loxostege sticticalis L. in North America. *Parasitology* 38, 150–156.

Spasojevic, T., Broad, G.R., Sääksjärvi, I.E., Schwarz, M., Ito, M., Korenko, S., Klopfstein, S. and Ho, S. (2021) Mind the outgroup and bare branches in total-evidence dating: a case study of pimpliform Darwin wasps (Hymenoptera, Ichneumonidae). *Systematic Biology* 70, 322–339.

Spradbery, J.P. (1968) The biology of *Pseudorhyssa sternata* Merrill (Hym., Ichneumonidae), a cleptoparasite of siricid woodwasps. *Bulletin of Entomological Research* 59, 291–297.

Steiner, S., Kropf, C., Graber, W., Nentwig, W. and Klopfstein, S. (2010) Antennal courtship and functional morphology of tyloids in the parasitoid wasp *Syrphoctonus tarsatorius* (Hymenoptera: Ichneumonidae: Diplazontinae). *Arthropod Structure & Development* 39, 33–40.

Takasuka, K. and Matsumoto, R. (2011) Lying on the dorsum: unique host-attacking behaviour of *Zatypota albicoxa* (Hymenoptera, Ichneumonidae). *Journal of Ethology* 29, 203–207.

Takasuka, K. and Watanabe, K. (2012) New records of *Zatypota albicoxa* (Hymenoptera, Ichneumonidae) and its potential host spider *Parasteatoda tepidariorum* (Araneae, Theridiidae) from Vietnam. *Japanese Journal of Systematic Entomology* 18, 447–450.

Takasuka, K., Matsumoto, R. and Ohbayashi, N. (2009) Oviposition behavior of *Zatypota albicoxa* (Hymenoptera, Ichneumonidae), an ectoparasitoid of *Achaearanea tepidariorum* (Araneae, Theridiidae). *Entomological Science* 12, 232–237.

Takasuka, K., Watanabe, K. and Konishi, K. (2011a) Genus *Cryptopimpla* Taschenberg new to Sulawesi, Indonesia, with description of a new species (Hymenoptera, Ichneumonidae, Banchinae). *Journal of Hymenoptera Research* 23, 65–75. doi: 10.3897/jhr.23.1595

Takasuka, K., Yoshida, H., Nugroho, P. and Matsumoto, R. (2011b) A new record of *Zatypota albicoxa* (Hymenoptera: Ichneumonidae) from Indonesia, with description of a new species of its host spider (Araneae: Theridiidae). *Zootaxa* 2910, 63–68.

Talekar, N.S. (2004) Biological control of diamondback moth in Asia. Improving Biocontrol of *Plutella xylostella*. CIRAD, Montpellier, France, pp.103–113.

Tereshkin, A. (2009) Illustrated key to the tribes of subfamilia Ichneumoninae and genera of the tribe Platylabini of world fauna (Hymenoptera, Ichneumonidae). *Linzer Biologische Beiträge* 41, 1317–1608.

Townes, H.K. (1969) The genera of Ichneumonidae, Part 1. *Memoirs of the American Entomological Institute* 11, 1–300.

Townes, H.K. (1970a) The genera of Ichneumonidae, Part 2. Gelinae. *Memoirs of the American Entomological Institute* 12, 1–537.

Townes, H.K (1970b) The genera of Ichneumonidae, Part 3. Banchinae, Scolobatinae & Porizontinae. *Memoirs of the American Entomological Institute* 13, 1–307.

Townes, H.K (1971) The genera of Ichneumonidae, Part 4. Cremastinae to Diplazontinae *Memoirs of the American Entomological Institute* 17, 1–372.

Townes, H.K. and Chiu, S.C. (1970) The Indo-Australian species of *Xanthopimpla* (Ichneumonidae). *Memoirs of the American Entomological Institute* 14, 1–372.

Townes, H.K. and Townes, M. (1951) Family Ichneumonidae. In Muesebeck, C.F.W., Krombein, K.V. and Townes, H.K. (eds) *Hymenoptera of America North of Mexico - Synoptic Catalog*. USDA. Agriculture Monograph. No. 2, pp. 184–409.

Townes, H.K. and Townes, M. (1960) Ichneumon-flies of America north of Mexico. Subfamilies Ephialtinae, Xoridinae, Acaentinae. *United States National Museum Bulletin* 216, 1–676.

Townes, H.K. and Townes, M. (1966) A catalogue and reclassification of Neotropic Ichneumonidae. *Memoirs of the American Entomological Institute* 8, 1–367.

Townes, H.K., Townes, M. and Gupta, V.K. (1961) A catalogue and reclassification of the Indo-Australian Ichneumonidae. *Memoirs of the American Entomological Institute* 1, 1–522.

Tripp, H.A. (1961) The biology of a hyperparasite, *Euceros frigidus* Cress. (Ichneumonidae) and description of the planidial stage. *The Canadian Entomologist* 00, 40–50.

Tripp, H.A. (1962) The biology of *Perilampus hyalinus* Say (Hymenoptera: Perilampidae), a primary parasite of *Neodiprion swainei* Midd. (Hymenoptera: Diprionidae) in Quebec, with descriptions of the egg and larval stages. *The Canadian Entomologist* 94, 1250–1270.

Ubaidillah, R., Yamaguchi, G. and Kojima, J.-I. (2009) A new *Arthula* Cameron (Ichneumonidae, Cryptinae) parasitoid of *Ropalidia plebeiana* Richards (Vespidae) and host of *Amoturoides breviscapus* Girault (Torymidae) (Hymenoptera). *Zootaxa* 2274, 45–50.

van Achterberg, C. (1988) Revision of the subfamily Blacinae Foerster (Hymenoptera: Braconidae). *Zoologische Verhandelingen, Leiden* 249, 1–324.

van Baarlen, P., Topping, O.J., and Sunderland, K.D. (1996) Host location by *Gelis festinans*, an eggsac parasitoid of the linyphiid spider *Erigone atra*. *Entomologia Experimentalis et Applicata* 81, 155–163.

van Rossem, G. (1990) Key to the genera of the Palaearctic Oxytorinae, with the description of three new genera (Hymenoptera: Ichneumonidae). *Zoologische Mededelingen* 63, 309–323.

Varga, O. and Reshchikov, A. (2015a) New records of the genus *Polysphincta* Gravenhorst, 1829 (Hymenoptera: Ichneumonidae: Pimplinae) from the Oriental region. *Zootaxa* 3955, 435–443. doi: 10.11646/zootaxa.3955.3.10

Varga, O. and Reshchikov, A. (2015b) First record of the genus *Acropimpla* Townes, 1960 (Hymenoptera: Ichneumonidae: Pimplinae) from Thailand, with descriptions of three new species. *Zootaxa* 4013, 556–570.

Varga, O., Reshchikov, A. and Broad, G.R. (2014) First record of the genus *Aplomerus* Provancher, 1886 (Hymenoptera: Ichneumonidae: Xoridinae) from the Oriental region, with descriptions of two new species. *Zootaxa* 3815, 591–599.

Varley, G.C., (1964) A note on the life history of the ichneumon fly *Euceros unifasciatus* (Voll.) with a description of its planidium-larvae. *Entomologist's Monthly Magazine* 100, 113–116.

Vas, Z. (2017) Data to the Vietnamese ichneumon wasp fauna with description of a new *Teleutaea* species (Hymenoptera: Ichneumonidae). *Folia Entomologica Hungarica* 78, 101–110.

Verkerk, R.H.J. and Wright, D.J. (1997) Field-based studies with the diamondback moth tritrophic system in Cameron Highlands of Malaysia: Implications for pest management. *International Journal of Pest Management* 43, 27–33.

Vilhelmsen, L., Isidoro, N., Romani, R., Basibuyuk, H.H. and Quicke, D.L.J. (2001) Host location and oviposition in a basal group of parasitic wasps: the subgenual organ, ovipositor apparatus, and associated structures in the Orussidae (Hymenoptera, Insecta). *Zoomorphology* 121, 63–84.

Wäckers, F.L., Mitter, E. and Dorn, S. (1998) Vibrational sounding by the pupal parasitoid *Pimpla* (*Coccygomimus*) *turionellae*: an additional solution to the reliability-detectability problem. *Biological Control* 11, 141–146.

Wahl, D.B. (1986) Larval structures of oxytorines and their significance for the higher classification of some Ichneumonidae (Hymenoptera). *Systematic Entomology* 11, 117–127. doi: 10.1111/j.1365-3113.1986.tb00171.x

Wahl, D.B. (1991) The status of *Rhimphoctona*, with special reference to the higher categories within Campopleginae and the relationships of the subfamily (Hymenoptera: Ichneumonidae). *Transactions of the American Entomological Society* 117, 193–213.

Wahl, D.B. (1993a) Family Ichneumonidae. In: Goulet, H. and Huber, J. (Eds) *Hymenoptera of the World: An Identification Guide to Families (Vol. vii)*. Ottawa, Canada: Research Branch, Agriculture Canada, pp. 395–448, 478–509 (Figs 154–183).

Wahl, D.B. (1993b) Cladistics of the genera of Mesochorinae (Hymenoptera: Ichneumonidae). *Systematic Entomology* 18, 371–387.

Wahl, D.B. and Gauld, I.D. (1998) The cladistics and higher classification of the Pimpliformes (Hymenoptera: Ichneumonidae). *Systematic Entomology* 23, 299–303.

Stephanids are most diverse in the Oriental, S.E. Asia and Afrotropical Regions; only a few species have been reported from Nearctic and only four are known from Europe (Hilszczański, 2011), of which *Foenatopus turcomanorum* and *Afromegischus gigas* are known in Europe only on the island of Crete (Hilszczański, 2011).

Stephanid wasps are solitary idiobiont ectoparasitoids of wood-boring beetle larvae but virtually all biological data concern only a handful of extralimital species (e.g. Rodd, 1951; Blüthgen, 1953; Mateu, 1972; Benoit, 1984; Halstead, 1986; Jansen et al., 1988; Aguiar et al., 2010; Tan et al., 2015), most records being from Buprestidae. A few other families are also attacked (such as Bostrychidae, Buprestidae and Cerambycidae) and one species is an important larval parasitoid of siricid wasps and a few solitary bees (Taylor, 1967; Aguiar, 2004). The extralimital *Schlettererius cinctipes*, from North America, is a parasitoid of horntail wasps (Hymenoptera: Siricidae) and has been introduced to Tasmania as a biological control agent (Hong et al., 2011).

Collectively, the few papers with some biological data indicate that different species are associated with wood of different ages of attack, from still-living trees to fallen trunks long after the bark has rotted off.

The fore leg and the large, distally swollen hind tibia contain a large vibration-detecting subgenual organ whose structure was described in detail by Vilhelmsen et al. (2008). In addition, the hind tarsi have only three segments and are highly modified. These assumed to be adaptations for detecting vibrations produced by moving/chewing hosts, often deep within wood. Unlike Orussidae and many cryptine Ichneumonidae, the antennae are not modified and so their host-searching does not involve active vibrational sounding (see *Echolocation* in Chapter 3, this volume).

Wing Venation

Stephanid wings are narrow (Fig. 11.1, 11.2) and rather short relative to the wasps' elongate bodies, never extending beyond the apex of the metasoma (Fig. 11.2). An important point to note is that the fore wing costal cell is not enclosed by a costal vein (C) anteriorly, but the cell is often folded at right angles to the rest of the wing and can be dark-pigmented, such that at a quick glance it appears as if there is no costal cell, which might lead beginners to key them to Ichneumonoidea. Also, vein 3RS does not reach the wing margin as a tubular vein.

Recognition

Stephanids are very uniform in general appearance (Fig. 11.2), and having seen one or two specimens there should be no need to use a key to identify them – they are that distinctive. The only group that might reasonably be confused with them are the Gasteruptiidae, but the latter lack the 'crown of thorns' on the head.

Unique characteristics of the Stephanidae are 'crown' or ocellar corona on their heads (Fig. 11.3 c); only stephanids and

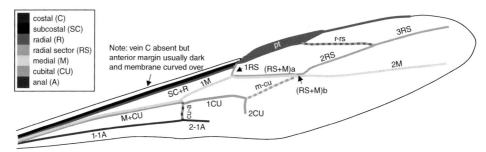

Fig. 11.1. Stylised stephanid wing venation using the same naming system used throughout this book.

Fig. 11.2. Unidentified stephanid female from Singapore. (Source: photograph by and © Zestin W.W. Soh (Singapore Botanic Gardens), reproduced with permission.)

Orussidae have this character. Medium-size to large, ranging from 2 mm to 20 mm (some species are larger than this), the largest species are in the genus *Megischus*, with body length nearly to 35 mm (Hong *et al.*, 2011), modified pronotum, mainly black or dark brown with head and legs, yellowish-, orange- or reddish-brown, the presence of ivory streaks on the frons or temple and ivory or whitish subapical band of the ovipositor sheath in some genera (van Achterberg, 2002; van Achterberg and Yang, 2004), modified hind legs with a swollen hind femur with teeth on the underside (Fig. 11.3 b).

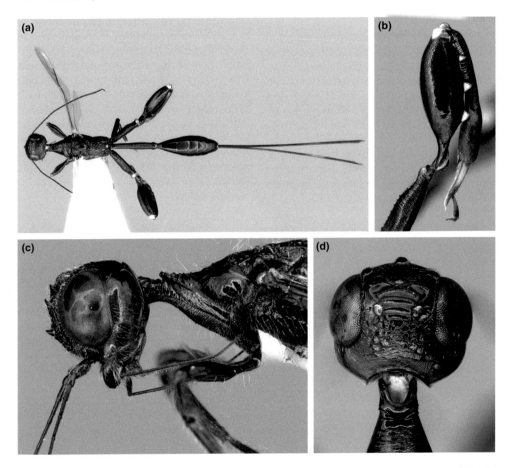

Fig. 11.3. Characters of *Foenatopus* sp., Stephaniidae: **(a)** habitus dorsal view, note the characteristic way that the hind coxae and femora are splayed out relative to the body; **(b)** detail of hind leg showing swollen and ventrally dentate femur and short, two-segmented tarsus; **(c)** head and pronotum lateral view showing the thorn-like projections of the frons; **(d)** head dorsal view showing characteristic sculpture and rather globular shape.

Identification

There are 13 recognised genera of stephanid wasps, of which only five are known from S.E. Asia, viz. *Foenatopus*, *Megischus*, *Parastephanellus*, *Pseudomegischus* and *Stephanus*. Van Achterberg (2002) provided a key to the world genera and Aguiar (2004) catalogued the world species. Hong et al. (2011) revised the Chinese Stephanidae.

S.E. Asian Genera

Foenatopus

The members of this genus can be found in tropical and subtropical areas of the Afrotropical and Indo-Australian regions. Unique characteristics of *Foenatopus* are its very slender neck with fine striations (Fig. 11.3 c, d) (van Achterberg, 2002). The S.E. Asian species are in urgent need of revision. Van Achterberg (2002) noted that there is considerable intrageneric variation and that it may, in the future, need to be split into a number of smaller genera. More than 50 species have been described from S.E. Asia (Aguiar, 2004).

Megischus

There are about 50 described species from around the world. They are mainly found in the Indo-Australian and Neotropical Regions, but one species has been recorded from central and southern Europe (van Achterberg, 2002). Approximately 30 species have been described from S.E. Asia. Van Achterberg (2002) revised the Old World species. Ge et al. (2023) provide a key to the four species known from Vietnam as well as the description of the new species.

Parastephanellus

This genus can be found in the Oriental and Australian regions. Characteristics of the genus are medium-sized to rather small species with a pale yellowish streak behind eyes and with an Indo-Australian distribution; neck is not separated from the pronotum; ovipositor sheath without subapical ivory band (van Achterberg, 2002).

Pseudomegischus

This genus was erected by van Achterberg (2002), and named because of its general resemblance to *Megischus*. It can be found in the Old World, including in Indonesia and the Philippines. Its distinguishing characteristics include an ovipositor sheath without ivory subapical band, and temple with pale yellowish streak behind eye. Species have been reared in China from wood infested with both siricid woodwasps and cerambycids, but the precise host is not known. Van Achterberg (2002) revised the Old World species, and this was followed by Tan et al. (2015), who provided an updated key to world species.

Stephanus

This is a small genus with only three known species (Aguiar, 2004). This genus can be differentiated from the others by having hind tarsus with five segments; hind femur with three large ventral teeth; legs with ivory patches; propodeum with fine irregular sculpture. Members of this genus are recorded from west Palaearctic (*S. serrator*) and from the Oriental regions (*S. borneensis* from Java, and 'Borneo', Indonesia, and *S. soror* from W. Malaysia). Van Achterberg (2002) revised the Old World species.

References

Aguiar, A.P. (2004) World catalog of the Stephanidae (Hymenoptera: Stephanoidea). *Zootaxa* 753, 1–120.
Aguiar, A.P., Jennings, J.T. and Turrisi, G.G. (2010) Three new Middle-Eastern species of *Foenatopus* Smith (Hymenoptera: Stephanidae) with a new host record and key to species with two spots on the metasoma. *Zootaxa* 2714, 40–58.

Aguiar, A.P., Deans, A.R., Engel, M.S., Forshage, M., Huber, J.T., Jennings, J.T., Johnson, N.F., Lelej, A.S., Longino, J.T., Lohrmann, V., Mikó, I., Ohl, M., Rasmussen, C., Taeger, A. and Yu, D.S.K. (2013) Order Hymenoptera. *Zootaxa* 3703, 51–62. doi: 10.11646/zootaxa.3703.1.12

Benoit, P.L.G. (1949) Les Stephanoidae (Hym.) du Congo belge. *Revue de Zoologie et de Botanique Africaines* 42, 285–294.

Benoit, P.L.G. (1984) Stephanidae du Sahara (Hymenoptera). *Revue de Zoologie Africaine* 98, 434–439.

Binoy, C., van Achterberg, C., Girish Kumar, P., Santhosh, S. and Sheela, S. (2020) A review of Stephanidae (Hymenoptera: Stephanoidea) from India, with the description of five new species. *Zootaxa* 4383, 1–51.

Blüthgen, P. (1953) Zur Biologie von *Stephanus serrator* F. (Hym., Stephanidae). *Zoologischer Anzeiger* 150, 229–234.

Engel, M.S. and Huang, D. (2017) A new crown wasp in Cretaceous amber from Myanmar (Hymenoptera: Stephanidae). *Cretaceous Research* 69, 56–61.

Engel, M.S., Grimaldi, D.A. and Ortega-Blanco, J. (2013) A stephanid wasp in mid-Cretaceous Burmese amber (Hymenoptera: Stephanidae), with comments on the antiquity of the Hymenopteran radiation. *Journal of the Kansas Entomological Society* 86(3), 244–252.

Ge, S.-X., Ren, L.-L. and Tan, J.L. (2023) A new species and new record species of *Megischus* Brullé (Hymenoptera, Stephanidae) from Vietnam. *Journal of Hymenoptera Research* 96, 723–734.

Gupta, A. and Oawao, S.M. (2020) A new species of the genus *Foenatopus* Smith (Hymenoptera: Stephanoidea: Stephanidae) from India. *Zootaxa* 4801, 389–394.

Halstead, J.A. (1986) Distribution and seasonality of *Megischus* spp. (Hymenoptera: Stephanidae) in California. *Entomological News* 97, 101–103.

Hilszczański, J. (2011) New data on the occurrence of stephanids (Hymenoptera: Stephanidae) in Turkey and Greece. *Opole Scientific Society Nature Journal* 44, 192–196.

Hong, C.D., van Achterberg, C. and Xu, Z.F. (2011) A revision of the Chinese Stephanidae (Hymenoptera, Stephanoidea). *Zookeys* 110, 1–108.

Jansen, E., Bense, J. and Schrameyer, K. (1988) *Stephanus serrator* (Fabricius, 1978) in der Bundesrepublik Deutschland (Hymenoptera, Stephanidae). *Entomofauna* 9, 421–428.

Mateu, J. (1972) *Les insectes xylophages des Acacia dans les régions sahariennes*. Publicações do Instituto de Zoologia 'Dr. Augusto Nobre'. Faculdade de Ciências do Porto. No. 116. 714 pp.

Rasnitsyn, A.P. (1969) [Origin and evolution of the lower Hymenoptera]. *Trudy Paleontologicheskogo Instituta. Akademiya Nauk SSSR* 123, 1–195. [In Russian: English translation, 1979. United States Department of Agriculture, Washington, DC]

Rodd, N.W. (1951) Some observations on the biology of Stephanidae and Megalyridae (Hym.). *Australian Zoologist* 11, 341–346.

Tan, J.L., Fan, X.L., van Achterberg, C. and Li, T. (2015) A new species of *Pseudomegischus* van Achterberg from China, with a key to the species (Hymenoptera, Stephanidae). *ZooKeys* 537, 103–110. doi: 10.3897/zookeys.537.6592

Taylor, K.L. (1967) Parasitism on *Sirex noctilio* F. by *Schlettererius cinctipes* (Cresson) (Hymenoptera: Stephanidae). *Journal of the Australian Entomological Society* 6, 13–19. doi: 10.1111/j.1440-6055.1967.tb02132.x

van Achterberg, C. (2002) A revision of the Old World species of *Megischus* Brullé, *Stephanus* Jurine and *Pseudomegischus* gen. nov., with a key to the genera of the family Stephanidae (Hymenoptera: Stephanoidea). *Zoologische Verhandelingen Leiden* 339, 3–204.

van Achterberg C. and Yang Z.Q. (2004) New species of the genera *Megischus* Brullé and *Stephanus* Jurine from China (Hymenoptera: Stephanoidea: Stephanidae), with a key to world species of the genus *Stephanus*. *Zoologische Medelingen Leiden* 78, 101–117.

Vilhelmsen L., Turrisi, G.F. and Beutel, R.G. (2008) Distal leg morphology, subgenual organs and host detection in Stephanidae (Insecta, Hymenoptera). *Journal of Natural History* 42, 1649–1663.

Vilhelmsen L., Mikó, I. and Krogmann, L. (2010) Beyond the wasp-waist: structural diversity and phylogenetic significance of the mesosoma in apocritan wasps (Insecta: Hymenoptera). *Zoological Journal of the Linnean Society* 159, 22–194.

The *Pristaulacus comptipennis* species group (Fig. 12.2) is endemic to S.E. Asia and ranges from China to Japan, Thailand, Laos and Vietnam (Turrisi and Smith, 2011). In other members of *Pristaulacus* posterior margin of the head is straight or weakly concave; however, in the *P. comptipennis* species group, occipital carinae deeply emarginate in the head which can be easily noticed in the dorsal view (Turrisi and Smith, 2011). Extensive searching of museum collections, along with many international collecting projects, has revealed many new species (Turrisi and Smith, 2011, 2020; Turrisi and Watanabe, 2011; Turrisi and Madl, 2013; Turrisi, 2014; Turrisi and Nobile, 2016; Kuroda, 2018). Turrisi (2013) revised the Oriental species of *Aulacus* to which Smith (2017a) added and differentiated a new species from the Philippines. Smith (2017b) provided a key to the Philippines species of *Pristaulacus*.

Evaniidae (Ensign Wasps)

Evaniids are the most speciose among the three subfamilies in Evanioidea. These wasps can be easily recognised by their short metasoma which is flag-shaped and attached to a long and thin petiole and often pumped up and down in a chopping motion; therefore they are called ensign wasps or hatchet wasps (Deans *et al.*, 2006). Both mesosoma and metasoma of evaniids are used to putatively simulate the shape and size of the oothecae within which the wasp larvae develop (Basibuyuk *et al.*, 2002; Deans and Huben, 2003; Vilhelmsen *et al.*, 2010). Moreover, ensign wasps share the following characteristics: (i) jugal lobes in fore and hind wing are distinctly separated (Fig. 12.3 a, b); (ii) distal part of fore wing without a tubular cross-vein; (iii) similar number of antennal flagellomeres in males and females; (iv) head is rather fixed and attached to the thorax on a short neck; and (v) ovipositor is usually short and thin (Li *et al.*, 2018). Currently, there are 21 extant genera with over 449 species described (Shih *et al.* 2020), including additional 17 extinct genera with 35 species from later Mesozoic and Paleogene deposits (Deans *et al.*, 2004; Deans, 2005; Li *et al.*, 2018). Evaniid wasps are frequently collected, but these wasps have been overlooked by ecologists and systematists since evaniid research was decreased in the 20th century. Deans (2002) reported that only one extant genus had been described since 1953. Of the 35 extinct species, 11 species have been recorded from Myanmar (Shih *et al.*, 2020). They are usuallt

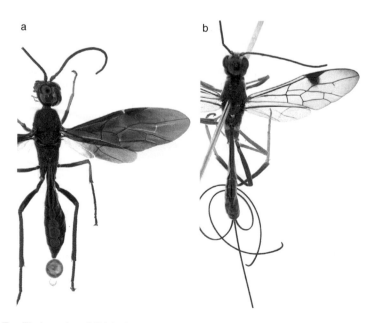

Fig. 12.2. Two Thai species of *Pristaulacus*.

black and/or orange-red, some with white on legs and antennae.

Ensign wasps are solitary egg predators of cockroach ootheca, therefore they can be used as potential natural enemies to control pestiferous cockroaches. Not much has been done to test their effectiveness (Deans and Huben, 2003) and one of the main reasons for this could be lack of evaniid classification, which is the main obstacle for further research (Deans et al., 2006).

So far, four genera have been reported from S.E. Asia as follows: *Evania, Parevania, Prosevania* and *Brachygaster* (Fig. 12.3 b). In addition, Deans (2002) described an apterous genus, *Papatuka*, from Papua New Guinea. Within the region, *Brachygaster* was known from Singapore. Elliott (2005) described a species from Australia as a major range extension and we have recently collected it in Thailand (Fig. 12.3 b). They may be distinguished using the following key.

Key to Evaniidae Genera of S.E. Asia (modified from Deans and Huben, 2003)

1	Fore wing with 1–3 cells enclosed by tubular veins (Fig. 12.3 b); legs relatively short; entire body densely foveate (Fig. 12.3 b) .... ***Brachygaster***
-	Fore wing with at least 6 cells enclosed by tubular veins (Fig. 12.3 c); not as above 2
2(1)	Distance between mid and hind coxa nearly equal to distance between fore and mid coxa (Fig. 12.3 c); mid coxa never touching hind coxa when projected posteriorly; fore wing RS+M present separating 1st subdiscal cell and 1st submarginal cell; head in lateral view slightly compressed; antennae arising on upper third of head (Fig. 12.3 d); ovipositor long and usually visible; female metasoma in lateral view triangular with metasomal tergite 8 expanded dorsally............................***Evania***
-	Distance between mid and hind coxa 0.6× or less than distance between fore and mid coxa; mid coxa often touching hind coxa when projected posteriorly; not as above 3
3(2) -	Gena costate or striate; sometimes irregular; legs relative long > 3× as long as mesosoma height...***Prosevania***
-	Gena smooth and shiny, setose, punctate and foveate; not as above.........***Parevania***

When an evaniid female finds a cockroach's ootheca, she will vibrate her antennae over it. Then the female lies on her side with her body parallel to the long axis of the egg case with the legs braced against the latter and the body against a substrate. Then she has to penetrate the tough cuticle of ootheca by wriggling the abdomen and then she inserts her ovipositor. This process usually takes from 15 to 30 minutes, with only one egg per ootheca. Pupation occurs within the egg case without a cocoon. The adult parasitoid emerges though a hole which it cuts near the end of one of the long sides of the egg case. The adults live for two to three weeks (Cameron, 1957). However, to be able to use evaniids in biological control one must consider a few problems, such as the mass production of evaniids which depends on a plentiful supply of a suitable host cockroach's ootheca; the parasitoid wasps must oviposit an egg at just the right time before the ootheca hardens. Many studies have focused on *Evania appendigaster* as a parasitoid of the American cockroach, *Periplaneta americana* (Tee and Lee, 2013; Fox and Bressan-Nascimento, 2006). In this case, a further complication is that the preferred host oothecal age of the *Evania* is also the age most liable to be cannibalised by the cockroach (Tee and Lee, 2017).

Gasteruptiidae (Carrot Wasps)

These constitute a distinctive family of parasitoid wasps, easily recognized by their elongate neck-like propleuron (Fig. 12.4),

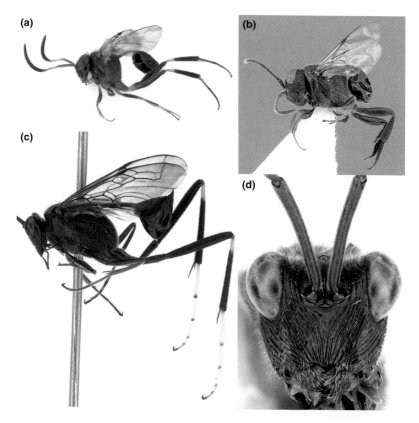

Fig. 12.3. Exemplar Thai evaniids: **(a)** *Parevania* sp.; **(b)** *Brachygaster* sp.; **(c, d)** *Evania* sp., possibly *E. postfurcalis*.

slender and posteriorly somewhat swollen (subclavate) and metasoma inserted high on the mesosoma, and the hind tibiae which are strongly distally swollen (clavate) (Fig. 12.4) (Jennings and Austin, 2002; Mikó *et al.*, 2019). The ovipositor is always quite long, and female antennae have 14 flagellomeres whilst those of males have only 13. The eye is relatively long, extending almost to the mandible (Zhao *et al.*, 2012). Fore wing vein (RS+M)a originates very close to M+CU far from the anterior wing margin (Fig. 12.5) and consequently the 1st discal cell is very small.

The Gasterutiidae is divided into two subfamilies: Gasteruptiinae and Hyptiogastrinae. The Hyptiogastrinae comprises two genera, *Hyptiogaster* and *Pseudofoenus*, but only the latter occurs outside of Australia and is recorded from New Guinea (and thus might occur in south-east Indonesia). The Gasteruptiinae includes four genera but only *Gasteruption* occurs in S.E. Asia. More than 500 species of *Gasteruption* have been described worldwide (Parslow *et al.*, 2020a, b), and it is quite diverse in S.E. Asia but there are no taxonomic revisions or keys for the region.

Little is known about biology of gasteruptiids except that they are predator-inquilines on larvae of bees nesting in soil, stems and in tunnels in wood, including members of the Apinae, Colletinae, Halictinae and Megachilinae (Zhao *et al.*, 2012) as well as masarine vespid wasps, a group that also feed their larvae on pollen. In addition, a few species have been reported from nests of some entomophagous provisioning Sphecidae and Eumeninae (Vespidae) (Parslow *et al.*, 2020b). However, Bogusch *et al.* (2018)

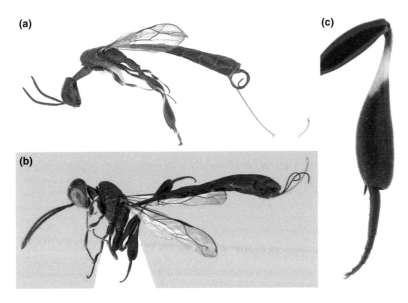

Fig. 12.4. Exemplar Thai *Gasteruption* species: **(a, b)** showing some variation in ovipositor length; **(c)** detail of clavate hind leg.

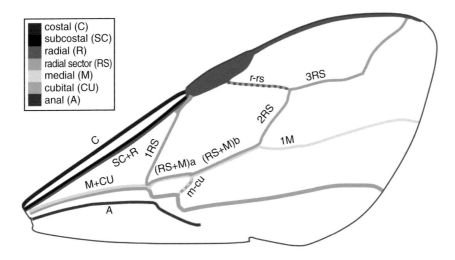

Fig. 12.5. Stylised *Gasteruption* wing venation.

pointed out that none of the purported observations of attacking nests of entomophagous aculeates have been confirmed. The gasteruptiid larvae usually predate the host egg or developing bee larva before consuming their provisions which the host has provided for its developing young, i.e. their pollen stores in the case of bees and masarines, and possibly prey insects in the case of crabronids, sphecids and other vespids but see above (Höppner, 1904; Crosskey, 1962; Bogusch *et al.*, 2018).

There appear to be no published host records for S.E. Asia. In temperate regions at least, adult gasteruptiids normally feed on flowers with easily accessible nectar (especially families Apiaceae, Asteraceae and Euphorbiaceae), but it is likely that at least some *Gasteruption* species feed on both nectar and pollen (Jennings and Austin, 2004 a, b).

The ovipositor is often distinctly upcurved (Fig. 12.4 a, b) and this is associated with the effects of an ovipositor steering mechanism (Quicke and Fitton, 1995). There are opposing bosses on the lower valves and stops on the upper valve near the tip and attempts to force the lower valves posteriorly force the upper valve (and the linked lower valves) to bend dorsally.

References

Barriga, J.E.T. (1990) Parásitos y depredadores de larvas de Cerambycidae y Buprestidae (Coleoptera) de Chile. *Revista Chilena de Entomologia* 18, 57–59.

Basibuyuk, H.H., Rasnitsyn, A.P., Fitton, M.G. and Quicke, D.L.J. (2002) The limits of the family Evaniidae (Insecta: Hymenoptera) and a new genus from Lebanese amber. *Insect Systematics and Evolution* 33, 23–34.

Bogusch, P., van Achterberg, C., Šilhán, K., Astapenková, A. and Heneberg, P. (2018) Description of mature larvae and ecological notes on *Gasteruption* Latreille (Hymenoptera, Evanioidea, Gasteruptiidae) parasitizing hymenopterans nesting in reed galls. *Journal of Hymenoptera Research* 65, 1–21.

Cameron, E. (1957) On the parasites and predators of the cockroach. II. – *Evania appendigaster* (L.). *Bulletin of Entomological Research* 48, 199–209.

Crosskey, R.W. (1962) The classification of the Gasteruptiidae (Hymenoptera). *Transactions of the Royal Entomological Society of London* 114, 377–402.

Deans, A.R. (2002) *Papatuka alamunyiga* Deans, a new genus and species of apterous ensign wasp (Hymenoptera: Evaniidae). *Zootaxa* 95, 1–8.

Deans, A.R. (2005) Annotated catalog of the world's ensign wasp species (Hymenoptera: Evaniidae). *Contributions of the American Entomological institute* 34, 1–164.

Deans, A.R. and Huben, M. (2003) Annotated key to the ensign wasp (Hymenoptera: Evaniidae) genera of the world, with descriptions of three new genera. *Proceedings of the Entomological Society of Washington* 105, 859–875.

Deans, A.R., Basibuyuk, H.H., Azar, D. and Nel, A. (2004) Descriptions of two new Early Cretaceous (Hauterivian) ensign wasp genera (Hymenoptera: Evaniidae) from Lebanese amber. *Cretaceous Research* 25, 509–516.

Deans, A.R., Gillespie, J.J. and Yoder, M.J. (2006) An evaluation of ensign wasp classification (Hymenoptera: Evaniidae) based on molecular data and insights from ribosomal RNA secondary structure. *Systematic Entomology* 31, 517–528.

Deyrup, M.A. (1984) A maple wood wasp, *Xiphydria maculata*, and its insect enemies (Hymenoptera: Xiphydriidae). *The Great Lakes Entomologist* 17, 17–28.

Elliott, M.G. (2005) First record of the genus *Brachygaster* Leach (Hymenoptera: Evaniidae) from Australia with the description of a new species. *Australian Journal of Entomology* 44, 2–5.

Fox, E.G.P. and Bressan-Nascimento, S. (2006) Biological characteristics of *Evania appendigaster* (L.) (Hymenoptera: Evaniidae) in different densities of *Periplaneta americana* (L.) oothecae (Blattodea: Blattidae). *Biological Control* 36, 183–188.

Gauld, I.D. and Hanson, B. (1995) The evaniomorph parasitoid families. In: Hanson, P.E. and Gauld, I.D. (eds) *The Hymenoptera of Costa Rica*. Oxford University Press, Oxford, pp. 185–208.

Goulet, H. and Huber, J.T. (1993) Hymenoptera of the world: an identification guide to families. Research Branch, Agriculture Canada, Ottawa, publ. 1894/E, 668 pp.

Grimaldi, D. and Engel, M.S. (2005) *Evolution of the Insects*. Cambridge University Press, New York, 755 pp.

Höppner, H. (1904) Zur Biologie der *Rubus*-bewohner. *Allgemeine Zeitschrift für Entomologie* 5, 97–103.

Jennings, J.T. and Austin, A.D. (1997) Revision of the Australian endemic genus *Hyptiogaster* Kieffer (Hymenoptera: Gasteruptiidae), with descriptions of seven new species. *Journal of Natural History* 31, 1533–1562.

Jennings J.T. and Austin A.D. (2002) Systematics and distribution of world hyptiogastrine wasps (Hymenoptera: Gasteruptiidae). *Invertebrate Systematics* 16, 735–811.

Jennings, J.T. and Austin, A.D. (2004a) Biology and host relationships of aulacid and gasteruptiid wasps (Hymenoptera: Evanioidea). In Austin, A.D. and Dowton, M. (eds) *Hymenoptera, Evolution, Biodiversity and Biological Control*. CSIRO Publishing, Collingwood, Australia, pp. 154–164.

Jennings, J.T. and Austin, A.D. (2004b) Biology and host relationships of aulacid and gasteruptiid wasps (Hymenoptera: Evanioidea): a review. In: Rajmohana, K., Sudheer, K., Girish Kumar, P. and Santhosh, S. (eds) *Perspectives on Biosystematics and Biodiversity*. University of Calicut, Kerala, India, pp. 187–215.

Jouault, C. Maréchal, A., Condamine, F.L., Wang, B., Nel, A., Legendre, F. and Perrichot, V. (2022) Including fossils in phylogeny: a glimpse into the evolution of the superfamily Evanioidea (Hymenoptera: Apocrita) under tip-dating and the fossilized birth–death process. *Zoological Journal of the Linnean Society* 194, 1396–1423.

Kuroda K. (2018) Discovery of the male of a rare aulacid wasp, *Pristaulacus emarginaticeps* Turner, 1922 (Hymenoptera: Aulacidae) from Vietnam and Laos. *Biodiversity Data Journal* 6, e26198.

Li, L., Rasnitsyn, A.P., Shih, C., Labandeira, C.C., Buffington, M., Li, D. and Ren, D. (2018) Phylogeny of Evanioidea (Hymenoptera, Apocrita), with descriptions of new Mesozoic species from China and Myanmar. *Systematic Entomology* 43, 810–842.

Mikó, I., Rahman, S.R., Anzaldo, S.S., Van de Kamp, T., Parslow, B.A., Tatarnic, N.J., Wetherington, M.T., Anderson, J., Schilder, R.J. and Ulmer, J.M. (2019) Fat in the leg: function of the expanded hind leg in gasteruptiid wasps (Hymenoptera: Gasteruptiidae). *Insect Systematics and Diversity* 3, 1–16.

Parslow, B.A., Jennings, J.T., Schwarz, M.P. and Stevens, M.I. (2020a) Phylogeny and divergence estimates for the gasteruptiid wasps (Hymenoptera: Evanioidea) reveals a correlation with hosts. *Invertebrate Systematics* 34, 319–327.

Parslow, B.A., Schwarz, M.P. and Stevens, M.I. (2020b) Review of the biology and host associations of the wasp genus *Gasteruption* (Evanioidea: Gasteruptiidae). *Zoological Journal of the Linnean Society* 189, 1105–1122.

Quicke, D.L.J. and Fitton, M.G. (1995) Ovipositor steering mechanisms in parasitic wasps of the families Gasteruptiidae and Aulacidae (Hymenoptera). *Proceedings of the Royal Society, London, Series B* 261, 99–103.

Rau, P. (1943) How the cockroach deposits its egg-case; a study in insect behavior. *Annals of the Entomological Society of America* 36, 221–226.

Sharanowski, B.J., Peixoto, L., Dal Molin, A. and Deans, A.R. (2019) Multi-gene phylogeny and divergence estimations for Evaniidae (Hymenoptera). *PeerJ* 7, e6689.

Shih, P.J.M., Li, L., Li, D. and Ren, D. (2020) Application of geometric morphometric analyses to confirm three new wasps of Evaniidae (Hymenoptera: Evanioidea) from mid-Cretaceous Myanmar amber. *Cretaceous Research* 109, 104249.

Skinner, E.R. and Thompson, G.H. (1960) *The Alder Woodwasp and Its Insect Enemies*. [Film.] World Educational Films, University of Oxford Department of Forestry.

Smith, D.R. (2001) World catalog of the family Aulacidae (Hymenoptera). *Contribution on Entomology, International* 4, 261–319.

Smith, D.R. (2017a) A new *Aulacus* (Hymenoptera: Aulacidae) from the Philippines. *Proceedings of the Entomological Society of Washington* 119, 112–115.

Smith, D.R. (2017b) *Pristaulacus* Kieffer (Hymenoptera: Aulacidae) of the Philippines, and New Records from Malaysia. *Proceedings of the Entomological Society of Washington* 119, 481–490.

Tee, H.S. and Lee, C.W. (2013) Feasibility of cold-stored *Periplaneta americana* (Dictyoptera: Blattidae) oothecae for rearing the oothecal parasitoids *Aprostocetus hagenowii* and *Evania appendigaster* (Hymenoptera: Eulophidae; Evaniidae): effect of ootheca age and storage duration. *Biological Control* 67, 530–538.

Tee, H.S. and Lee, C.W. (2017) Cockroach oothecal parasitoid, *Evania appendigaster* (Hymenoptera: Evaniidae) exhibits oviposition preference towards oothecal age most vulnerable to host cannibalism. *Journal of Economic Entomology* 110, 2504–2511.

Turrisi, G.F. (2013) Contribution to the revision of Oriental *Aulacus* Jurine, 1807 (Hymenoptera: Aulacidae): description of *A. ceciliae* sp. nov. from Laos and redescription of *A. bituberculatus* Cameron, 1899 from India. *Entomological Science* 16, 326–334.

Turrisi, G.F. (2014) A new species of *Pristaulacus* Kieffer, 1900 from Laos (Hymenoptera: Aulacidae). *Natura Somogyiensis* 24, 165–172.

Turrisi, G.F. (2017) The parasitoid wasp family Aulacidae with a revised World checklist (Hymenoptera, Evanioidea). *Proceedings of the Entomological Society of Washington* 119 (special issue), 931–939.

Turrisi, G.F. and Madl, M. (2013) Addition to the revision of the *Pristaulacus comptipennis* species-group: Description of two new species from Laos and Thailand (Hymenoptera: Aulacidae). *Journal of Asia-Pacific Entomology* 16, 237–243.

Turrisi, G.F. and Nobile, V. (2016) Description of *Pristaulacus leleji* sp.n. (Hymenoptera: Aulacidae) from Thailand. *Euroasian Entomological Journal* 15, 152–155.

Turrisi, G.F. and Smith, D.R. (2011) Systematic revision and phylogeny of the endemic southeastern Asian *Pristaulacus comptipennis* species group (Hymenoptera: Aulacidae). *Zootaxa* 2959, 1–72.

Turrisi, G.F. and Smith, D.R. (2020) Three new species of Aulacidae (Hymenoptera: Evanioidea) with additional records from Thailand and Laos. *Proceedings of the Entomological Society of America* 122, 197–210.

Turrisi, G.F. and Vilhelmsen, L. (2010) Into the wood and back: morphological adaptations to the wood-boring parasitoid lifestyle in adult aulacid wasps (Hymenoptera: Aulacidae). *Journal of Hymenoptera Research* 19, 244–258.

Turrisi, G.F. and Watanabe, K. (2011) Description of two new Asian *Pristaulacus* Kieffer 1900 (Hymenoptera: Aulacidae). *Zootaxa* 2895, 35–46.

Turrisi, G.F., Jennings, J.T. and Vilhelmsen, L. (2009) Phylogeny and generic concepts of the parasitoid wasp family Aulacidae (Hymenoptera: Evanioidea). *Invertebrate Systematics* 23, 27–59.

Vilhelmsen, L., Mikó, I. and Krogmann, L. (2010) Beyond the wasp-waist: structural diversity and phylogenetic significance of the mesosoma in apocritan wasps (Insecta: Hymenoptera). *Zoological Journal of the Linnean Society* 159, 22–194.

Visitpanich, J. (1994) The parasitoid wasps of the coffee stem borer, *Xylotrechus quadripes* Chevrolat (Coleoptera, Cerambycidae) in Northern Thailand. *Japanese Journal of Entomology* 62, 597–606.

Yeh, C.C. (1995) Oviposition concealment behavior of *Periplaneta americana* L. and its application on the oothecal trap in the laboratory. *Chinese Journal of Entomology* 15, 153–161.

Zhao, K., van Achterberg, C. and Xu, Z. (2012) A revision of the Chinese Gasteruptiidae (Hymenoptera, Evanioidea). *ZooKeys* 237, 1–123.

13

Ceraphronoidea

Abstract

The Ceraphronoidea are rather diverse and frequently collected, but almost nothing is known of their taxonomy in S.E. Asia; there are no recent keys to either the genera or species of any genera. A key to the two families Ceraphronidae and Megaspilidae is presented. What is known of their biology is described, but this is almost entirely based on extralimital species, some species possibly being economically important. The few relevant identification works are cited.

Common, small (0.5–4.0 mm) wasps but extremely difficult to identify, indeed for the tropics probably totally impossible because the vast majority of species are undescribed. Recent literature seems to have more papers concerning fossil taxa than on the extant but much under-investigated S.E. Asian fauna.

These are the only apocritan group that have two fore tibial spurs (Fig. 13.1 a), sometimes regarded as a putatively plesiomorphic character. However, their small size makes this difficult to see, needing a good microscope. In addition, fore wing veins C and R are fused to form a united C+R vein that runs along the anterior margin, thus with no membranous edge until after the pterostigma, although there is a break between C+R and the pterostigma. Vein RS is well developed and usually strongly curving anteriorly in a very distinctive fashion. Metasomal syntergite is very large and appears to abut directly and quite to very broadly onto the propodeum, because segment 1 is very small.

In addition, in dorsal view, the metasoma is narrowed anteriorly in Megaspilidae to about a third of its widest width, whereas in Ceraphronidae there is virtually no neck-like constriction anteriorly. Both families include brachypterous and apterous species as well as fully winged ones.

The separation of the two families is generally clear and Mikó and Deans (2009) listed the traits that enable their identification. *Trassedia*, which includes the S.E. Asian species, *T. yanegai* (Thailand), is problematic in that it displays features of both families (Mikó et al., 2018). It has a large triangular pterostigma characteristic of Megaspilidae, but also a Waterston's evaporatorium, a structure unique to Ceraphronidae (Fig. 13.2). The latter is an area of modified cuticle associated with one or two type III exocrine glands and the presence of a posterior abdominal pulsatory organ (Fig. 13.2 a, b) (Ulmer et al., 2021). Waterston's organ is sexually dimorphic in *Trassedia*; it is unpaired

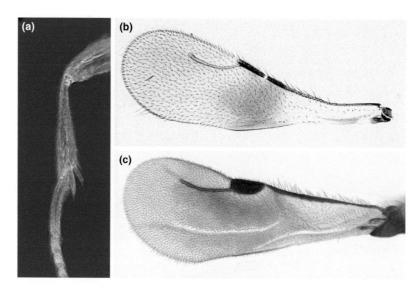

Fig. 13.1. Features of Ceraphronoidea and its families: **(a)** megaspilid sp., fore leg showing two spurs; **(b, c)** fore wings of a ceraphronid and a megaspilid, respectively.

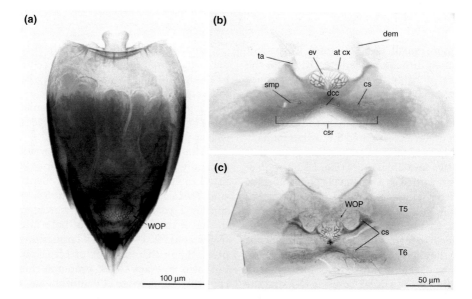

Fig. 13.2. Waterston's evaporatorium and surrounding tergites of a *Ceraphron* sp.: **(a)** metasoma in dorsal view, with evaporatorium visible on metasomal tergite 6; **(b)** metasomal tergum 6 and Waterston's evaporatorium detail; **(c)** metasomal terga 5 and 6. Abbreviations: at cx, acrotergal calyx; cs, campaniform sensilla; csr, caudal setal row; dcc, distal crenulate carina; ev, evaporatorium; ite, intertergal extensor muscle; smp, submedial patches; sr ta, sclerotized ridge of tergal apodeme; ta, tergal apodeme. (Source: reproduced from Ulmer *et al.* (2021) under Creative Commons Attribution Licence CC-BY-4.0.)

in males and paired in females (Mikó et al., 2018). The genus *Masner* from Australia and the Fiji Islands also displays an intermediate mix of character states and may be sister group to the remaining Ceraphronidae (Mikó and Deans, 2009).

Ernst et al. (2013) described the interesting ovipositor system of the superfamily using confocal laser scanning microscopy. They found considerable variation between the studied species and some of the features might be associated with the need for rapid oviposition (e.g. in *Trassedia*). Unlike other Hymenoptera, ceraphronoids lack an ovipositor retractor muscle and instead rely on movement of the seventh sternite.

Johnson and Musetti (2004) catalogued all the world literature on the superfamily up until that date, and for every species gave author, date, synonyms and broad geographical region. Their Oriental Region was taken to include S.E. Asia plus India, presumably Sri Lanka, and all of China, but excluding New Guinea, which was included in their Australian region. Thus, it is simply not possible to get from this the numbers of described species for S.E. Asia alone in the sense of this book. However, it allows us to estimate as follows: (i) Ceraphronidae is represented by approximately 20 species of *Ceraphron*, ten of *Aphanogmus* and one or two of *Cyoceraphron*; and (ii) Megaspilidae is represented by fewer than ten species of *Lagynodes* (see also Dessart, 1999), approximately ten of *Conostigmus* and a few more of *Dendrocerus*. All of these genera are cosmopolitan, with the vast majority of described species being from the Palaearctic and Nearctic regions.

Almost all publications mentioning members of the superfamily from S.E. Asia, in relation to either biocontrol or ecological/diversity studies, simply give numbers of morphospecies, with no named species, or, usually, not even genera, and sometimes not even family. This is clearly one of the most taxonomically neglected groups in the region.

Key to the Families of Ceraphronoidea

1   Tibia of mid leg with 1 spur; pterostigma usually small and narrow (exceptions are *Trassedia* and *Masner*; Fig. 13.1 b); notauli nearly always absent (Fig. 13.3 a), synsternite is not subdivided into an anterior short and a posterior longer sclerite..............**Ceraphronidae**
-   Tibia of mid leg with 2 spurs; pterostigma usually large and triangular (exception Lagynodinae; Fig. 13.1 c); notauli almost always present, at least anteriorly (Fig. 13.4 a), synsternite is subdivided into an anterior short and a posterior longer sclerite......................**Megaspilidae**

Ceraphronidae

Ceraphronids are typically brown or black, sometimes yellow, never metallic (Fig. 13.3). Most species are fully winged. The elbowed (geniculate) antennae have a long scape and are inserted very close to the mouth (Fig. 13.3); those of females usually have eight flagellomeres, males have nine.

Very little is known about their biology apart from simple host records. However, it is clear that the host range is very large and includes Diptera, Hemiptera, Hymenoptera, Neuroptera, Thysanoptera and several others.

Some *Aphanogmus* species could be considered as pests because they parasitise various cecidomyiid fly predators of pest Auchenorhyncha and spider mites. Some also attack prepupae of microgastrine braconids and of Bethylidae (Dessart, 1988; Kamarudin et al., 1996). Evans et al. (2005) described an *Aphanogmus* species from material that emerged in quarantine from the specimens of cybocephalid (formerly Nitidulidae) beetle *Cybocephalus nipponicus* that had been collected in Thailand and were in quarantine in Florida, USA, prior to potential use in the biocontrol of the introduced armoured scale insect pest, *Aulacaspis yasumatsui* (Diaspididae). Several S.E. Asian species belong to the *Aphanogmus hakonensis* species complex, member of which, as far as is known, are hyperparasitoids of lepidopteran larvae via various hymenopteran or dipteran primary parasitoids (Polaszek and Dessart, 1996). At least one Aphanogmus species is a primary endoparasitoid of a pest, Gelechiidae leaf-miner, in India (Youssef *et al.*, 2022). Some of these species are extremely difficult to distinguish and require detailed examination of the male genitalia.

This family was called Calliceratidae by some earlier workers. The genera can be identified using the key by Dessart and Cacemi (1986) and probably many also by the key to Indian genera by Bijoy and Rajmohana (2021). The genus *Ceraphron* is further subdivided into a number of subgenera (Dessart, 1981). Almost the most recent relevant literature on the family is Baltazar's (1966) catalogue which records a single species, *Ceraphron manilae*, from the Philippines. What little recent taxonomic literature there is regarding the Oriental and S.E. Asian faunas mostly involves just the descriptions of single species. Alekseev (2004) described a new species of *Ceraphron* from Vietnam.

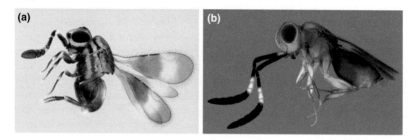

Fig. 13.3. Two rather uncommonly brightly marked Ceraphronidae from Thailand: **(a)** probably *Elysoceraphron* sp.; **(b)** *Ceraphron* (*Allomicrops*) sp.

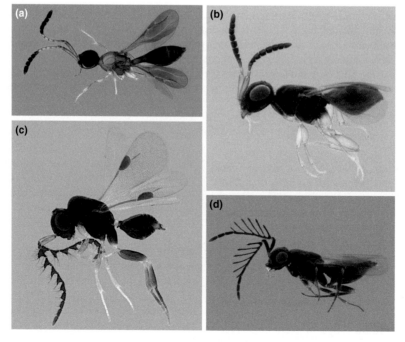

Fig. 13.4. Four undetermined exemplars of Thai Megaspilidae: **(a, b)** females; **(c, d)** males, but note that male flagellomeres are not always modified such as in these. **(a)** *Conostigmus* sp. nr *ampullaceus*; **(b)** *Dendrocerus* sp.; **(c)** *Dendrocerus* sp. nr *indicus*; **(d)** *Dendrocerus* sp. nr *katmandu*.

Megaspilidae

Megaspilids (Fig. 13.4) are generally very similar to ceraphronids. Most are dark; a few have a very slight metallic blue/violetness but we have not encountered any such species in Thailand.

From the little that is known on a global scale, megaspilids have diverse host–parasitoid relationships. It must be emphasised that virtually all biological data refer to European or North American species. Some are primary parasitoids of Coccoidea (Hemiptera), Neuroptera (lacewings), Mecoptera or dipteran puparia. Others are hyperparasitoids of aphids and scale insects (Hemiptera) via primary parasitoid aphidiine Braconidae (but also, less commonly, aphelinid chalcidoids, encyrtids, pteromalids and figitids). Megaspilids that attack Diptera pupae are still ectophagous, feeding on the true pupa externally though concealed within the protective host puparium. Haviland (1920) provided a fairly detailed study of the biology of *Dendrocerus carpenteri*, a pseudohyperparasitoid of aphids, but as aphids are uncommon in S.E. Asia the biology of this species just gives a hint as to what the local megaspilids might do. Interestingly, even though these wasps are small, the adults, provided with water and hosts, can live remarkably long, up to 75 days (Walker and Cameron, 1981).

There are 14 genera divided into two subfamilies: Megaspilinae (12) and Lagynodinae (2). The genera can be identified using the key by Dessart and Cacemi (1986). Female *Lagynodes* are always brachypterous whereas *Dendrocerus* are macropterous. Some *Dendrocerus* males have highly pectinate (ramose) antennae (Dessart, 1999).

References

Alekseev, V.K. (2004) A new species from subgenus *Larsoceraphron* of the genus *Ceraphron* Jurine from Vietnam (Hymenoptera: Ceraphronidae). *Proceedings of the Russian Entomological Society, St Petersburg* 75, 191–193.

Baltazar, C.R. (1966) A catalogue of Philippine Hymenoptera. *Pacific Insects Monographs* 8, 1–488.

Bijoy, C. and Rajmohana, K. (2021) First report of rare genera, *Pteroceraphron* Dessart 1981, *Elysoceraphron* Szelenyi 1936 and *Cyoceraphron* Dessart 1975 (Ceraphronidae: Hymenoptera) from India with new species descriptions. *Journal of Asia-Pacific Entomology* 24, 1326–1333.

Dessart, P. (1981) Definition de quelques sous-genres de *Ceraphron* Jurine, 1807 (Hymenoptera: Ceraphronoidea: Ceraphronidae). *Bulletin de l'Institut Royal des Sciences Naturelles de Belgique* 53, 1–29.

Dessart, P. (1988) *Aphanogmus goniozi* sp. n., hyperparasite d'un béthylidé au Sri Lanka (Hymenoptera Ceraphronoidea Ceraphronidae). *Bulletin & Annales de la Société Royale Belge d'Entomologie* 124, 99–104.

Dessart, P. (1999) Revision des *Dendrocerus* du groupe «*halidayi*» (Hymenoptera Ceraphronoidea Megaspilidae). *Belgian Journal of Entomology* 1, 169–265.

Dessart, P. and Cacemi, P. (1986) Tableau dichotomique des genres de Ceraphronoidea (Hymenoptera) avec commentaires et nouvelles espèces. *Frustula Entomologica, Nouvelles Series* VII–VIII (XX–XXI), 307–372.

Ernst, A.F., Mikó, I. and Deans, A.R. (2013) Morphology and function of the ovipositor mechanism in Ceraphronoidea (Hymenoptera, Apocrita). *Journal of Hymenoptera Research* 33, 25–61. doi: 10.3897/JHR.33.5204

Evans, G.A., Dessart, P. and Glenn, H. (2005) Two new species of *Aphanogmus* (Hymenoptera: Ceraphronidae) of economic importance reared from *Cybocephalus nipponicus* (Coleoptera: Cybocephalidae). *Zootaxa* 1018, 47–54.

Haviland, M.D. (1920) On the bionomics and development of *Lygocerus testaceimanus* Kieffer, and *Lygocerus cameroni* Kieffer (Proctotrypoidea-Ceraphronidae), parasites of *Aphidius* (Braconidae). *Quarterly Journal of Microscopical Science* 65, 361–372.

Johnson, N.F. and Musetti, L. (2004) Catalog of systematic literature of the superfamily Ceraphronoidea (Hymenoptera). *Contributions of the American Entomological Institute* 33, 1–149.

Kamarudin, N., Walker, A.K., Wahid, M.B., LaSalle, J. and Polaszek, A. (1996) Hymenopterous parasitoids associated with the bagworms *Metisa plana* and *Mahasena corbetti* (Lepidoptera: Psychidae) on oil palms in Peninsular Malaysia. *Bulletin of Entomological Research* 86, 423–439. doi:10.1017/S000748530003501X

Masner, L. (1993) Superfamily Ceraphronoidea. 566–569. In: Goulet, H., and Huber, J. (eds) *Hymenoptera of the World: an Identification Guide to Families*. Centre for Land and Biological Resources Research. Issued by Research Branch, Agriculture Canada.

Mikó, I. and Deans, A.R. (2009) *Masner*, a new genus of Ceraphronidae (Hymenoptera, Ceraphronoidea) described using controlled vocabularies. *ZooKeys* 20, 127–153. doi: 10.3897/zookeys.20.119

Mikó, I., Trietsch, C., van de Kamp, T., Masner, L, Ulmer, J.M., Yoder, M.J., Zuber, M., Sandall, E.L., Baumbach, T. and Deans, A.R. (2018) Revision of *Trassedia* (Hymenoptera: Ceraphronidae), an evolutionary relict with an unusual distribution. *Insect Systematics and Diversity* 2, 1–29. doi: 10.1093/isd/ixy015 Research

Polaszek, A. and Dessart, P. (1996). Taxonomic problems in the *Aphanogmus hakonensis* species complex; (Hymenoptera: Ceraphronidae) common hyperparasitoids in biocontrol programmes against lepidopterous pests in the tropics. *Bulletin of Entomological Research* 86, 419–422. doi:10.1017/s0007485300035008

Ulmer, J.M., Mikó, I. Deans, A.R. and Krogmann, L. (2021) The Waterston's evaporatorium of Ceraphronidae (Ceraphronoidea, Hymenoptera): a morphological barcode to a cryptic taxon. *Journal of Hymenoptera Research* 85, 29–56. doi: 10.3897/jhr.85.67165

Walker, G.P. and Cameron, P.J. (1981) The biology of *Dendrocerus carpenteri* (Hymenoptera: Ceraphronidae), a parasite of *Aphidius* species, and field observations of *Dendrocerus* species as hyperparasites of *Acyrthosiphon* species. *New Zealand Journal of Zoology* 8, 531–538. doi: 10.1080/03014223.1981.10427979

Youssef, R.M., Kaf, N.H.A., Abboud, R. and Al Tawaha, A.R.M. (2022) New record of *Aphanogmus clavicornis* Thomson (Hymenoptera: Ceraphronidae) as a larval parasitoid of tomato leaf miner *Tuta absoluta* (Meyrick) in Syria. *International Journal of Biology Sciences* 24(2), 190–194.

14

Megalyroidea

Abstract
The Megalyroidea comprises a single extant family, Megalyridae, members of which are associated with hosts in dead wood where they are normally collected. The distributions of the three genera (*Carminator*, *Ettchellsia* and *Megalyra*) and 13 species known from S.E. Asia are tabulated. What is known of their biology is entirely based on extralimital species. The few relevant identification works are cited.

This superfamily comprises the single extant family Megalyridae. It is cosmopolitan and had long been considered to be one of the most basal (primitive) superfamilies of parasitoid wasps, although they now appear to be sister group to Trigonalyoidea and somewhat derived within the Evaniomorpha+Aculeata clade (Blaimer *et al.*, 2023). Their fossil history goes back to the early Jurassic period (Rasnitsyn, 2002).

Megalyridae

Usually uncommon, the megalyrids comprise eight extant genera worldwide of which three are known from S.E. Asia: *Carminator*, *Ettchellsia* and *Megalyra*, based on 13 species (Table 14.1). The common name 'long-tailed wasps' has been applied to the Australian *Megalyra* species in which the ovipositor can be eight times longer than the body, but that name is inappropriate for most because many other species have rather short ovipositors.

Many of the described species are represented in collections by one or only a very few specimens. S.E. Asian megalyrids are mostly rather small, 'hairy' wasps with body lengths of a few millimetres (Fig. 14.1). They are characterised by a combination of features rather than any one synapomorphy. Most notable is the possession of distinct subantennal grooves, 14-segmented antennae and highly reduced hind-wing venation (Vilhelmsen *et al.*, 2010).

The few host records indicate that they are parasitoids of wood-boring beetle larvae or pupae of the families Buprestidae, Bostrichidae and Cerambycidae (Froggatt, 1906; Hacker, 1913, 1915; Rodd, 1951; Douglas, 1954; Hadlington and Gardner, 1960; Gess, 1964; Moore, 1961). They are presumed to be idiobionts but their biology has not been studied in detail. One might notice that all of these host records are for Australian species, nothing is known of the hosts of any S.E. Asian taxa. Quite a few of these records are also more than 100 years old which must say something about a lack

Table 14.1. The species of Megalyridae occurring in S.E. Asia (New Guinea is included as it is contiguous with Irian Jaya and species might occur in both countries).

Tribe	Species	Known distribution	Reference
Dinapsini	*Ettchellsia ignita*	'Borneo', Malaysia (Peninsular)	Mita and Shaw, 2012
	Ettchellsia nigripes	Indonesia (Sulawesi)	Mita and Shaw, 2012
	Ettchellsia philippinensis	Philippines	Mita and Shaw, 2012
	Ettchellsia reidi	Indonesia (Kalimantan)	Mita and Shaw, 2012
Megalyrini	*Carminator affinis*	Malaysia (Sabah)	Shaw, 1988
	Carminator ater	Thailand	Shaw, 1988; Basibuyuk et al., 2000
	Carminator coronatus	Malaysia, Indonesia (Kalimantan)	Mita and Konishi, 2011
	Carminator gracilis	Vietnam	Mita and Konishi, 2011
	Megalyra rufiventris	Papua New Guinea	Shaw, 1990a
	Megalyra sedlaceki	New Guinea	Shaw, 1990a
	Megalyra longiseta	Indonesia (Ceram), Papua New Guinea	Shaw, 1990a
	Megalyra spectabilis	Indonesia (Irian Jaya), Papua New Guinea	Shaw, 1990a
	Megalyra tawiensis	Malaysia (Sabah), Indonesia (Sumatra), Philippines (Tawi Tawi)	Shaw, 1990a

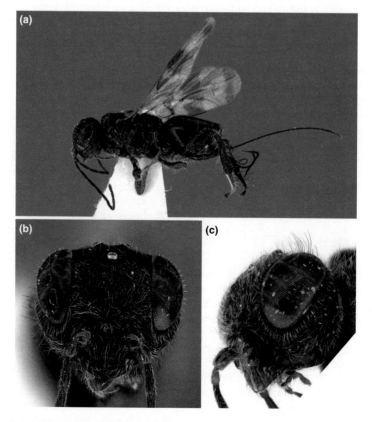

Fig. 14.1. *Ettchellsia* sp. nr *ignita* (but metasoma black) from Thailand: **(a)** habitus; **(b)** face; **(c)** head, oblique ventro-lateral view.

of practical natural history these days. Mesaglio and Shaw (2022) describe the oviposition behaviour od an Australian species with long ovipositor, and speculate that they may employ passive vibration detection to find hosts inside wood.

Shaw's (1990b) morphological phylogenetic hypothesis indicated the S.E. Asian megalyrids belonged to two separate tribes, each genus with its closest relatives coming from outside of our region. This was followed by Vilhelmsen et al. (2010) who performed a combined analysis of extant and fossil Megaylridae.

The genera may be recognised using the key by Vilhelmsen et al. (2010). Shaw (1990a) revised *Megalyra*, Mita and Konishi (2011) carried out a morphological phylogenetic analysis of species relationships within *Carminator* and provided a key to the species, and Mita and Shaw (2012) revised *Ettchellsia* and provided a key to the species.

References

Basibuyuk, H.H., Quicke, D.L.J., Rasnitsyn, A.P. and Fitton, M.G. (2000) Morphology and sensilla of the orbicula, a sclerite between the tarsal claws, in the Hymenoptera. *Annals of the Entomological Society of America* 93, 625–636.

Blaimer, B.B., Santos, B.F., Cruaud, A., Gates, M.W., Kula, R.R., Mikó, I., Rasplus, J.-Y., Smith, D.R., Talamas, E.J., Brady, S.G. and Buffington, M.L. (2023) Key innovations and the diversification of Hymenoptera. *Nature Communications* 14, 1212.

Douglas, A. (1954) Observations on a long-tailed wasp, *Megalyra shuckardi* Westwood. *Western Australian Naturalist* 4, 145–146.

Froggatt, W.W. (1906) Notes on the hymenopterous genus *Megalyra* Westwood, with descriptions of new species. *Proceedings of the Linnaean Society of New South Wales* 31, 399–407.

Gess, F.W. (1964) The discovery of a parasite of the *Phoracantha* beetle (Coleoptera: Cerambycidae) in the Western Cape. *Journal of the Entomological Society of South Africa* 27, 152.

Hacker, H. (1913) Some field notes on Queensland insects. *Memoirs of the Queensland Museum* 2, 96–100.

Hacker, H. (1915) Notes on the genus *Megachile* and some rare insects collected during 1913–14. *Memoirs of the Queensland Museum* 3, 137–141.

Hadlington, P. and Gardner, M.J. (1960) *Diadoxus erythrurus* (White) (Coleoptera—Buprestidae), attack of fire-damaged *Callitris* spp. *Proceedings of the Linnean Society of New South Wales* (1959) 84, 325–332.

Mesaglio, T. and Shaw, S.R. (2022) Observations of oviposition behaviour in the long-tailed wasp *Megalyra fasciipennis* Westwood, 1832 (Hymenoptera: Megalyridae). *Austral Ecology* 47(4), 889–893.

Mita, T. and Konishi, K. (2011) Phylogeny and biogeography of *Carminator* (Hymenoptera: Megalyridae). *Systematic Entomology* 36, 104–114. doi: 10.1111/j.1365–3113.2010.00548.x

Mita, T. and Shaw, S.R. (2012) A taxonomic study on the genus *Ettchellsia* Cameron, with descriptions of three new species (Hymenoptera, Megalyridae, Dinapsini). *ZooKeys* 254, 99–108. doi: 10.3897/zookeys.254.4182

Moore, K.M. (1961) Observations on some Australian forest insects. *Proceedings of the Royal Zoological Society of New South Wales* (1958–59), 87–95.

Rasnitsyn, A.P. (2002) Superorder Vespidea Laicharting, 1781. Order Hymenoptera Linne, 1758 (= Vespida Laicharting, 1781). In: Rasnitsyn, A.P. and Quicke, D.L.J. (eds) *History of Insects*. Kluwer Academic, Dordrecht, pp. 242–254.

Rodd, N.W. (1951) Some observations on the biology of Stephanidae and Megalyridae (Hymenoptera). *The Australian Zoologist* 11, 341–346.

Shaw, S.R. (1988) *Carminator*, a new genus of Megalyridae (Hymenoptera) from the Oriental and Australian regions, with a commentary on the definition of the family. *Systematic Entomology* 13, 101–113. doi: 10.1111/j.1365-3113.1988.tb00233.x

Shaw, S.R. (1990a) Taxonomic revision of the long-tailed wasps of the genus *Megalyra* Westwood (Hymenoptera: Megalyridae). *Invertebrate Taxonomy* 4, 1005–1052.

Shaw, S.R. (1990b) Phylogeny and biogeography of the parasitoid wasp family Megalyridae (Hymenoptera). *Journal of Biogeography* 17, 569–581.

Vilhelmsen, L., Perrichot, V. and Shaw, S.R. (2010) Past and present diversity and distribution in the parasitic wasp family Megalyridae (Hymenoptera). *Systematic Entomology* 35, 58–677. doi: 10.1111/j.1365-3113.2010.00537.x

15

Trigonalyoidea

Abstract
The fascinating biology of this uncommon group of obligate indirect hyperparasitoids is described, although this is based entirely on extralimital species. An identification key is provided to the five genera occurring in the region: *Taeniogonalos*, *Pseudonomadina*, *Bareogonalos*, *Lycogaster* and *Bakeronymus*. The family is very poorly recorded, and most records are based on only one or a very few individuals, so little can be said of their biogeography in the regions. The few relevant identification works are cited.

This is a small monotypic cosmopolitan superfamily with only the family Trigonalyidae. The literature is inconsistent about the spelling of the family group name. Sometimes it is spelt Trigonalyidae and sometimes Trigonalidae. This is discussed by Aguiar *et al.* (2013) who concluded that the original spelling without the 'y' is correct, but the majority of workers nevertheless still spell it with an 'i'. With the most complete wing venation of any extant, non-sawfly Hymenoptera, trigonalyids have often been considered to be very 'primitive' (basally derived) apocritan wasps. The adult wasps are relatively large, 5–15 mm, and often found visiting flowers.

Nearly all trigonalyids are obligate indirect (sequential) hyperparasitoids. Their specialised and complex biologies suggest that they are very far from primitive. Female trigonalyids lay vast numbers (3000–10,000) of microtype eggs on leaves (Clausen, 1931, 1940), and these must be consumed by a folivorous insect, basically a moth or butterfly caterpillar, or sometimes a sawfly larva. The trigonalyid egg then hatches but the 1st instar larva remains in that instar until something else happens. This is because trigonalyids are not lepidopteran parasitoids, but rather they rely on a 'parasitised' caterpillar subsequently being consumed by another insect, typically a vespid wasp, but also other entomophagous insects. The vespid then regurgitates the chewed-up caterpillar, including the tough 1st instar trigonalid larva, and feeds it to one of the vespid larvae in her nest. The trigonalyid larva then bores through the gut wall of the larval vespid and starts its truly parasitoid development in the immature social wasp, which represents its definitive host (Carmean, 1991; Smith, 1996). Having said this, Weinstein and Austin (1995) described the biology of an Australian *Taeniogonalos* species that is a primary koinobiont endoparasitoid of a pergid sawfly, so the host in this case is also a hymenopteran. A Russian species of *Taeniogonalos* is now known to be a primary larval-pupal endoparasitoid of a papilionid butterfly caterpillar (Marchiori, 2022).

Other primary hosts of trigonalyids include various ichneumonid parasitoids and also tachinid Diptera which subsequently parasitise a trigonalyid-attacked caterpillar (Gauld and Bolton, 1988). One extralimital species is known to be thelytokous and only produces males exceedingly rarely (Weinstein and Austin, 1996).

Weinstein and Austin (1991) reviewed known host associations of the family and catalogued the world species, but this is rather out of date.

Morphology

Of all the parasitoid Hymenoptera, members of this family have the most complete wing venation, with four submarginal cells in the fore wing (Fig. 15.1). Another key character is that the tarsal segments except for the telotarsus all have fleshy, finger-like protuberances from the apicoventral ends, called plantar lobes (Mason, 1993). The mandibles are usually asymmetric with three teeth on the left and four on the right. Males usually possess elongate oval tyloids on some of the middle segments.

S.E. Asian Trigonalyid Fauna

Five genera are known from the region: *Taeniogonalos* (Myanmar, Philippines, Indonesia), *Pseudonomadina* (Philippines), *Bareogonalos* (Indonesia), *Lycogaster* (Indonesia and Malaysia) and *Bakeronymus* (Philippines) (Figs 15.2, 15.3). The occurrence of additional genera in nearby countries suggests that additional genera probably also occur.

Carmean and Kimsey (1998) provided separate generic keys for various geographical regions, including one for the 'Eurasian and Indo-Australian' genera.

Chen *et al.* (2014) revised the Chinese species and provided a key to all the Oriental and Palaearctic genera, including all of those occurring in S.E. Asia. Smith and Tripotin (2015) recorded seven species from Thailand and four from Laos. Each species was illustrated photographically in colour but no identification key was presented. In earlier literature there is often mention of the genus *Poecilogonalos*, but this was synonymised with *Taeniogonalos* by Carmean and Kimsey (1998).

Poecilogonalos thwaitesi gestroi Schulz has been recorded from Myanmar (as Burma) and an extralimital subspecies has been shown to be an hyperparasitoid of a large *Enicospilus* sp. (Ichneumonidae: Ophioninae) (Clausen, 1929).

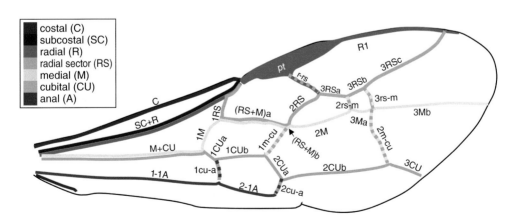

Fig. 15.1. Stylised fore wing venation of Trigonalyidae with nomenclature used in this book.

Fig. 15.2. Specimen of Trigonalyidae from Singapore. (Source: photograph by and © Wendy Wang (Lee Kong Chian Natural History Museum, LKCNHM, National University of Singapore), reproduced with permission.)

Fig. 15.3. Females of two genera and four species of Trigonalyidae from Thailand showing some of the range of colour pattern variation. **(a)** *Lycogaster flavonigrata*; **(b)** *L. rufiventris*; **(c)** *Taeniogonalos fasciata*; **(d)** *T. tricolor* (Source: from Smith and Tripotin, P. (2015), reproduced under Creative Commons Attribution Licence CC-BY-4.0.)

Simplified Key to the Genera of Trigonalyidae Occurring in S.E. Asia

1 Antenna with 13–15 segments; vertex with medio-longitudinal depression dorsally; vein 1RS of fore wing short; mandibles narrow in anterior view; apical segment of labial palp slender and tapered; ♀ third sternite with protuberance that is much larger than that of second sternite; hind tarsus strongly modified 2
- Antenna with 17–32 segments; vertex normal, at most with slight median depression dorsally; vein 1RS of fore wing medium-sized to long; mandibles wide in anterior view and sublaterally attached to head; apical segment of labial palp widened and obtuse, more or less triangular; ♀ third sternite without protuberance 3
2(1) Maxillary palp 2-segmented; first metasomal tergite comparatively wide anteriorly; third antennal segment widened, distinctly longer than fourth segment; apical half of antenna of ♀ with distinctly moniliform and asymmetric segments .... ***Pseudonomadina***
- Maxillary palp 4-segmented; first tergite subpetiolate and comparatively narrow anteriorly; third antennal segment slender, 0.8 times as long as fourth; apical half of antenna of ♀ hardly moniliform and with symmetric segments ... ***Bakeronymus***
3(1) Mandible base far removed from eyes; hind trochanter simple, dorsal triangular part not separated from basal part; metanotum nearly flat and smooth medially; fore trochanter distinctly widened apically and slightly longer than hind trochanter .... ***Bareogonalos***
- Mandible base close to level of eyes; apicodorsal triangular part of hind trochanter separated from basal part by oblique groove; metanotum usually convex and sculptured medially; fore trochanter subparallel-sided and distinctly longer than hind trochanter 4
4(3) Fifth sternite of ♀ distinctly emarginate medio-posteriorly; second sternite with pair of small triangular teeth on apical protuberance; metanotum smooth, shiny and only weakly convex... ***Lycogaster***
- Fifth sternite of ♀ straight or slightly emarginate medio-posteriorly; second sternite without pair of small teeth medio-apically; metanotum often sculptured, matt and distinctly convex.. ***Taeniogonalos***

References

Aguiar, A.P., Deans, A.R., Engel, M.S., Forshage, M., Huber, J.T., Jennings, J.T., Johnson, N.J., Lelej, A.S., Longino, J.T., Lohrmann, V., Mikó, I., Ohl, M., Rasmussen, C., Taeger, A. and Yu, D.S.K. (2013) Order Hymenoptera. *Zootaxa* 3703, 051–062.

Carmean, D. (1991) Biology of the Trigonalyidae (Hymenoptera), with notes on the vespine parasitoid *Bareogonalos canadensis*. *New Zealand Journal of Zoology* 18, 209–214.

Carmean, D. and Kimsey, L. (1998) Phylogenetic revision of the parasitoid wasp family Trigonalidae (Hymenoptera). *Systematic Entomology* 23, 35–76.

Chen, H.Y., van Achterberg, C., He, J.H. and Xu, Z.F. (2014). A revision of the Chinese Trigonalyidae (Hymenoptera, Trigonalyoidea). *ZooKeys* 385, 1–207.

Clausen, C.P. (1929) Biological studies on *Poecilogonalos thwaitesi* (Westw.), parasitic in the cocoons of *Henicospilus* (Hymen.: Trigonalidae). *Proceedings of the Entomological Society of Washington* 31, 67–79.

Clausen, C.P. (1931) Biological notes on the Trigonalidae (Hymenoptera). *Proceedings of the Entomological Society of Washington* 33, 72–81.

Clausen, C.P. (1940) *Entomophagous Insects*. McGraw-Hill, New York, 688 pp.

Gauld, I.D. and Bolton, B. (1988) *The Hymenoptera*. British Museum (Natural History), London, and Oxford University Press, Oxford, 332 pp.

Marchiori, C.H. (2022) Inventory on the family Trigonalidae (Insecta: Hymenoptera). *Open Access Research Journal of Science and Technology* 04(02), 82–104.

Key to the Families of Chrysidoidea in S.E. Asia

1   Antennae with 8 flagellomeres (Figs 16.9 a, d, e, 16.11) 2
-      Antennae with 9 or more flagellomeres (Fig. 16.2 a, b, d) 3
2(1)   Antenna arising from a small protuberance far from mouth (Fig. 16.11); apex of fore tarsus never modified to form a pincer .... **Embolemidae**
-      Antenna arising close to mouth (Fig. 16.9 a, d); females often with apical fore tarsal segment modified into a pincer (Fig. 16.9 c, d, f) .... **Dryinidae**
3(1)   Femora and antennal scape with transparent flanges (Fig. 16.7 a, d); vertex strongly produced posteriorly and with a flange that has the appearance of dense gold-silver associated specialised setosity (Fig. 16.7 a, c); never metallic .... **Chrysididae (Loboscelidiinae)**
-      Femora, and antennal scape without transparent flanges (Fig. 16.3 a, d); vertex rounded (Figs 16.4–6); colour variable, but sometimes with mesosoma metallic green, purple, blues and or red (Figs 16.5, 16.6) 4
4(3)   Antennae with 13–37 flagellomeres; antennae inserted under a shelf-like lamella (Fig. 16.12 a, b, d); head prognathous; (uncommon) .... **Sclerogibbidae**
-      Antennae with 9–11 flagellomeres; antennae not normally inserted under a shelf; head shape variable 5
5(4)   Metasoma with 3–5 clearly exposed tergites (Figs 16.3–16.6); pronotum usually separated from tegula; colour variable .... **Chrysididae**
-      Metasoma with 6–7 clearly exposed tergites; pronotum usually reaching tegulae; no metallic green or copper coloration 6
6(5)   Propleuron not visible dorsally; fore trochanter arising from apex of fore coxa; prosternum variable, if visible usually small and transverse; legs variable, not usually swollen; (very common) .... **Bethylidae**
-      Propleuron extended as a neck and therefore visible dorsally; fore trochanter arising laterally closer to base of fore coxa than its apex; prosternum large and diamond-shaped and clearly visible in ventral view; femora swollen; (extremely rare) .... **Scolebythidae**

Bethylidae (Flat Wasps)

The described world fauna of bethylids amounts to more than 2900 species, which is probably about half the real total, or less, and these are classified in just over 100 genera (Azevedo et al., 2018; Vargas et al., 2020). The family is divided into five extant subfamilies which can be identified using the key by Azevedo et al. (2018). A few bethylids completely lack fore wing vein C, complicating higher group keys, and Pristocerinae and Scleroderminae females may be completely apterous. A number of species are quite markedly dorsoventrally compressed (e.g. Fig. 16.2 d), hence their common name. Phylogenetic relationships within the family were investigated by Terayama (2003 a, b).

Bethylids are idiobiont ectoparasitoids of holometabolous insect larvae in concealed situations, the majority (except Bethylinae) attacking beetle larvae. Several species are frequently encountered in domestic situations, where their populations can become quite large as they attack associated domestic pests such as furniture beetles (Anobiidae) or carpet beetles (Dermestidae). This can sometimes have relatively minor medical consequences, as their frequent stings can cause dermatitis (Mohar et al., 1985). Bethylid stings can be somewhat painful to humans, though their main role is host paralysis, their hosts often being far larger than the female wasp. Their paralysing venom acts rapidly; they are usually gregarious and females often remain with the host guarding their broods. It is even known that in some species an unmated female will lay a brood of haploid males, stay with them until her sons emerge and then mate with one or more of those, and then can go on to produce daughters (Gauld and Bolton, 1988). Siblicide is very uncommon in the family (Mayhew and Hardy, 1998).

Some bethylids are borderline when it comes to the definition of a parasitoid because they display some pre-oviposition physical host manipulation, as was noted by Wheeler (1928) long ago. This has been noted in some species that are important parasitoids of stored product pests which are easy both to study in the laboratory and to obtain. For example, the female of the extralimital but extremely well-studied bethylids *Cephalonomia tarsalis* and *C. waterstoni*, parasitoids of free-living grain beetles, paralyses a host larva and then drags it along the surface of the infested substrate to a crevice or cavity where she oviposits (Finlayson, 1950; Howard et al., 1998; Amante et al., 2017a). Some species also display destructive host-feeding (Amante et al., 2017b).

Gordh and Móczár (1990) catalogued the world species of the family. Excluding fossil taxa, that work covers just over 1700 species, of which only 121 are from S.E. Asia, highlighting the paucity of regional systematic work on the group. Further, of these 121, the great majority of records/descriptions are from the Philippines dating back to three works by Kieffer (1913a, b, 1914). Indeed, more species of *Seriola* are known from the Hawai'ian Islands, where that genus underwent a large radiation. As noted by Azevedo et al. (2018), Gordh and Móczár's catalogue is very out of date, with many taxonomic changes having been made since it was published.

Their generally small size and rather uniform morphology does not make their systematic and taxonomic study easy. Many species have been described over the past 30 or so years, mostly by Mamoru Terayama from Japan and Celso Azevedo from Brazil. Lanes et al. (2020) provided an enormously useful review of morphology and standardised terminology. Accurate taxonomy often requires study of male genitalia.

Key to Subfamilies of Fully Winged Bethylidae*

(*based on Azevedo et al., 2018)

1		Fore wing with vein RS&M showing a sharp angulation where 1M becomes 1RS (Fig. 16.1 e) often with at least as stub of tubular vein RS+M................ **Bethylinae**
-		Fore wing with RS+M vein absent and vein RS&M smoothly curved or straight (Fig. 16.1 f).. 2
2(1)		Metanotum well developed, medially emarginate and usually with protruding submedial lobes overlapping scutellum posteriorly (Fig. 16.1 a,d) .... **Pristocerinae**
-		Metanotum not developed medially, usually absent (Fig. 16.1 c) or occasionally present as a narrow transverse strip (Fig. 16.1 b) 3
3(2)		Propodeum depressed medially with posterolateral spines (Fig. 16.2 a); body usually strongly sculptured (foveolate) (Fig. 16.2 a) .... **Mesitiinae**
-		Propodeum not depressed medially, never with posterior spines (Figs 16.2 b–f); body not strongly foveolate 4
4(3)		Mesopleuron with transepisternal line; fore wing of macropterous forms with anterior broad straight, costal vein always present; occipital carina always present........ **Epyrinae**
-		Mesopleuron without transepisternal line; fore wing of macropterous forms with anterior board angularly incurved, anterior to pterostigma, costal vein usually absent; occipital carina often absent.. **Scleroderminae**

Apterous females occur in the Pristocerinae and Scleroderminae, various degrees of brachyptery in the other subfamilies.

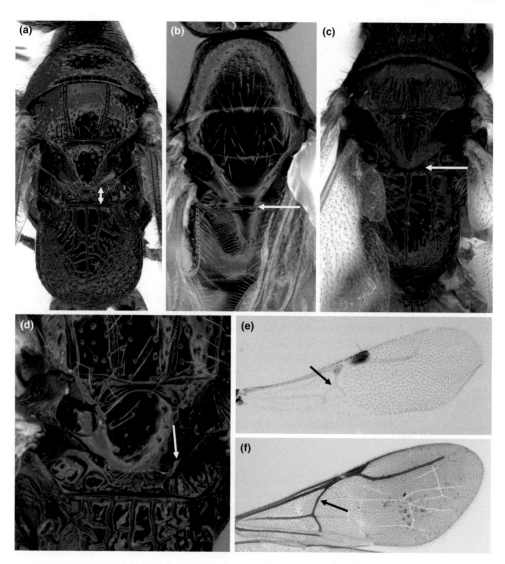

Fig. 16.1. Bethylidae details of dorsal mesosoma and fore wings: **(a, d)** Pristocerinae gen. sp., showing well developed metanotum (arrow) and in **(d)** the edge of the medial metanotal emargination actually overlapping posterior of scutellum; **(b)** Epyrinae gen. sp., showing very reduced strip-like metonum (arrow); **(c)** Epyrinae gen. sp., absence of external metanotal structure; **(e)** Bethylinae gen. sp., showing angulation between 1-M and 1-RS (arrow) and tubular vein RS+M; **(f)** Pristocerinae gen. sp. with smoothly curved vein RS&M with no RS+M spur present.

Bethylinae

Following a morphological cladistic analysis, Polaszek and Krombein (1994) concluded that there were seven valid genera and provided an illustrated identification key to these. Ramos and Azevedo (2020) presented an updated morphological phylogeny of the subfamily. Most bethylines are gregarious ectoparasitoids of larval Lepidoptera. It comprises some 540 species in 11 genera worldwide. Lim and Lee (2013) provided a key to the six species of *Odontepyris* known from Cambodia.

Epyrinae

This is a rather small cosmopolitan group. The tribe Epyrini contains approximately

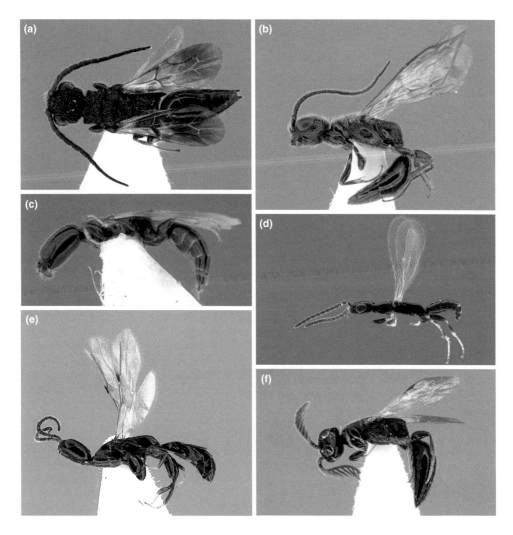

Fig. 16.2. Exemplar Bethylidae from Thailand: **(a)** Mesitiinae gen. sp.; **(b)** male *Pristepyris* sp. (Pristocerinae); **(c)** *Allobethylus* sp. (Sclerodarminae); **(d)** *Plastanoxus* sp. (Sclerodarminae); **(e)** male *Foenobethylus* sp. (Pristocerinae); **(f)** male *Calyoza* sp. (Epyrinae).

20 recognised genera but has more than 750 described species. Males of some genera have pectianate antennae (Fig. 16.2 f). The taxonomic overview and generic identification key for Epyrini by Azevedo (2011) is now somewhat out of date. This was followed by Alencar and Azevedo (2013), who conducted a detailed cladistic analysis of 391 morphological characters. This resulted in the resurrection of the Sclerodarminae as a subfamily separate from Epyrinae. Colombo *et al.* (2022) used an integrated taxonomic approach to validate, or otherwise, the genera, which has resulted in quite a few changes.

Mesitiinae

A rather small (c. 200 spp.), entirely Old World subfamily. Most species are black, sometimes with paler reddish or white markings. They are easy to recognise because the propodeum is medially concave and possesses a pair of lateral spines (Fig. 16.2 a).

Barbosa et al. (2022) presented a morphological phylogeny of the group which showed some genera were polyphyletic and some were paraphyletic. The genera are obviously treated by Azevedo et al. (2018) but Móczár's (1984) key might still be a useful adjunct and it also includes keys to species of many of the genera as known at that time. Separation of males into genera is much more straightforward than for females. Of the 17 included genera, eight are known from S.E. Asia, most of the rest being Palaearctic and/or Afrotropical.

Pristocerinae (= Afgoiogfinae)

Pristocerines (Fig. 16.2 b, e) are a diverse group (23 genera) of parasitoids of beetle larvae that bore into wood, live in the soil and some that live in ant nests. It includes by far the largest number of known species. Some species can reach up to 30 mm in body length. Females of this subfamily are apterous, have no eyes (or ones reduced to just a few facets) and lack ocelli. Their extreme sexual dimorphism makes associations between the sexes extremely difficult and will probably mostly require DNA data for most species, although occasional instances of phoretic copulation or rearing mixed-sex broods can also be used. At present, most species are known only from one sex.

Azevedo et al. (2018) necessarily provided separate generic keys for the two sexes, and many couplets in the key to males require genitalia preparations (see Fig. 23.13). Terayama (1995) described *Caloapenesia* and *Neoapenesia* based on specimens from the Philippines and Thailand, and provided a key to the two known species of the former. Lanes and Azevedo (2007) redescribed the Indonesian monotypic genus *Scaphepyris* and transferred it from Sclerodermini to Pristocerinae. Azevedo and Alencar (2009) synonymised *Pristepyris*, based on rediscovery of the holotype of its Malaysian type species, with *Acrepyris* and so transferred it to the Pristocerinae from the Epyrinae. Liu et al. (2011a) revised the species of the rare genus *Foenobethylus* (Fig. 16.2 e). Mugrabi and Azevedo (2013) revised the Thai species of *Dissomphalus*, including 24 that were newly described. Gobbi and Azevedo (2014) revised the S.E. Asian genus *Caloapenesia* and provided a key to males, including 16 new species that they described from Indonesia, Thailand and Vietnam.

Scleroderminae

This subfamily is cosmopolitan with 227 currently recognised species in 30 genera (Fig. 16.2 c, d). Females of the subfamily can be apterous, but unlike Pristocerinae they have eyes with many facets. For a long while they were considered as a tribe of Epyrinae along with Cephalonomiini, but Alencar and Azevedo (2013) resurrected them to subfamily status. Vargas et al. (2020) presented a morphological phylogenetic analysis of the Scleroderminae which necessitated descriptions of four new genera, including *Pilocutis* from Thailand, all of which display strongly divergent morphology. Barbosa and Azevedo (2011) reviewed the S.E. Asian species of *Allobethylus*, including description of new species from Myanmar and Thailand, and provided a key to the nine known species (most extralimital). Following a morphological cladistic analysis (Lanes and Azevedo, 2008), Azevedo and Lanes (2009) placed the aberrant genus *Galodoxa* from the Philippines in the Sclerodermini (then treated as part of the Epyrinae). Lim and Lee (2011) described a new species of *Prorops* from Cambodia and provided a key to world species of the genus. Azevedo and Barbosa (2010) recorded *Pararhabdepyris* from S.E. Asia for the first time and described a new species from Thailand.

Lanes and Azevedo (2008) provided a key to the world genera, both males and females, of Scleroderminae (as Sclerodermini).

Chrysididae

This family comprises four subfamilies: Amiseginae, Chrysidinae, Cleptinae and Loboscidiinae. Some authors treat the chrysidine tribe Panporpini as a separate subfamily too (Kimsey, 1987); however, they probably do not occur in S.E. Asia and appear to be restricted to the Holarctic, Near East and Madagascar. Members of the four subfamilies each have very different biologies, and the Loboscelidiinae are so morphologically aberrant that they were for a long while considered not to be aculeate hymenopterans at all, but rather to be proctotrupoids and treated as a separate family.

Simplified Key to Chrysidid Subfamilies Occurring in S.E. Asia

1	Femora, and antennal scape with transparent flanges (Fig. 16.7 a, d); vertex strongly produced posteriorly and with dense gold-silver associated specialised setosity (Fig. 16.7 a, c); fore wing costal vein absent (Fig. 16.7 e); never metallic .... **Loboscelidiinae**	
-	Femora, and antennal scape without transparent flanges (Fig. 16.3 a, d); vertex rounded (Figs 16.4–6); colour variable, but sometimes with mesosoma metallic green, purple, blues and or red (Figs 16.5, 16.6); fore wing costal vein present...... 2	
2(1)	Metasoma with only three visible tergites (Fig. 16.5); metasoma concave ventrally (wasp capable of rolling into a tight, protective ball), always metallic green, purple, blues and or red (Fig. 16.5) .... **Chrysidinae**	
-	Metasoma with more than three visible tergites (Figs 16.3, 16.6); metasoma convex ventrally (wasp cannot roll into a ball); usually non-metallic.................................. 3	
3(2)	Combined midlengths of metasomal tergites 1+2 shorter than tergites 3+4 .... **Cleptinae**	
-	Combined midlengths of metasomal tergites 1+2 longer than tergite 3 +4 (Fig. 16.3 a, c) .... **Amiseginae**	

Amiseginae

Amisegines have a strange disjunct distribution comprising parts of South America, southern Africa and the region from India to Australia. These are very commonly collected by Malaise traps throughout the tropics. Thirty-two extant genera are known globally. They are all believed to be endoparasitoids of stick insect eggs (Kimsey and Bohart, 1991). Krombein (1956) keyed out the world genera, and since then one more genus has been described, *Nipponosega*, which includes a species in Thailand (Mita, 2021).

Sexual dimorphism in those genera with fully winged females is very slight, exhibited only in shape of mandibles, and characters of the retracted metasomal segments (Krombein, 1956). A small selection of S.E. Asian species is shown in Figs 16.3 and 16.4.

Relatively common S.E. Asian genera include *Cladobethylus*, which was revised by Kimsey (2019). It includes eight S.E. Asian species plus two more from Papua New Guinea. The genus *Mahinda* includes some of the largest species, whose distinctive females are brachypterous with pad-like fore wings and large, conical or tooth-like propodeal angulations (Fig. 16.4) (Kimsey et al., 2016). The rather brightly coloured genus *Atoposega* was revised by Kimsey (2014).

Chrysidinae

All species are metallic in coloration (Fig. 16.5); these are the true 'cuckoo wasps' or 'emerald wasps', and in Western Europe many are considered rare and some are endangered. Almost all are cleptoparasitoids in solitary bee nests; that is, the female chrysidine enters the nest and lays her egg in the host cell, usually before the host female has sealed it and completed provisioning it. The newly hatched chrysidine larva then kills the host egg or early-instar host larva and then continues to feed on the host's food store.

Evolution has led to chrysidids having a very hard cuticle as well as the ability

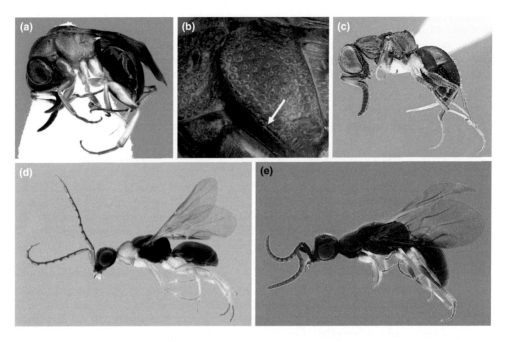

Fig. 16.3. Exemplar Thai amisegines: **(a, b)** *Atoposega* sp. nr *lineata*, showing detail of the carina-like omaulus (arrow); **(c)** *Nipponosega lineata*; **(d)** *Cladobethylus* sp.; **(e)** *Bupon* sp.

Fig. 16.4. Living apterous female of *Mahinda bo* from Vietnam. (Source: from Kimsey *et al.* (2016) reproduced under terms of Creative Commons Attribution License CC-BY-4.0.)

to roll their bodies into a protective ball, with no obvious soft gaps presenting themselves for attacked by their bee or wasp hosts. Not only do many chrysidoids roll themselves into a ball when disturbed, but their wings are often somewhat curved over the metasoma and it may not be easy to flatten them out. Their tough cuticle also makes them quite difficult to push a pin through.

In Vietnam an unidentified *Chrysis* species has been recorded as frequent parasitoid of the stem-nesting eumenine vespid, *Pareumenes quadrispinosus* (Dang and Fateryga, 2021).

Specimens collected into ethanol, e.g. using Malaise traps often change colour thus, blue metallic colour turns into dark blue, green turns into blue, yellow turns into greenish and red becomes more yellowish (Rosa *et al.*, 2016).

Most species of *Praestochrysis* are idiobiont ectoparasitoids of large, heavily cocooned moth prepupae, particularly of Limacodidae (Kimsey and Bohart, 1991). One such example is the Chinese (and Japanese), *Praestochrysis shanghaiensis*, whose biology was described by Yamada (1987), which attacks *Monema flavescens* (Lepidoptera: Limacodidae). The female wasp chews a small hole through the cocoon wall, approximately 0.5 mm in diameter, then inserted her delicate ovipositor apparatus and laid one egg on the inside wall of the host cocoon, and finally she plugged the hole with chewed material. The whole process taking about one hour, and the chrysidid normally only laid one egg (maximum three) per day but due to long adult life, mean lifetime fecundity is nearly 30 eggs

S.E. Asian species are classified into two tribes: Chrysidini and Elampini. Elampini are a cosmopolitan group of approximately 21 genera. They are characterised by having dentate tarsal claws, and lack of a pit row or row of sublateral foveae on metasomal

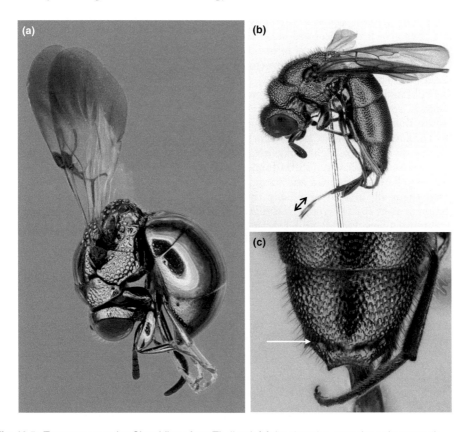

Fig. 16.5. Two representative Chrysidinae from Thailand: **(a)** An elampine, note the rather smooth metasomal tergites and the protruded (death artefact) telescopic 'ovipositor' composed of the more posterior segments, the true ovipositor indicated by arrows; **(b, c)** a chrysidine, note that the arrow points towards the transverse pit row of metasomal tergite 3.

tergite 3. The Chrysidinae have simple claws and possess a pit row on metasomal tergite 3 (Fig. 16.5 c, arrow). The phylogenetics of the lineages was investigated by Lucena and Almeida (2022) which robustly confirmed monophyly of the family and each tribe, with Elampini being basal.

The world Chrysididae were revised by Kimsey and Bohart (1991). Wiśniowski et al. (2020) described four new species of *Trichrysis* from Vietnam, and Nguyen and Wiśniowski (2021) revised Oriental species of *Holophris* (Elampini), which includes two S.E. Asian species.

Cleptinae

Cleptines are principally Holarctic as their hosts are tenthredinid sawflies ('Symphyta'), and they attack the cocooned stage of the host. From the little biological information available, cleptines are idiobiont ectoparasitoids of their pre-pupal hosts.

The subfamily includes only three genera, *Cleptes*, *Cleptidea* and *Lustrina*, and a key to separate these is provided by Rosa et al. (2020). The largest genus is *Cleptes*, which used to be divided into eight subgenera; however, Kimsey and Bohart (1991) found these to be indistinctly defined and so synonymised them all, instead proposing the recognition of eight species-groups. *Cleptidea* is extralimital (New World) and *Lustrina* includes a single species originally described from India, but recently discovered in Vietnam (Rosa et al., 2020) and here we report it from Thailand (Fig. 16.6).

Sexual dimorphism is relatively slight. Females have four visible metasomal tergites, males have five. Coloration of Cleptinae varies considerably. Some are entirely metallic, others have only the head and mesosoma metallic, and some are largely or entirely black – see Wei et al. (2013) for colour photographs showing the huge range of variation among the Chinese species.

There have been no systematic studies of Cleptinae outside of the Holarctic region. Móczár (1998) revised the world subgenera and species groups of *Cleptes*. To date only three species are known from S.E. Asia. In addition to *Lustrina*, Tsuneki (1982) described a new species (*Cleptes thaiensis*) from Thailand, and Móczár (2000) in his revision of the *Cleptes asianus* and *townesi* groups described *C. humeralis* from Malaysia. Given that 17 species are now known from China (Wei et al., 2013), it seems likely that more will be discovered in northern S.E. Asia, and these are most likely to be discovered at higher elevations, where sawfly hosts are abundant and diverse.

Loboscelidiinae

These are highly distinctive and aberrant, mostly very dark brown or orange-brown species, approximately 2–3 mm long (males). One can easily say 'once seen,

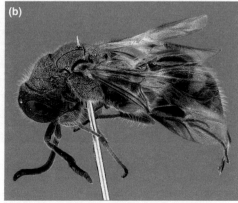

Fig. 16.6. *Lustrina assamensis* from Thailand: **(a)** dorsal habitus; **(b)** oblique lateral habitus.

never forgotten' about them. Most specimens in collections are males, which has hampered taxonomy, but although they are sometimes referred to as uncommon, we have encountered hundreds of specimens in Thai Malaise trap samples from a variety of habitats. Although an affinity with the Chrysidoidea had been suggested when they were first described as a separate family (Maa and Yoshimoto, 1961) it was not really until Day (1978) that these wasps were generally accepted as chrysidoid aculeates (Kimsey, 2012).

Loboscelidiines are characterised by the antennae being inserted horizontally on to a shelf-like projection at the middle of the face, the vertex being produced posteriorly to form a neck-like projection (Fig. 16.7 a, c); the pronotum has an anteroventral flange appearing like a short line of ribbon-like setae, which is overlapped by the flange at the back of the head (although both may be reduced in some species); the fore wing lacks a stigma, costal or subcostal veins (Fig. 16.7 e); the tegulae are very large, covering both wing bases (Fig. 16.7 a) and held in place by a mesopleural ridge.

The Loboscelidiinae comprises just two genera, *Loboscelidia* and *Rhadinoscelidia* (Fig. 16.7) (Kimsey, 1988), and is largely restricted to S.E. Asia, with a few species known from both China and Australia. *Loboscelidea* includes approximately 45 described

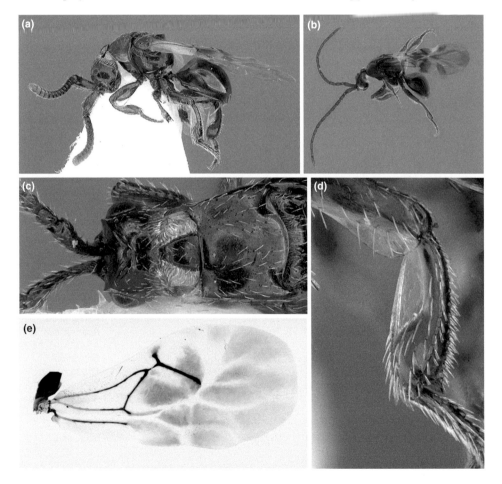

Fig. 16.7. Thai *Loboscelidea* spp.: **(a, b)** lateral habitus of two species; **(c)** head and pronotum dorsal aspect showing the posterior cephalic protrusion flange, and dense modified golden setal patches; **(d)** transparent flanges on hind legs and tibia; **(e)** highly derived fore wing venation.

species to which Kimsey (2012) provided a key. Females are virtually unknown and the taxonomy is based only on males. Sexual dimorphism is extreme and in fact one new genus was erected based on a female, but that turned out to be just a female of *Loboscelidia*. Females probably spend most of their adult lives in cryptic situations such as among leaf litter or under bark (Kimsey, 2012).

Little is known of their biology. Kimsey (2012) suggested that the peculiar structural modifications of the group, including flanges on the head, legs and antennae, large tegula and the tegular clip, may be adaptations enabling the females to search for stick-insect eggs in ant nests. *Rhadinoscelidia lixa* has been found on an ant nest in Thailand (Hisasue and Mita, 2020) but tropical ants often carry stick-insect eggs into their nests as part of a mimicry system (Quicke, 2017). Whether the same behaviour is displayed by Asian ants is not known.

Rhadinoscelidia is rarer and even less well known, though it has now been recorded from China (Liu *et al.*, 2011b). Kojima and Ubaidillah (2003) recorded both genera from Indonesia. Liu *et al.* (2011b) illustrated and described a few new taxa of both genera from China. The Vietnamese Loboscelidia species were revised by Hisasue *et al.* (2023).

Dryinidae (Pincer Wasps)

The Dryinidae is divided into 11 subfamilies, seven of which occur in S.E. Asia: Anteoninae (Fig. 16.9 a, b), Aphelopinae (Fig. 16.8), Bocchinae (Fig. 16.9 c, d, e), Conganteoninae, Dryininae (Fig. 16.9 e, f, g), Gonatopodinae (Fig. 16.10 a) and Thaumatodryininae (the latter sometimes included in Dryininae, e.g. Olmi, 1993). A key to the world subfamilies was provided by Finnamore and Brothers (1995), although it is not always easy to see some of the critical characters. In Table 16.1 we summarise the major character combinations for females. Tribull (2015) investigated the relationships between the subfamilies, using sequence data from two mitochondrial gene fragments (cytochrome oxidase I and cytochrome b).

Dryinids are koinobiont parasitoids of nymphal and sometimes adult auchenorrynchan Hemiptera in the Cicadelloidea and Fulgoroidea. Anteoninae are parasitoids of Cicadellidae; Aphelopidae mainly attack Membracidae but some attack Cicadellidae (Typhlocybinae); whereas Bocchinae are parasitoids of Cicadellidae and Issidae; Gonatopodinae attack various Fulgoromorpha and Cicadomorpha; Thaumatodryinidae are parasitoids of Flatidae (Finnamore and Brothers, 1995).

Females of most dryinids grasp their prospective hosts, usually nymphal stages,

Table 16.1. Summary of female subfamily-level characters of dryinid tribes occurring in S.E. Asia.

Character	Present	Absent
Fore tibia chelate (modified into pincer)	The rest	Aphelopinae
Enlarged protarsal claw sub-apical teeth or lammelum	The rest	Anteoninae
Occipital carina	The rest	Anteoninae
Middle tibia without apical spur	Gonatopodinae	The rest
Flagellomeres 3–8 with tufts of long setae	Thaumatodryininae	The rest
Brachypterous or apterous	Anteoninae (some, but always with a scale-like vestige), Bocchinae (some), Gonatopodinae (most)	Anteoninae (most)
Mandible with four large teeth (alternative is either fewer teeth or with three large teeth and one reduced tooth)	Dryininae	Bocchinae
In fully winged forms three enclosed cells present (alternative is fewer)	Anteoninae, Bocchinae	The rest
Chela with rudimentary claw (often difficult to see)	Anteoninae	The rest

firmly with their chelate claws and then usually deliver a sting that causes temporary host paralysis. However, members of the dryinine genus *Aphelopus* (Fig. 16.8) lack chelate claws. Females may be fully winged (Fig. 16.8), brachypterous or apterous, often rather ant-like (see Fig. 16.10 a), whereas males are always fully winged. Females of most species can be readily distinguished from males because they have the fore tarsus highly modified as a raptorial grasper (Fig. 16.9 a-f), hence the common name pincer wasps is increasingly being used for them.

Although males of most species look quite different from females and lack the modified fore tarsi, they are more commonly collected but are relatively easy to recognise as dryinids based on their general body shape and their 8-segmented antennal flagellums. Association of the sexes is difficult because of the high level of dimorphism. Mita and Matsumoto (2012) were able to associate the sexes of *Gonatopus javanus* with the use of DNA barcode data. Some species are capable of thelytokous parthenogenesis and males are either unknown or exceedingly rare. Whether or not this is caused by *Wolbachia* does not seem to be known.

The year 2013 was particularly active for publications on the family. Xu *et al.* (2013) re-revised the Oriental Dryinidae with strong emphasis on the Chinese fauna. This monumental monograph nevertheless covers many species from the whole of S.E. Asia (Table 16.2). Kim *et al.* (2013) published a checklist of Cambodian dryinids, which added several species to the records in Xu *et al.* (2013), and also included a key to the 14 known species from that country. Many species have wide distributions, extending beyond S.E. Asia as far as Australia, China, India, Pakistan and Japan. For example, *Gonatopus nearcticus* is a Holarctic species that occurs also in the oriental and

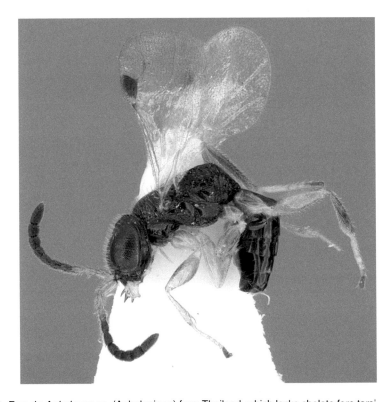

Fig. 16.8. Female *Aphelopus* sp. (Aphelopinae) from Thailand, which lacks chelate fore tarsi.

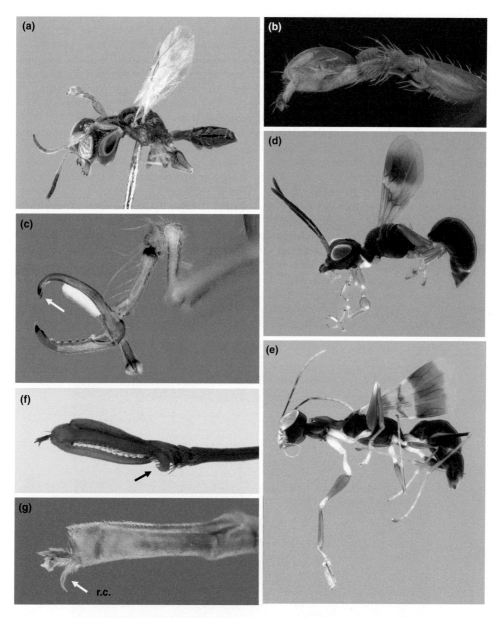

Fig. 16.9. Exemplar female dryinids from Thailand: **(a, b)** *Anteon* sp. (Anteoninae), note absence of rudimentary claw; **(c, d)** *Bocchus* sp. (Bocchinae), with detail of raptorial fore leg forced open post mortem (in death in alcohol they usually lock very rigid), with arrow pointing to single flattened lamella at apex of telotarsus; **(e, f, g)** *Dryinus* sp. (Dryininae), and detail of raptorial fore leg in normal death state showing multiple flattened lamellae at apex of telotarsus (arrow) and with tarsus rotated showing rudimentary claw next to aroleum (r.c.).

S.E. Asian regions. Very few taxonomists study Dryinidae, so it is hard to tell how complete it is and undoubtedly many undescribed species still exist in the region.

Since Xu *et al.*'s (2013) revision there have been a number of descriptions, mostly of isolated species, from S.E. Asia (Guglielmino and Olmi 2013; Olmi *et al.*, 2015, 2016a;

Table 16.2. Distribution of dryinid species in S.E. Asia (data from Mita and Okajima, 2011; Kim *et al.*, 2013; Olmi *et al.*, 2013, 2015, 2016a; Xu *et al.*, 2013; and Guglielmino *et al.*, 2017).

Country	Species recorded
Brunei	*Anteon* (*chui, diaoluoshanense, expolitum, peterseni, sabahnum, sarawaki, thai*); *Aphelopus* (*borneanus, malayanus, spadiceus*); *Bocchus pedunculatus; Dryinus* (*bruneianus, pyrillivorus*); *Neodryinus diffusus, Pseudodryinus beckeri; Thaumatodryinus* (*pasohensis, philippinus*)
Cambodia	*Aphelopus alebroides; Anteon fidum, hilare, ingenuum, laminatum, metuendum, thai, yangi, yasumatsui; Dryinus browni, indianus, stantoni; Gonatopus medius*
Indonesia	*Anteon* (*atrum, borneanum, cerberum, clavatum, devriesi, expolitum, gauldi, hilare, insertum, krombeini, laminatum, muiri, nanlingense, pahanganum, parapriscum, priscum, quatei, subdignum, sulawesianum, supaibum, thai, yasiri, yasumatsui*); *Aphelopus* (*achtbergeri, borneanus, maculiceps, ochreus, orientalis, penanganus, taiwanensis*); *Bocchus* (*achtbergeri, beckeri, pedunculatus, pedunculatus*); *Deinodryinus philippinus; Dryinus* (*alboniger, bellicus, browni, indicus, mansus, parvulus, sinicus*); *Echthrodelphax fairchildii; Gonatopus* (*asiae, iarensis, insulae, javanus, lucens, muiri, nigricans, pajanensis, perpolitus, plebeius*); *Neodryinus* (*chelatus, leptopus, pseudodiffusus, sumatranus*); *Pseudodryinus piceus; Thaumatodryinus* (*noyesi, philippinus*)
Laos	*Anteon* (*expolitum, hilare, holzschuhi, kresli, laotianum, munitum, striolaticeps, thai, yoshimoto*); *Aphelopus* (*malayanus, ochreus, orientalis, taiwanensis*) *Bocchus* (*banianus, laotianus, levis, viet*) *Deinodryinus asiaticus, Dryinus* (*barbarus, chenae, choui, exilis, expolitus, irregularis, krombeini, latus, parvulus, pyrillae, sinicus, trifasciatus*) *Echthrodelphax* (*laotianus, rufus*); *Neodryinus* (*chelatus, javanus, karaentensis, leptopus, pseudodiffusus, robustus*)
Malaysia	*Anteon* (*adebratti, austini, borneanum, flaccum, gauldi, hilare, hirashimai, maai, munitum, naduense, nemorale, pahanganum, parapriscum, peterseni, quatei, sabahnum, sarawaki, silvestre, sinicum, songyangense, taiwanense, thai, wushense, yasumatsui, yoshimotoi, zoroastrum*); *Aphelopus* (*albiclypeus, borneanus, malayanus, ochreus, penanganus, philippinus, sabahnus, taiwanensis*); *Bocchus* (*adebratti, beckeri, levis, laotianus, levis, muluensis, pedunculatus, thai*); *Dryinus* (*asiaticus, bellicus, browni, indicus, krombeini, lucens, parvulus, pyrillae, trifasciatus*); *Echthrodelphax fairchildii; Gonatopus* (*achterbergi, asiaticus, beaveri, borneanus, iarensis, lucens, maurus, malesiae, nudus, nigricans, plebeius, rufoniger, sarawakensis, validus, yasumatsui*); *Haplogonatopus apicalis; Neodryinus* (*diffusus, javanus, malayanus, pseudodiffusus, reticulatus, sumatranus*); *Pseudodryinus* (*beckeri*); *Thaumatodryinus* (*alienus, asiaticus, beckeri, malayanus, pasohensis, philippinus, sharkeyi*)
Myanmar	*Anteon* (*fidum, gauldi, hilare, hirashimai, malaisei, meifenganum, munitum, peterseni, quatei, sulawesianum*); *Aphelopus* (*birmanus, orientalis, taiwanensis*) *Deinodryinus malaisei; Gonatopus* (*cristatus, malesiae*); *Lonchodryinus sinensis*
Philippines	*Anteon* (*abatanense, achterbergi, autumnale, borneanum, expolitum, gauldi, hilare, insertum, luzonense, munitum, parapriscum, peterseni, philippinum, provinciale, thai*); *Aphelopus* (*malayanus, philippinus, wushensis*) *Bocchus* (*rubricus, pedunculatus*); *Deinodryinus philippinus; Dryinus* (*achterbergi, browni, gibbosoides, kiefferi, krombeini, latus, peterseni, praeclarus, pyrillae, pyrillivorus, trifasciatus*); *Echthrodelphax fairchildii; Gonatopus* (*attenuatus, bicuspis, lucens, nudus, yasumatsui, philippinus*); *Haplogonatopus apicalis; Neodryinus* (*chelatus, javanus, pseudodiffusus, reticulatus, sumatranus*); *Thaumatodryinus* (*asiaticus, philippinus*)
Singapore	*Pseudodryinus sinensis; Thaumatodryinus malayanus*
Thailand	*Anteon* (*austini, borneanum, chui, dignum, doiense, expolitum, fidum, gauldi, hilare, huettingeri, insertum, khaokhoense, laminatum, meifenganum, munitum, mysorense, naduense, papillatum, parapriscum, phetchabunense, phuphayonense, raptor, semipolitum, sulawesianum, thai, viraktamathi, yasumatsui*); *Aphelopus* (*albiclypeus, fuscoflavus, malayanus, orientalis, penanganus, spadiceus, taiwanensis, tha*); *Bocchus* (*banianus, thai, viet*); *Deinodryinus* (*asiaticus, constrictus*); *Dryinus* (*browni, fulvus, indicus, irregularis, krombeini, latus, longipes, parvulus, pyrillae, pyrillivorus, trifasciatus, viet*); *Echthrodelphax* (*fairchildii, laotianus, rufus*); *Gonatopus* (*cantonensis, cristatus, nearcticus, nigricans, nudus, thai, yasumatsui*); *Haplogonatopus apicalis; Neodryinus* (*beaveri, diffusus, lohmani, phuphayonensis, pseudodiffusus, sumatranus*); *Pseudodryinus thai; Thaumatodryinus* (*asiaticus, philippinus*)

Continued

Table 16.2. Continued.

Country	Species recorded
Vietnam	*Anteon* (*acre, debile, fyanense, gauldi, hirashimai, expolitum, peterseni, silvicolum, spenceri, thai, yoshimotoi*); *Aphelopus* (*albiclypeus, maculiceps*); *Bocchus* (*banianus, beckeri, viet*); *Conganteon richardsi*; *Dryinus* (*asiaticus, fulvus, indianus, krombeini, latus, lini, viet*); *Gonatopus* (*achterbergi, fyanensis, nearcticus, philippinus, viet, yasumatsui*); *Neodryinus leptopus*

Guglielmino et al., 2017). Most are based on males and include modified key sections from previous publications to accommodate the new taxa.

Despite the economic importance of some species, and the relative ease with which some can be cultured, the developmental biology, especially early development, is surprisingly poorly known (Carcupino et al., 1998). Details of egg placement and development differ a great deal between different subfamilies, and for a good and detailed summary see Olmi (1994). The egg is laid between a pair of the host's abdominal sclerites or under the wing buds, depending on species. The eggs of some species have a hook-like process (Giri and Freytag, 1989) that is thought to secure the egg between the tergite and intersegmental membrane of its host (Giri and Freytag, 1989; Quicke, 1997), but the exact mechanism appears to be unknown. In some other species, the egg is apparently placed fully internally; for example, Mangione and Virla (2004) clearly stated in reference to the South American species *Gonatopus bonaerensis* [translated from Spanish]:

'The females place the eggs in the abdomen of the hosts, between two sclerites and through the intersegmental membrane, below the integument and with only their posterior end protruding slightly from the wound produced by the ovipositor.'

The head end of the larva is thus able to feed internally whilst the rest of its body remains and grows on the outside of the host's body but actually within a sac or cyst (Fig. 16.10 b). The larval stages of several species have been described in detail for quite a lot of species (e.g. Guglielmino et al., 2015). Species of the extralimital genus *Crovettia* (Aphelopinae) are polyembryonic and hence totally endoparasitoid (Askew, 1971; Carcupino et al., 1998). Eggs of Gonatopodinae are small and lack yolk; they may be laid either internally or externally and are glued on to intersegmental membranes of the host's thorax or abdomen. In the case of external ovipositing species the 1st instar larva waits until the host moults and enters part way into the host via the ovipositor wound site in the soft new cuticle. The posterior part remains external, where it grows much larger as development progresses. The cast larval skins produced by each moult remain attached and form an external protective sac called a thylacium (Fig. 16.10 b).

Aphelopus (Aphelopinae) species lay their small, alecythal eggs inside the host's haemocoel. Development of the egg forms an embryo surrounded by an extra-embryonic cellular membrane called the trophamnion which isolates the 1st instar larva from the host's haemocytes. The 2nd instar remains near the oviposition wound against which it is pressed and its posterior body is pushed outside between two overlapping host abdominal sclerites but the larval head remains inside the haemocoel, where it feeds (Olmi, 1994).

Several dryinid species are important parasitoids of pest species of rice hoppers (Homoptera Cicadellidae) (Omar et al., 1996). Guglielmino et al. (2013) catalogued all known host records for the family based on data from 38 different countries. Some play important roles in regulating host populations in agroecosystems (Guglielmino et al., 2013). In addition to their parasitisation of hosts, many (maybe most) are also active predators (host feeders) on plant hoppers (Sahragard et al., 1991). Adult females of the brown rice hopper (*Nilaparvata lugens*) parasitoid *Gonatopus flavifemur* live for

Fig. 16.10. Living Thai dryinids: **(a)** apterous adult female *Gonatypus* sp. (Gonatopodinae); **(b)** nymph of delphacid plant hopper with protruding black larva of a dryinid inside its thylacium. (Source: photographs by and © Jean-Yves Rasplus, reproduced with permission.)

an average of nearly 20 days and during this time they can parasitise more than 400 host individuals, and may additionally prey on another 50 or so (Chua and Dyck, 1982; He et al., 2020).

Embolemidae

These are small and rather uncommon wasps, often with a high degree of sexual dimorphism. Males are always fully winged (Fig. 16.11); females may be fully winged, brachypterous or apterous. They are easy to recognise based on their having only eight flagellomeres, which arise from a small conical protuberance above the middle of the head. Female *Embolemus* are apterous, male *Embolemus* are usually fully winged, and both sexes of *Ampulicomorpha* are fully winged.

Xu et al. (2001) revised the six species in two genera (*Ampulicomorpha* and *Embolemus*) known from China, and this provides a helpful start. Three of the species covered occur in S.E. Asia: *Ampulicomorpha collinsi*, *Embolemus krombeini* and *E. pecki*. Most work on S.E. Asian species comprises isolated descriptions (Guglielmino and Olmi, 2014). However, it should be noted that van Achterberg and van Kats (2000)

Fig. 16.11. Male embolemmid from Thailand (with a parasitic mite attached); note the way the antennae are inserted on a protuberance far from the mandibles.

considered *Ampulicomorpha* to be a synonym of *Embolemus*, the key difference for males being very slight.

Very little is known of their biology (Wharton, 1989), which appears to be similar to that of Dryinidae, parasitising and perhaps specialising on ground-dwelling Fulgoroidea.

Sclerogibbidae

This is a small, relatively rarely collected family, with only four extant genera (Olmi, 2005; Olmi *et al.*, 2016b).

Olmi recognised only three extant genera, a great reduction from some earlier classifications, but later added a fourth from South Africa (Olmi *et al.*, 2016b). *Sclerogibba* is widespread in S.E. Asia, and *Caenosclerogibba* is known from two species, *C. rossi* from Malaysia and Singapore, and *C. longiceps* recently discovered in the Philippines (Lucañas and Olmi, 2017) and Japan (Mita and Olmi, 2018). Both species have wide distributions, *C. rossi* being known from Bangladesh and *C. longiceps* occurring in North and South America, Yemen, Africa (including Madagascar), India and Japan. Lim *et al.* (2013) recorded three species from Cambodia and provided diagnoses and photographs for each. Lucañas *et al.* (2016) recorded *Caenosclerogibba* from the Philippines for the first time.

Sclerogibbids may be recognised by the combination of prognathous head with antennae inserted under a shelf-like lower part of face near the mouth (Fig. 16.12 a, b, d), robust femora and the insertion of the fore trochanter onto the apico-lateral part of the fore coxae (Fig. 16.12 d). Olmi (2005) noted 'the extreme difficulty of this family composed of very uniform species' as the reason for the relatively little attention it has received.

Fig. 16.12. A sclerogibbid male from Thailand. **(a)** habitus; **(b)** head, dorsal view, arrow pointing to shelf over antennal sockets; **(c)** fore wing; **(d)** head and anterior mesosoma, ventral view, arrow indicating the lateral insertion of fore trochanter on coxa.

The majority of species have enormous distributions, some being virtually cosmopolitan, or at least incorporating several biogeographical regions. Several species of *Sclerogibba* are known from the region: *S. africana* from Myanmar, *S. berlandi* from Malaysia, *S. impressa* from the Philippines; *S. madegassa* from Thailand, *S. rapax* from Malaysia, Laos, Vietnam and Thailand, *S. rossi* from Myanmar and Malaysia, *S. rugosa* from Thailand and *S. talpiformis* from Myanmar.

Sclerogibbids are gregarious koinobiont ectoparasitoids of Embioptera (webspinners) nymphs and adults (Ananthasubramanian and Ananthakrishnan, 1959); see also photograph in Mita and Olmi (2018).

Scolebythidae

This family was erected as recently as 1962 by Howard Evans on the basis of their unusual combination of characters, and indeed they had been thought to show some affinities to the vespoid family Scoliidae (Evans, 1963). Scolebythids are rare and known mostly from the New World, Africa (including Madagascar) and Australia, but with specimens also known from north China, Thailand (Azevedo et al., 2011), and Fiji, as well as Cretaceous fossils in Burmese amber (Zhang et al., 2020). The Thai (and Chinese) species, *Pristapenesia asiatica*, is the only known extant Asian scolebythid with records from the region. It is 'dark castaneous' (i.e. black with a chestnut tinge) and a little over 5 mm long. The only other known member of the genus is from Brazil.

Melo (2000) provided the first biological information of the family based on a New World species, which was at that time placed in *Dominibythus* but is now regarded as a junior synonym of *Pristapensia* (Brothers and Janzen, 1999). As previously suspected, based on circumstantial evidence for species of *Ycaploca* and *Scolebythus*, the species was proved to be a gregarious idiobiont ectoparasitoid of wood boring beetle larvae, in this case Anobiidae.

References

Alencar, I.D.C.C. and Azevedo, C.O. (2013) Reclassification of Epyrini (Hymenoptera: Bethylidae): a tribal approach with commentary on their genera. *Systematic Entomology* 38, 45–80.
Amante, M., Schöller, M., Suma, P. and Russo, A. (2017a) Bethylids attacking stored-product pests: an overview. *Entomologia Experimentalis et Applicata* 163, 251–264.
Amante, M., Schöller, M., Hardy, I.C.W. and Russo, A. (2017b) Reproductive biology of *Holepyris sylvanidis* (Hymenoptera: Bethylidae). *Biological Control* 106, 1–8.
Ananthasubramanian, K.S. and Ananthakrishnan, T.N. (1959) The biology of *Sclerogibba longiceps* Richards and *Sclerogibba embiidarum* (Kieff.) (Sclerogibbidae: Hymenoptera) parasitic on Embioptera. *Journal of the Bombay Natural History Society* 56, 101–113. Available at http://www.biodiversitylibrary.org/part/153240
Askew, R.R. (1971) *Parasitic Insects*. Heinemann, London, 316 pp.
Azevedo, C.O. (2011) Synopsis of *Aspidepyris* Evans, 1964 (Hymenoptera, Bethylidae). *Zootaxa* 3016, 63–68.
Azevedo, C.O. and Alencar, I.D.C.C. (2009) Rediscovery of *Pristepyris* Kieffer (Hymenoptera, Bethylidae), a new synonym of *Acrepyris* Kieffer. *Zootaxa* 2287, 45–54.
Azevedo, C.O. and Barbosa, D.N. (2010) Two new species *Pararhabdepyris* Gorbatovskii (Hymenoptera, Bethylidae) from Australia and Thailand. *Zootaxa* 2668, 55–62.
Azevedo, C.O. and Lanes, G.O. (2009) Cladistic assessment and redescription of *Galodoxa torquata* Nagy (Hymenoptera, Bethylidae), a striking species with swallow tailed metasomal sternite. *Zoologische Mededelingen, Leiden* 83, 841–851.
Azevedo, C.O., Xu, Z. and Beaver, R.A. (2011) A new species of *Pristapenesia* Brues (Hymenoptera, Scolebythidae) from Asia. *Zootaxa* 2750, 60–64.
Azevedo, C.O., Alencar, I.D.C.C., Ramos, M.S., Barbosa, D.N., Colombo, W.D., Vargas, J.M.R. and Lim, J. (2018) Global guide of the flat wasps (Hymenoptera, Bethylidae). *Zootaxa* 4489, 1–294. https://doi.org/10.11646/zootaxa.4489.1.1
Barbosa, D.N. and Azevedo, C.O. (2011) Taxonomy of *Allobethylus* Kieffer (Hymenoptera: Bethylidae) from southeastern Asia. *Journal of Asia-Pacific Entomology* 14, 89–94.

Barbosa, D.N., Vilhelmsen, L. and Azevedo, C.O. (2021) Morphology of sting apparatus of Chrysidoidea (Hymenoptera, Aculeata). *Arthropod Structure & Development* 60, 100999. doi: 10.1016/j.asd.2020.100999

Barbosa, D.N., Hermes, M.G. and Lepeco, A. (2022) Phylogeny of Mesitiinae (Hymenoptera: Bethylidae): assessing their classification, character evolution and diversification. *Arthropod Systematics & Phylogeny* 80, 603–625.

Brothers, D.J. and Janzen, J.-W. (1999) New generic synonymy in Scolebythidae, with redescription of both sexes of *Pristapenesia primaeva* Brues from Baltic amber (Hymenoptera: Chrysidoidea). In Vršanský, P. (ed.) *Proceedings of the First Palaeoentomological Conference, AMBA, Moscow*, pp. 17–26.

Carcupino, M., Guglielmino, A., Olmi, M. and Mazzini, M. (1998) Morphology and ultrastructure of the cephalic vesicles in two species of the *Gonatopus* genus: *Gonatopus camelinus* Kieffer and *Gonatopus clavipes* (Thunberg) (Hymenoptera, Dryinidae, Gonatopodinae). *Invertebrate Reproduction and Development* 34, 177–186.

Chua, T.H. and Dyck, V.A. (1982) Assessment of *Pseudogonatopus flavifemur* E. & H. (Dryinidae: Hymenoptera) as a biocontrol agent of the rice brown planthopper. In: *Proceedings of the International Conference on Plant Protection in the Tropics*, Kuala Lumpur, 1–4 March 1982. Malaysian Plant Protection Society, Kuala Lumpur, Malaysia, pp. 253–265.

Colombo, W.D., Tribull, C.M., Waichert, C. and Azevedo, C.O. (2022) Integrative taxonomy solves taxonomic impasses: a case study from Epyrinae (Hymenoptera, Bethylidae). *Systematic Entomology* 47, 504–529. doi: 10.1111/syen.12544

Dang, H.T. and Fateryga, A.V. (2021) Nesting biology of *Pareumenes quadrispinosus* (de Saussure, 1855) (Hymenoptera: Vespidae: Eumeninae) in trap nests in North Vietnam. *Journal of Asia-Pacific Entomology* 24, 1276–1285.

Day, M.C. (1978) The affinities of *Loboscelidia* Westwood. *Systematic Entomology* 4, 21–30.

de Brito, C.D., de O. Lanes, G. and Azevedo, C.O. (2022) Morphology and evolution of the mesopleuron in Bethylidae (Hymenoptera: Chrysidoidea) mapped on a molecular phylogeny. *Arthropod Structure & Development* 71, 101214.

Evans, H.E. (1963) A new family of wasps. *Psyche* 1963, 7–16.

Finlayson, L.H. (1950) The biology of *Cephalonomia watestoni* Gahan (Hymenoptera: Bethylidae), a parasite of *Laemophloeus* (Coleoptera: Cucujidae). *Bulletin of Entomological Research* 41, 79–97.

Finnamore, A.T. and Brothers, D.J. (1995) Superfamily Chrysidoidea. In: Goulet, H. and Huber, J. (eds) *Hymenoptera of the World: An Identification Guide to Families* (Vol. vii). Research Branch, Agriculture Canada, Ottawa, pp. 130–160.

Gauld, I.D. and Bolton, B. (1988) *The Hymenoptera*. British Museum (Natural History), London, and Oxford University Press, Oxford, 332 pp.

Giri, M.K. and Freytag, P.H. (1989) Development of *Dicondylus americanus* (Hymenoptera: Dryinidae). *Frustula Entomologica, N.S.* 9, 215–222.

Gobbi, F.T. and Azevedo, C.O. (2014) Revision of *Caloapenesia* (Hymenoptera, Bethylidae), with description of sixteen new species. *Zootaxa* 3860, 501–535.

Gordh, G. and Móczár, L. (1990) A catalog of the world Bethylidae (Hymenoptera: Aculeata). *Memoirs of the American Entomological Institute* 42, 1–364.

Goulet, H. and Mason, W.R.M. (1993) Superfamily Cynipoidea. In: Goulet, H. and Huber, J. (eds) *Hymenoptera of the World: An Identification Guide to Families* (Vol. vii). Research Branch, Agriculture Canada, Ottawa, pp. 60–64.

Guglielmino, A. and Olmi, M. (2013) Description of *Anteon seramense* (Hymenoptera: Dryinidae), a new species from Indonesia. *Florida Entomologist* 96, 598–601. doi: 10.1653/024.096.0226

Guglielmino, A. and Olmi, M. (2014) Description of two new species of *Ampulicomorpha* Ashmead from Indonesia and Madagascar (Hymenoptera: Embolemidae). *Journal of the Kansas Entomological Society* 87, 234-241. doi: 10.2317/JKES130808.1

Guglielmino, A., Olmi, M. and Bückle, C. (2013) An updated host–parasite catalogue of world Dryinidae (Hymenoptera: Chrysidoidea). *Zootaxa* 3740, 1–113. doi: 10.11646/zootaxa.3740.1.1

Guglielmino, A., Parise, G. and Bückle, C. (2015) Description of larval instars of *Dryinus tarraconensis* Marshall, 1868 and *Gonatopus baeticus* (Ceballos, 1927) (Hymenoptera: Chrysidoidea: Dryinidae), parasitoids of the genus *Dictyophara* Germar (Hemiptera: Auchenorrhyncha: Dictyopharidae). *Zootaxa* 4032, 42–54. doi: 10.11646/zootaxa.4032.1.2

Guglielmino, A. Olmi, M. Marletta, A. and Xu, Z.-F. (2017) Description of *Aphelopus fuscoflavus*, a new species of Dryinidae from Thailand (Hymenoptera, Chrysidoidea). *Journal of Hymenoptera Research* 57, 115–121. doi: 10.3897/jhr.57.12462

He, J., He, Y., Lai, F., Chen, X. and Fu, Q. (2020) Biological traits of the pincer wasp *Gonatopus flavifemur* (Esaki & Hashimoto) associated with different stages of its host, the brown planthopper, *Nilaparvata lugens* (Stål). *Insects* 11, 279. doi: 10.3390/insects11050279

Hisasue, Y. and Mita, T. (2020) *Rhadinoscelidia lixa* sp. nov. (Hymenoptera, Chrysididae, Loboscelidiinae) found on an ant nest in Thailand. *ZooKeys* 975, 1–9.

Hisasue, Y., Pham, T.H. and Mita, T., (2023) Taxonomic revision of the genus *Loboscelidia* Westwood, 1874 (Hymenoptera: Chrysididae: Loboscelidiinae) from Vietnam. *European Journal of Taxonomy* 887, 1–68.

Howard, R.W., Charlton, M. and Charlton R.E. (1998) Host-finding, host-recognition, and host-acceptance behavior of *Cephalonomia tarsalis* (Hymenoptera: Bethylidae). *Annals of the Entomological Society of America* 91, 879–889.

Kieffer, J.J. (1913a) Serphides des îles Philippines. *Insecta* 3, 253–260.

Kieffer, J.J. (1913b) Serphides des îles Philippines. (Suite). *Insecta* 3, 317–324.

Kieffer, J.J. (1914) Énumération des Serphides (Proctotrupides) des Îles Philippines avec description du genres nouveaux et d'espèces nouvelles. *Philippine Journal of Science, Series D* 9, 285–311.

Kim, C.-J., Olmi, M., Lee, S., Lim, J., Choi, G.W. and Lee, J.-W. (2013) A checklist of Dryinidae (Hymenoptera: Chysidoidea) from Cambodia with new records. *Journal of Asia-Pacific Entomology* 16, 485–488.

Kimsey, L.S. (1987) Review of the Subfamily Parnopinae (Hymenoptera, Chrysididae). *Journal of the Kansas Entomological Society* 1987, 83–91.

Kimsey, L.S. (1988) Loboscelidiinae, new species and a new genus from Malaysia (Hymenoptera: Chrysididae). *Psyche* 95, 67–79. doi: 10.1155/1988/16535

Kimsey, L.S. (2012) Review of the odd chrysidid genus *Loboscelidia* Westwood, 1874 (Hymenoptera, Chrysididae, Loboscelidiinae), *ZooKeys* 213, 1–40.

Kimsey, L.S. (2014) Reevaluation of the odd chrysidid genus *Atoposega* Krombein (Hymenoptera, Chrysididae, Amiseginae). *ZooKeys* 409, 35–47. doi: 10.3897/zookeys.409.7414

Kimsey, L.S. (2019) Revision of the south Asian amisegine genus *Cladobethylus* Kieffer, 1922 (Hymenoptera, Chrysididae, Amiseginae). *Journal of Hymenoptera Research* 70, 41–64. doi: 10.3897/jhr.70.34206

Kimsey, L.S. and Bohart, R.M. [1990] (1991) *The Chrysidid Wasps of the World*. Oxford University Press, Oxford, 652 pp.

Kimsey, L.S., Mita, T. and Pham, H.T. (2016) New species of the genus *Mahinda* Krombein, 1983 (Hymenoptera, Chrysididae, Amiseginae). *ZooKeys* 551, 145–154. doi: 10.3897/zookeys.551.6168

Kojima, J.I. and Ubaidillah, R. (2003) Two new species of the cryptic chrysidid parasitoid subfamily Loboscelidiinae: the second species in *Rhadinoscelidia* and the first *Loboscelidia* for the Indonesian fauna. *Entomological Science* 6, 199–207. doi: 10.1046/j.1343-8786.2003.00023.x

Krombein, K.V. (1956) A generic review of the Amiseginae, a group of phasmatid egg parasites, and notes on the Adelphinae (Hymenoptera, Bethyloidea, Chrysididae). *Transactions of the American Entomological Society* 82, 147–215.

Lanes, G.O. and Azevedo, C.O. (2007) Redescription and placement of the Oriental *Scaphepyris rufus* Kieffer (Hymenoptera: Bethylidae). *Zootaxa* 1654, 55–60.

Lanes, G.O. and Azevedo, C.O. (2008) Phylogeny and taxonomy of Sclerodermini (Hymenoptera: Bethylidae, Epyrinae). *Insect Systematics and Evolution* 39, 55–86.

Lanes, G.O., Kawada, R., Azevedo, C.O. and Brothers, D.J. (2020) Revisited morphology applied for systematics of flat wasps (Hymenoptera, Bethylidae). *Zootaxa* 4752, 1–127.

Lim, J. and Lee S. (2011) A new species of *Prorops* Waterston 1923 (Hymenoptera: Bethylidae) from Cambodia with a key to world species. *Zootaxa* 3040, 25–28. doi:10.11646/zootaxa.3040.1.3

Lim, J. and Lee, S. (2013) Taxonomy of the family Bethylidae (Hymenoptera: Chrysidoidea) from Cambodia and adjacent countries. I. Genus *Odontepyris* Kieffer (Bethylidae: Bethylinae) with four new species and two new records. *Journal of Natural History* 47, 2017–2038. doi: 10.1080/00222933.2012.763057

Lim, J., Olmi, M., Kim, I.-K. and Lee, S. (2013) First record of Sclerogibbidae (Hymenoptera: Chrysidoidea), ectoparasitoids of Embiidina (Insecta: Neoptera) from Cambodia. *Korean Society of Applied Entomology* 4, 273.

Liu, J., Chen, H. and Xu, F. (2011a) Two new species of the genus *Foenobethylus* Kieffer 1913 (Hymenoptera: Bethylidae) from China with a key to the known species. *Zootaxa* 2806, 53–59.

Liu, J., Yao, J. and Xu, Z. (2011b) A new species of the rare chrysidid subfamily Loboscelidiinae from China: the third species of *Rhadinoscelidia* Kimsey, 1988. *Zookeys* 87, 11–17. doi: 10.3897/zookeys.87.1295

Lucañas, C.C. and Olmi, M. (2017) First record of the genus *Caenosclerogibba* Yasumatsu, 1958 (Hymenoptera: Sclerogibbidae) from the Philippines. *Check List* 13, 2100. doi: 10.15560/13.3.2100

Lucañas, C.C., Olmi, M.D. and Lit, I.L. Jr (2016) Sclerogibbidae (Hymenoptera): wasps ectoparasitic on webspinners (Embioptera) in the Philippines. *Philippine Entomologist* 30, 180.

Lucena, D.A.A. and Almeida, E.A.B. (2022) Morphology and Bayesian tip-dating recover deep Cretaceous-age divergences among major chrysidid lineages (Hymenoptera: Chrysididae). *Zoological Journal of the Linnean Society* 194, 36–79. doi: 10.1093/zoolinnean/zlab010

Maa, T.C. and Yoshimoto, C.M. (1961) Loboscelididae, a new family of Hymenoptera. *Pacific Insects* 3, 523–548.

Mangione, S. and Virla, E.G. (2004) Morfología de los preimaginales de *Gonatypus bonaerensis*, y consideraciones sobre la morfología interna de sus larva inmaduras (Hymenoptera, Dryinidae). *Acta Zoológica Lilloana* 48, 91–202.

Mayhew, P.J. and Hardy I.C.W. (1998) Nonsiblicidal behavior and the evolution of clutch size in bethylid wasps. *The American Naturalist* 151, 409–424.

Melo, G.A.E. (2000) Biology of an extant species of the scolebythid genus *Dominibythus* (Hymenoptera: Chrysidoidea: Scolebythidae), with description of its mature larva. In: Austin, A.D. and Dowton, M. (eds) *Hymenoptera: Evolution, Biodiversity and Biological Control*. CSIRO, Canberra, pp. 281–284.

Mita, T. (2021) Taxonomic study of *Baeosega* and its allies, with description of a new species of *Nipponosega* (Hymenoptera, Chrysididae, Amiseginae). *ZooKeys* 1041, 1–25. doi: 10.3897/zookeys.1041.66267

Mita, T. and Matsumoto, Y. (2012) First description of the male of *Gonatopus javanus* (R.C.L. Perkins) determined with mitochondrial COI sequence (Hymenoptera: Dryinidae). *Entomological Science* 15, 214–218. doi: 10.1111/j.1479-8298.2011.00502.x

Mita, T. and Okajima, S. (2011) *Dryinus* species collected from Laos (Hymenoptera, Dryinidae, Dryininae). *Japanese Journal of Systematic Entomology* 17, 153–154.

Mita, T. and Olmi, M.S. (2018) Taxonomic additions of Embolemidae and Sclerogibbidae (Hymenoptera: Chrysidoidea) from Japan, with description of a new species of *Trogloembolemus*. *Zootaxa* 4497, 586–592.

Móczár, L. (1984) Oriental Mesitiinae (Hymenoptera: Bethylidae). *Folia Entomologica Hungarica* 45, 109–150.

Móczár, L. (1998) Revision of the Cleptinae of the World. Genus *Cleptes* subgenera and species groups (Hymenoptera, Chrysididae). *Entomofauna* 19, 501–516.

Móczár, L. (2000) Revision of the *Cleptes asianus* and *townesi* groups of the World (Hymenoptera, Chrysididae, Cleptinae). *Acta Zoologica Academiae Scientiarum Hungaricae* 46, 319–331.

Mohar, N., Dujmović, R. and Uremović, V. (1985) Season dermatitis ex insectis caused by stings of *Sclerodermus domesticus*. *Dermatologica* 171, 446–449. doi: 10.1159/000249471

Mugrabi, D.F. and Azevedo, C.O. (2013) Revision of Thai *Dissomphalus* Ashmead, 1893 (Hymenoptera: Bethylidae), with description of twenty-four new species. *Zootaxa* 3662, 1–73. doi: 10.11646/zootaxa.3662.1.1

Nguyen, L.T.P. and Wiśniowski, B. (2021) Review of *Holophris* Mocsáry (Hymenoptera: Chrysididae) from Vietnam, with description of a new species. *Zootaxa* 4963, 393–399

Olmi, M.S. (1993) A new generic classification for Thaumatodryininae, Dryininae, and Gonatopodinae, with descriptions of new species (Hymenoptera Dryinidae). *Bollettino di Zoologia agraria e di Bachicoltura, Ser. II* 25, 57–89.

Olmi, M.S. (1994) The Dryinidae and Embolemidae (Hymenoptera: Chrysidoidea) of Fennoscandia and Denmark. *Fauna Entomologica Scandinavica* 30. E.J. Brill, Leiden.

Olmi, M.S. (2005) A revision of the world Sclerogibbidae (Hymenoptera Chrysidoidea). *Frustula Entomologica* 26–27 (2003-2004), 46–193.

Olmi, M., Xu, Z. and Guglielmino, A. (2013) First supplement to the monograph of the Oriental Dryinidae (Hymenoptera: Chrysidoidea): description of *Neodryinus lohmani*, a new species from Thailand. *Florida Entomologist* 96, 1556–1558.

Olmi, M.S., Xu, Z. and Guglielmino, A. (2015) A new species of the genus *Anteon* Jurine (Hymenoptera, Dryinidae) from Thailand. *ZooKeys* 504, 141–147. doi: 10.3897/zookeys.504.9333

Olmi, M.S., Xu, Z.-F. Guglielmino, A. and Sparanza, S. (2016a) A new species of the genus *Anteon* Jurine (Hymenoptera: Dryinidae) from Laos. *ZooKeys* 561, 31–38. doi: 10.3897/zookeys.561.7417

Olmi, M.S., Marletta, A., Guglielmino, A. and Speranza, S. (2016b) *Protosclerogibba australis* gen. et sp. nov., new genus and species of sclerogibbid wasp (Hymenoptera: Sclerogibbidae) from South Africa. *Zootaxa* 4085, 127–134. doi: 10.11646/zootaxa.4085.1.6

Omar, M.Y., Azman, A. and Olmi, M. (1996) *Anteon yasumatsui* Olmi, parasitoid of *Nephotettix nigropictus* (Stål) and *N. malayanus* Ishihara and Kawase in Malaysia (Hymenoptera Dryinidae and Homoptera Cicadellidae). *Frustula Entomologica* 19, 82–188.

Paukkunen, J., Berg, A., Soon, V., Ødegaard, F. and Rosa, P. (2015) An illustrated key to the cuckoo wasps (Hymenoptera, Chrysididae) of the Nordic and Baltic countries, with description of a new species. *ZooKeys* 548, 1–116.

Perkins, J.F. (1976) Hymenoptera Bethyloidea (excluding Chrysididae). *Handbook of Royal Entomological Society of London* Vol. VI, Part 3(a). Handbooks for the Identification of British Insects. Royal Entomological Society of London.

Polaszek, A. and Krombein, K.V. (1994) The genera of Bethylinae (Hymenoptera: Bethylidae). *Journal of Hymenoptera Research* 3, 91–105.

Quicke, D.L.J. (1997) *Parasitic Wasps*. Chapman & Hall, London, 470 pp.

Quicke, D.L.J. (2017) *Mimicry, Crypsis, Masquerade and other Adaptive Resemblances*. Wiley, Oxford, 576 pp.

Ramos, M.S. and Azevedo, C.O. (2020) Revisited phylogeny of Bethylinae (Hymenoptera, Bethylidae) solves basal polytomy. *Insect Systematics & Evolution* 51(3), 296–346. https://doi.org/10.1163/187631 2X-00002202

Rosa, P., Na-sen, W., Feng, I and Zai fu, X. (2016) Revision of the genus *Trichrysis* Lichtenstein, 1876 from China, with description of three new species (Hymenoptera, Chrysididae). *Deutsche Entomologische Zeitschrift* 63(1), 109.

Rosa, P., Pham, H.T. and Mita, T. (2020) Rediscovery of *Lustrina* Kurian (Hymenoptera, Chrysididae), with redescription of *L. assamensis* Kurian. *Zootaxa* 4718, 285–291. doi: 10.11646/zootaxa.4718.2.10

Sahragard, A., Jervis, M.A. and Kidd, N.A.C. (1991) Influence of host availability on rates of oviposition and host-feeding, and on longevity in *Dicondylus indianus* Olmi (Hym., Dryinidae), a parasitoid of the rice brown planthopper, *Nilaparvata lugens* Stål (Hem., Delphacidae). *Journal of Applied Entomology* 112, 153–162.

Terayama, M. (1995) *Calvapenesia* and *Neoapenesia*, new genera of the family Bethylidae (Hymenoptera, Chrysidoidea), from the Oriental region, with proposals of two new synonymies of genera. *Japanese Journal of Entomology* 64, 881–891.

Terayama, M. (2003a) Phylogenetic systematics of the family Bethylidae (Insecta: Hymenoptera) Part I. Higher classification. *Academic Reports of the Faculty of Engineering of Tokyo Polytechnic University* 26, 1–14.

Terayama, M. (2003b) Phylogenetic systematics of the family Bethylidae (Insecta: Hymenoptera) Part II. Keys to subfamilies, tribes and genera in the world. *Academic Reports of the Faculty of Engineering of Tokyo Polytechnic University* 26, 16–29.

Tribull, C.M. (2015) Phylogenetic relationships among the subfamilies of Dryinidae (Hymenoptera, Chrysidoidea) as reconstructed by molecular sequencing. *Journal of Hymenoptera Research* 45, 15–29. doi: 10.3897/JHR.45.5010

Tsuneki, K. (1982) Two new species of *Cleptes* from Thailand and Formosa (Hymenoptera, Chrysididae). *Special Publications of the Japan Hymenopterists Associations* 23, 1–2.

van Achterberg, C. and van Kats, R.J.M. (2000) Revision of the Palaearctic Embolemidae (Hymenoptera). *Zoologische Mededelingen* 74 (1–17), 251–269.

Vargas, J.M., Colombo, W.D. and Azevedo, C.O. (2020) Revisited phylogeny of Scleroderminae (Hymenoptera: Bethylidae) reveals a plastic evolutionary history. *Arthropod Systematics & Phylogeny* 78, 217–243.

Wei, N.-S., Rosa, P. and Xu, Z.-F (2013) Revision of the Chinese *Cleptes* (Hymenoptera, Chrysididae) with description of new species. *ZooKeys* 362, 55–96. doi: 10.3897/zookeys.362.6175

Wharton, R.A. (1989) Final instar larva of the embolemid wasp, *Ampulicomorpha confusa* (Hymenoptera). *Proceedings of the Entomological Society of Washington* 91, 509–512.

Wheeler, W.M. (1928) *The Social Insects. Their Origin and Evolution*. Kegan Paul, Trench, Trubner, London.

Wiśniowski, B., Nguyen, L.T.P. and Nguyen, C.Q. (2020) Discovery of four new species of the genus *Trichrysis* Lichtenstein, *cyanea* species group (Hymenoptera, Chrysididae) from Vietnam. *Zootaxa* 4881, 165–178. doi: 10.11646/zootaxa.4881.1.10

Xu, Z.-F., He, J.-H. and Olmi, M. (2001) The Embolemidae (Hymenoptera: Chrysidoidea) from China. *Entomologia Sinica* 8, 213–217. doi: 10.1111/j.1744-7917.2001.tb00444.x

Xu, Z., Olmi, M. and He, J. (2013) Dryinidae of the Oriental region (Hymenoptera: Chrysidoidea). *Zootaxa* 3614, 1–460. doi: 10.11646/zootaxa.3614.1.1

Yamada, Y. (1987) Characteristics of the oviposition of a parasitoid, *Chrysis shanghaiensis* (Hymenoptera: Chrysididae). *Applied Entomology and Zoology* 22, 456–464.

Zhang, Q., Rasnitsyn, A.P., Olmi, M., Martynova, K.V. and Perkovsky, E.E. (2020) First scolebythid wasp (Hymenoptera: Chrysidoidea, Scolebythidae) in the mid-Cretaceous Burmese amber. *Palaeoentomology* 3, 41–45. doi: 10.11646/palaeoentomology.3.1.5

17

Parasitoid Aculeates – Vespoidea *sensu lato*

Abstract

Recent systematics changes to the Vespoidea *sensu lato* are followed. Although most people are familiar with the stinging, social wasps belonging to the family Vespidae, several related groups, now considered as separate superfamilies based on molecular phylogenetic analyses, comprise parasitoids. Even a few spider wasps Pompilidae are technically parasitoids. A simplified key to the fully winged and some wingless families is provided. Representatives of most families are illustrated photographically and their biologies are described. Relevant identification works are cited.

The proposal of Brothers (1975, 1999) that the aculeate Hymenoptera was best classified into just three superfamilies gradually became accepted, with the Chrysidoidea being composed entirely of parasitoid families, the Vespoidea s.l. containing some parasitoid groups and the Apoidea none. This concept of the Vespoidea contained the familiar entirely social groups stinging wasps (yellow-jackets and allies, Vespidae) and ants (Formicidae, also often treated as Formicoidea), plus the spider-hunting wasps (Pompilidae) and seven other families that are all strictly parasitoids, i.e., the Bradynobaenidae, Mutillidae (velvet ants), Rhopalosomatidae, Sapygidae, Scoliidae, Sierolomorphidae and Tiphiidae (e.g. Brothers, 1975; Gauld and Bolton, 1988; Brothers and Finnamore, 1993). With the exceptions of the Bradynobaenidae (but see below) and the Sierolomorphidae, all of these smaller families are represented to some degree in S.E. Asia. A few of the spider-hunting wasps are also known strictly to be parasitoids in the sense of this book, i.e. with no physical host manipulation other than stinging and ovipositing on it, whereas most others are provisioning predators (Gonzaga, 2013).

A number of fairly recent molecular phylogenetic studies (Pilgrim *et al.*, 2008; Debevec *et al.*, 2012; Johnson *et al.*, 2013; Branstetter *et al.*, 2017; Zheng *et al.*, 2021) have resulted in radical changes to the concept of 'Vespoidea' and many workers have now agreed that it should be regarded as comprising six (or even seven) superfamilies: Formicoidea (ants, not discussed here), Pompiloidea, Scolioidea, Thynnoidea, Tiphioidea and Vespoidea. Branstetter *et al.*'s (2017) own analyses (Fig. 17.1 d) additionally treated the extralimital Sierolomorphidae as a separate superfamily. This arrangement was followed in the latest ultra-conserved element (UCE) study of Blaimer *et al.* (2023). The main groups are thus:

- Pompiloidea – defined as including Mutillidae, Myrmosidae, Pompilidae and Sapygidae
- Scolioidea – defined to contain Scoliidae (including Proscoliinae) and Bradynobaenidae (subfamilies Apterogyninae and Bradynobaeninae only)
- Sierolomorphoidea – containing only Sierolomorphidae
- Thynnoidea – defined to contain Chyphotidae (former bradynobaenid subfamilies Typhoctinae and Chyphotinae) and Thynnidae (former tiphiid subfamilies Anthoboscinae, Diamminae, Thynninae, Myzininae and Methochinae)
- Tiphioidea – defined to contain either only Tiphiidae (Tiphiinae and Brachycistidinae) or also Sierolomorphidae by Pilgrim et al. (2008). (Tiphiinae and Brachycistidinae only)
- Vespoidea – redefined to contain only Vespidae and Rhopalosomatidae.

Some hypothesised phylogenetic relationships between these and other aculeate groups are shown in Fig. 17.1. Although Tiphiidae were long considered to include Thynninae as a subfamily based on morphology (e.g. Brothers, 1993a), the molecular studies cited consistently recover it as paraphyletic with respect to several other families (e.g., Sapygidae, Pompilidae, Mutillidae), so that various components have been split between two superfamilies; a similar fate has befallen components of the previously recognised Bradynobaenidae. The ants are now generally accepted as not being close to the Vespidae as the morphological study of Brothers (1975, 1999) had suggested, but rather being closer to the bees (Fig. 17.1 a, c, d).

There is obviously a trade-off between amount of sequence data and number of taxa sequenced (taxonomic coverage) due to cost, time and sample-availability constraints. Therefore, the new taxonomic arrangement may well be subject to further changes. However, it should be noted that the affinities of various groups have not been totally set in stone (see for example, Myrmosidae below) and that in Blaimer et al.'s (2023) tree, many of the nodes near the base of the Aculeata had rather low support despite the wealth of data. Brothers (2021) provided a balanced discussion of the proposed changes and actually concluded that it would be premature to adopt the classificatory changes of Pilgrim et al. (2008) and Branstetter et al. (2017). Here we have nevertheless adopted a split approach which, as with lumper versus splitter species concepts, may allow more precise information to be given about various groups. It is always easier to re-lump taxa if they are found to be the same than to extract individual data if they were previously lumped.

Separating some of the 'vespoid' families can be very tricky. The comprehensive key to males and females, and winged and wingless forms, of the ten families in the previous unsplit concept of Vespoidea s.l. in the *Hymenoptera of the World* (Brothers and Finnamore, 1993) runs to 27 couplets, and even then, some of them are not that easy for the beginner. Male ants, in particular, can sometimes be very confusing.

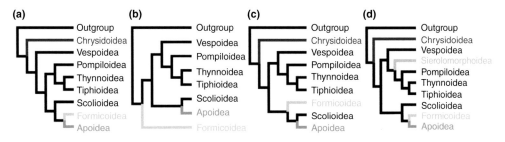

Fig. 17.1. Some proposed relationships between major groups of aculeate wasps including: **(a)** Johnson et al., 2013; **(b, c)** Faircloth et al. (2015) 45 and 56 taxon ultraconserved elements (UCE) datasets respectively; **(d)** Branstetter et al.'s (2015) results from combining targeted enrichment of UCEs with multiplexed next-generation sequencing (NGS) using a Hymenoptera-specific bait set designed to enrich 1510 UCE loci. (Source: from Branstetter et al., 2017, reproduced under terms of Creative Commons Attribution licence 4.0.)

The morphological similarity between members of many of the groups now recognised as separate families and superfamilies is likely continued at lower classificatory levels. Recent barcoding of European specimens of several of the smaller families suggests that cryptic species could be rather widespread (Schmid-Egger and Schmidt, 2021).

Simplified Key to the Fully Winged and Some Wingless Vespoidea (*s.l.*) in S.E. Asia

Note: if wingless, proceed to next couplet until characters fit.

1		Fore wings longitudinally folded (in life and usually in death unless they are specifically set) **AND** eyes with deep notch opposite antennal sockets (but see also couplet 5); pronotum extending and slightly past tegula medially, posteriorly rectangular or angulate common; not parasitoids) ....**Vespidae**
-		Fore wings not longitudinally folded; eyes variable; pronotum posteriorly pointed or rounded and just reaching tegula, not extending beyond tegula on median side tegula medially but may extend beyond tegula laterally.. 2
2(1)		Metasomal tergite 1 or tergites 1 and 2 forming clearly defined nodes with narrow connections to next posterior segment(see Fig. 17.1 a, b) .... ants, **Formicidae**
-		Metasomal tergite 1 or tergites 1 and 2 not node-like but broadly connected to subsequent tergites, or at most only with slightly narrowed connection (a few male ants might run here too, e.g. Amblyoponinae, see Fig. 7.1 d) 3
3(2)		Apical part of wing membrane finely longitudinally corrugated (Fig. 17.8 a); posterior part of metasternum protruding as a plate concealing (protecting) bases of hind coxae which are widely separated (Fig. 17.8 b); males with three-pronged hypopygium (Fig. 17.8 c) ....**Scoliidae**
-		Apical wing membrane not finely corrugated (Fig. 17.9 a, b) (although there are fine sub-apical corrugations in some bees, Apoidea); posterior part of metasternum not protruding to cover bases of hind coxae, and the latter not so widely separated (Fig. 17.9 c); males without hypopygial spines or with just one 4
4(3)		Mesopleuron with straight diagonal groove; hind legs long (end of femur often level with apex of metasoma), hind tibia spinose; fully winged; (mostly provisioning predators but a few are idiobiont parasitoids, and one or two are koinobiont parasitoids) ....**Pompilidae**
-		Mesopleuron without straight diagonal groove; hind legs variable but usually end of femur near midlength of metasoma; fully winged or wingless 5
5(4)		Eyes deeply notched opposite antennal sockets; fully winged 6
-		Eyes convex, straight or only shallowly and broadly concave next to antennal sockets; fully winged or wingless 8
6(5)		Basal flagellomeres with 1 or 2 conspicuous setae near distal edge (Fig. 17.10 b); costal cell obliterated broadly medially by secondary fusion of veins C and SC+R (Fig. 17.10 c); (uncommon except sometimes at light) .... **Rhopalosomatidae**
-		Basal flagellomeres simple; costal cell present along whole length although sometimes narrow... 7
7(6)		First sternite scarcely or not depressed posteriorly, overlapping second, no deep constriction between them; venation of fore wing reaching wing apex; (rare) ......**Sapygidae**
-		First sternite strongly depressed posteriorly, not overlapping second, a strong constriction between them; venation of fore wing not reaching wing apex .... some male **Mutillidae**
8(5)		Fully winged; males and females 9
-		Wingless; females only... 11

9(8) Metasomal tergite 2 and/or sternite 2 with felt-line (line of closely appressed setae) laterally (Fig. 17.2 a, b); hypopygium simple, not produced into a median spine (Fig. 17.2 a, b) .... some male **Mutillidae**
- Metasomal tergite 2 and sternite 2 without felt-line, although tergite often with bare longitudinal lateral groove; male hypopygium often produced medially into an upcurved spine (Fig. 17.9 b) but sometimes simple or reduced 10
10(9) Male hypopygium produced medially into an upcurved spine (Fig. 17.8 b); female mid and hind tibiae stout and strongly spinose ....
.... **Tiphiidae** (Tiphiinae) and **Thynnidae** (Myzininae and Methochinae)
- Male hypopygium simple or much reduced, not forming a median spine; males only .... **Myrmosidae**
11(8) Metasomal tergite 2 and/or sternite 2 with felt-line (line of closely appressed setae) laterally (Fig. 17.2 a, b); mesosomal dorsum entirely fused, without any obvious subdivisions (Fig. 17.2 c–f) .... **Mutillidae**
- Metasomal tergite 2 and sternite 2 without felt-line, although tergite often with bare longitudinal lateral groove; mesosomal dorsum distinctly subdivided into two or more unfused subdivisions 12
12(11) Mesosomal dorsum subdivided into two articulated regions, pronotum and the remainder (Fig. 17.5) .... **Myrmosidae**
- Mesosomal dorsum subdivided into three articulated regions, pronotum, mesonotum and metanotum+propodeum .... **Thynnidae**

Pompiloidea

This superfamily now embraces Mutillidae, Myrmosidae and Sapygidae as well as the spider-hunting wasps, Pompilidae (Pilgrim *et al.*, 2008).

Mutillidae (Velvet Ants)

Much of what is known about S.E. Asian Mutillidae is due to the efforts of Arkady S. Lelej, Denis J. Brothers, Kevin Williams and Juriya Okayasu. This family has actually received more attention than almost any other in the region, probably because of their often very conspicuous colour patterns (Fig. 17.2) and often very visible females searching over the ground for host nests. Several generic revisions and distributional summaries have been published. Ultra-conserved elements (UCEs) were recently employed to elucidate phylogenetic relationships within the family (Waldren *et al.*, 2022) and this resulted in several changes at tribe level.

The female wasps are ant-like except for conspicuous setosity (hence the common English name), usually seen running over the ground. Some have very long and ferocious stings (Fig. 17.3) and since the sting shaft is coiled within the metasoma, the tip can be extruded a long way (Kumpanenko *et al.*, 2022). The Mutillidae are a rather large group with eight recognised subfamilies (Brothers and Lelej, 2017) and some 4300 described species worldwide. There is extreme sexual dimorphism and some species perform phoretic copulation (Fig. 17.4); that is, the winged males copulate with females which are far smaller and are sufficiently firmly attached by their genitalia that the male can fly with them to sites where they can seek hosts (see Evans, 1969). Vivallo (2020) reviewed this behaviour across the aculeate Hymenoptera (see Thynnoidea below for further examples). Phoretic copulation is primarily found in Mutillinae (Ctenotillini, Ephutini, Smicromyrmini, and Trogaspidiini), and Rhopalomutillinae but has been found in others, and, not surprisingly given the morphological adaptations, appears largely conserved at tribe and subfamily level (Waldren *et al.*, 2020). In other groups, mating occurs *in situ* where the male locates the female.

Finding *in copulo* pairs is obviously an ideal way of associating the sexes of species,

Fig. 17.2. Exemplar S.E. Asian Mutillidae: **(a)** male, *Bischoffitilla* sp. (Myrmillinae), Thailand, insert arrow indicating felt-line; **(b)** male *Smicromyrme dardanus* (Mutillinae: Smicromyrmini) (often misidentified as *Sinotilla boheana*), Thailand; **(c)** female *Mickelomyrme puttasoki* (Mutillinae: Smicromyrmini), Vietnam; **(d)** female *Trogaspidia doricha* (Mutillinae: Trogaspidini), Indonesia (Irian Jaya); **(e)** *Bischoffitilla lamellata*, Laos; **(f)** *Karlidia peterseni* (Mutillinae, Mutillini, Ephutina), Thailand. (Source: (c–f) photographs by and © Juriya Okayasu, reproduced with permission.)

but it seems likely that for the majority it will have to rely on molecular methodology, and whilst DNA barcoding is widely used for this sort of thing, Pilgrim and Pitts (2006) demonstrated that the internal transcribed spacer regions (ITS1 and ITS2) can also be used.

Even though some mutillids are quite common, surprisingly little is known about their biology. Mutillids are mainly parasitoids

Fig. 17.3. Exemplar female S.E. Asian Mutillidae showing the considerable length of their oiocrted stingers: **(a)** *Orientilla vietnamica* (Dasylabrinae), Vietnam; **(b)** *Sinotilla calopoda*, (Mutillinae, Smicromyrmini), Indonesia. (Source: photographs by and © Juriya Okayasu, reproduced with permission.)

of the larvae of various solitary and, less commonly, eusocial Hymenoptera (Brothers et al., 2000). Despite the size of the family, there are rather few reliable published host records, and Ronchetti and Polidori (2020) in their phylogenetic and food web study could only include information on 132 species (representing 49 genera) which collectively included 305 host associations. Host associations appeared to reflect ecological host traits much more than host taxonomy (Brothers, 1989).

The host stage attacked by mutillids is no longer a feeding one, and is cocooned or in a cell. Hosts mostly belong to the Aculeata but also include beetles, flies, moths or even cockroach oothecae. The social hosts involved include halictine bees, bumble-bees (extralimital association) and a few occasionally attack honey bees. A few mutillid species are parasitoids of commensals or symbionts of ants. In the case of the species attacking bumble-bees, there is an old assertion that they are endoparasitoids, but this needs checking; in other cases, they are certainly ectoparasitoids. *Sinotilla decorata* has been reared from nests of a widely distributed, mud-dauber wasp *Pison argentatum* (Hymenoptera: Crabronidae) in Malaysia (Pagden, 1934; Okayasu, 2017).

Brothers (1975, 1999) recognised seven subfamilies, but Lelej and Nemkov (1997) recognised ten subfamilies with rather different relationships. Brothers and Lelej (2017) collaborated on a much broader analysis to reconcile their earlier differences, and recognised eight subfamilies: Dasylabrinae, Mutillinae, Myrmillinae, Myrmosinae, Pseudophotopsidinae, Rhopalomutillinae, Sphaeropthalminae and Ticoplinae; all of these studies agreed in concluding that the Myrmosinae were the sister group to all remaining Mutillidae. That placement was also found by most molecular studies, but the Myrmosinae has now been recognised as its own family (e.g., Peters et al., 2017).

Brothers (2022) assessed all the type specimens of Mutillidae from the Australasian region, essentially east of Wallace's line. This 215-page work illustrates all the types photographically and presents several new combinations. Okayasu et al. (2018) reviewed the colour patterns exhibited by female mutillids from S.E. Asia and the southern half of China, and presented colour dorsal photographs of many species.

Brothers (1993b) provided a key to world subfamilies. Lelej (2005) catalogued the Oriental species of the family and provided keys to the subfamilies, tribes and genera. Williams et al. (2019a) provided a key to species for female mutillids known from southern Thailand, and a key to the genera occurring in S.E. Asia. Lelej (1996a) described a new genus and several species from Borneo; Lelej (1996b) described a new species of *Promecidia* from Sarawak; and Lelej (2012) described a new species of *Rhopalomutilla* from Vietnam. Lelej et al. (2016) provided a key to the species of *Promecidia*. Lelej and Krombein (1999) described remarkable genera of Sphaeropthalminae (Pseudomethocini) from Thailand. Lelej and Brothers (2008) catalogued genus group names in the family, providing details of type species and lectotypifications. The genera of Rhopalomutilinae

Fig. 17.4. Mating pair, phoretic copulation, in the extralimitial *Pherotilla rufitincta* (Rhopalomutillinae), male above; the S.E. Asian representatives are very similar. Note extreme sexual dimorphism. (Source: from Brothers (2015) reproduced under terms of Creative Commons Attribution Licence CC-BY 4.0.)

were keyed by Brothers (2015); of these, only *Pherotilla* (Fig. 17.4) occurs in S.E. Asia. Sutanto *et al.* (2021) revised the eight *Bischoffitilla* (Fig. 17.2 a, e) species occurring in Indonesia. Terayama and Mita (2015) provided a key to the genera and tribes of Japanese Mutillidae. Lelej *et al.* (2017) revised *Zeugomutilla*, with several species from the northern part of S.E. Asia. Zhou *et al.* (2018) revised Oriental *Zavatilla* species which included one from Vietnam. Okayasu (2017) described a new species of *Sinotilla* (Fig. 17.3 b) from West Java. Okayasu *et al.* (2021a) redefined *Andreimyrme*, provided a key to females and described nine new species from S.E. Asia. Okayasu *et al.* (2021b) reviewed the trogaspidiine genus *Eotrogaspidia* and some related taxa. Okayasu (2021) erected a new genus, *Arkaditilla*, for several S.E. Asian species that were previously assigned to *Trogaspidia*. Lelej (2021) extended the distribution of *Petersenidia* into Wallacea. Okayasu (2022) described a new species of *Nordeniella* from Sumbawa, thus extending the known range of the genus across Wallace's line. Thaochan *et al.* (2022) described three new species belonging to the genera *Mickelomyrme* (Fig. 17.2 c), *Nordeniella* and *Smicromyrme* from Thailand. Lelej *et al.* (2023) revised the Oriental species of Mutillini (the genera *Kurzenkotilla*, *Standfussidia* and *Storozhenkotilla*) which includes species from Laos, Thailand and Vietnam. Lelej (2023) provides a key to the Oriental genera and species of Ctenotillini. Okayasu (2003) recorded the genus *Orientilla* from Laos for the first time based on two species.

Myrmosidae

As a subfamily, the Myrmosinae were shifted from the Mutillidae (or separate from but close to that family), to the Tiphiidae and then back to the Mutillidae once more (Brothers, 1975, 1978), as well as being treated as a full family, Myrmosidae, in their own right (Peters *et al.*, 2017), although they appear to be close to Mutillidae (Blaimer *et al.*, 2023) (Fig. 17.5).

As with several mutillids, some myrmosids perform phoretic copulation (Waldren

Fig. 17.5. *Taimyrmosa nigrofasciata*, specimen from Japan (Source: photograph by and © Juriya Okayasu, reproduced with permission.)

et al., 2020). Female myrmosids have been observed to enter the nest burrows of their host bees, not only to oviposit, but additionally to act as predators on the host's larvae and prepupae (Brothers, 1978; Brothers and Lelej, 2017).

Williams *et al.* (2019b) provided keys to Old World genera for males and females of the family (as subfamily) and to males of the species of *Kudakrumia* and *Krombeinella*, both of which include at least one species in S.E. Asia. *Erimyrmosa burmanensis* (Myrmosini) was described from Myanmar.

Pompilidae (Spider-hunting Wasps)

The Pompilidae, which were referred to as the Psammocharidae through much of the first half of the 20th century, are commonly called spider-hunting wasps. It is a large family including some of the largest of all wasps, and these are specialists at attacking large mygalomorph spiders. The females have powerful stings. The family is cosmopolitan except for arctic regions, but most species-rich in the tropics. The basic biology of the majority of species is that of provisioning predator: the female pompilid hunts for host spiders typically by running over the ground erratically, searching for clues such as silk threads.

Pompilids are mostly idiobionts provisioners, building their nests in pre-existing holes in the ground or canopy, digging new burrows in soil or manufacturing nests with leaves or mud (e.g. Harris, 1987; Grout and Brothers, 1993; Harris, 1999; Kurczewski *et al.*, 2017). A few, however, are true parasitoids and at the moment all such examples are extralimital, we will briefly discuss those here since it is quite probable that some S.E. Asian species are too, but simply have not been investigated yet. Indeed, the behaviour of pompilids in S.E. Asia is very poorly investigated, with most observations concerning large-bodied species that attack 'tarantulas' (Araneae: Mygalomorpha).

Most of the strictly parasitoid Pompilidae are idiobiont ectoparasitoids. In these, the female wasp locates a spider in its retreat, stings it, paralyses it and oviposits on to it in situ where her larva completes development; this strategy is exemplified by European species of Aporus, Homonotus and some Arachnospila (Day, 1988). A few, mostly Neotropical, species are actually koinobionts, that is their larva stays attached to and feeding on a still-mobile spider host (da Souza et al., 2015; Villanueva-Bonilla et al., 2018; Benamu et al., 2020). Koinobiontism has been documented only in some genera such as *Paracyphononyx* (da Souza et al., 2015), *Notocyphus* (Martins et al., 2016; Reiban et al., 2021), *Epipompilus* (Harris, 1987; Villanueva Bonilla et al., 2018) and *Homonotus* (Evans, 1953). Townes (1957) had actually suggested a koinobiont lifestyle might be employed by the genus *Minagenia* and this was confirmed by Benamu et al. (2020).

Minagenia occurs from Australia to Taiwan, including Indonesia and mainland Asia, and a key to the species was provided by Decker et al. (2020). Although the examples of pompilid koinobiont are mostly from elsewhere, they each involve different pompilid genera and therefore this developmental strategy could conceivably also be present in some S.E. Asian species.

Sapygidae

Sapygids are nearly always rather uncommon, even in the Holarctic where they are commonest (Huber, 2009); in the Oriental Region they are exceedingly rare. Only three genera and five species have been discovered thus far, out of a world total of approximately 75 species in 12 genera (Kurzenko, 1996). The known S.E. Asian Sapygidae comprise *Sapyga* (one species) and *Parasapyga* (three species) (van Achterberg, 2014).

Unlike nearly all other Aculeata, sapygids have a fully developed ovipositor with apical serrations (Fig. 17.6 b). The larvae are kleptoparasitoids or ectoparasitoids of megachilid and anthophorid bees as well as of eumenine vespid wasps.

There are rather few investigations of sapygid biology. Rozen and Kemel (2009) described the hospicidal behaviour of the extralimital species *Sapyga luteomaculata* in Egypt. The hosts are species of megachilid bee. The *Sapyga* lays its egg next to that of the host bee in a provisioned nest cell. Upon hatching, the 1st instar sapygid attacks any additional sapygid eggs in the same cell, and also attacks the host bee's egg; in each case it is presumed that the successful sapygid larva consumes the contents of both conspecific and host eggs. The 1st instar sapygid larva is very agile, and after consuming the available egg(s) it easily moves around the

Fig. 17.6. Two S.E. Asian species of *Parasapyga*: **(a)** *P. yvonnae*, holotype female from Indonesia (Sumatra); **(b)** *P. boschi*, holotype female from Cát Tien National Park, S. Vietnam. (Source: from van Achterberg (2014) reproduced under terms of Creative Commons Attribution Licence CC-BY 4.0.)

bee's cell feeding on the food provisions that are stored there.

Kurzenko (1996) provided a key to the world genera, and Parasapyga from India and S.E. Asia was revised by van Achterberg (2014).

Scolioidea

This superfamily includes two families, of which only the Scoliidae are known to occur in S.E. Asia, where they are quite common.

Bradynobaenidae (extralimital)

A very poorly known, seldom collected and principally arid tropical and Mediterranean family with apterous females and fully winged males, superficially similar to Mutillidae. The closest published record to S.E. Asia is a species from Pakistan (Pagliano and Romano, 2017). There is a single host record from a solfugid arachnid. As of 2017 the family comprised 11 genera and 227 species worldwide. There are two subfamilies: Apterogyninae and Bradynobaeninae. The Typhoctinae and Chyphotinae have been removed to the Thynnoidea as the family Chyphotidae.

Scoliidae (Digger Wasps, Mammoth Wasps)

The Scoliidae comprises approximately 300 predominantly tropical or warm-temperate species in 22 genera from around the world. It is divided into two subfamilies: the Proscoliinae, with just two eastern Mediterranean species, and the Scoliinae, which includes all the rest. The latter is further divided into the tribes Scoliini and Campsomerini. The former have fore wing vein 2m-cu whereas the latter do not.

They are generally large, robust, setose wasps (Fig. 17.7). Unlike many thynnids, scoliid females and males are generally

Fig. 17.7. Live scoliids from Thailand: **(a)** *Scolia* sp., male; **(b)** *Megascolia* sp., female; **(c)** *Megascolia azurea*, female; **(d)** unidentified scoliid, female. (Source: photographs by and © Ian Dugdale (Baan Maka Nature Lodge), reproduced with permission.)

similar in appearance, although males tend to be thinner with relatively longer antennae, but in some species there is marked dimorphism. Females have ten and males have 11 flagellomeres. In addition to their roles as parasitoids of some pest beetles, they are frequent flower visitors and may be important incidental pollinators. Males are generally easily recognised by the hypopygium (metasomal sternite 8) which is produced into three posterior prongs (Fig. 17.8 c).

In general, Scoliidae, Thynnidae and Tiphiidae have very similar developmental biology (Day *et al.*, 1981). They are solitary ectoparasitoids of beetle larvae, usually in the soil, but also some in rotting wood or leaves. As with the pompilids, there are some very large species of scoliid, and most are between 2 and 3 cm long. To locate hosts, females are adapted for a fossorial life with robust, spiny legs. Males may dig themselves underground to sleep or avoid inclement weather (Kurczewski and Spofford, 1986). Scoliids are protandrous (Spradbery, 1973), and mating takes place underground soon after the female emerges from her cocoon, the males locating where they are (presumably by pheromones) and digging down to meet them (Barratt, 2003). It appears that females need to feed before they can lay eggs. Early reports that some species are parthenogenetic are inconclusive and possibly erroneous (Barratt, 2003).

Technically, some scoliids at least may not be true parasitoids in the sense of this book because apparently some species, after having stung and paralysed their host, dig down further, moving it into a new chamber that the scoliid constructs (Askew, 1971). However, so little detail is known about their biology that it is impossible to say how widespread this behaviour might be.

Phylogenetics relationships within the family have recently been investigated by Khouri *et al.* (2022) as ultraconserved-element data (UCE) data. Their results supported the

Fig. 17.8. Male *Campsomeris* sp. from Thailand: **(a)** habitus, note fine pleating of distal wing membrane; **(b)** posterior mesosoma, ventral view showing wide separation of hind coxae (HC) and lamella of metasternum (MS) partly overlapping coxal base (arrow); **(c)** posterior metasoma, dorsal view, showing three straight spines originating from hypopygium (8th metasomal sternite).

basal position of Proscoliinae and a sister relationship between the other two subfamilies, with the exclusion of *Colpa* from the Campsomerinae. *Colpa* has long been considered anomalous but now appears to be a basal member of the Scoliinae, although possibly placing them in their own subfamily might be justified (as did Argaman, 1996; see below).

At this point we need to mention the work of the infamous Hungarian–Romanian–Israeli entomologist Qabir Argaman, who also published under his birth name of Carol Nagy before emigrating to Israel and adopting a Hebrew name. It is almost an understatement that Argaman was a 'splitter' and, as regards Scoliidae, he published several papers culminating in his generic synopsis of the group (Argaman, 1996) in which he not only provided a key to world genera but described 21 new tribes and 62 new genera. It may well be that some of the larger currently accepted genera will need to be redefined and split in the light of new detailed studies, but 62 new genera were excessive, and the characters used to differentiate them seem unlikely to be very phylogenetically helpful. Therefore, Argaman's key might be better regarded as a practical aid but one requiring very careful consideration of the actual applied names. Kimsey and Brothers (2016) provided an overview of Nagy/Argaman's life's work. Refer also to section 'Perilampidae' in Chapter 22, this volume.

Osten (2005) provided a checklist to the approximately 560 valid world species and 220 subspecies as well as a comprehensive bibliography. Most of the taxonomic work on the family relevant to S.E. Asia was carried out more than 50 years ago, and the most important nearly 100 years ago: Betrem's (1928) revision of the Indo-Australian species. Species identification is sometimes challenging, especially for males (Liu *et al.*, 2021a). Schulten *et al.* (2011) re-diagnosed and illustrated *Bellimeris*, which includes two species: one from Myanmar, the other from China. Liu *et al.* (2021b) revised the Chinese fauna.

Thynnoidea and Tiphioidea (Flower Wasps)

These two superfamilies were long considered to be members of a monophyletic Tiphiidae until molecular data showed that this was actually a grade taxon, and consequently, to preserve equivalent taxonomic status of clades, they were raised in status. Brothers (1993a) provided a key that will enable separation of the various groups. Thynnidae (Methocinae and Myzininae) and Tiphiidae occur in S.E. Asia. The other recognised subdivisions are absent from the region, viz. Thynnidae (Anthoboscinae from Australia, South Africa, South America and Sri Lanka, and Diamminae from Australia only), Chyphotidae (New World) and Tiphiidae (Brachycistidinae in the Nearctic only). Larger species can appear superficially rather similar to some male scoliids, but most preserved specimens can be readily identified by the presence of posteriorly directed mesosternal lamellae that cover the bases of the mesocoxae (Fig. 17.9 c), although the character is much reduced in Methochinae.

Kimsey (1991) re-investigated the phylogenetic relations between 'tiphiid' subfamilies (including ones now placed in Thynnoidea) based on morphological evidence, following on from Brothers (1975). The Anthoboscinae were discovered as sister group to the remainder of the family in both cases, suggesting a Gondwanan origin.

The sexes may be distinguished by the number of visible metasomal tergites and number of flagellar segments: 7 and 11 in males, 6 and 10 in females. Keys to the subfamilies are provided by Kimsey (1991) and by Brothers and Finnamore (1993).

Thynnoidea (Thynnidae)

Thynninae is the largest subfamily with more than 50 genera, being predominantly Neotropical and Australian but with some species known from S.E. Asia. The thynnine tribe Rhagigasterini are a Gondwanan group (Chile and Australia) but extend into the Island of New Guinea (Kimsey, 1996a). The *Ariphron* generic group is predominantly Australian, but *Takyomyia* includes species from New Guinea and Borneo (Kimsey, 1996b).

The Myzininae are distributed worldwide, with 17 valid genera divided into four tribes: Austromyzinini (Australia), Myzinini (New World), Mesini (Old World) and Meriini (Old World, except Oriental Region) (Boni Bartalucci, 2004). Females are stouter than

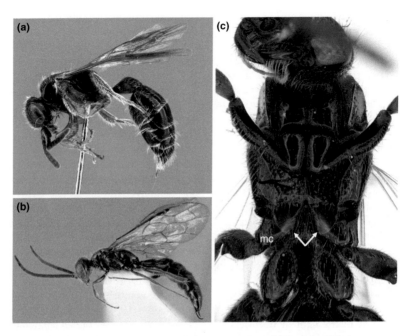

Fig. 17.9. Exemplar male S.E. Asian Thynnoidea and Tiphioidea, and characters: **(a)** female *Tiphia* sp., (Tiphioidea: Tiphiidae: Tiphiinae); **(b)** male *Methocha granulosa* Methochinae (Thynnoidea: Thynnidae) from Laos, note the single curved pseudosting formed from the hypopygium; **(c)** *Tiphia* sp. (Tiphioidea), ventral view of mesosoma showing the lamellar mesopleural plates overlapping the mid-coxae bases (arrows) (Source: (b) from Narita and Mita (2018) reproduced under terms of Creative Commons Attribution Licence CC-BY 4.0.)

the slender males and have swollen spiny legs used for burrowing; some groups of females have reduced wings or are even wingless.

Members of the subfamily Methocinae have wingless females and are unusual in that they are parasitoids of tiger beetle larvae (Cicindellidae). The genus *Methocha* is cosmopolitan except for Australia. Pagden (1949) described a species, *M. malayana*, from Malaysia, and Narita and Mita (2018) recently described a species from Laos (Fig. 17.9 b) – in earlier literature it was often spelled *Methoca*. Tiger beetle larvae live in burrows (in the ground or in trees) and are ferocious predators of ants that may inadvertently pass over their entrances, grasping them with their large mandibles (Burdick and Wasbauer (1959) after Williams, 1919). *Methocha* females may allow this to happen to themselves, but rapidly sting and paralyse the beetle larva. If the host escapes, they may chase and resting it and then drag it back to the burrow (Iwata, 1936), where they oviposit on it.

Boni Bartalucci (2004) provided a key to the tribe groups of Myzininae with particular emphasis on Palaearctic fauna, and Boni Bartalucci (2011) surveyed the S.E. Asian 'Tiphiidae'. Brown (2005) further partially revised the *Ariphron* genus group of which three species of *Tachyphron* are known from Indonesia (Halmahera, Morotai and Waigeo). The genera *Epactiothynnus*, *Rhagigaster*, *Tachynomyia* and *Thynnus* are known from the Aru Islands of Wallacea (Indonesia) (Kimsey and Brown, 1993). Most recently, Liao *et al.* (2022) revised the Chinese species of Myzininae and included colour photographs of the S.E. Asian species *Hylomesa bakeri* and *H. longiceps*.

Tiphioidea (Tiphiidae)

The new classification of Vespoidea s.l. places two families in the Tiphioidea, the well-known Tiphiidae (Tiphiinae) and the rare Bradynobaenidae (northern Hemisphere but absent from S.E. Asia).

The Tiphiidae is a cosmopolitan family with approximately 1500 described species. Many are superficially like smaller versions of Scoliidae. Virtually all species are thought to be idiobiont ectoparasitoids of soil-inhabiting beetle larvae, notably of the Scarabaeoidea. The females of many groups are poorly known, spending most of their time searching for soil-dwelling beetle-larvae hosts. Therefore, for these, much of the taxonomy is based on males which are generally more easily collected, often visiting flowers.

Two subfamilies are recognised in the Tiphiidae: the entirely Nearctic Brachycistidinae and the cosmopolitan Tiphiinae. Nine to 12 genera or subgenera are recognised globally, but only two genera are known from S.E. Asia: *Cyanotiphia* and *Tiphia*, the latter being by far the largest genus with more than 500 described species worldwide. The monotypic genus *Cyanotiphia*, known now from Malaysia (Borneo), Java and Sumatra, is unique in the family in having metallic blue coloration (Saini *et al.*, 2022). Interestingly, since its original description by Cameron (1907) it was not seen again for another 114 years.

Tiphia intrudens was originally described from Mysol (now Misool) Island, Indonesia, and is very widespread, extending from India to S.E. Queensland, Australia (Brown, 1995). As long ago as the 1920s *Tiphia* species were being investigated as possible classical biocontrol agents against the Japanese beetle, *Popillia japonica* (Scarabaeidae), an important introduced pest in the USA and elsewhere, which prompted research on their taxonomy in the east Palaearctic and S.W. Asia (Allen and Jaynes, 1930).

A major aid to the study of Tiphiidae was the redescription of the types of the S.E. Asian (and other) species in the collections of the Natural History Museum London and Oxford University Museum by Allen (1969), who also gave a key to type specimens of *Tiphia* and *Cyanotyphia* of S.E. Asia. Little work has been conducted on the regional fauna but Hanima *et al.* (2022) provided a key to the subgenera of *Tiphia* occurring in India as well as to the Indian species, which include several that occur in peninsular S.E. Asia, and therefore provides a start. Han *et al.* (2021) provided a key to the four subgenera, and a key to all known species of *Tiphia* (*Jaynesia*) although, as yet, that subgenus is not recorded from the Region.

Vespoidea

The largest family of Vespoidea by far are the well known Vespidae comprising the social wasps (Vespinae), paper wasps (Polistinae), potter wasps (Eumeninae) and a few other subfamilies, totalling approximately 5000 described species. None of these include any parasitoids. However, the Rhopalosomatidae are true parasitoids.

Rhopalosomatidae

Rhopalosomatidae are a small, worldwide family of ant-like wasps with about 72 extant species in four genera (Aguiar *et al.*, 2013; Lohrmann and Engel, 2017). They are entirely subtropical or tropical in distribution. Two genera are recorded from S.E. Asia: six species of *Paniscomima* and one species of the predominantly New World genus *Liosphex*, *L. trichopleurum*, which occurs in the Philippines and Indonesia. The genus *Olixon* (26 spp.) is distributed through Africa and Australia and the New World and includes one undescribed species from India (Lohrmann and Engel, 2017), which may therefore also be present in S.E. Asia.

They are solitary, and the larvae are ectoparasitic on orthopteran nymphs (Grylloidea) (Gurney, 1953). This makes them different from other hymenopteran cricket ectoparasitoids (Melo *et al.*, 2011). They are yellowish with red or brown markings, but may be all brown in colour. Many of the nocturnal species bear a considerable resemblance to nocturnal 'ophionoid-like' Ichneumonidae (Fig. 17.10). Some diurnal ones resemble spider-hunting Pompilidae, and the wingless species are a bit similar to Mutillidae. Winged species are usually nocturnal, while wingless or reduced-wing species are mainly diurnal. Unlike nearly all other aculeates, the fore wing costal cell largely

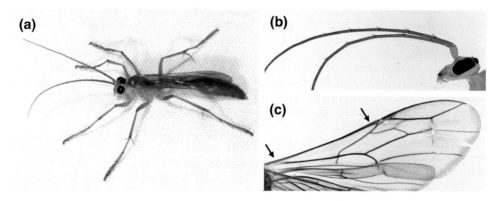

Fig. 17.10. Rhopalosomatid features: **(a)** extralimital nocturnal *Rhopalosoma* sp. attracted to a light sheet, but it would be hard to distinguish from Asian ones; **(b, c)** Thai *Paniscomima* sp., from Thailand, fore wing and flagellum, respectively. Note in **(b)** the well-developed spine-like setae apico-dorsally on some of the basal flagellar segments. Note in **(c)** that the costal cell is almost entirely closed except for small basal and apical regions (seen more easily by viewing from the anterior). (Source: from Miller *et al.* (2019) reproduced under terms of Creative Commons Attribution Licence CC-BY 4.0.)

obliterated medially by veins C and SC+R being fused over much of its length, but the cell is narrowly open both basally and apically (Fig. 17.10 c).

Townes (1977) revised the family and much more recently *Paniscomima* was revised by Guidotti (2007). *Paniscomima* males and females are both fully winged and superficially similar to the New World species illustrated in Fig. 17.10 a.

Even in the absence of rearing, DNA barcoding can lead to the recognition of host relationships when parasitised hosts are found (Miller *et al.*, 2019).

References

Aguiar, A.P., Deans, A.R., Engel, M.S., Forshage, M., Huber, J.T., Jennings, J.T., Johnson, N.F., Lelej, A.S., Longino, J.T., Lohrmann, V., Mikó, I., Ohl, M., Rasmussen, C., Taeger, A. and Yu, D.S.K. (2013) Order Hymenoptera. *Zootaxa* 3703, 51–62. doi: 10.11646/zootaxa.3703.1.12
Allen, H.W. (1969) Redescriptions of types of Tiphiinae from Asia, Africa, Oceania in the British Museum (NH) and at Oxford University. *Transactions of the American Entomological Society* 95, 353–438.
Allen, H.W. and Jaynes, H.A. (1930) Contributions to the taxonomy of Asiatic wasps of the genus *Tiphia* (Scoliidae). *Proceedings of the United States National Museum* 76, 1–105.
Argaman, Q. (1996) Generic synopsis of Scoliidae (Hymenoptera, Scolioidea). *Annales Historico-Naturales Musei Nationalis Hungarici* 88, 171–222.
Askew, R.R. (1971) *Parasitic Insects*. Elsevier, New York, 316 pp.
Barratt, B.I.P. (2003) Aspects of Reproductive Biology and Behaviour of Scoliid Wasps. DOC Science Internal Series 147. Department of Conservation, Wellington, New Zealand, 11pp.
Benamu, M., García, L.F., Viera, C., Lacava, M. and Korenko, S. (2020) Koinobiont life style of the spider wasp *Minagenia* (Hymenoptera, Pompilidae) and its consequences for host selection and sex allocation. *Zoology* 140, 125797.
Betrem, J.G. (1928) Monographie der Indo-Australischen Scoliiden (Hym., Acul.) mit zoogeographischen Betrachtungen. *Treubia* 9, 1–399 + plates I–IV.
Blaimer, B.B., Santos, B.F., Cruaud, A., Gates, M.W., Kula, R.R., Mikó, I., Rasplus, J.-Y., Smith, D.R., Talamas, E.J., Brady, S.G. and Buffington, M.L. (2023) Key innovations and the diversification of Hymenoptera. *Nature Communications* 14, 1212.
Boni Bartalucci, M. (2004) Tribe-groups of the Myzininae with special regard to the Palaearctic taxa of the tribe Meriini (Hymenoptera, Tiphiidae). *Linzer biologische Beiträge* 36, 1205–1308.

Boni Bartalucci, M. (2011) Tiphiidae from South East Asia (Hymenoptera). *Onychium* 8, 101–144.

Branstetter, M.G., Danforth, B.N., Pitts, J.P., Faircloth, B.C., Ward, P.S., Buffington, M.L., Gates, M.W., Kula, R.R. and Brady, S.G. (2017) Phylogenomic insights into the evolution of stinging wasps and the origins of ants and bees. *Current Biology* 27, 1019–1025. doi: 10.1016/j.cub.2017.03.027

Brothers, D.J. (1975) The phylogeny and classification of the aculeate Hymenoptera, with special reference to Mutillidae. *University of Kansas Science Bulletin* 50, 483–648.

Brothers, D.J. (1978) Biology and immature stages of *Myrmosula parvula* (Hymenoptera: Mutillidae). *Journal of the Kansas Entomological Society* 51, 698–671.

Brothers, D.J. (1989) Alternative life-history styles of mutillid wasps (Insecta, Hymenoptera). In: Bruton, M.N. (ed.) *Alternative Life-History Styles of Animals. Perspectives in Vertebrate Science* 6, 279–291.

Brothers, D.J. (1993a) Key to subfamilies of Tiphiidae. In: Goulet, H. and Huber, J.T. (eds) *Hymenoptera of the World: An Identification Guide to Families*. Publication I894/E, Research Branch Agriculture Canada, Ottawa, pp. 178–185.

Brothers, D.J. (1993b) Key to subfamilies of Mutillidae. In: Goulet, H. and Huber, J.T. (eds) *Hymenoptera of the World: An Identification Guide to Families*. Publication I894/E, Research Branch Agriculture Canada, Ottawa, pp. 188–201.

Brothers, D.J. (1999) Phylogeny and evolution of wasps, ants and bees (Hymenoptera, Chrysidoidea, Vespoidea and Apoidea). *Zoologica Scripta* 28(1), 233–249.

Brothers, D.J. (2015) Revision of the Rhopalomutillinae (Hymenoptera, Mutillidae): 1, generic review with descriptions of three new genera. *Journal of Hymenoptera Research* 46, 1–24. doi: 10.3897/JHR.46.5733

Brothers, D.J. (2021) Aculeate Hymenoptera: Phylogeny and Classification. In: Starr, C.K. (ed.) *Encyclopedia of Social Insects*. Springer Nature Switzerland AG, Basel, pp. 3–11.

Brothers, D.J. (2022) Critical analysis of the type material of Mutillidae described from the Australasian Region (Hymenoptera). *Zootaxa* 5140, 1–215.

Brothers, D.J. and Finnamore, A.T. (1993). Superfamily Vespoidea. In: Goulet H. and Huber, J.T. (eds) *Hymenoptera of the World: An Identification Guide to Families*. Publication I894/E, Research Branch Agriculture Canada, Ottawa, pp. 161–278.

Brothers, D.J. and Lelej, A.S. (2017) Phylogeny and higher classification of Mutillidae (Hymenoptera) based on morphological reanalyses. *Journal of Hymenoptera Research* 60, 1–97. doi: 10.3897/jhr.60.20091

Brothers, D.J., Tschuch, G. and Burger, F. (2000) Associations of mutillid wasps (Hymenoptera, Mutillidae) with eusocial insects. *Insectes Sociaux* 47, 201–211.

Brown, G.R. (1995) Notes on the distribution and identity of '*Tiphia intrudens*' Smith (Hymenoptera: Tiphiidae). *Australian Entomologist* 22, 1–4.

Brown, G.R. (2005) A revision of *Tachyphron* Brown and description of two new genera within the *Ariphron* group (Hymenoptera: Tiphiidae). *Journal of Natural History* 39, 197–239.

Burdick, D.J. and Wasbauer, M.S., (1959) Biology of *Methocha californica* Westwood (Hymenoptera: Tiphiidae). *Wasmann Journal of Biology* 17, 75. Available at: https://digitalcommons.usu.edu/bee_lab_bu/9

Cameron, P. (1907) On the Bornean Tiphiidae, including a new genus. *The Entomologist* 40, 287–289.

da Souza, H.S., Messas, Y.F., Masago, F., dos Santos, E.D. and Vasconcellos-Neto, J. (2015) *Paracyphononyx scapulatus* (Hymenoptera: Pompilidae), a koinobiont ectoparasitoid of *Trochosa* sp. (Araneae: Lycosidae). *Journal of Hymenoptera Research* 46, 165–172. doi:

Day, M.C. (1988) Spider wasps. Hymenoptera: Pompilidae. *Handbooks for the Identification of British Insects* 6 (part 4), 1–60.

Day, M.C., Else, G.R. and Morgan, D. (1981) The most primitive Scoliidae (Hymenoptera), *Journal of Natural History* 15, 671–684. doi: 10.1080/00222938100770471

Debevec, A.H., Cardinal, S. and Danforth, B.N. (2012) Identifying the sister group to the bees: a molecular phylogeny of Aculeata with an emphasis on the superfamily Apoidea. *Zoologica Scripta* 41, 527–535.

Decker, B.L., Pitts, J.P., Yuan, D. and Rodriguez, J. (2020) Re-examination of Australian and Oriental species of *Minagenia* Banks, 1934 (Hymenoptera: Pompilidae), with a new record for the genus in Australia and a new species description. *Zootaxa* 4768(3), 383–394.

Evans, H.E. (1953) Comparative ethology and the systematics of spider wasps. *Systematic Zoology* 2, 155–172.

Evans, H.E. (1969) Phoretic copulation in Hymenoptera. *Entomological News* 80, 113–124.

Gauld, I.D. and Bolton, B. (1988) *The Hymenoptera*. British Museum (Natural History), London, and Oxford University Press, Oxford.

Gonzaga, O. (2013) Natural history of tropical parasitoid wasps. In: Del Claro, K., Oliveira, P.S. and Rico-Gray, V. (eds) *Tropical Biology and Conservation Management. Case Studies*. Vol. XI. Encyclopedia of Life Support Systems (EOLSS). Available at http://www.eolss.net/Sample-Chapters/C20/E6-142-CS-11.pdf (accessed 26 August 2022)

Grout, T.G. and Brothers, D.J. (1983) Behaviour of a parasitic pompilid wasp (Hymenoptera). *Journal of Entomological Society of Africa* 45, 217–220.

Guidotti, A.E. (2007) A revision of the wasp genus *Paniscomima* (Hymenoptera: Rhopalosomatidae) and a proposal of phylogenetic relationships among species. *Invertebrate Systematics* 21, 297–309. doi: 10.1071/is04027

Gurney, A.B. (1953) Notes on the biology and immature stages of a cricket parasite of the genus *Rhopalosoma*. *Proceedings of the U.S. National Museum* 103, 19–34.

Han, Q., Chen, B. and Li, T.-J. (2021) Three new species of the subgenus *Jaynesia* Allen, 1969 of the genus *Tiphia* Fabricius, 1775 (Hymenoptera: Tiphiidae: Tiphiinae) from China, with a key to all known species. *Zootaxa* 4970, 313–324.

Hanima, R.K.P., Kumar, P.G. and Hegde, V.D. (2022) Additions to the knowledge on the genus *Tiphia* Fabricius (Hymenoptera: Tiphiidae: Tiphiinae) from India with the description of ten new species. *Zootaxa* 5204, 1–106.

Harris, A.C. (1987) Pompilidae (Insecta: Hymenoptera). *Fauna of New Zealand*, Vol. 12. DSIR Science Information Publishing Centre, Wellington, New Zealand. doi: 10.7931/J2/FNZ.12.

Harris, A.C. (1999) The life histories and nesting behaviour of the Pompilidae (Hymenoptera) in New Zealand: a comparative study. *Species Diversity* 4(1), 143–235.

Huber, J.T. (2009) Biodiversity of Hymenoptera. In: Foottit, R.G. and Adler, P.H. (eds) *Insect Biodiversity, Science and Society i–xxi + 1–632*. Wiley-Blackwell, Oxford, pp. 303–323.

Iwata, K. (1936) Biology of two Japanese species of *Methoca* with the description of a new species (Hymenoptera, Thynnidae). *Kontyu* 10, 57–89.

Johnson, B.R., Borowiec, M.L., Chiu, J.C., Lee, E.K., Atallah, J. and Ward, P.S. (2013) Phylogenomics resolves evolutionary relationships among ants, bees, and wasps. *Current Biology* 23, 2058–2062. doi: 10.1016/j.cub.2013.08.050

Khouri, Z., Gillung, J.P. and Kimsey, L.S. (2022) The evolutionary history of mammoth wasps (Hymenoptera: Scoliidae). *bioRxiv* preprint. doi: 10.1101/2022.01.24.474473

Kimsey, L.S. (1991) Relationships among the tiphiid wasp subfamilies (Hymenoptera). *Systematic Entomology* 16, 427–438. doi: 10.1111/j.1365-3113.1991.tb00677.x

Kimsey, L.S. (1996a) Phylogenetic relationships of the thynnine wasp tribe Rhagigasterini (Hymenoptera: Tiphiidae). *Journal of Hymenoptera Research* 5, 80–99.

Kimsey, L.S. (1996b) Re-examination of the *Ariphron* generic group of thynnine wasps (Hymenoptera: Tiphiidae). *Australian Journal of Entomology* 35, 303–311.

Kimsey, L.S. and Brothers, D.J. (2016) The life, publications and new taxa of Qabir Argaman (Carol Nagy). *Journal of Hymenoptera Research* 50, 141–178. doi: 10.3897/JHR.50.7973

Kimsey, L.S. and Brown, G.R. (1993) Lectotype designations within the subfamily Thynninae (Hymenoptera: Tiphiidae). *Journal of the Australian Entomological Society* 32, 317–326.

Kumpanenko, A., Gladun, D. and Vilhelmsen, L. (2022) Morphology of the sting apparatus in velvet ants of the subfamilies Myrmosinae, Dasylabrinae, Myrmillinae and Mutillinae (Hymenoptera: Mutillidae). *Zoomorphology* 141, 81–94. doi: 10.1007/s00435-022-00552-w

Kurczewski, F.E. and Spofford, M.G. (1986) Observations on the behaviours of some Scoliidae and Pompilidae (Hymenoptera) in Florida. *Florida Entomologist* 69, 636–644.

Kurczewski, F.E., Edwards, G.B. and Pitts, J.P. (2017) Hosts, nesting behavior, and ecology of some North American spider wasps (Hymenoptera: Pompilidae), II.*Southeastern Naturalist* 16(m9), 1–82.

Kurzenko, N.V. (1996) A new Nearctic genus of Sapygidae with a key to the Nearctic and Palaearctic genera (Hymenoptera, Sapygidae). *Memoirs of the Entomological Society of Washington* 17, 89–94.

Lelej, A.S. (1996a) Mutillid wasps collected in Malaysia and Indonesia by Dr. Sk. Yamane (Hymenoptera, Mutillidae). *Tropics* 6, 91–104. doi: 10.3759/tropics.6.91

Lelej, A.S. (1996b) To the knowledge of the East Asian species of tribe Trogaspidiini Bischoff, 1920 (Hymenoptera, Mutillidae) with description of eight new genera and two new species. *Far Eastern Entomologist* 30, 1–24.

Lelej, A.S. (2005) *Catalogue of the Mutillidae (Hymenoptera) of the Oriental Region*. Dalnauka, Vladivostok, 252 pp.

Lelej, A.S. (2012) A new species of *Rhopalomutilla* André, 1901 (Hymenoptera: Mutillidae) from South Vietnam. *Russian Entomological Journal* 21, 143–145.

Lelej, A.S. (2021) A new species of the genus *Petersenidia* Lelej (Hymenoptera: Mutillidae) from Indonesia, eastward of Wallaces line. *Zootaxa* 5006, 101–105.

Lelej, A.S. (2023) Review of the tribe Ctenotillini (Hymenoptera: Mutillidae) from Oriental and Palaearctic regions. *Far Eastern Entomologist* 480, 1–22.

Lelej, A.S. and Krombein, K.V. (1999) Two remarkable new genera of mutillid wasps (Hymenoptera: Mutillidae, Sphaeropthalminae, Pseudomethocini) from Thailand. *Far Eastern Entomologist* 79, 1–8.

Lelej, A.S. and Brothers, D.J. (2008) The genus-group names of Mutillidae (Hymenoptera) and their type species, with a new genus, new name, new synonymies, new combinations and lecto-typifications. *Zootaxa* 1889, 1–79.

Lelej, A.S. and Nemkov, P.G. (1997) Phylogeny, evolution and classification of Mutillidae (Hymenoptera). *Far Eastern Entomologist* 46, 1–24.

Lelej, A.S., Zhou, H.-T., Loktionov, V. and Xu, Z.-F. (2016) Review of the genus *Promecidia* Lelej, 1996, with description of two new species from China (Hymenoptera, Mutillidae, Trogaspidiini). *ZooKeys* 641, 103–120. doi: 10.3897/zookeys.641.10765

Lelej, A.S., Williams, K.A., Loktionov, V.M., Pang, H. and Xu, Z.-F. (2017) Review of the genus *Zeugomutilla* Chen, 1957 (Hymenoptera, Mutillidae, Mutillini), with description of two new species. *Zootaxa* 4247, 1–15.

Lelej, A.S., Williams, K.A., Terine, J.B., Okayasu, J., Parikh, G.R. and Kumar, G.P. (2023) Review of the tribe Mutillini (Hymenoptera: Mutillidae) from the Oriental Region. *Zootaxa* 5228, 455–476.

Liao, X.-P., Chen, B. and Li, T.-J. (2022) A taxonomic revision of the subfamily Myzininae from China, with a key to the Chinese species (Hymenoptera: Tiphiidae). *Zootaxa* 5154, 152–174. doi: 10.11646/zootaxa.5154.2.3

Liu, Z., van Achterberg, C., He, J.-H., Chen, X.-X. and Chen, H.Y. (2021a) Illustrated keys to Scoliidae (Insecta, Hymenoptera, Scolioidea) from China. *ZooKeys* 1025, 139–175.

Liu, Z., Yang, S.-J., Wang, Y.-Y., Peng, Y.-Q., Chen, H.-Y. and Luo, S.-X. (2021b) Tackling the taxonomic challenges in the family Scoliidae (Insecta, Hymenoptera) using an integrative approach: a case study from Southern China. *Insects* 12, 802. doi: 10.3390/insects12100802

Lohrmann, V. and S. Engel, M.S. (2017) The wasp larva's last supper: 100 million years of evolutionary stasis in the larval development of rhopalosomatid wasps (Hymenoptera: Rhopalosomatidae). *Fossil Record* 20, 239–244. doi: 10.5194/fr-20-239-2017

Martins, A.L., Gallão, J.E., Bichuette, M.E. and Santos E.F. (2016) The first record of *Notocyphus tyrannicus* Smith, (Hymenoptera: Pompilidae) as parasitoid of *Acanthoscurria* Ausserer, 1871 (Teraphosidae: Teraphosinae). *Brazilian Journal of Biology* 76, 3. doi: 10.1590/1519-6984.07515

Melo, G.A., Hermes, M.G. and Garcete-Barrett, B.R. (2011) Origin and occurrence of predation among Hymenoptera: a phylogenetic perspective. In: Polidori, C. (ed.) *Predation in the Hymenoptera: An Evolutionary Perspective*. Transworld Research Network, Trivandrum, pp. 1–22.

Miller, L.A., Benefield, T.D., Lounsbury, S.A., Lohrmann, V. and Blaschke, J.D. (2019) DNA barcoding of rhopalosomatid larvae reveals a new host record and genetic evidence of a second species of *Rhopalosoma* Cresson (Hymenoptera, Rhopalosomatidae) in America north of Mexico. *Journal of Hymenoptera Research* 74, 35–46. doi: 10.3897/jhr.74.38276

Narita, K., and Mita, T. (2018) Two new species of the genus *Methocha* from Laos (Hymenoptera, Tiphiidae). *ZooKeys* 775, 59–68. doi: 10.3897/zookeys.775.24945

Okayasu, J. (2017) Two new species of the genus *Sinotilla* Lelej, 1995 (Hymenoptera: Mutillidae), with notes on taxonomic characters. *Zootaxa* 4294, 151–169.

Okayasu, J. (2021) A new velvet ant genus *Arkaditilla* (Hymenoptera, Mutillidae, Trogaspidiini) from the Oriental Region, with review of species. *Journal of Hymenoptera Research* 84, 145–168. doi: 10.3897/jhr.84.68709

Okayasu, J. (2022) Remarkable range extension of the genus *Nordeniella* Lelej, 2005 (Hymenoptera: Mutillidae: Smicromyrmini): description of a new species from Sumbawa, Indonesia. *Journal of Insect Biodiversity* 32, 56–63.

Okayasu, J. (2023) Discovery of the velvet ant genus *Orientilla* Lelej from Laos (Hymenoptera, Mutillidae, Dasylabrinae), with description of a related new species from India. *Journal of Hymenoptera Research* 96, 817–834. https://doi.org/10.3897/jhr.96.110590

Okayasu, J., Williams, K.A. and Lelej, A.S. (2018) A remarkable new species of *Sinotilla* Lelej (Hymenoptera: Mutillidae: Smicromyrmini) from Taiwan and an overview of color diversity in East Asian mutillid females. *Zootaxa* 4446, 301–324.

Okayasu, J., Williams, K.A., Lelej, A.S. and Pham, T.H. (2021a) Review of female *Andreimyrme* Lelej (Hymenoptera: Mutillidae: Smicromyrmini). *Zootaxa* 5061, 1–38.

Okayasu, J., Lelej, A.S. and Williams, K.A. (2021b) Review of *Eotrogaspidia* Lelej (Hymenoptera: Mutillidae: Trogaspidiini). *Zootaxa* 4920, 56–90.

Osten, T. (2005) *Checkliste der Dolchwespen der Welt (Insecta: Hymenoptera, Scoliidae)*. Supplement. Bericht der Naturforschenden Gesellschaft, Augsburg, 62 pp.

Pagden, H.T. (1934) Biological notes on some Malayan aculeate Hymenoptera I. (Sphecoidea and Vespoidea). *Journal of the Federated Malay States Museums* 17, 458–466.

Pagden, H.T. (1949) Descriptions and records of Austro-Malaysian Methocidae and Mutillidae (Hymenoptera). *Transactions of the Royal Entomological Society of London* 100, 191–231.

Pagliano, G. and Romano, M. (2017) World list of all known species of Bradynobaenidae (Hymenoptera). *Il Naturalista Valtellinese* 28, 15–46.

Peters, R.S., Krogmann, L., Mayer, Ch., Donath, A., Gunkel, S., Meusemann, K., Kozlov, A., Podsiadlowski, L., Petersen, M., Lanfear, R., Diez, P.A., Heraty, J., Kjer, K.M., Klopfstein, S., Meier, R., Polidori, C., Schmitt, T., Liu, S., Zhou, X., Wappler, T., Rust. J., Misof, B. and Niehuis, O. (2017) Evolutionary history of the Hymenoptera. *Current Biology* 27, 1–6. doi: 10.1016/j.cub.2017.01.027

Pilgrim, E.M. and Pitts, J.P. (2006) A molecular method for associating the dimorphic sexes of velvet ants. *Journal of the Kansas Entomological Society* 79, 222–230.

Pilgrim, E.M., Von Dohlen, C.D. and Pitts, J.P. (2008) Molecular phylogenetics of Vespoidea indicate paraphyly of the superfamily and novel relationships of its component families and subfamilies. *Zoologica Scripta* 41, 527–535.

Reibán, J.M.F., Rentería, R.P.P. and Torres, J.D.F. (2021) First record of parasitism in *Avicularia purpurea* Kirk, 1990 (Araneae: Theraphosidae: Aviculariinae) by *Notocyphus* aff. *tyrannicus* Smith, 1855 (Hymenoptera: Notocyphinae) in the Ecuadorian Amazon. *Revista Chilena de Entomología* 47, 157–164.

Ronchetti, F. and Polidori, C. (2020) A sting affair: a global quantitative exploration of bee, wasp and ant hosts of velvet ants. *PLoS ONE* 15, e0238888. doi: 10.1371/journal.pone.0238888

Rozen, J.G. Jr and Kemel, S.M. (2009) Hospicidal behavior of the cleptoparasitic wasp *Sapyga luteomaculata* and investigation into ontogenetic changes in its larval anatomy (Hymenoptera: Vespoidea: Sapygidae). *American Museum Novitates* 3644, 1–24.

Saini, J., Gupta, D., Chandra, K. and Gupta, S.K. (2022) Taxonomic notes on the little-known genus *Cyanotiphia* Cameron, 1907 (Hymenoptera: Tiphidae: Tiphinae). *Zootaxa* 5154, 87–92.

Schmid-Egger, C. and Schmidt, S. (2021) Unexpected diversity in Central European Vespoidea (Hymenoptera, Mutillidae, Myrmosidae, Sapygidae, Scoliidae, Tiphiidae, Thynnidae, Vespidae), with description of two species of *Smicromyrme* Thomson, 1870. *ZooKeys* 1062, 49–72.

Schulten, G.G.M, Feijen, H.R, and Feijen, C. (2011) The genus *Bellimeris* Betrem (Hymenoptera, Scoliidae, Campsomerinae). *Zoologische Mededelingen* 85, 887–903.

Spradbery, J.P. (1973) *Wasps: An Account of the Biology and Natural History of Social and Solitary Wasps.* Sidgewick and Jackson. London. 408 pp.

Sutanto, D.D., Williams, K.A., Nugroho, H. and Lelej, A.S. (2021) To the knowledge of the velvet ant genus *Bischoffitilla* Lelej (Hymenoptera: Mutillidae) in Indonesia. *Treubia* 48, 1–12.

Terayama, M. and Mita, T. (2015) New species of the genera *Methocha* Latreille and *Hylomesa* Krombein from Japan (Hymenoptera: Tiphiidae). *Japanese Journal of Systematic Entomology* 21, 373–380.

Thaochan, N., Williams, K.A., Thoawan, K., Jeenthong, T. and Sittichaya, W. (2022) Three new species and one new country record of velvet ants (Hymenoptera, Mutillidae) from Thailand. *Journal of Hymenoptera Research* 93, 151–165.

Townes, H.K. (1957) Nearctic wasps of the subfamilies Pepsinae and Cerapalinae. *United States National Museum* 209, 1–286.

Townes, H.K. (1977) A revision of the Rhopalosomatidae (Hymenoptera). *Contributions of the American Entomological Institute* 15, 1–34.

van Achterberg, C. (2014) Revision of the genus *Parasapyga* Turner (Hymenoptera, Sapygidae), with the description of two new species. *ZooKeys* 369, 61–77.

Villanueva-Bonilla, G.A., Brescovit, A.D., dos Santos, E.F. and Vasconcellos-Neto, J. (2018) First record of *Epipompilus excelsus* (Bradley, 1944) (Hymenoptera, Pompilidae) as a koinobiont ectoparasitoid of *Ariadna mollis* (Holmberg, 1876) (Araneae, Segestriidae). *Journal of Hymenoptera Research* 66, 15–21. doi: 10.3897/jhr.66.28915

Vivallo, F. (2020) Phoretic copulation in Aculeata (Insecta: Hymenoptera): a review. *Zoological Journal of the Linnean Society* 191, 627–636.

Waldren, G.C., Roberts, J.D. and Pitts, J.P. (2020) Phoretic copulation in the velvet ant *Sphaeropthalma pensylvanica* (Lepeletier) (Hymenoptera, Mutillidae): a novel behavior for Sphaeropthalminae with a synthesis of mating strategies in Mutillidae. *Journal of Hymenoptera Research* 78, 69–89. doi: 10.3897/jhr.78.55762

Waldren, G.C., Sadler, E.A., Murray, E.A., Bossert, S., Danforth, B.N. and Pitts, J.P. (2023) Phylogenomic inference of the higher classification of velvet ants (Hymenoptera: Mutillidae). *Systematic Entomology* 48(3), 463–487.

Williams, F.X. (1919) Philippine wasp studies. Part II. Descriptions of new species and life history studies. *Bulletin of the Experimental Station of the Hawaiian Sugar Planters' Association, Entomological Series* 14, 19–186.

Williams, K.A., Lelej, A.S., Okayasu, J., Borkent, C.J., Malee, R., Thoawan, K. and Thaochan, N. (2019a) The female velvet ants (aka modkhong) of southern Thailand (Hymenoptera: Mutillidae), with a key to the genera of southeast Asia. *Zootaxa* 4602, 1–69. doi: 10.5281/zenodo.2669927

Williams, K.A., Lelej, A.S. and Thaochan, N. (2019b) New species of Myrmosinae (Hymenoptera: Mutillidae) from Southeastern Asia. *Zootaxa* 4656, 525–534 doi: 10.11646/zootaxa.4656.3.9

Zheng, X.-Y., Cao, L.-J., Chen, P.-Y., Chen, X.-X., van Achterberg, K.[C.], Hoffmann, A.A., Liu, J.-X. and Wei, S.-J. (2021) Comparative mitogenomics and phylogenetics of the stinging wasps (Hymenoptera: Aculeata). *Molecular Phylogenetics and Evolution* 159, 107119. doi: 10.1016/j.ympev.2021.107119

Zhou, H.T., Lelej, A.S., Williams, K.A. and Liu, J.X. (2018) Revision of the Oriental genus *Zavatilla* Tsuneki (Hymenoptera, Mutillidae, Trogaspidiini), with descriptions of two new species. *Zootaxa* 4418, 101–120.

18

Platygastroidea

Abstract

The superfamily Platygastroidea comprises five families in the region, of which only the Platygastridae and Scelionidae are massively species-rich. Special morphological terminology is described. An identification key to the families is provided, as well as ones to the subfamily of the Platygastridae and to the traditional genus clusters of the Scelionidae. Their biologies are described and representative species of all groups are illustrated photographically. Relevant identification works are cited.

The platygastroids are generally small wasps, with some 4500 described species worldwide (Austin et al., 2005). There really has never been any doubt that the platygastrids and scelionids were very closely related. The question was really only about whether they were mutually monophyletic. For many years the Platygastroidea comprised just the two families: the Platygastridae (often erroneously written Platygasteridae) and Scelionidae. Then there was a period when all the species were treated as belonging to just the one family Platygastridae, and now the Platygastroidea is considered as comprising the Platygastridae, Scelionidae, again, plus some additional isolated families (Chen et al., 2021).

Murphy et al. (2007) published the first substantial molecular phylogenetic study of the superfamily using sequence data from three genes. They recovered Platygastridae and Scelionidae *sensu stricto* as sister groups and Sparasionini as sister to the remaining scelionids. With the exception of both Teleasinae and Telenominae, which were monophyletic, the other subfamilies were mostly recovered as polyphyletic. Chen et al. (2021) moved things forward and changed things quite a lot based on a considerably larger study combining data from morphology, four Sanger-sequenced genes and, for a subset of taxa, phylogenomic data from 4371 single-copy protein-coding genes. However, they rejected the morphological data as it was evident that any phylogenetic signal was 'swamped by noise', i.e. homoplasy was far too pervasive. As a result, they recognised a total of seven extant families: Geoscelionidae, Janzenellidae, Neuroscelionidae, Nixoniidae, Platygastridae, Scelionidae and Sparasionidae. Five of these (all except Geoscelionidae and Janzenellidae) occur in S.E. Asia, Platygastridae and Scelionidae being by far the commonest.

Since these are mostly small and often dull-coloured wasps their taxonomy is still rather poorly known, especially in tropical regions, but also in other countries where work has mostly concerned those species of agricultural relevance.

Collectively, platygastroids attack members of nine insect orders.

Talamus and Pham (2017) provided an online photographic resource of all the Platygastroidea (Hymenoptera) in the Institute of Ecology and Biological Resources in Hanoi (Vietnam) including all the holotypes (which are illustrated in the 'printed' journal article). It has to be said that the conditions of most of the specimens are truly atrocious and, in many cases, useless for taxonomic study.

Morphological Terminology

Mikó et al. (2007) provided a detailed anatomical description of the head and thorax of the superfamily, including all the internal musculature and their homologies to those of other groups. The terminology is broadly similar to that used for Chalcidoidea.

Mesepimeral Sulcus

Keys often use the term mesepimeral sulcus, which corresponds to what in many other groups is called the mesopleural furrow or suture (Talamas et al., 2019; Mikó et al., 2021). It is a groove, often sculptured, or row of pits parallel and close to the posterior margin of the mesopleuron.

Skaphion

A differentiated anterior band of the mesonotum defined posteriorly by the transverse skaphial carina is present in many taxa.

Metasomal Horn

The internalised telescopic ovipositor system of platygastroids (Austin and Field, 1997) is often longer than the metasoma in which it is contained, and many taxa have independently evolved a horn like structure protruding from the 1st metasomal tergite to accommodate this extra length (see Fig. 18.5 b, d). In some such species, such as the cosmopolitan platygastrid genus *Inostemma*, the horn extends far anteriorly over the whole mesosoma and head.

Key to S.E. Asian Families of Platygastroidea*

(*modified after Chen et al. (2021)).

1. Mesepimeral sulcus absent (Fig. 18.1 b, c); antenna with 10 or fewer segments...... **Platygastridae**
- Mesepimeral sulcus usually present, indicated by a line of foveae parallel to mesopleural-metapleural suture (Fig. 18.1 a, a'); antennal with 7–15 segments............. 2
2(1) Antenna with 14 or 15 segments; metasomal segments with broad, paired depressions along anterior margin (Fig. 18.3); pronotal cervical sulcus setose (Fig. 18.3) . ... **Nixoniidae**
- Antenna with 12 or fewer segments (Figs 18.2, 18.7); metasomal segments either simple (no foveae or depressions), or with line of foveae along anterior margins; pronotal cervical sulcus not setose.. 3
3(2) Middle tibia with one apical spur .... **Scelionidae**
- Middle tibia with two apical spurs 4
4(3) Fore wing with bulla, i.e., a gap between the apex of the submarginal vein and the more distal venation (Fig. 18.9); maxillary palp elongate and easily visible, 5-segmented (Fig. 18.9)... **Sparasionidae**
- Fore wing without bulla, submarginal vein continuous to costal margin apically; sutures between metasomal segments 3–5 simple, without line of foveae (Fig. 18.2); maxillary palp minute, 2-segmented.............................. **Neuroscelionidae**

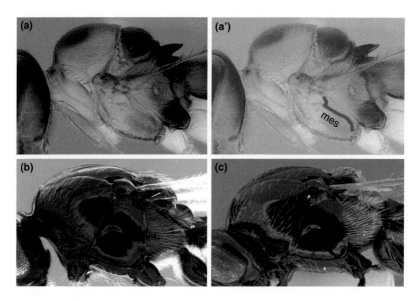

Fig. 18.1. Lateral views of mesosomas: **(a, a')** scelionid wasp, showing the mesipemeral suture (mes); **(b, c)** two platygastrids lacking the suture.

Neuroscelionidae

Galloway *et al.* (1992) placed it as a basal member of the Gryonini of the Scelionidae. However, it was removed from there but left unplaced by Valerio *et al.* (2009) who recognised it as being a rather isolated group within the superfamily, most similar to some fossil genera with relatively complete wing venation. The family was described by Chen *et al.* (2021). *Neuroscelio*, the only extant genus of Neuroscelionidae, for a long time was thought to be endemic to Australia. Species are typically collected in forests and have now been collected from Thailand, Vietnam, Malaysia (Sarawak) and Australia. These are medium-sized to large platygastroids, approximately 5 mm long (Fig. 18.2). Their biology is unknown.

Valerio *et al.* (2009) provided a key to the species of *Neuroscelio*.

Nixoniidae

Nixonia, the only known genus of Nixoniidae, is widespread, occurring in S.E. Asia as well as in sub-Saharan Africa, Egypt, Somalia, India and Sri Lanka. *Nixonia krombeini* (Fig. 18.3) is recorded from India, Laos, Sri Lanka, Thailand and Vietnam.

The only host record is for an extralimital species, *N. watshami* in southern Africa, which

Fig. 18.2. *Neuroscelio doddi* female, dorsal view. (Source: from Chen *et al.* (2021), reproduced under terms of Creative Commons Attribution Licence CC-BY 4.0.)

Fig. 18.3. *Nixonia krombeini* female, dorsal view. (Source: from Chen *et al.* (2021), reproduced under terms of Creative Commons Attribution Licence CC-BY 4.0.)

has been reared from the eggs of the armoured bush cricket, *Acanthoplus discocidalis* (Orthoptera: Tettigoniida, Hetrodinae). However, since Hetrodinae are endemic to sub-Saharan Africa, those species that occur in other regions must have different host relationships but very likely to be orthopteran eggs.

Johnson and Masner (2006) revised the world species of *Nixonia*, and van Noort and Johnson (2009) provided another key to world species.

Platygastridae

Platygastrids *sensu stricto* are minute insects, essentially all about 1 mm long or less but a few reach 3.0 mm. It is an enormously species-rich and extremely common family but it is taxonomically poorly known. They are characterised by the absence of a mesepimeral sulcus, which distinguishes them from all other S.E. Asian platygastroids. In addition, the antennae of females having only eight or fewer flagellomeres (but this overlaps with Scelionidae) and the absence of a frontal depression behind antennal sockets. In winged species the venation is reduced to a single tubular vein along the basal part of the fore wing, or this vein entirely absent, Thus the 'stigmal' vein (r-rs) is absent, or if this is present then it is very short. Nearly all species are smooth and shiny. A very few species are brachypterous or apterous.

The family comprises two subfamilies: Platygastrinae and Sceliotrachelinae. Earlier works also recognised Inostemmatinae. Masner and Huggert (1989) revised the world genera that were up until then included in the Inostemmatinae and reassigned 15 of them to the Platygastrinae and the remaining 26 to the Sceliotrachelinae. This was because the traditional Inostemmatinae was a polyphyletic assemblage based on plesiomorphic traits. The genera of Inostemmatinae were reclassified variously into each of the two subfamilies, but the subfamily name is now formally a junior synonym of Platygastrinae. If specimens can be recognised as belonging to the former Inostemmatinae (primarily by the presence of a submarginal vein), then one can use the key to the 41 genera in Masner and Huggert (1989).

Key to Subfamilies of Platygastridae

1 Antennal clava of female abruptly defined, usually 3-segmented but sometimes 1-segmented; male antennae often similar to females but may be thread-like; laterotergites large.. **Sceliotrachelinae**
- Antennal clava of female not so abruptly defined and composed of 4 or 5 segments; male antennae usually threadlike, rarely clavate; laterotergites narrow ....**Platygastrinae**

Almost all Platygastrinae are parasitoids of gall-forming Cedidomyiidae (Diptera) whereas sceliotrachelines attack a variety of Hemiptera (Fulgoroidea, Aleyrodidae, Pseudococcidae) and also Coleoptera (i.e. members of the tribe Fidobiini) (Austin et al., 2005; Polaszek, 2009); a very few platygastrids are parasitoids of gall-inducing psyllids (Veenakumari et al., 2018). Unlike scelionids, platygastrids are koinobiont endoparasitoids; most studied species are solitary parasitoids but a few are gregarious. Oviposition by Platygastrini is into either the host egg or an early-instar larva, and the adult wasp emerges from the pre-pupa or pupa.

Males have the second flagellomere (occasionally the first) modified and contain a gland. Isidoro and Bin (1995) demonstrated that these glands are involved in courtship in the genus *Amitus*, and this is likely to be true for all other species.

Sceliotrachelines are minute and infrequently collected, and are most diverse in regions where they have been studied the least (Lahey et al., 2021a).

Many species are of economic importance (Polaszek, 2009). *Platygaster oryzae*, and to a lesser extent another species usually called *P. foersteri*, is an important contributor to control of the serious pest rice gall midge *Orseolia oryzae* throughout the region (Hummelen and Soenarjo, 1977; Kobayashi and Kadkao, 1981, 1984, 1986, 1991; Leu *et al.*, 1982; Saran *et al.*, 2016).

To say that knowledge of S.E. Asian Platygastridae was abysmal would be an understatement of high order. In the 1995 calalogue of world species, only two are listed from the region (Vlug, 1995).

Getting started on the taxonomy of Platygastridae is very difficult because of the lack of even good generic keys. Fortunately, nearly all of the genera are cosmopolitan and therefore keys to the genera of one region might work reasonably well for other regions. Kozlov's (1971) key to the fauna of the former Soviet Union is probably the soundest starting point, but he has been considered a splitter (which in our view is safer than being a lumper, because named taxa can always be combined again later) (Talamas and Buffington, 2014). More accessible is Buhl and Choi's (2006) key to the genera known from Korea, which includes 15 genera. It is worth noting that the great majority of species belong to a small minority of genera, notably *Platygaster* (Fig. 18.4 a, b), *Synopeas* (Fig. 18.4 c) and *Leptacis* (Fig. 18.4 d) (Polaszek, 2009).

Synopeas is one of the larger genera of Platygastridae with some 400 described species, and Jackson (1969) discussed characters that help separate it from *Leptacis*, another very large cosmopolitan genus. The species of *Synopeas* from New Guinea were revised by Awad *et al.* (2021). Buhl (2002, 2008, 2009, 2021) described rather a large number of new platygastrid species in a number of genera from Indonesia, Malaysia and Vietnam, but unfortunately does not provide any comprehensive keys, making his papers really rather hard to use. Lahey *et al.* (2019) revised *Aleyroctonus* (Sceliotrachelinae), which includes one described species in the region, but they also provide a well-illustrated (photographically) key to the genera of the *Aphanomerus* group of genera.

Scelionidae

The Scelionidae are morphologically diverse. However, Chen *et al.* (2021) noted that diagnosing the family is not easy as no single

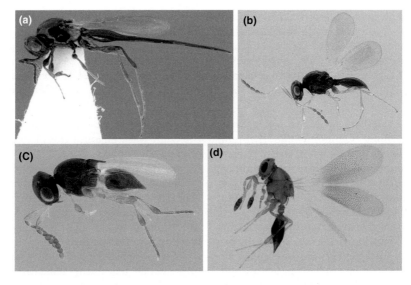

Fig. 18.4. Exemplar common Platygastridae from Thailand: **(a, b)** *Platygaster* spp.; **(c)** *Synopaeus* sp.; **(d)** *Leptacis* sp.

character, or even any simple combination of characters, will allow every species to be separated from those of other families, and that they 'are probably most easily identified by a process of elimination'.

All scelionids are solitary idiobiont endoparasitoids within insect or spider eggs. Collectively many insect orders are attacked, although not termites or ants. Host associations are, however, rather conserved at tribe/genus level.

Chen et al.'s (2021) analysis found that two of the traditional subfamilies, Telenominae and Teleasinae, rendered the Scelioninae paraphyletic, although both Telenominae and Teleasinae were recovered as monophyletic. Because of the historical use of these names, we deal with these separately but put them in inverted commas to indicate that they may not be subfamilies in the strict sense. Their analysis recovered a monophyletic tribe of spider egg parasitoids, the Baeini, comprising the genera *Baeus*, *Odontacolus* and *Mirobaeoides*. Interestingly, the represented Embidobiini which are parasitoids of webspinners (Embioptera) were recovered as a grade leading to the Baeini, thus suggesting that the use of silk in host-location in the Scelionidae may have had a single origin.

Males have flagellomere 3 modified (see Fig. 18.7) and this is associated with a specialised antennal gland.

Adult *Mantibaria* females attach themselves between the abdominal tergites of female praying mantises (Mantodea) and feed on their haemolymph, thus acting as ectoparasites. When the mantis female lays eggs which are encased in an ootheca, the *Mantibaria* moves to the egg mass and oviposits in the mantis eggs before the ootheca hardens. This genus is yet to be recorded from S.E. Asia but since it occurs in Africa, India (Veenakumari et al., 2012) and Australia (as well as the Palaearctic and New World) it seems possible that it could occur here.

Masner (1976) provided the most recent key to world genera, but since many genera have wide distributions, it is also worth trying his key to the Holarctic genera (Masner, 1980), obviously with appropriate caution.

Key to Clusters of Scelionidae*

*modified after Masner (1976).

1	Metasoma with wide laterotergites that are attached loosely to their sternites, impressed submarginal ridge absent; T2 by far the largest of all metasomatic tergites; female antennae 11-segmented (rarely fewer); male antennae 12-segmented.......... **'Telenominae'**
-	Metasoma with narrow laterotergites closely attached to their sternites to form an impressed submarginal ridge **OR** if ridge absent (rare condition) female antennae not 11-segmented and T2 not the largest, metasoma segments more similar in length to one another, female antennae usually 12-segmented (rarely only 6–11-segmented); male antennae usually 12-segmented, rarely with fewer 2
2(1)	Lateral ocelli much closer to median ocellus than to eye; T3 by far the largest segment; largest of all tergites; marginal vein several times longer than stigmal vein, postmarginal vein absent.. **'Teleasinae'**
-	Lateral ocelli usually closer to eye than to median ocellus **OR** if closer to median ocellus then **EITHER** T3 is not the largest **OR** marginal vein shorter than stigmal vein and postmarginal long, or wing venation absent .... **Scelioninae**

Note that the internal relationships within the family are far from settled (Chen et al., 2021) and that, probably, each of the above groups should be down classified as tribes. It is likely that, in the future, several new tribes will be created to render all groups monophyletic.

Within the Scelionidae there are two different types of ovipositor system, the simplest being referred to as the *Ceratobaeus*-type found in Baeini and Gryonini and relying on associated musculature to protract it; and a more complex telescopic system, called the *Scelio*-type which relies on metasomal hydrostatic pressure (Field and Austin, 1994; Austin and Field, 1997). The *Scelio*-type ovipositor itself is short, but can be extruded far beyond the tip of the metasomal segments. In some, the retracted part is more than three times longer than the main body of the metasoma, as sometimes happens when they die in alcohol (Fig. 18.5 a, c, e).

Masner (1976) provided a key to the world genera but unfortunately none of the characters are illustrated. Lê (2000) provided a key to genera and some species of the Vietnamese fauna in Vietnamese.

Scelioninae *sensu* Authors

This is a diverse group, in terms of numbers of genera and species and morphology. The world species were catalogued by Johnson (1992), but this is obviously dated.

Members of the type genus *Scelio* are virtually ubiquitous parasitoids of grasshopper (Orthoptera: Acrididae) eggs, and often play an important role in host population regulation. Their biology, ecology and collection were described in detail by Dangerfield *et al.* (2001) who also listed all species-level host associations known at the time. Although that list is extensive, only five such associations are known for S.E. Asia (most summarised by Siddiqui *et al.*, 1986). Qodir *et al.* (2017) described the life cycle and natural history of *S. pembertoni* in Indonesia (Java).

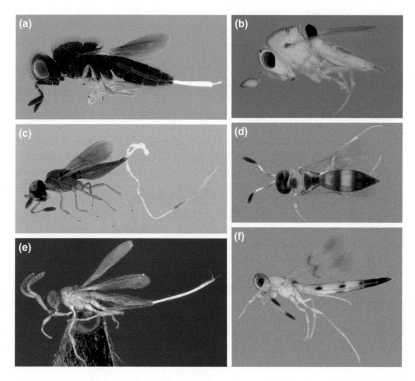

Fig. 18.5. Exemplar Scelionidae with (**a, c, e**) showing partially or completely extruded, telescopic ovipositor systems: (**a**) *Heptascelio* sp.; (**b**) *Idris* sp., note the humped 1st metasomal tergum into which the internal telescopic ovipositor structure extends; (**c**) *Dicroscelio* sp. (dry mounted so the extruded ovipositor system has partially collapsed and distorted); (**d**) ? *Opisthacantha* sp.; (**e**) *Macroteleia* sp.; (**f**) *Calliscelio* sp.

The Baeini such as the genus *Idris* (Fig. 18.5 b) are typically very small, sub-spherical parasitoids of spider eggs (Fig. 18.6). Parasitoid–host co-evolution of baeines was investigated by Austin (1985) and more recently a morphologically based phylogeny was investigated by Iqbal and Austin (2000a). Members of *Baeus* have been referred to as 'micro-flea' wasps because of their small size (mostly less than 1 mm), habitus and extreme wing reduction in females (Stevens and Austin, 2007). Baeines can exert strong effects on their host spider populations, often attacking a large proportion of the eggs in a batch (e.g. Fig. 18.6 a).

Carey *et al.* (2006) presented a two-gene molecular phylogeny of the spider egg-parasitising scelionines and found that they did not form a monophyletic group. They also showed that there was much convergent evolution in the possession of an anterior metasomal horn and in wing reduction, both apparently associated with searching for host egg masses in crytobiotic habitats, and/or penetrating the silk egg sacs of spiders.

There are four very large genera of Baeini which account for the great majority of the species: *Baeus* (53 species), *Ceratobaeus* (165 species), *Idris* (160 species) and *Odontacolus* (55 species) (Johnson, 1992; Johnson *et al.*, 2018). The Australasian species of *Ceratobaeus* were revised by Iqbal and Austin (2000b). This genus is recognisable (and distinguished from *Idris*) by the horn-like projection of the 1st metasomal tergite (reflecting the internal housing of the retracted telescopic ovipositor system). Valerio *et al.* (2010) described a new *Cyphacolus* species that occurs in Thailand as well as Africa, the genus later being synonymised with *Odontacolus* (Valerio *et al.*, 2013). Talamas *et al.* (2011) revised the *Paridris nephta* species group and provided a key to recognise *Paridris* from some related genera.

Members of at least eight scelionid genera worldwide are parasitoids of aquatic insects with submerged eggs, specifically those of water-skaters, water scorpions (Heteroptera: Gerridae and Nepidae, respectively) and dragonflies (Odonata). They access these by crawling down underwater on the stones or plants on which their host eggs are attached.

Thoron species that parasitise the submerged eggs of water scorpions (Hemiptera: Nepidae) include one S.E. Asian species,

Fig. 18.6. Pholcid spiders from S.E. Asia with their baeine (*Idris* spp.) egg parasitoids: **(a)** *Nipisa phyllicola* from Malaysia carrying her eggs, only two of which (the pale greenish-white ones) have escaped parasitisation; **(b, c)** female *Tissahamia gombak* and *Belisana khaosok*, from Malaysia and Thailand, respectively, carrying eggs from which a parasitoid has just recently emerged (arrows). (Source: reproduced and slightly modified from Johnson *et al.* (2018) under terms of Creative Commons Attribution Licence CC-BY 4.0.)

T. dayi (from Seram island, Indonesia) (Johnson and Masner, 2004). Three other genera of Thoronini are represented in S.E. Asia: *Microthoron*, *Tanaodytes* and *Tiphodytes* (Masner, 1972; Mineo *et al.*, 2009). Masner (1972) provided a key to the genera and a diagnosis. The biology of North American *Tiphodytes* species, parasitoids of Gerridae, has been extensively investigated. Other members of the tribe have been found as phoretic on dragonflies (Carlow, 1992).

The Gryonini is a small, rather homogeneous tribe of chubby parasitoids of Heteroptera eggs. In Sarawak, *Gryon flavipes* was reported to be the only common parasitoid of *Leptocorisa oratorius* (Heteroptera: Alydidae), which is the most important rice earbug pest in S.E. Asia (Rothschild, 1970). One species from Singapore and India is a parasitoid of the blood-sucking reduviid bugs (*Triatoma rubrofasciata* and *Linshcosteus* sp.) (Masner, 1995).

Several species play important roles in regulating pest populations in S.E. Asia. For example, *Gryon ancinla* parasitises egg masses of the leaf-footed bug, *Acanthocoris scaber* (Hemiptera: Coreidae), which is a pest that feeds on chilli, potato, tomato and aubergine and other economically important vegetables, and is distributed in China and mainland S.E. Asia (Chen *et al.*, 2020). *Calliscelio elegans* is extremely widespread, occurring in all tropical continents as well as Oceania and Japan. It is frequently collected in sugar-cane plantations; its host remains unknown, but it is speculated to be a cricket (or crickets) associated with cane plantations. Galloway *et al.* (1992) described the ovipositor system.

Lubomir Masner (pers. comm.) investigated how to collect platygastroid parasitoids of submerged aquatic insect eggs and invented a floating yellow pan trap (with anchor and expanded polystyrene flotation ring) that could be deployed just downstream of emergent rocks in a stream on whose underwater surface pondskaters (Hemiptera: Gerridae) had oviposited. Downstream was important, because this is where the host kairomones will be strongest. Given that this is not the normal mode of collecting for hymenopterists, it is not surprising that very little is known about the biology and taxonomy of aquatic scelionids (Johnson and Masner, 2004).

For generic identification apart from Masner's (1976) key for the whole 'family', Galloway and Austin (1984) is probably the most useful work for the subfamily in its restricted sense, even though it only deals with the Australian fauna and includes many Australian endemic elements. Caleca and Bin (1995) revised the world species of the highly distinctive genus *Encyrtoscelio* (Gryonini) but, whilst noting that a species had been described from Vietnam by Lê (1986), they did not include it in their key. Regarding the S.E. Asian fauna, Johnson *et al.* (2008a) revised the world species of *Heptascelio*, a genus closely related to *Scelio*, which includes several representatives in the region, Taekul *et al.* (2008) revised *Fusicornia* and Taekul *et al.* (2010) revised *Platyscelio*. *Calliscelio* has been expanded to incorporate all taxa with a special type of mandibular sensilla (Talamus *et al.*, 2016a). The Australian genus *Nyleta* was recently discovered in the Andaman and Nicobar Islands and Vietnam (Kamalanathan and Mohanraj, 2016). Chen *et al.* (2013) revised the Chinese species of *Macrotelia* and these include quite a few species that also occur in adjacent parts of S.E. Asia. They also provided a key to separate *Habroteleia*, *Macroteleia* and *Triteleia*. Valerio *et al.* (2013) revised the Old World *Odontacolus* and provided a key to females. Chen *et al.* (2018) revised the world species of *Habroteleia*.

'Teleasinae'

This is a moderately large, clearly defined group, predominantly known from temperate regions, that includes specialist parasitoids of carabid beetle eggs – indeed the only known carabid egg parasitoids (Mikó *et al.*, 2021). Worldwide it comprises more than 450 species in 11 genera (Johnson, 1992). Consequently, some species are important in various agro-ecosystems. Several species display considerable wing reduction, and probably wing development polymorphism. *Triteleia* species are particularly elongate, and large, some species reaching 12 mm. *Trimorus* is a huge genus with well over 300 currently recognised species and with many more undescribed ones.

Obtaining reliable host records for this group is challenging because of where and how their hosts deposit their eggs, for example in mud cells. Mikó et al. (2021) reviewed the known host records.

Mikó *et al.* (2010) revised the large Old World genus *Xenomerus* and gave a character allowing it to be distinguished from other 'teleasines'. Talamas *et al.* (2016b) provided a key to the 'teleasine' genera with scutellar spines and revised the teleasine genus *Dvivarnus*, of which the distinctive *D. agamades* (Fig. 18.7) is commonly collected in mainland S.E. Asia.

'Telenominae'

Telenomiines are apparently derived scelionines but the group is readily recognisable by their large laterotergites that overlap the sternites in side view. The genera *Trissolcus* and *Telenomus* are particularly large, and most of their tropical species are undescribed.

Trissolcus and *Telenomus* have been particularly intensively studied because of their importance as biological control agents. These two genera are closely related (Chen *et al.*, 2021) with very similar appearance and biology (Fig. 18.8). *Trissolcus japonicus* (Fig. 18.8 a) has been given the common name of 'samurai wasp' in reference to its Japanese origin and lethality. It is native to east Asia but now is found in Europe, North America and Chile. It is chiefly known for parasitising *Halyomorpha halys*, the invasive brown marmorated stink bug (Hemiptera: Pentatonidae). *Telenomus remus* (Fig. 18.8 b-d), a native of Indonesia (Sarawak) and New Guinea (Wojcik *et al.,* 1976), is an important parasitoid of the fall armyworm, *Spodoptera frugiperda* (Lepidoptera: Noctuidae), a serious pest of economically important cereal crops including maize, rice, sorghum and sugar cane (Liao *et al.*, 2019). Most famous perhaps is *Trissolcus basalis*, an egg parasitoid of the invasive pest, green stink bug, *Nezara viridula* (Hemiptera: Pentatomidae) (Jones, 1988), a host–parasitoid association that has become a much-studied model system in research on insect ecology, behaviour and physiology. Its complete genome has been published (Lahey *et al.*, 2023), and the authors were able to identify 174 rapidly evolving gene families including olfactory receptors and pheromone/general odorant binding proteins, which are important in its host detection.

Bin *et al.* (1989) discovered that the multiporous plate sensillae on the antenna of *Trissolcus basalis* are associated with glands, and are probably gustatory rather than olfactory (Isidoro *et al.*, 1996). These gustatory antennal sensillae are present only in females, and they display several unique features found only in the Platygastroidea (Isidoro *et al.*, 2001). However, those in Scelioninae and Platygastridae are slightly different.

Trissolcus species are parasitoids of the eggs of pentatomomorph Hemiptera, i.e., members of the superfamilies Aradoidea, Coreoidea, Lygaeoidea, Pentatomoidea and Pyrrhocoroidea. Several species of *Trissolcus* occur in S.E. Asia (Talamas *et al.*, 2017).

Several species of 'Telenominae' are phoretic (e.g. Clausen, 1976; Rajmohana *et al.*, 2019) using their host species' sex pheromones (Arakaki *et al.*, 2011) to locate recently emerged adult females on which they ride until the female commences egg-laying. Interestingly, phoresy in scelionids has evolved on multiple independent occasions (Yan *et al.*, 2022).

Fig. 18.7. Male of the distinctive teleasine, *Dvivarnus agamades*, from Thailand; note the sexually modified 3rd flagellomere and the large marginal to stigmal vein ratio.

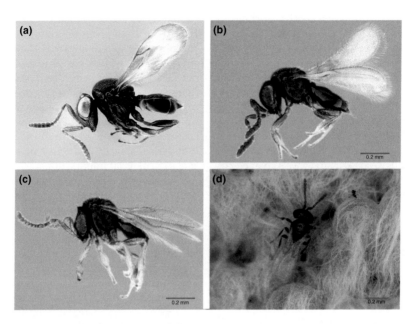

Fig. 18.8. Two telenomine scelionids of considerable biocontrol importance: **(a)** *Trissolcus japonicus*; **(b-d)** *Telenomus remus*, female, male and female on egg mass of *Spodoptera frugiperda* (respectively). (Source **(a)** by Elijah Talamas, public domain via Wikimedia Commons; **(b-d)** from Liao *et al.* (2019) reproduced under terms of Creative Commons Attribution Licence CC-BY 4.0.)

Taekul *et al.* (2014) presented a molecular phylogeny, based on four genes, of the traditional subfamily (*sensu* Masner, 1976) and concluded that it was non-monophyletic, with the *Psix* group of genera (*Psix* and *Paratelenomus*) actually belonging to the Scelioninae close to Gryonini, and so excluded them from Telenominae. However, the situation is slightly different in Chen *et al.*'s (2021) trees with 'Telenominae' including the *Psix* group being well supported but collectively being the sister group to *Gryon*. Therefore, in a sense, both results were right.

From the above molecular investigations it appears that the ground-plan biology of the superfamily is parasitism of Orthoptera (Chen *et al.*, 2021), and with a single shift in 'Telenominae'+Gryonini to being parasitism of terrestrial Hemiptera, and from them to Lepidopteran eggs, etc. (Taekul *et al.*, 2014).

There are not many revisions of genera occurring in S.E. Asia. Johnson (1991) revised Australasian (including New Guinea) species of *Trissolcus*. Johnson (1996) revised *Paratelenomus*, which occurs throughout the region. Johnson and Masner (1985) revised *Psix*, which has one species in S.E. Asia (Indonesia).

Lahey *et al.* (2021b) revised *Phoenoteleia*, principally from material from Indonesia, Malaysia and Philippines, and provided a key to separate it from some closely related genera. *Phanuromyia*, a very large genus in the Neotropics, is represented in the region by only three species, one of which is a parasitoid of the agricultural and forestry pest *Pochazia shantungensis* (Hemiptera: Auchenorrhyncha: Ricaniidae) (Johnson and Musetti, 2003; Nam *et al.*, 2020). Separation of the two common genera, *Trissolcus* and *Telenomus*, is not always straightforward, even in the Palaearctic. Most specimens can be assigned to genus using the key by Talamas *et al.* (2015) and further help is provided by Talamas *et al.* (2017).

Sparasionidae

Sparasionids include some of the largest species in the superfamily (up to 12 mm). Johnson *et al.* (2008b) reviewed the four extant world genera but only *Sparasion* occurs in the Old World, and is particularly species-rich

Fig. 18.9. *Sparasion* sp. from Thailand. Arrow indicates the bulla (note the small clear break) between fore wing vein R and more distal venation.

in the Palaearctic region. Several members, especially *Sparasion* species from S.E. Asia, have metallic coloration (Fig. 18.9).

Sparasion species are parasitoids of bush cricket (Tettigoniidae) and ordinary cricket (Grylloidea) eggs (Grissell, 1997).

A few species are brachypterous but otherwise they are identifiable from all other platygastroids by the presence of a bulla in the fore wing, a gap at the apex of the submarginal vein (R) separating it from more distal venation (Fig. 18.9). Popovici *et al.* (2014) presented a detailed anatomical description of the labio-maxillary complex for *Sparasion*.

References

Arakaki, N., Yamazawa, H. and Wakamura, S. (2011) The egg parasitoid *Telenomus euproctidis* (Hymenoptera: Scelionidae) uses sex pheromone released by immobile female tussock moth *Orgyia postica* (Lepidoptera: Lymantriidae) as kairomone. *Applied Entomology and Zoology* 46, 195–200.

Austin, A.D. (1985) The function of spider egg sacs in relation to parasitoids and predators, with special reference to the Australian fauna. *Journal of Natural History* 19(2), 359–376.

Austin, A.D. and Field, S.A. (1997) The ovipositor system of scelionid and platygastrid wasps (Hymenoptera: Platygastroidea): comparative morphology and phylogenetic implications. *Invertebrate Taxonomy* 11, 1–87.

Austin, A.D., Johnson, N.F., and Dowton, M. (2005) Systematics, evolution, and biology of scelionid and platygastrid wasps. *Annual Review of Entomology* 50, 553–582. doi: 10.1146/annurev.ento.50.071803.130500

Awad, J., Bremer, J.S., Butterill, P.T., Moore, M.R. and Talamas, E.J. (2021) A taxonomic treatment of *Synopeas* Förster (Platygastridae, Platygastrinae) from the island of New Guinea. *Journal of Hymenoptera Research* 87, 5–65. doi: 10.3897/jhr.87.65563

Bin, F., Strand, M.R. and Vison, S.B. (1986) Antennal structures and mating behavior in *Trissolcus basalis* (Woll.) (Hym.: Scelionidae), an egg parasitoid of the green stink bug, *Nezara viridis* (Hemiptera: Pentatomidae). *International Journal of Insect Morphology and Embryology* 15, 129–138.

Bin, F., Colazza, S., Isidoro, N., Solinas, M. and Vinson, S.B. (1989) Antennal chemosensilla and glands, and their possible meaning in the reproductive behaviour of *Trissolcus basalis* (Woll.) (Hym.: Scelionidae). *Entomologica* 24, 33–97.

Buhl, P.N. (2002) New species of *Leptacis* Förster, 1856 from Malaysia (Hymenoptera, Platygastridae). *Entomofauna* 23, 13–25.

Buhl, P.N. (2008) New and little known Platygastridae from Indonesia and Malaysia (Hymenoptera: Platygastroidea). *Zoologische Mededelingen, Leiden* 82, 515–579.

Buhl, P.N. (2009) New species of Platygastridae from Vietnam (Hymenoptera: Platygastroidea). *Zoologische Mededelingen, Leiden* 83, 877–918.

Buhl, P.N. (2021) Two new species of Platygastrinae (Hymenoptera, Platygastridae) from Vietnam. *Entomologist's Monthly Magazine* 157, 104–108.

Buhl, P.N. and Choi, J.Y. (2006) Taxonomic Review of the family Platygastridae (Hymenoptera: Platygastroidea) from the Korean Peninsula. *Journal of Asia-Pacific Entomology* 9, 121–137.

Caleca, V. and Bin, F. (1995) World revision of the genus *Encyrtoscelio* (Hymenoptera: Scelionidae). *Invertebrate Taxonomy* 9, 1021–1045.

Carey, D., Murphy N.P. and Austin A.D. (2006) Molecular phylogenetics and the evolution of wing reduction in the Baeini (Hymenoptera: Scelionidae): parasitoids of spider eggs. *Invertebrate Systematics* 20, 489–501.

Carlow, T. (1992) *Thoronella* sp. (Hymenoptera: Scolionidae) discovered on the thorax of an Aeshnidae (Anisoptera). *Notulae Odonatologicae* 3, 149–150.

Chen, H.-Y., Johnson, N.F., Masner, L. and Xu, Z.-F. (2013) The genus *Macroteleia* Westwood (Hymenoptera, Platygastridae *s.l.*, Scelioninae) from China. *ZooKeys* 300, 1–98. doi: 10.3897/zookeys.300.4934

Chen, H.-Y., Talamas, E.J., Masner, L. and Johnson, N.F. (2018) Revision of the world species of the genus *Habroteleia* Kieffer (Hymenoptera, Platygastridae, Scelioninae). *ZooKeys* 730, 87–122. doi: 10.3897/zookeys.730.21846

Chen, H., Talamas, E.J., Bon, M.-C. and Moore, M.R. (2020) *Gryon ancinla* Kozlov & Lê (Hymenoptera: Scelionidae): host association, expanded distribution, redescription and a new synonymy. *Biodiversity Data Journal*, 8, e47687. doi: 10.3897/BDJ.8.e4768

Chen, H., Lahey, Z., Talamas, E.J., Valerio, A.A., Popovici O.A., Musetti, L., Klompen, H., Polaszek, A., Masner, L., Austi, A.D. and Johnson, N.F. (2021) An integrated phylogenetic reassessment of the parasitoid superfamily Platygastroidea (Hymenoptera: Proctotrupomorpha) results in a revised familial classification. *Systematic Entomology* 46, 1088–1113.

Clausen, C.P. (1976) Phoresy among entomophagous insects. *Annual Review of Entomology* 21, 343–368. doi: 10.1146/annurev.en.21.010176.002015

Dangerfield, P.A., Austin, A. and Baker, G. (2001) *Biology, Ecology and Systematics of Australian Scelio, wasp parasitoids of locust and grasshopper eggs*. CSIRO Publishing, Melbourne, 254 pp. doi: 10.1071/9780643100763

Field, S.A. and Austin, A.D. (1994) Anatomy and mechanics of the telescopic ovipositor system of *Scelio* Latreille (Hymenoptera: Scelionidae) and related genera. *International Journal of Insect Morphology and Embryology* 23, 135–158. doi: 10.1016/0020-7322(94)90007-8

Galloway, I.D. and Austin, A.D. (1984) Revision of the Scelioninae (Hymenoptera: Scelionidae) in Australia. *Australian Journal of Zoology Supplementary Series* 32, 1–138.

Galloway, I.D., Austin, A.D. and Masner, L. (1992) Revision of the genus *Neuroscelio* Dodd, primitive Scelionids (Hymenoptera: Scelionidae) from Australia, with a discussion of the ovipositor system of the tribe Gryonini. *Invertebrate Taxonomy* 6, 523–545.

Grissell, E. (1997) Biological notes on *Sparasion* Latreille (Hymenoptera: Scelionidae), an egg parasitoid of *Atlanticus gibbosus* Scudder (Orthoptera: Tettigoniidae). *Proceedings of the Entomological Society of Washington* 99, 693–696.

Hummelen, P.J. and Soenarjo, E. (1977) Notes on the biology of *Platygaster oryzae, Obtusiclava oryzae*, and *Neanastatus oryzae*, parasites of the rice gall midge, *Orseolia oryzae*. *Contributions from the Central Research Institute for Agriculture, Bogor, Indonesia* 31, 1–18.

Iqbal, M. and Austin, A.D. (2000a) A preliminary phylogeny for the Baeini (Hymenoptera: Scelionidae) endoparasitoids of spider eggs. In: Austin, A.D. and Dowton, M. (eds) *The Hymenoptera: Evolution, Biodiversity and Biological Control*. CSIRO, Melbourne, pp. 178–191.

Iqbal, M., and Austin, A.D. (2000b) Systematics of the wasp genus *Ceratobaeus* Ashmead (Hymenoptera: Scelionidae) from Australasia: parasitoids of spider eggs. *Records of the South Australian Museum Monograph Series* 6, 1–164.

Isidoro, N. and Bin, F. (1995) Male antennal gland of *Amitus spiniferus* (Brethes) (Hymenoptera: Platygastridae), likely involved in courtship behavior. *International Journal of Insect Morphology and Embryology* 24, 365–373. doi: 10.1016/0020-7322(95)00014-U.

Isidoro, N., Bin, F., Colazza, S. and Vinson, S.B. (1996) Morphology of the antennal gustatory sensilla and glands in some parasitic Hymenoptera with hypothesis on their role in sex and host recognition. *Journal of Hymenoptera Research* 5, 206–239.

Isidoro, N., Romani, R. and Bin, F. (2001) Antennal multiporous sensilla: their gustatory features for host recognitions in female parasitic wasps (Insecta, Hymenoptera: Platygastroidea). *Microscopy Research & Technology* 55, 350–358.

Jackson, D.L. (1969) New characters for generic separation in the *Synopeas-Leptacis* complex. *Proceedings of the Entomological Society of Washington* 71, 400–404.

Johnson, N.F. (1991) Revision of Australian *Trissolcus* species (Hymenoptera: Scelionidae). *Invertebrate Taxonomy* 5, 211–239.

Johnson, N.F. (1992) Catalog of world species of Proctotrupoidea, exclusive of Platygastridae (Hymenoptera). *Memoirs of the American Entomological Institute* 51, 1–825.

Johnson, N.F. (1996) Revision of world species of *Paratelenomus* Dodd (Hymenoptera: Scelionidae). *The Canadian Entomologist* 128, 273–291. doi:10.4039/Ent128273-2

Johnson, N.F. and Masner, L. (1985) Revision of the genus *Psix* Kozlov & Lê (Hymenoptera: Scelionidae). *Systematic Entomology* 10, 33–58.

Johnson, N.F. and Masner, L. (2004) The genus *Thoron* Haliday (Hymenoptera: Scelionidae), egg parasitoids of waterscorpions (Hemiptera: Nepidae) with key to world species *American Museum Novitates* 3452, 1–16. doi: 10.1206/0003-0082(2004)452<0001:TGTHHS>2.0.CO;2

Johnson, N.F. and Masner, L. (2006) Revision of world species of the genus *Nixonia* Masner (Hymenoptera: Platygastroidea, Scelionidae). *American Museum Novitates* 3518, 1–32. doi: 10.1206/0003-0082(2006)3518[1:ROWSOT]2.0.CO;2

Johnson, N.F. and Musetti, L. (2003) Redefinition of the genus *Phanuromyia* Dodd (Hymenoptera: Scelionidae). *Journal of New York Entomological Society* 111, 138–144. doi: 10.1664/0028-71 99(2003)111[0138:ROTGPD]2.0.CO;2

Johnson, N.F., Masner, L., Musetti, L., van Noort, S., Rajmohana, K., Darling, C.D. and Quicutti, A. (2008a) Revision of world species of the genus *Hoptoscelio* Kieffer (Hymenoptera: Platygastroidea, Platygastridae). *Zootaxa* 1776, 1–51

Johnson, N.F., Masner, L. and Musetti, L. (2008b) Review of genera of the tribe Sparasionini (Hymenoptera: Platygastroidea, Scelionidae), and description of two new genera from the New World. *American Museum Novitates* 3629, 1–24. doi: 10.1206/578.1

Johnson, N.F., Chen, H. and Huber, B.A. (2018) New species of *Idris* Förster (Hymenoptera, Platygastroidea) from southeast Asia, parasitoids of the eggs of pholcid spiders (Araneae, Pholcidae). *ZooKeys* 811, 65–80. doi: 10.3897/zookeys.811.29725

Jones, W.A. (1988) World review of the parasitoids of the southern green stink bug, *Nezara viridula* (L.) (Heteroptera: Pentatomidae). *Annals of the Entomological Society of America* 81, 262–273. doi: 10.1093/aesa/81.2.262

Kamalanathan, V. and Mohanraj, P. (2016) A new species of *Nyleta* Dodd (Hymenoptera: Scelionidae) from Southeast Asia. *Journal of Insect Biodiversity* 4, 1–9.

Kobayashi, M. and Kadkao, S. (1981) Developmental biologies of *Platygaster oryzae* (Cameron) and *Platygaster foersteri* (Gahan) (Hymenoptera, Platygasteridae), parasitoids of the rice gall midge, *Orseolia oryzae* (Wood-Mason) in Thailand. *Kontyu* 49, 506–519.

Kobayashi, M. and Kadkao, S. (1984) Biological characteristics of *Platygaster oryzae* and *Platygaster foersteri* (Hymenoptera, Platygasteridae) parasitoids of the rice gall midge *Orseolia oryzae* in Thailand. *Kontyu* 52, 128–136.

Kobayashi, M. and Kadkao, S. (1986) Interrelation of *Platygaster oryzae* (Cameron) and *P. foersteri* (Gahan) (Hymenoptera, Platygastridae), parasitoids of the rice gall midge, *Orseolia oryzae* (Wood-Mason) in Thailand. *Kontyu* 54, 225–233.

Kobayashi, M. and Kadkao, S. (1991) Seasonal occurrence of the rice gall midge, *Orseolia oryzae* (Wood-Mason), in a wild rice field in Thailand and parasitism by its parasitoids, *Platygaster oryzae* (Cameron) and *P. foersteri* (Gahan) (Hymenoptera, Platygasteridae). *Japanese Journal of Entomology* 59, 449–466.

Kozlov, M.A. (1971) [Proctotrupoids (Hymenoptera, Proctotrupoidea) of the USSR]. *Trudy Vsesoyuznogo Entomologicheskogo Obshchestva*, 54, 3–67.

Lahey, Z., Masner, L., Johnson, N.F. and Polaszek, A (2019) Revision of *Aleyroctonus* Masner & Huggert (Hymenoptera, Platygastridae, Sceliotrachelinae). *Journal of Hymenoptera Research* 73, 73–93. doi: 10.3897/jhr.73.38383

Lahey, Z., Talamas, E., Masner, L. and Johnson, N.F. (2021a) Revision of the Australian genus *Alfredella* Masner & Huggert (Hymenoptera, Platygastridae, Sceliotrachelinae). *Journal of Hymenoptera Research* 87, 81–113. doi: 10.3897/jhr.87.58368

Lahey, Z., Musetti, L., Masner, L. and Johnson, N.F. (2021b) Revision of *Phoenoteleia* Kieffer (Hymenoptera, Scelionidae, Scelioninae). *Journal of Hymenoptera Research* 87, 575–611. doi: 10.3897/jhr.87.59794

Lahey, Z., Chen, H., Dowton, M., Austin, A.D. and Johnson, N.F. (2023) The genome of the egg parasitoid *Trissolcus basalis* (Wollaston) (Hymenoptera, Scelionidae), a model organism and biocontrol agent of stink bugs. *Journal of Hymenoptera Research* 95, 31–44.

Lê, X.-H. (1986) A new species of the genus *Encyrtoscelio* Dodd 1914 (Hym. Scelionidae, Scelioninae) from Vietnam. *Tap Chi Sihn Hoc* 8, 40–41 [in Vietnamese].

Lê, X.-H. (2000) Egg-parasites of family Scelionidae (Hymenoptera). *Fauna of Vietnam*, Vol. 3. Science and Technics Publishing House, Hanoi, 386 pp. [in Vietnamese].

Leu, S.K., Sin-Wei, G. and Ye-min, L. (1982) Ontogeny and biology of *Platygaster oryzae*, a gregarious endoparasite of the rice gall midge. *Acta Entomologia Sinica* 25, 373–381.

Liao Y.-L., Yang, B., Xu, M.-F., Lin, W., Wang, D.-S., Chen, K.-W. and Chen, H.-Y. (2019) First report of *Telenomus remus* parasitizing *Spodoptera frugiperda* and its field parasitism in southern China. *Journal of Hymenoptera Research* 73, 95–102. doi: 10.3897/jhr.73.39136

Masner, L. (1972) The classification and interrelationships of Thoronini (Hymenoptera: Proctotrupoidea, Scelionidae). *The Canadian Entomologist* 104, 833–849. doi: 10.4039/Ent104833-6

Masner, L. (1976) Revisionary notes and keys to world genera of Scelionidae (Hymenoptera: Proctotrupoidea). *Memoirs of the Entomological Society of Canada* 97, 1–87.

Masner, L. (1980) Key to genera of Scelionidae of the Holarctic Region, with descriptions of new genera and species (Hymenoptera: Proctotrupoidea). *Memoirs of the Entomological Society of Canada* 113, 1–54.

Masner, L. (1975) Two new sibling species of *Gryon* Haliday (Hymenoptera, Scelionidae), egg parasites of blood-sucking Reduviidae (Heteroptera). *Bulletin of Entomological Research* 65, 209–213.

Masner, L. and Huggert, L. (1989). World review and keys to genera of the subfamily Inostemmatinae with reassignment of the taxa to the Platygastrinae and Sceliotrachelinae (Hymenoptera: Platygastridae). *Memoirs of the Entomological Society of Canada* 121(147), 1–216. doi: 10.4039/entm121147fv

Mikó, I., Vilhelmsen, L., Johnson, N.F., Masner, L. and Pénzes, Z. (2007) Morphology of Scelionidae (Hymenoptera: Platygastroidea): head and mesosoma. *Zootaxa* 1571, 1–78.

Mikó, I., Masner, L. and Deans, A.R. (2010) World revision of *Xenomerus* Walker (Hymenoptera: Platygastroidea, Platygastridae). *Zootaxa* 2708, 1–73.

Mikó, I., Masner, L., Ulmer, J.M., Raymond, M., Hobbie, J., Tarasov, S., Margaría, C.B., Seltmann, K.C. and Talamas, E.J. (2021) A semantically enriched taxonomic revision of *Gryonoides* Dodd, 1920 (Hymenoptera, Scelionidae), with a review of the hosts of Teleasinae. *Journal of Hymenoptera Research* 87, 523–573. doi: 10.3897/jhr.87.72931

Mikó, I., Raymond, M. and Talamas, E.J. (2021) New family-level characters for Platygastroidea. In: Lahey, Z. and Talamas, E. (Eds) Advances in the Systematics of Platygastroidea III. *Journal of Hymenoptera Research* 87, 235–249. https://doi.org/10.3897/jhr.87.72906

Mineo, G., O'Connor, J.P. and Ashe, P. (2009) A world revision of *Tiphodytes* Bradley, 1902 (Hym. Platygastroidea: Scelionidae). *Entomologist's Monthly Magazine* 145, 227–245.

Murphy, N.P., Carey, D., Castro, L.R., Dowton, M. and Austin, A.D. (2007) Phylogeny of the platygastroid wasps (Hymenoptera) based on sequences from the 18S rRNA, 28S rRNA and cytochrome oxidase I genes: implications for the evolution of the ovipositor system and host relationships. *Biological Journal of the Linnean Society* 91, 653–669.

Nam, S., Chen, H.Y., Talamas, E.J., Lee, G.S., Dong, W., Sun, L.J. and Lee, S. (2020) *Phanuromyia ricaniae* Nam, Lee and Talamas sp. nov. (Hymenoptera: Scelionidae) reared from the eggs of *Ricania shantungensis* Chou & Lu (Hemiptera: Ricaniidae) in Asia. *Zootaxa* 4890, 109–118. doi: 10.11646/ zootaxa.4890.1.6

Polaszek, A. (2009) *Masnerium wellsae* gen. nov., sp. nov. (Hymenoptera, Platygastridae, Sceliotrachelinae) a parasitoid of *Aleuroduplidens wellsae* Martin (Hemiptera, Aleyrodidae) in Australia. *ZooKeys* 20, 119–125. doi: 10.3897/ zookeys.20.189

Popovici, O.A., Mikó, I., Seltmann, C.K. and Deans, A.R. (2014) The maxillo-labial complex of *Sparasion* (Hymenoptera, Platygastroidea). *Journal of Hymenoptera Research* 37, 77–111.

Qodir, H.A., Maryana, N. and Pudjianto (2017) Biology of *Scelio pembertoni* Timberlake (Hymenoptera: Scelionidae) on eggs of *Oxya japonica* (Thunberg) (Orthoptera: Acrididae). *Indonesian Journal of Entomology* 14, 58–68 [in Indonesian].

Rajmohana. K., Sachin, J.P., Talamas, E.J., Shamyasree, M.S., Jalali, S.K. and Rakshit, O. (2019) *Paratelenomus anu* Rajmohana, Sachin & Talamas (Hymenoptera, Scelionidae): description and biology of a new species of phoretic egg parasitoid of *Megacopta cribraria* (Fab.) (Hemiptera, Plataspidae). *Journal of Hymenoptera Research* 73, 103–123. doi: 10.3897/jhr.73.34262

Rothschild, G.H.L. (1970) *Gryon flavipes* (Ashmead) [Hymenoptera, Scelionidae], an egg-parasite of the rice earbug *Leptocorisa oratorius* (Fabricius) [Hem. Alydidae]. *Entomophaga* 15, 15–10.

Saran, R., Rani, W.B., Kalyanasundaram, M, and Vanniarajan, C. (2016) Varietal influence and off-season occurrence of Asian rice gall midge *Orseolia oryzae* (Wood-Mason) and its parasitoid *Platygaster oryzae* (Cameron). *Advances in Life Sciences* 5, 3519–3523.

Siddiqui, R.K., Irshad, M. and Mohyuddin, A.I. (1986) Digest: *Scelio* spp. as biocontrol agents of acridids. *Biocontrol News and Information* 7, 69–76.

Stevens, N.B. and Austin, A.D. (2007) Systematics, distribution and biology of the Australian 'micro-flea' wasps, *Baeus* spp. (Hymenoptera: Scelionidae): parasitoids of spider eggs. *Zootaxa* 1499, 1–45.

Taekul, C., Johnson, N.F., Masner, L., Rajmohana, K. and Chen, S.-P. (2008) Revision of the world species of the genus *Fusicornia* Risbec (Hymenoptera: Platygastridae, Scelioninae). *Zootaxa* 1966, 1–52.

Taekul, C., Johnson, N.F., Masner, L., Polaszek, A. and Rajmohana, K. (2010) World species of the genus *Platyscelio* Kieffer (Hymenoptera, Platygastridae). *ZooKeys* 50, 97–126. doi: 10.3897/zookeys.50.485

Taekul, C., Valeri, A.A., Austin, A.D., Klompen, H. and Johnston, N.F. (2014) Molecular phylogeny of telenomine egg parasitoids (Hymenoptera: Platygastridae s.l.: Telenominae): evolution of host shifts and implications for classification. *Systematic Entomology* 39, 24–35.

Talamas, E.J., Johnson, N.F., Shih, C., Ren, D. (2019) Proteroscelliopoidae. A new family of Platygastroidea from Cretaceous amber. In: Talamas, E. (Ed.) Advances in the Systematics of Platygastroidea II. *Journal of Hymenoptera Research* 73, 3–38. https://doi.org/10.3897/jhr.73.32256

Talamas, E.J. and Pham, H.-T. (2017) An online photographic catalog of Platygastroidea (Hymenoptera) in the Institute of Ecology and Biological Resources (Hanoi, Vietnam), with some taxonomic notes. *Journal of Hymenoptera Research* 56, 225–239. doi.: 10.3897/jhr.56.10214

Talamas, E.J., Masner, L. and Johnson, N.F. (2011) Revision of the *Paridris nephta* species group (Hymenoptera, Platygastroidea, Platygastridae). *ZooKeys* 133, 49–94. doi: 10.3897/zookeys.133.1613

Talamas, E.J. and Buffington, M. (2014) Updates to the nomenclature of Platygastroidea in the Zoological Institute of the Russian Academy of Sciences. *Journal of Hymenoptera Research* 39, 99–117. doi: 10.3897/JHR.39.7698

Talamas, E.J., Johnson, N.F. and Buffington ML (2015) Key to Nearctic species of *Trissolcus* Ashmead (Hymenoptera, Scelionidae), natural enemies of native and invasive stink bugs (Hemiptera, Pentatomidae). *Journal of Hymenoptera Research* 43, 45–110. doi: 10.3897/JHR.43.4661

Talamas, E.J., Johnston-Jordan, D. and Buffington, M.E. (2016a) *Calliscelio* Ashmead expands (Hymenoptera: Scelionidae). *Proceedings of the Entomological Society of Washington* 118, 404–423.

Talamas, E.J., Mikó, I. and Copeland, R.S. (2016b) Revision of *Dvivarnus* (Scelionidae, Teleasinae). *Journal of Hymenoptera Research* 49, 1–23. doi: 10.3897/JHR.49.7714

Talamas, E.J., Buffington, M.L. and Hoelmer, K. (2017) Revision of Palearctic *Trissolcus* Ashmead (Hymenoptera, Scelionidae). *Journal of Hymenoptera Research* 56, 3–185. doi: 10.3897/jhr.56.10158

Valerio, A.A., Masner, L., Austin, A.D. and Johnson, N.F. (2009) The genus *Neuroscelio* Dodd (Hymenoptera: Platygastridae s.l.) reviewed: new species, distributional update, and discussion of relationships. *Zootaxa* 2306, 29–43.

Valerio, A.A., Masner, L. and Austin A.D. (2010). Systematics of *Cyphacolus* Priesner (Hymenoptera: Platygastridae s.l.), an Old World genus of spider egg parasitoid. *Zootaxa* 2645, 1–48. doi:10.11646/zootaxa.2645.1.1

Valerio, A.A., Austin A.D., Masner, L. and Johnson, N.F. (2013) Systematics of Old World *Odontacolus* Kieffer s.l. (Hymenoptera, Platygastridae s.l.): parasitoids of spider eggs. *ZooKeys* 314, 1–151.

van Noort, S. and Johnson, N.F. (2009) New species of the plesiomorphic genus *Nixonia* Masner (Hymenoptera, Platygastroidea, Platygastridae, Scelioninae) from South Africa. *ZooKeys* 20, 31–51.

Veenakumari, K., Rajmohana, K. and Prashanth, M. (2012) Studies on phoretic Scelioninae (Hymenoptera: Platygastridae) from India along with description of a new species of *Mantibaria* Kirby. *Linzer Biologische Beiträge* 44, 1715–1725.

Veenakumari, K., Buhl, P.N. and Mohanraj, P. (2018) A new species of *Synopeas* (Hymenoptera: Platygastridae) parasitizing *Pauropsylla* cf. *depressa* (Psylloidea: Triozidae) in India. *Acta Entomologica Musei Nationalis Pragae* 58, 137–141. doi: 10.2478/aemnp-2018-0011

Vlug, H.J. (1995) Catalogue of the Platygastridae (Platygastroidea) of the world (Insecta: Hymenoptera). *Hymenopterorum Catalogus* 19, 1–168.

Wojcik, B., Whitcomb, W.H. and Habeck, D.H. (1976) Host range testing of *Telenomus remus* (Hymenoptera: Scelionidae). *Florida Entomologist* 59, 195–198. doi: 10.2307/3493972.

Yan, C.-J., Talamas, E., Lahey, Z. and Chen, H.-Y. (2022) *Protelenomus* Kieffer is a derived lineage of *Trissolcus* Ashmead (Hymenoptera, Scelionidae), with comments on the evolution of phoresy in Scelionidae. *Journal of Hymenoptera Research* 94, 121–137.

Yoder, M.J., Valerio, A.A., Masner, L. and Johnson, N.F. (2009) Identity and synonymy of *Dicroscelio* Kieffer and description of Axea, a new genus from tropical Africa and Asia (Hymenoptera: Platygastroidea: Platygastridae). *Zootaxa* 2003, 1–45.

19

Cynipoidea – Gall Wasps and Their Kin

Abstract
The superfamily Cynipoidea includes phytophagous, gall-forming species, their inquilines and a number of families that are strictly parasitoids. A simplified key of females of major groups of S.E. Asian Cynipoidea is provided and the biology of all groups are described. Representatives of most groups are illustrated photographically. Relevant identification works are cited.

Familiar to many in the temperate zone are the gall-forming Cynipidae and the superfamily is often thought of as 'the gall wasps'. However, only the Cynipidae form plant galls, and not even all of those. Although all are phytophagous some are inquilines in galls. The remaining families (Austrocynipidae, Figitidae, Ibaliidae and Liopteridae) are all parasitoids. Members of the Ibaliidae, Liopteridae and the rare, entirely Australian Austrocynipidae are often referred to as macrocynipoids because most of them are markedly larger than the remainder. Cynipidae themselves are not common in the tropics, their main host plants being oaks (Fagaceae) and roses (Rosaceae).

Identification to all families, subfamilies and tribes of the world fauna are keyed, illustrated in colour and discussed by Buffington *et al.* (2020). Earlier keys, e.g. Gauld and Bolton (1988) and Ritchie (1993), included Charipidae ad Eucoilidae as separate families, whereas they are now treated as clades within an enlarged Figitidae. For various reasons, it is often easier to identify specimens direct to lower taxa (e.g. subfamilies and tribes) rather than attempting highly complicated couplets allowing for all the variation within Figitidae and Cynipidae. Hence, Buffington *et al.*'s (2020) key runs to 34 couplets compared with just seven in Ritchie (1993). Beginners might therefore want to try running specimens through both (hopefully getting compatible answers). Van Noort *et al.* (2015) noted that quite a few genera and even species of figitids are shared between the Afrotropical and Oriental Regions, but the fauna of S.E. Asia is still very poorly studied. However, this shared component means that van Noort *et al.*'s nicely illustrated keys should also be useful in the region, with the obvious caveats.

Morphological Terminology

There are not many specific morphological terms used for this superfamily. Liu *et al.* (2007) provided a fairly detailed set of morphological drawings for the relatively basal Liopteridae. Features of the scutellar sulcus and scutellum are often important, as are the relative lengths of the metasomal tergites medially. In the traditional 'Eucoilidae' the scutellum has a cup-like structure medially. The scutellar sulcus is usually in the form of a pair of foveae (depressions) whose form is important for lower-level identification.

Vein RS&M is often referred to as the basal vein. The wing venation of most of the smaller species is fairly conserved and distinctive (Fig. 19.1); there is no true pterostigma (except in the extralimital Austrocynipidae), and in most, fore wing veins r-rs and 3RS form a conspicuous triangle with the fore wing margin (Fig. 19.1 b; see also Fig. 19.4); the radial vein is normally absent at the anterior of this cell (Fig. 19.1 b). Buffington and Sandler (2011) showed that, except in large species and ones with dark wings (Fig. 19.1 a). The wings show very distinct and phylogenetically informative interference patterns (see Fig. 19.5 g) – a nice source of taxonomic characters but not often used.

The metasoma is nearly always rather strongly laterally compressed. The first segment is usually referred to as the petiole. The ovipositor is seldom seen exserted and is largely internalised, often coiled inside metasoma (Fergusson, 1988).

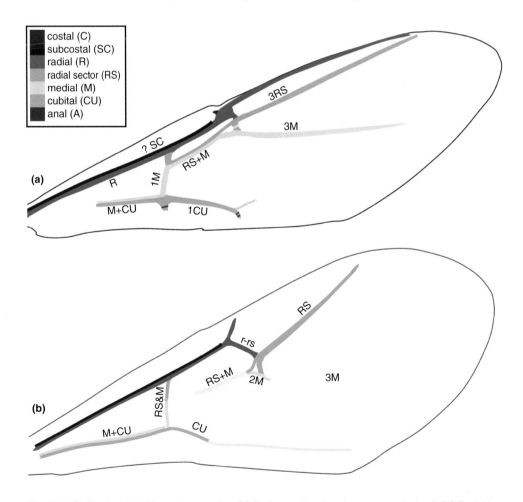

Fig. 19.1. Stylised cynipoid fore wing venation: **(a)** *Ibalia* sp., showing elongate marginal cell; **(b)** Cynipidae (Cynipini) showing rather triangular marginal cell.

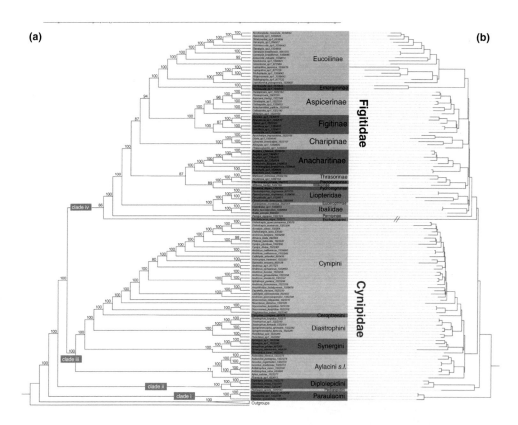

Fig. 19.2. Maximum-likelihood tree for Cynipoidea based on phylogenomic analysis. Current subfamily (for Figitidae) and tribe (for Cynipidae) assignments are indicated. (Source: from Blaimer *et al.* (2020) reproduced under terms of Creative Commons Attribution Licence CC-BY-4.0.)

Phylogeny

There have been several phylogenetic analyses of the Cynipoidea, starting with purely morphological ones (Ronquist, 1994, 1995a, 1999), then total evidence combining morphology with Sanger-sequenced gene sequences (Ronquist *et al.*, 2015), ones based on Sanger-sequenced genes and, most recently, a detailed phylogenomic study (Fig. 19.2) (Blaimer *et al.*, 2020).

Simplified Key to Females of Major Groups of S.E. Asian Cynipoidea

1	Fore wing marginal cell very elongate, approximately 9× longer than wide (Fig. 19.1 a); (very rare) ....**Ibaliidae**	
-	Fore wing marginal cell very elongate, less than 6× longer than wide (Figs 19.1 b, 19.4 b, f).. 2	
2(1)	Apparent metasomal tergite 4 or 5 the longest mediodorsally (Fig. 19.5) ....**Liopteridae**	
-	Metasomal tergites 2 or 3 the longest mediodorsally (Fig. 19.4) 3	
3(2)	Fore wing apically bilobed (indented) (Fig. 19.4 f) (rare).. ....**Figitidae: Emargininae**	
-	Fore wing without apical incision (Fig. 19.4 a, b) 4	

4(3)		Scutellum with a dorsal teardrop-shaped, posteriorly rounded protuberance which has a central depression which is often smooth, never with a median carina (Figs 19.4 d, e, 19.5 a, b) .... **Figitidae: Eucoilinae**
-		Scutellum without such a structure, sometimes protruding, if with a median area it is defined then this has a median occasionally as a spine (Fig. 19.5 e) 5
5(4)		Fore wing vein RS+M present and pointing basally towards the middle of vein RS&M (Fig. 19.3 b) 6
-		Fore wing vein RS+M sometimes absent or just represented by a narrow ridge in the wing membrane, when visibly indicated pointing basally towards the posterior end of RS&M (Fig. 19.5 f, g) 7
6(5)		Head, pronotum and mesoscutum variously sculptured, and notauli present (very rare, except mountains where host trees occur) ....**Cynipidae**
-		Head, pronotum and mesoscutum shiny, more or less smooth, and notauli............. .... **Figitidae: Charipinae**
7(5)		First metasomal segment (petiole) elongate, considerably longer than wide, nearly as long as the second or longer (Fig. 19.4 a); mouth region narrow and mandibles broadly overlapping; scutellum sometimes formed into spine projecting dorsally... ....**Figitidae: Anacharatinae**
-		First metasomal segment (petiole) short, usually quadrate or shorter; mouth region wider and mandibles only slightly overlapping; scutellum sometimes formed into spine projecting posteriorly (Fig. 19.5 e) ....**Figitidae: Aspicerinae**

Cynipidae

Most of the approximately 1400 described species of Cynipidae are gall inducers; however, around 15% develop as inquilines inside galls of other cynipids and these belong to the tribe Synergini (Pujade-Villar *et al.*, 2003; Csóka *et al.* 2005). The 'true' gall-forming cynipids are divided among 15 tribes, the great majority of which are predominantly Palaearctic, Nearctic or Neotropical (Pénzes *et al.*, 2018; Buffington *et al.*, 2020). Pénzes *et al.* (2018) reviewed data on the occurrence of the family in the Oriental Region where it is represented by the tribes Cynipini and Synergini.

Cynipini

The Cynipini is associated with oaks and chestnuts (Fagaceae) and very diverse in the East Palaearctic (Abe *et al.* 2007). Lack of sampling in S.E. Asia meant that the first species of Cynipini in the region were not discovered until 2014, with the description of two new species from Vietnam (Abe *et al.*, 2014): *Plagiotrochus indochinensis* and *Dryocosmus okajimai*. As expected, these gall wasps are associated with the tropical oak genus *Cyclobalanopsis*. Subsequently, Ide *et al.* (2022) recorded *Andricus mukaigawae* as a gall-former on *Quercus griffithii* in the Chin Hills of north-western Myanmar. *A. mukaigawae* was originally described from Japan but also occurs in the Russian Far East, Korea, mainland China and north-eastern India. We have collected a species *Dryocosmus* from northern Thailand (Fig. 19.3). As northern Myanmar is home to the gall wasp host tree genera *Castanea* (one species), *Castanopsis* (seven species), *Lithocarpus* (17 species) and *Quercus* (21 species), it would seem likely that other species of Cynipini must also occur there. Many Cynipini have the hypopygium ventromedially extended into a spine.

With only a couple of known exceptions, all Cynipini go through alternating sexual and asexual generations; that is, a generation of all females (agamic generation) lays unfertilised eggs that develop to produce both males and females of the sexual generation, which mate, producing the next generation of agamic females. The different generations are usually morphologically quite different and so associating them nowadays is normally done with the aid of DNA barcoding. The different generations usually produce different types of galls. Very

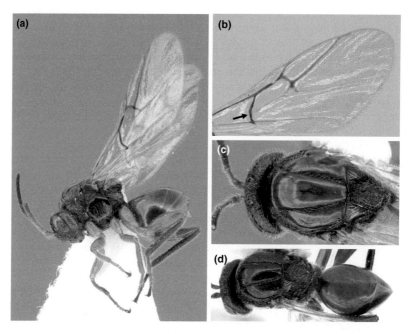

Fig. 19.3. Female *Dryocosmus* sp. (Cynipini) from Doi Phu Kha National Park, northern Thailand: **(a)** habitus; **(b)** fore wing, arrow pointing to where vein RS+M would intersect RS&M (basal vein); **(c)** top of head to scutellum detail; **(d)** dorsal habitus showing large 2nd metasomal tergite 2.

little is known of the biology of most species, including basically all the ones likely to be found in S.E. Asia.

The plant gall is actually not induced by female venom or ovarian secretions, but rather by secretions (saliva) of the larva, so the gall only starts to develop once the gall wasp egg has hatched. However, recent consideration of venom gland morphology does suggest that the female's venom plays some role in gall formation (Guiguet *et al.*, 2023).

Tang *et al.* (2016) provided a key to the majority of *Dryocosmus* species known from Asia, including *D. okajimai*.

Synergini

Six species of Synergini have been recorded from S.E. Asia including Papua New Guinea (Weld, 1926; Nieves-Aldrey and Butterill, 2014; Ide *et al.* 2020; Lobato-Vila *et al.*, 2021, 2022): two species of ?*Saphonecrus* from the Philippines, *Lithosaphonecrus mindatus* (Myanmar), *L. papuanus* (Papua New Guinea), *L. ?serratus* (Philippines), and *L. vietnamensis* (Vietnam). Undoubtedly more are present but as yet unrecorded. These species were mostly previously placed in the genus *Saphonecrus* but that has now been shown to be polyphyletic and most of the East Asian species are transferred to *Lithosaphonecrus* (Tang *et al.*, 2015). The majority of *Lithosaphonecrus* species are extralimital, especially in China, Taiwan and Japan. An updated key to the species was provided by Melika *et al.* (2021).

Synergines are obligate inquilines of other gall wasps. This is a form of cleptoparasitism (also called agastoparasitism) (Ronquist, 1994); whilst the benefits are entirely unilateral, there appears to be no detrimental effects on the gall-forming host/partner (Pénzes *et al.*, 2012). They may be distinguished from Cynipini in having the anteromedial part of the pronotum being relatively longer in dorsal view.

Figitidae

In terms of number of species, the Figitidae is by far the largest family of Cynipoidea.

There are more than 1700 described species classified among some 157 genera. Buffington et al. (2007) presented a molecular phylogeny of the family based on three gene regions. Many earlier publications treated the Eucoilinae and Charipinae as separate families (e.g. Gauld and Bolton, 1988), but it is now clear that they are derived clades within an enlarged Figitidae. This family now comprises a total of 11 (or 12) subfamilies, but of those, only five have been recorded from S.E. Asia. Figitids usually have the anterior portion of the pronotum differentiated (sharply defined laterally, sometimes by a carina) (Figs 19.4 a, b, 19.5 c, d), called the pronotal plate, but this can be rather weak in some groups.

All species, as far as it is known, are endoparasitoids of various holometabolous insects. Most are primary parasitoids but some, especially among Charipinae, are hyperparasitoids. All reliable host records were summarised by Buffington et al. (2012).

Anacharitinae

The Anacharitinae is characterised by these following characters: rounded and continuous pronotal plate, mandibles broadly overlapping and triangular-shaped head in frontal view (Mata-Casanova et al., 2014). Currently, there are nine genera belonging to this subfamily: *Acanthaegilips*, *Acanthaegilopsis*, *Aegilips*, *Anacharis* (Fig. 19.4 a), *Calofigites*, *Hexacharis*, *Proanacharis*, *Solenofigites* and *Xyalaspis*. Of these, *Anacharis* and *Xyalaspis* are represented in the region.

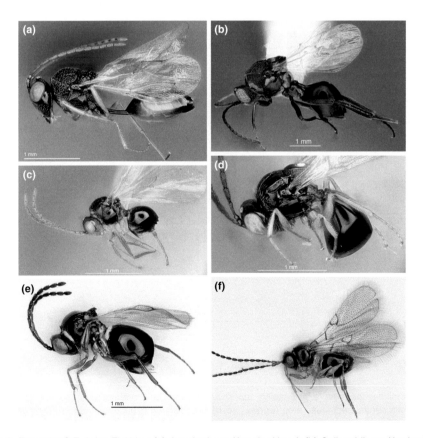

Fig. 19.4. Exemplar S.E. Asian Figitidae: **(a)** *Anacharis* sp. (Anacharitinae); **(b)** *Callaspidia* sp. (Aspicerinae); **(c)** *Alloxysta* (Charipiinae); **(d)** *Afrostilba* sp. (Eucoilinae); **(e)** *Gastraspis* sp. (Eucoilinae); **(f)** *Thoreauella* sp., from Thailand (Emargininae). (Source: a-e photographs by Matt Buffington, reproduced with permission.)

The biology of anacharitines is very different from that of other figitids in that they are larval parasitoids of Hemerobiidae (Neuroptera) that feed on aphid-feeding hosts (Díaz, 1979; Miller and Lambdin, 1985; Cave and Miller, 1987; Mata-Casanova *et al.*, 2014).

Mata-Casanova *et al.* (2014) revised the Asian species of genus *Xyalaspis*; of the six species reported from Asia (mostly Japan, China and India), two are discovered in Thailand. Mata-Casanova *et al.* (2014) revised Chinese Anachatarinae and included a key to the genera and species which would probably be helpful.

Aspicerinae

This subfamily is considered to be monophyletic, characterized by a facial depression, strongly ligulate (strap-shaped) third abdominal tergum, and a unique pronotal plate formed by lateral fusion of protruding dorsal and ventral elements (Fig. 19.5 d). They are

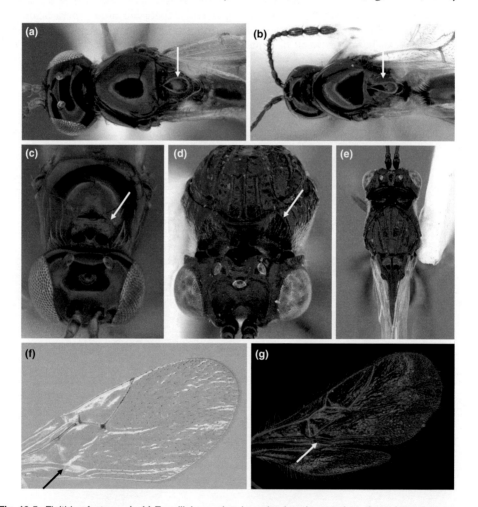

Fig. 19.5. Figitidae features: **(a, b)** Eucoilini spp. showing raised oval to teardrop-shaped structure on scutellum with central depressed area (arrows); **(c, d)** Eucoilini and Aspecerini, respectively, views of anterior mesosoma showing pronotal plate (arrows); **(e)** *Aspicera* sp. (Aspicerini) showing posteriorly directed scutellar spine; **(f, g)** fore wings of Eucoilini spp. showing line of missing vein RS+M indicated by crease in wing membrane, arrows indicating intercept with posterior end of vein RS&M (basal vein), and note interference pattern in **(g)** which was imaged against a black background.

parasitoids of aphid-predating Diptera larvae of Syrphidae and Chamaemyiidae (Ronquist, 1999). Since aphids are uncommon in the tropics, so are they. Currently there are about 99 species in eight genera, viz. *Anacharoides, Aspicera, Balna, Callaspidia* (Fig. 19.4 b), *Xyalophora* (= *Ceraspidia*), *Omalaspis, Paraspicera* and *Prosaspicera* plus, *Melanips* (previously in Figitinae) was consistently recovered as sister group to Aspicerinae by Blaimer *et al.* (2020). Of these, *Aspicera, Calluspidea, Melanips* and *Prosaspicera* occur in S.E. Asia. Some members of the group have large scutellar spines. *Aspicera* species have a particularly large scutellar spine (Fig. 19.5 e).

Ros-Farré (2007) provided a key to all aspicerine genera. Ros-Farré and Pujade-Villar (2006) revised *Prosaspicera*; Ros-Farré and Pujade-Villar (2009) revised *Callaspidia*.

Charipinae

In older literature, usually given family level status. The Charipinae is a small group of hyperparasitoids of aphids and psyllids (Hemiptera) (Menke and Evenhuis, 1991), including many pest species, and are therefore of economic importance, mostly in temperate regions.

A total of 281 species of Charipinae have been described worldwide in eight genera (Ferrer-Suay *et al.*, 2014a,b). Ferrer-Suay *et al.* (2012) catalogued the world species and provided a key to genera. The S.E. Asian species of Charipinae known to date are given in Table 19.1. However, various other species occur in adjacent regions, and since their hosts also occur, albeit relatively uncommonly or at altitude,

Table 19.1. Country distributions of Charipinae in S.E. Asia.

Country	S.E. Asian species records
Indonesia	*Dilyta orientalis*
Malaysia	*Alloxysta castanea, A. sawoniewiczi*
Thailand	*Alloxysta arcuata, A. asiatica, A. areeluckator, A. brevis, A. melanogaster, A. paretasmartinezi, A. petchabunensis, A. pilosa, A. pusilla, A. sawoniewiczi, Phaenoglyphis chiangmaiensis, P. wongchaiensis*

we must expect others to occur. *Dilyta* was recorded from S.E. Asia (Indonesia: Java) for the first time by Ferrer-Suay *et al.* (2011).

Species of *Alloxysta* (Fig. 19.4 c), *Phaenoglyphis* and *Lytoxysta* previously treated as Alloxystini are endoparasitic hyperparasitoids. Their primary hosts are aphidiine braconids and members of the chalcidoid genus *Aphelinus*, which are all endoparasitoids of aphids (Hemiptera: Aphididae). *Dilyta* and *Apocharips* are endoparasitic hyperparasitoids of Encyrtidae (Chalcidoidea) that are themselves endoparasitoids of psyllid hosts (Hemiptera: Psyllidae) (Fergusson, 1986; Menke and Evenhuis, 1991).

Ferrer-Suay *et al.* (2013, 2014a, b) described, recorded and keyed a number of species from S.E. Asia. Ferrer Suay *et al.* (2019) provided a key to world species of the whole group; however, identification of Charipinae to species level is extremely difficult because of their generally small size and a dearth of interspecifically varying morphological characters. Many 'species' have been recognised largely on the basis of colour characters and these are unreliable for species recognition (Ferrer-Suay *et al.*, 2012). Because of this it is simply impossible to identify most *Alloxysta* up to the species level, and as a consequence most ecologists are forced to stop at a generic level (e.g. El-Heneidy *et al.*, 1987; Müller *et al.*, 1997; Suay *et al.*, 1998; Talebi *et al.*, 2009).

Emargininae

Rather uncommon but distinctive group with the apical margin of the fore wing indented (Fig. 19.4 f) (Paretas-Martínez *et al.*, 2012). They are believed to be associated with ant nests probably as parasitoids of myrmecophilous Diptera larvae. Adults of extralimital species have been collected from refuse deposits of army ants, using Berlese funnels (Weld, 1960), and also found in *Camponotus* ant nests (Díaz, 1978).

Eucoilinae

Eucoilines for which biology is known are koinobiont endoparasitoids of cyclorrhaphous

Diptera larvae, sometimes ovipositing in early instars but the offspring emerging from the host puparium (Buffington and Forshage, 2014). The Eucoilinae is the most species-rich subfamily of Figitidae with almost 1000 described species. However, despite their economic importance they are a generally poorly known group everywhere, especially in the tropics and perhaps particularly in S.E. Asia. It has been estimated that only 5% to (at most) 20% of species have currently been described (Nordlander, 1984). Their generic classification is not well studied and numerous new genera await description. Many genera have wide geographical distributions and probably most instances of apparent endemism are just consequences of having failed to find and identify the genera elsewhere where they occur.

For a period, the extralimital *Leptopilina heterotoma* and *L. boulardi*, both parasitoids of *Drosophila*, were investigated quite a lot in European laboratories, and in particular their mechanisms of host immuno-suppression (Rizki and Rizki, 1984, 1990a, b, 1991). Without doubt application of modern molecular techniques would provide vastly improved understanding.

A few eucoilines have been investigated in terms of their potential to control pest species of Diptera and quite a few are important in controlling a wide range of leaf-mining pests, including Agromyzidae (particularly *Liriomyza* spp.) and Tephritidae (Nordlander, 1978; Johnson, 1987; Wharton *et al.*, 1998; Buffington, 2010a, 2010b).

A morphological phylogeny of the subfamily was presented by Fontal-Cazalla *et al.* (2002).

The regional fauna includes *Afrostilba* (Fig. 19.4 d), *Aganaspis* and *Gastraspis* (Fig. 19.4 e). Van Noort *et al.* (2015) noted that *Afrostilba* is often very abundant in bulk samples. *Aganaspis* is mainly a S.E. Asian genus. *A. daci*, originally from Malaysia and Borneo, was introduced into the Hawai'ian Islands as a potential biocontrol agent against the fruitfly pest *Bactrocera dorsalis* (Tephritidae) (Clausen *et al.*, 1965), and subsequently into Florida and Costa Rica to control *Anastrepha* spp. (Wharton *et al.*, 1981).

Figitinae

This is a largely Holarctic, Neotropical and Afrotropical subfamily. *Neralsia* and *Xyalophora* are both widely distributed but have not yet been recorded from the Oriental Region (Buffington *et al.*, 2020). They are parasitoids of cyclorhaphous Diptera.

Ibaliidae

This very distinctive, principally north temperate family comprises two genera, *Ibalia* and *Heteribalia* (Liu and Nordlander, 1994), one of each occurring in the region: *Heteribalia confluens* from northern Vietnam and *Ibalia kalimantanica* from Indonesia (Borneo) (Liu, 1998). The two genera are quite speciose in China, Japan and Taiwan so it is quite likely that other species may occur in the north of the region or at high elevation. In addition, *Eileeniella* is a very strange-looking ibaliid from New Guinea (Fergusson, 1992); they are quite weird-looking and so they might not often find their way into a cynipoid collection, but perhaps be placed among the ichneumonoids (M. Buffington, pers. comm.).

Ibaliids are rather large wasps with body lengths up to 3 cm, and strongly laterally compressed metasomas. Host records include larvae of several woodwasp species (Siricidae), which are uncommon in the tropics. What is known of their biology is largely based on European species (Chrystal, 1930; Spradbery, 1970). They are koinobiont endoparasitoids. The female inserts her ovipositor down the oviposition shaft bored by the host, and deposits her egg into the haemocoel of the host embryo in its egg or its early-instar larva. The 3rd instar larva chews its way out of the host, then continues feeding on it from the outside (see *External Feeding Phase*, Chapter 2, this volume). The parasitoid larva then moults to a fourth instar, which does not feed, and then pupates without making a cocoon. Interestingly, hosts that are parasitised make shorter burrows than unparasitised ones, and bore closer to the surface of the wood before they are killed. The strong mesoscutal rugae of the adult siricid are an adaptation to facilitate chewing its way out of the host substrate when it emerges, by helping to brace the thorax so that the wasp can gain more purchase for chewing.

Since *Sirex noctilio* is a serious forestry pest in many parts of the temperate world, Ibalia has been used in several classical biological control programmes (Fischbein and Corley, 2015). For the same reason its ecology has been quite well investigated (e.g. Fischbein *et al.*, 2012). Siricid wood wasps carry symbiotic fungi and inject fragments of them along with their eggs during oviposition. The fungi help to kill the tree and digest the wood. Both woodwasps and Ibaliidae are attracted to the volatiles emitted from infected wood (Spradbery, 1974; Martinez *et al.*, 2006), which has been termed a double-edged sword by foresters (Faal *et al.*, 2021). Given what is known about these temperate ibaliids and the importance of volatiles from *Amylostereum* and *Cerrena* fungi (in coniferous and broadleaf wood, respectively) it would be interesting to set out traps baited with these in S.E. Asian higher-altitude forests to see whether ibaliid diversity might be higher than currently appreciated.

Liopteridae

Liopterids have a rather distinctive appearance, are generally rather large, and some are brightly coloured, and are sometimes unrecognised as being cynipoids in museum Hymenoptera. There is only anecdotal evidence that they are parasitoids of wood-boring insect larvae (Ronquist, 1995b; Buffington *et al.*, 2012) but there are no definitive host records. As with some other xylophilic groups (see Ibaliidae above), the mesoscutum is transversely striate (Fig. 19.6 a), a presumed adaptation for bracing themselves against the walls of host borings to help them chew their way out.

Eleven genera and more than 70 species are known worldwide. Ronquist (1995b) presented a morphological phylogenetic analysis of the family. Four subfamilies are recognised: Dallatorrellinae, Mayrellinae, Oberthuerellinae and Liopterinae, of which only the first two occur in S.E. Asia.

Hedicke and Kerrich (1940) and Ronquist (1995b) revised the family. *Paramblynotus* (Fig. 19.6) is by far the most species-rich genus, and it has an enormous radiation in S.E. Asia. They appear to be quite seasonal and are generally collected by Malaise trapping. The genus was revised by Liu *et al.* (2007). The Dallatorrellinae includes two genera: *Mesocynips*, with one known species; and *Dallatorrella*, which is widely distributed in the region, being known in Australia, Borneo, China, Indonesia, Laos, West Malaysia, Papua New Guinea, Philippines, Sarawak and Singapore (Liu, 2009). Liu's phylogenetic analysis led to the conclusion that the Dallatorrellinae originated in S.E. Asia.

Liu *et al.* (2007) revised *Paramblynotus*, presented a morphological phylogeny, and provided a key to species groups and, for each, a key to species. Liu (2009) revised the Dallatorrellinae.

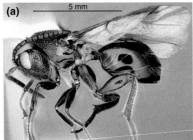

Fig. 19.6. Two species of *Paramblynotus* (Mayrellinae): **(a)** *P. grossus*; **(b)** sp. indet. from Thailand. (Source: (a) photograph by Matt Buffington, reproduced with permission.)

References

Abe, Y., Melika G. and Stone, G.N. (2007) The diversity and phylogeography of cynipid gallwasps (Hymenoptera: Cynipidae) of the Oriental and eastern Palearctic regions, and their associated communities, *Oriental Insects* 41, 169–212. doi: 10.1080/00305316.2007.10417504

Abe, Y., Ide, T., Konishi, K. and Ueno, T. (2014) Discovery of Cynipidae (Hymenoptera: Cynipoidea) from the Indochina Region, with description of three new species. *Annals of the Entomological Society of America* 107, 399–406. doi: 10.1603/AN13180

Blaimer, B.B., Gotzek, D., Brady, S.G. and Buffington, M.L. (2020) Comprehensive phylogenomic analyses re-write the evolution of parasitism within cynipoid wasps. *BMC Evolutionary Biology* 20, 155. doi: 10.1186/s12862-020-01716-2

Buffington, M.L. (2010a) A revision of *Ganaspidium* Weld, 1952 (Hymenoptera, Figitidae, Eucoilinae): new species, bionomics, and distribution. *Zookeys* 37, 81–101. doi: 10.3897/ zookeys.37.311

Buffington, M.L. (2010b) The description of *Banacuniculus* Buffington, new genus (Hymenoptera: Figitidae: Eucoilinae). *Journal of Hymenoptera Research* 19, 94–112.

Buffington, M.L. and Forshage, M. (2014) The description of *Garudella* Buffington and Forshage, new genus (Hymenoptera: Figitidae: Eucoilinae). *Proceedings of the Entomological Society of Washington* 116, 225–242.

Buffington, M.L. and Sandler, R.J. (2011) The occurrence and phylogenetic implications of wing interference patterns in Cynipoidea (Insecta : Hymenoptera). *Invertebrate Systematics* 25, 586–597.

Buffington, M.L. and Van Noort, S. (2012) Revision of the Afrotropical Oberthuerellinae (Cynipoidea, Liopteridae). *ZooKeys* 202, 1-154.

Buffington, M.L., Nylander, J.A.A. and Heraty, J.M. (2007) The phylogeny and evolution of Figitidae (Hymenoptera: Cynipoidea). *Cladistics* 2, 403–431. doi: 10.1111/j.1096-0031.2007.00153.x.

Buffington, M.L., Brady, S.G., Morita, S.I. and Van Noort, S. (2012) Divergence estimates and early evolutionary history of Figitidae (Hymenoptera: Cynipoidea). *Systematic Entomology* 37(2), 287–304.

Buffington, M.L., Forshage, M., Liljeblad, J., Tang, C.-T. and van Noort, S. (2020) World Cynipoidea (Hymenoptera): a key to higher-level groups. *Insect Systematics and Diversity* 4, 1–69.

Cave, R.D. and Miller, G.L. (1987) Notes on *Anacharis melanoneura* (Hymenoptera: Figitidae) and *Charitopes mellicornis* (Hymenoptera: Ichneumonidae) parasitizing *Micromus posticus* (Neuroptera: Hemerobiidae). *Entomologicak News* 98, 211–216.

Chrystal, R.N. (1930) Studies on *Sirex* parasites. The biology and post-embryonic development of *Ibalia leucospoides* Hochenw. (Hymenoptera-Cynipoidea). *Oxford Forestry Memoirs* 11, 1–63.

Clausen, C.P., Clancy, D.W. and Chock, Q.C. (1965) *Biological control of the oriental fruit fly (Dacus dorsalis Hendel) and other fruit flies in Hawaii*. United States Department of Agriculture Technical Bulletin 1322, 102 pp.

Csóka, G., Stone, G.N. and Melika, G. (2005) The biology, ecology and evolution of gall wasps. In: Raman, A., Schaeffer, C.W. and Withers, T.M. (eds) *Biology, Ecology and Evolution of Gall-inducing Arthropods*. Science Publishers, Inc., Enfield, New Hampshire, pp. 569–636.

Díaz, N.B. (1978) Estudio ecológico y sistemático de cinipoideos neotropicales IV (Hymenoptera: Cynipidae). *Neotropica* 24, 123–125.

Díaz, N.B. (1979) Neotropical Hymenoptera parasitizing Neuroptera and Cynipoidea (Hymenoptera). *Revista de la Sociedad Entomológica Argentina* 38, 21–28 [in Spanish].

El-Heneidy, A.H., Klausnitzer, B. and Richter, K. (1987) Beitrag zur Kenntnis der Parasitoide (Hymenoptera) von *Aphis fabae cirsiiacanthoides* Scop. im Gebiet von Leipzig. *Entomologische Nachrichten und Berichte* 31, 67–70.

Faal, H., Cha, D.H., Hajek, A.E. and Teale, S.A. (2021) A double-edged sword: *Amylostereum areolatum* odors attract both *Sirex noctilio* (Hymenoptera: Siricidae) and its parasitoid, *Ibalia leucospoides*. *Fungal Ecology* 54, 101108.

Fergusson, N.D.M. (1986) Charipidae, Ibaliidae & Figitidae. Hymenoptera: Cynipoidea. Charipidae, Ibaliidae & Figitidae. Hymenoptera: Cynipoidea. *Royal Entomological Society* 8(1c), 55.

Fergusson, N.D.M. (1988) A comparative study of the structures of phylogenetic significance of female genitalia of the Cynipoidea (Hymenoptera). *Systematic Entomology* 13, 12–30.

Fergusson, N. (1992) A remarkable new genus and species of Cynipoidea (Hymenoptera) from Papua New Guinea. *Journal of Natural History* 26, 659–662.

Ferrer-Suay, M., Paretas-Martínez, J., Selfa, J. and Pujade-Villar, J. (2011) First record of the genus *Dilyta* in Asia, with a description of a new species *Dilyta orientalis* (Hymenoptera: Cynipoidea: Figitidae: Charipinae). *Zoological Studies* 50, 230–234.

Ferrer-Suay, M., Paretas-Martínez, J., Selfa, J. and Pujade-Villar, J. (2012) Taxonomic and synonymic world catalogue of the Charipinae and notes about this subfamily (Hymenoptera: Cynipoidea: Figitidae). *Zootaxa* 3376, 1–92.

Ferrer-Suay, M., Selfa, J. and Pujade-Villar, J. (2013) Charipinae fauna (Hymenoptera: Figitidae) from Asia with a description of 11 new species. *Zoological Studies* 52, 41.

Ferrer-Suay, M., Selfa J. and Pujade-Villar, J. (2014a) New charipine wasps (Hymenoptera: Figitidae) from Thailand. *Oriental Insects* 48, 83–91. doi: 10.1080/00305316.2014.959784

Ferrer-Suay, M., Selfa, J. and Pujade-Villar, J. (2014b) New Australasian records of 'Alloxysta forster' (Hymenoptera: Cynipoidea: Figitidae: Charipinae) from the Canadian National Collection of Insects, Ottawa. *Australian Entomologist* 41, 91–106.

Ferrer-Suay, M., Selfa, J. and Pujade-Villar, J. (2019) Keys to world Charipinae (Hymenoptera, Cynipoidea, Figitidae). *ZooKeys* 822, 79–130.

Fischbein, D. and Corley, J. (2015) Classical biological control of an invasive forest pest: a world perspective of the management of *Sirex noctilio* using the parasitoid *Ibalia leucospoides* (Hymenoptera: Ibaliidae). *Bulletin of Entomological Research* 105, 1–12.

Fischbein, D., Bettinelli, J., Bernstein, C. and Corley, J. C. (2012) Patch choice from a distance and use of habitat information during foraging by the parasitoid *Ibalia leucospoides*. *Ecological Entomology* 37(3), 161–168.

Fontal-Cazalla, F.M., Buffington, M.L., Nordlander, G., Liljeblad, J., Ros-Farré, P., Nieves-Aldrey, J.L., Pujade-Villar, J. and Ronquist, F. (2002) Phylogeny of the Eucoilinae (Hymonoptera. Cynipoidea: Figitidae). *Cladistics* 10, 154–199.

Gauld, I.D. and Bolton, B. (1988) *The Hymenoptera*. British Museum (Natural History), London, and Oxford University Press, Oxford, 332 pp.

Guiguet, A., Tooker, A.F., Deans, A.R., Mikó, I., Ning, G., Schwéger, S. and Hines, H.M. (2023) Comparative anatomy of venom glands suggests a role of maternal secretions in gall induction by cynipid wasps (Hymenoptera: Cynipidae). *Insect Systematics and Diversity* 7(5), 3. https://doi.org/10.1093/isd/ixad022

Hedicke, H., and Kerrich, G. (1940) A revision of the family Liopteridae (Hymenopt., Cynipoidea). *Transactions of the Royal Entomological Society, London* 90, 177–228.

Ide, T., Aung, M.M. and Tanaka, N. (2020) First record of Cynipidae from Myanmar with description of a new species of *Lithosaphonecrus* (Hymenoptera: Cynipidae: Synergini). *Zootaxa* 4810, 344–350.

Ide T., Aung, M.M. and Tanaka, N. (2022) First record of the oak gall wasp (Hymenoptera, Cynipidae) in Myanmar. *Bulletin of the National Museum of Natural Science, Ser. A* 48, 89–95.

Johnson, M.W. (1987) Parasitization of *Liriomyza* spp. (Diptera: Agromyzidae) infesting commercial watermelon plantings in Hawaii. *Journal of Economic Entomology* 80, 56–61.

Liu, Z. (1998) A new species of *Ibalia* from Borneo, with a revised phylogeny and historical biogeography of Ibaliidae (Hymenoptera: Cynipoidea). *Journal of Hymenoptera Research* 7, 149–156.

Liu, Z. (2009) Phylogeny, biogeography, and revision of the subfamily Dallatorrellinae (Hymenoptera: Liopteridae). *American Museum Novitates* 3353, 1–23.

Liu, Z. and Nordlander, G. (1994) Review of the family Ibaliidae (Hymenoptera: Cynipoidea) with keys to genera and species of the World. *Insect Systematics & Evolution* 25, 377–392. doi: 10.1163/187631294X00153

Liu, Z., Ronquist, F. and Nordlander, G. (2007) The cynipoid genus *Parambynotus*: revision, phylogeny and historical biogeography (Hymemoptera: Liopteridae). *Bulletin of the American Museum of Natural History* 304, 1–151.

Lobato-Vila, I., Wang, Y., Melika, G., Guo, R., Ju, X.-X. and Pujade-Villar, J. (2021) A review of the species in the genus *Synergus* Hartig (Hymenoptera: Cynipidae: Synergini) from mainland China, with an updated key to the Eastern Palaearctic and Oriental species. *Journal of Asia-Pacific Entomology* 24, 341–362.

Lobato-Vila, I., Bae, J., Roca-Cusachs, M., Kang, M., Jung, S., Melika, G., Pénzes, Z. and Pujade-Villar, J. (2022) Global phylogeny of the inquilinous gall wasp tribe Synergini (Hymenoptera: Cynipoidea: Cynipidae): first insights and establishment of a new cynipid tribe. *Zoological Journal of the Linnean Society* 195, 1338–1354. doi: 10.1093/zoolinnean/zlab085

Martinez, A.S., Fernández-Arhex, V. and Corley, J.C. (2006) Chemical information from the fungus *Amylostereum areolatum* and host-foraging behaviour in the parasitoid *Ibalia leucospoides*. *Physiological Entomology* 31, 336–340.

Mata-Casanova, N., Selfa, J. and Pujade-Villar, J. (2014) Revision of the Asian species of genus *Xyalaspis* Hartig, 1843 (Hymenoptera: Figitidae: Anacharitinae). *Journal of Asia-Pacific Entomology* 17, 569–576.

Melika, G., Ranjith, A.P., Lobato-Vila, I., Priyadarsanan, D.R. and Pujade-Villar, J. (2021) A new cynipid inquiline of the genus *Lithosaphonecrus* (Hymenoptera: Cynipidae: Synergini) from India, with an updated key to all known species. *Zootaxa* 5060, 124–136. doi: 10.11646/zootaxa.5060.1.6

Menke, A.S. and Evenhuis, H.H. (1991) North American Charipidae: key to genera, nomenclature, species checklists, and a new species of *Dilyta* Forster (Hymenoptera: Cynipoidea). *Proceedings of the Entomological Society of Washington* 93, 136–158.

Miller, G.L. and Lambdin, P.L. (1985) Observations on *Anacharis melanoneura* (Hymenoptera: Figitidae), a parasite of *Hemerobius stigma* (Neuroptera: Hemerobiidae). *Entomological News* 96, 93–97.

Müller, C.B., Völkl, W. and Godfray, H.C.J. (1997) Are behavioural changes in parasitised aphids a protection against hyperparasitism? *European Journal of Entomology* 94, 221–234.

Nieves-Aldrey, J.L. and Butterill, P.T. (2014) First evidence of cynipids from the Oceanian Region: the description of *Lithonecrus papuanus* a new genus and species of cynipid inquiline from Papua New Guinea (Hymenoptera: Cynipidae, Synergini). *Zootaxa* 3846, 221–234.

Nordlander, G. (1978) Revision of the genus *Rhoptromeris* Förster, 1869 with reference to north-western European species studies on Eucoilidae (Hym.: Cynipoidea) II. *Entomologica Scandinavica* 9, 47–62. doi: 10.1163/187631278X00214

Nordlander, G. (1984) Vad vet vi om parasitiska Cynipoidea. *Entomologisk Tidskrift*, 105(1–2), 36–40.

Paretas-Martínez, J., Forshage, M., Buffington, M., Fisher, N., La Salle, J. and Pujade-Villar, J. (2012) Overview of Australian Cynipoidea (Hymenoptera). *Australian Journal of Entomology* 52, 73–86.

Pénzes, Z., Tang, Chang-Ti, Bihari, P., Bozsó, M., Schwéger, S. and Melika, G. (2012) Oak associated inquilines (Hymenoptera, Cynipidae, Synergini). Tiscia Monograph Series 11 (Szeged), 1–76. Available at http://expbio.bio.u-szeged.hu/ecology/tiscia/monograph/TISCIA-monograph11.pdf (accessed 17 February, 2023)

Pénzes, Z., Tang, C.T., Stone, G.N., Nicholls, J.A., Schwéger, S., Bozsó, M., and Melika, G. (2018) Current status of the oak gallwasp (Hymenoptera: Cynipidae: Cynipini) fauna of the Eastern Palaearctic and Oriental Regions. *Zootaxa* 4433, 245–289.

Pujade-Villar, J., Melika, G., Ros-Farré, P., Ács, Z. and Csóka, G., (2003) Cynipid inquiline wasps of Hungary, with taxonomic notes on the Western Palaearctic fauna (Hymenoptera: Cynipidae, Cynipinae, Synergini). *Folia Entomologica Hungarica* 64, 121–170.

Ritchie, A.J. (1993) Superfamily Cynipoidea. In: Goulet, H. and Huber, J. (eds) *Hymenoptera of the World: An Identification Guide to Families* (Vol. vii). Research Branch, Agriculture Canada, Ottawa, pp. 521–536.

Rizki, R.M. and Rizki, T.M. (1984) Selective destruction of a host blood cell type by a parasitoid wasp. *Proceedings of the National Academy of Science of the USA* 81, 6154–6158.

Rizki, R.M. and Rizki, T.M. (1990a) Parasitoid virus-like particles destroy *Drosophila* cellular immunity. *Proceedings of the National Academy of Science of the USA* 87, 8388–8392.

Rizki, R.M. and Rizki, T.M. (1990b) Microtubule inhibitors block morphological changes induced in *Drosophila* blood cells by a parasitoid wasp factor. *Experientia* 46, 311–315.

Rizki, R.M. amd Rizki, T.M. (1991) Effects of lamellolysin from a parasitoid wasp on *Drosophila* blood cells in vitro. *Journal of Experimental Zoology* 257, 236–244.

Ronquist, F. (1994) Evolution of parasitism among closely related species: phylogenetic relationships and the origin of inquilinism in gall wasps (Hymenoptera, Cynipidae). *Evolution* 48, 241–266.

Ronquist, F. (1995a) Phylogeny and early evolution of the Cynipoidea (Hymenoptera). *Systematic Entomology* 20, 309–335.

Ronquist, F. (1995b) Phylogeny and classification of the Liopteridae, an archaic group of cynipoid wasps (Hymenoptera). *Entomologica Scandinavica Supplement* 46, 1–74.

Ronquist, F. (1999) Phylogeny, classification and evolution of the Cynipoidea. *Zoologica Scripta* 28, 139–164.

Ronquist, F., Nieves-Aldrey, J.-L., Buffington, M.L., Liu, Z., Liljeblad, J. and Nylander, J.A.A. (2015) Phylogeny, evolution and classification of gall wasps: the plot thickens. *PLoS ONE* 10(5), e0123301. doi: 10.1371/journal.pone.0123301

Ros-Farré, P. and Pujade-Villar, J. (2006) Revision of the genus *Prosaspicera* Kieffer, 1907 (Hym, Figitidae: Aspicerinae). *Zootaxa* 1379, 1–102.

Ros-Farré, P. (2007) *Pujadella* Ros-Farré, a new genus from the Oriental Region, with a description of two new species (Hymenoptera: Figitidae: Aspicerinae). *Zoological Studies Taipei*, 46(2) 168–175.

Ros-Farré, P. and Pujade-Villar, J. (2009) Revision of the genus *Callaspidia* Dahlbom, 1842. (Hym.: Figitidae: Aspicerinae). *Zootaxa* 2105, 1–31.

Spradbery, J.B. (1970) The biology of *Ibalia drewseni* Borries (Hymenoptera: Ibaliidae) a parasite of siricid woodwasps. *Proceedings of the Royal Entomological Society of London (A)* 45, 104–113.

Spradbery, J.P. (1974) The responses of *Ibalia* species (Hymenoptera: Ibaliidae) to the fungal symbionts of siricid woodwasp hosts. *Journal of Entomology Series A, General Entomology* 48, 217–222.

Suay, V.A., Luna, F. and Michelena, J.M. (1998) Parasitoids not aphidiins of aphids (Chalcidoidea: Aphelinidae) and hyperparasitoids of the superfamilies Chalcidoidea, Ceraphronoidea and Cynipoidea (Hymenoptera: Apocrita: Parasitica) in the province of Valencia. *Bolentin de la Asociación Española de Entomologia* 22, 99–113.

Talebi, A.A., Rakhshani, E., Fathipour, Y., Starý, P., Tomanoviaé, Z. and Rajabi-Mazhar, N. (2009) Aphids and their parasitoids (Hym., Braconidae: Aphidiinae) associated with medicinal plants in Iran. *American-Eurasian Journal of Sustainable Agriculture* 3(2), 205–219.

Tang, C.-T., Yang, M.-M., Schwéger, S., Pujade-Villar, J., Melika, G., Bozsó, M., Pénzes, Z. and Bihari, P. (2015) A new genus of cynipid inquiline, *Lithosaphonecrus* Tang, Melika & Bozsó (Hymenoptera: Cynipidae: Synergini), with description of four new species from Taiwan and China. *Insect Systematics & Evolution* 46, 79–114. doi: 10.1163/1876312x-45032116

Tang, C.-T., Mikó, I., Nicholls, J.A., Schwéger, S., Yang, M.-M., Stone, G.N., Sinclair, F., Bozsó, M., Melika, G. and Pénzes, Z. (2016) New *Dryocosmus* Giraud species associated with *Cyclobalanopsis* and non-*Quercus* host plants from the Eastern Palaearctic (Hymenoptera, Cynipidae, Cynipini). *Journal of Hymenoptera Research* 53, 77–162.

van Noort, S., Buffington, M.L. and Forshage, M. (2015) Afrotropical Cynipoidea (Hymenoptera). *ZooKeys* 493, 1–176.

Weld, L.H. (1926) Field Notes on gall-inhabiting cynipid wasps with descriptions of new species. *Proceedings of the United States National Museum* 68, 1–131.

Weld, L.H. (1960) A new genus in Cynipoidea (Hymenoptera). *Proceedings of the Entomological Society of Washington* 62, 195–196.

Wharton, R.A., Gilstrap, F.E., Rhodei, R.H., Fischel, M.M. and Hart, W.G. (1981) Hymenopterous egg-pupal and larval-pupal parasitoids of *Ceratitis capitata* and *Anastrepha* spp. (Diptera: Tephritidae) in Costa Rica. *Entomophaga* 26, 285–290.

Wharton, R.A., Ovruski, S. and Gilstrap, F. (1998) Neotropical Eucoilidae (Cynipoidea) associated with fruit-infesting Tephritidae, with new records from Argentina, Bolivia and Costa Rica. *Journal of Hymenoptera Research* 7, 102–115.

20

Proctotrupoidea

Abstract
Only four families of the heterogeneous superfamily Proctotrupoidea have been found in S.E. Asia so far. Of these, only the Proctotrupidae are in any way common. An identification key to the four families is provided. Representatives of most groups are illustrated photographically, and their wing venations interpreted with diagrams. The biology of each group as far as is known is described but this is based entirely on extralimital records. Relevant identification works are cited.

The heterogeneous superfamily Proctotrupoidea has in the past contained additional taxa (see Diaprioidea). It currently comprises the families Austroniidae (restricted to Australia), Heloridae, Peradeniidae (two rare species from southern Australia), Pelecinidae (entirely New World), Proctotrupidae and Vanhorniidae.

Simplified Key to S.E. Asian Families of Proctotrupoidea

1		Metasomal segment 1 occupying approximately 90% of metasoma length (Fig. 20.5); mandibles exodont (i.e., not overlapping or touching when closed and usually pointed outwards); ovipositor directed forwards between legs, housed in a mid-ventral groove along metasomal sternites; (extremely rare) ..... **Vanhorniidae**
-		Metasomal segment 1 occupying not more than 60% of metasoma length (sometimes very short and T2 very large); mandibles endodont (i.e., their tips touching or overlapping when closed); ovipositor not directed anteriorly between legs............ 2
2(1)		Antenna with 11 flagellomeres; wing venation much reduced and characteristic (Fig. 20.2); fore wing vein RS+M absent, never tubular (Fig. 20.2), sometimes indicated as short slightly pigmented (nebulous) line; metasomal segment one far shorter and wider; (moderately common) .... **Proctotrupidae**
-		Antenna with 12 or more flagellomeres; wing venation more complete (Figs 20.1, 20.4); fore wing vein RS+M present and defining medial cell posteriorly [rare] 3
3(2)		Fore wing medial cell triangular, vein 1-RS absent (Fig. 20.1); flagellum with 13 well-developed segments (plus one reduced and ring-like); (extremely rare)............ .... **Heloridae**

- Fore wing medial cell rhombic or pentagonal, vein 1-RS present (Fig. 20.4); flagellum with only 12 flagellomeres of which none are ring-like; (extremely rare) ....**Roproniidae**

Heloridae

All extant species belong to the genus *Helorus*. This distinctive family is predominantly Holarctic, but species have been recorded from South America, India, China, New Guinea and Australia (New, 1975; Naumann, 1983), and a single species has recently been described from Indonesia (Sulawesi), filling the gap (van Achterberg, 2006). The triangular medial cell in the fore wing, unconnected to the anterior venation, serves to identify extant members (Fig. 20.1), although some fossil members have the connecting 1 RS vein.

All that is known of their biology concerns extralimital temperate species, much owing to Clancy (1946). They are koinobiont endoparasitoids of lacewing (Chrysopidae) larvae and if the host enters an overwintering diapause, there is a prolonged 1st instar phase. Further development starts after the host spins its cocoon and is completed before host pupation. As with other studied proctotrupoids, the full-grown parasitoid larva partially exits from the host but leaves its posterior inside, before it pupates. They appear to be very uncommon or even absent from most tropical lowland forest despite chrysopids occurring there.

Proctotrupidae

Proctotrupids are commonest in the temperate zone, and most tropical species occur at higher altitude. Their general habitus is easily recognised and the fore wing venation is characteristic with tubular venation confined to the anterior, costal cell present, and a pterostigma well-developed with vein RS running very close to its distal edge to the wing margin (Fig. 20.2).

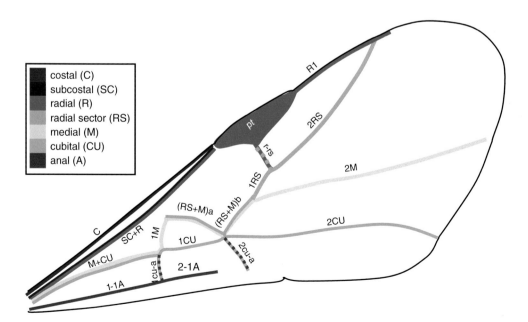

Fig. 20.1. Stylised fore wing venation of an extant *Helorus* showing terminology used here. Note the characteristic triangular medial cell (surrounded by veins 1M, 1CU and (RS+M)a.

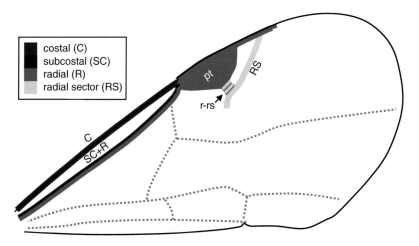

Fig. 20.2. Stylised drawing of characteristic proctotrupid fore wing venation showing well-developed pterostigma and small narrow marginal cell between it, r-rs and RS. The tubular veins are shown thick and coloured; other venation is usually indicated only by pigmented lines or wing folds.

Readers may wonder why Henry Townes used the name Serphidae rather than Proctotrupidae, which is now nearly universally applied. This is discussed in Johnson (1992) but essentially the genus name *Serphus* is a senior synonym of *Proctotrupes* but, because of usage, the International Commission of Zoological Nomenclature published an Opinion in 1946 that was intended to fix the name as Proctotrupidae. However, Townes argued that the Commission had not properly followed its own rules and that the Opinion was not valid.

With the removal of the Vanhorniidae to a family of their own, the Proctotrupidae contains only two subfamilies, the Austroserphinae (=Acanthoserphinae) and Proctotrupinae (Townes and Townes, 1981). The latter are by far the largest group with more than 20 genera worldwide and hundreds of species, whereas the latter includes only two mainly Gondwanan genera with a few species occurring in S.E. Asia in addition to Australia and New Guinea.

Townes and Townes (1981) summarised what was then known of proctotrupid biology and not much has changed since then, as this has been a terribly neglected group. When hosts are small-bodied they are solitary (e.g. Lee *et al.*, 1988; Hoebeke and Kovarik, 1988) but on larger hosts they are often gregarious They are koinobiont endoparasitoids but when they have completed development the larva usually exits partially from the host, leaving its posterior end still embedded in the host, before shedding its final instar larval skin to reveal a naked exarate pupa (Williams, 1932; Hoebeke and Kovarik, 1988). However, *Nothoserphus* (Fig. 20.3 a) appears to pupate completely internally (Lee *et al.*, 1988).

Hosts of most species inhabit ground layer, such as leaf litter larval (Carabidae, Elateridae, Staphylinidae) (Hoebeke, 1978), decaying wood on the ground (Anthribidae, Curculionidae, Erotylidae, Melandryidae, Phalacridae). In the same general habitat, *Cryptoserphus* appear to be specialist parasitoids of fungus gnat larvae (Diptera: Mycetophilidae) in fungal fruiting bodies (Huggert, 1979), and one species, *Phaneroserphus calcar*, has been reared from a lithobiid centipede (Chilopoda). At least one extralimital *Brachyserphus* species parasitises nitidulid (Coleoptera) larvae feeding on sap-fluxes, and in the laboratory can be reared successfully on a range of hosts (Williams *et al.*, 1992). A New Zealand species, *Fustiserphus intrudens*, has been reared from oecophorid moth larvae collected from leaf litter in *Nothofagus* forest (Early and Dugdale, 1994). Another specialised host association has been reported for

Nothoserphus species, from Korea (Lee *et al.*, 1988) and Pakistan (Bodlah *et al.*, 2019). The two species involved were reared from exposed coccinellid beetle larvae, and in the former case *N. afissae* on *Helosepilachna vigintioctomaculata* (the 28-spotted potato ladybird or Hadda beetle); the host is an important pest of Solanaceae crops, including potato and aubergines. Essentially nothing is known of host relationships for any S.E. Asian species. Pupation occurs with the posterior of the pupa still inside the host, the rest external and unprotected.

Some species are very widespread; for example, *Oxyserphus rossica*, originally described from Russia, is known from the west Palaearctic, New Zealand, New Caledonia, Indonesia, Philippines, Thailand, Laos, Taiwan, Japan, Mexico and Costa Rica (Kolyada, 2007). In the absence of molecular investigation, we cannot know whether such examples are genuinely widespread single species or a mix of geographically separated cryptic species. Kolyada and Mostovski (2017) showed that *Hormoserphus* is actually a synonym of *Oxyserphus*, a genus that is very speciose in the Oriental to Australian region. Kolyada and Mostovski (2017) also described *Trachyserphus* for some other species of *Hormoserphus*, and provided a key to the three known species. which are distributed from Nepal, through southern China and Taiwan, and Vietnam.

Johnson (1992) catalogued the known genera and species and summarised what information was given in the literature for each (biology, taxonomy, keys, etc.), as well as where type material was deposited (if known). However, for each only a broad geographical region was indicated rather than country.

The world species of the family were revised (under the name Serphidae) by Henry and Marjorie Townes of Ichneumonidae fame (Townes and Townes, 1981), although this should no longer be regarded as fully reliable and it certainly misses many S.E. Asian taxa. Since then a few extra genera have been described, some of which occur in S.E. Asia or may well do (Fan and He, 1993; He and Xu, 2007; Kolyada and Mostovski, 2017). Two further generic keys that may be useful, which although not including all genera in the region, do use more standard terminology, are: (i) those of Izadizadeh *et al.* (2022), who provided a key to the genera known from Iran which was based on Townes and Townes (1981) and Kolyada's (1998) key in Russian; and (ii) that of Park and Lee (2021) to the Korean fauna, again modified from Townes and Townes (1981). Currently the family comprises approximately 600 species in about 30 genera (Kolyada and Mostovski, 2017) (Fig. 20.3).

Roproniidae

Uncommon or very rare, and (usually) medium-sized (5.0–10 mm) rather distinctive parasitoid wasps (Fig. 20.4), mostly known from North America and the temperate Palaearctic, with species extending to China, Japan, northern Myanmar and Taiwan (Hedqvist, 1959; Lin, 1988). Most likely, it occurs in the north or all Indo-Chinese countries. There are only two extant genera. Most species belong to *Ropronia* and the second genus *Xiphyropronia* is only known from China (He and Chen, 1991). In the past, Ropronia was included in the Heloridae.

The species of *Ropronia* were revised by Townes (1948). A few extra have been added since then and Lin (1987) provided an updated key to the world species including the Burmese species, *Ropronia malaisei*, named after the inventor of the Malaise trap: the Swedish entomologist René Malaise, who carried out some field work and invented the prototype trap there in 1934 (Vårdal and Taeger, 2011).

Vanhorniidae

A small, highly distinctive family with a single genus, *Vanhornia*, that are all, as far as is known, larval parasitoids of false click beetles (Eucnemidae). They are medium-sized wasps with body length ranging from 4.0 to

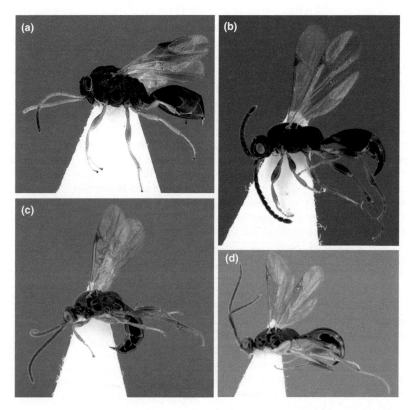

Fig. 20.3. Some representative Thai Proctotrupidae: **(a)** *Nothoserphus* sp.; **(b)** *Phaneroserphus* sp.; **(c)** *Exallonyx* sp.; **(d)** *Phaneroserphus* sp.

Fig. 20.4. *Ropronia*, probably an extralimital species from Israel.

nearly 7.0 mm. Muona (2021) questioned exactly how much was really known about the biology of vanhorniids, and pointed out that, although a number of publications mentioned host records, there were actually remarkably few original observations involved and that there may also be some problems with the taxonomy of the beetles mentioned.

The family comprises the following four species: *V. eucnemidarum* (Fig. 20.5) from North America, *V. leileri* from the Palaearctic, *V. quizhouensis* from southern China and Thailand (Artmann-Graf, 2016), and *V. yurii* from northeast Asia. Timokhov and Belokobylskij (2020) provided a key to the world species.

Mason (1983) described the unusual metasoma and ovipositor system in detail. The ovipositor itself is very thin-walled and it has been postulated that the ovipositor sheaths may play an important role in its support during the oviposition process (Quicke *et al.*, 1994).

Fig. 20.5. The extralimital vanhorniid species, *Vanhornia eucnemidarum*: **(a)** dorsal view; **(b)** habitus, lateral view. (Source: from Timokhov and Belokobylskij (2020), reproduced under terms of Creative Commons licence CC-BY-4.0.)

References

Artmann-Graf, G. (2016) *Vanhornia leileri* Hedqvist, 1976: eine für Mitteleuropa neue Art und Familie parasitischer Wespen in der Region Olten. *Oltner Neujahrsblätter* 74, 64–67.

Bodlah, I., Gull-E-Fareen, A., Rasheed, M.T., Amin, M. and Aihetasham, A. (2019) First record of *Nothoserphus mirabilis* Brues, 1940 (Hymenoptera: Proctotrupidae) from Pothwar, Pakistan. *Punjab University Journal of Zoology* 34, 47–49. doi: 10.17582/journal. pujz/2019.34.1.47.49

Clancy, D.W. (1946) The insect parasites of the Chrysopidae (Neuroptera). *University of California Publications in Entomology* 7, 403–496.

Early, J.W. and Dugdale, J.S. (1994) *Fustiserphus* (Hymenoptera: Proctotrupidae) parasitises Lepidoptera in leaf litter in New Zealand. *New Zealand Journal of Zoology* 21, 249–252. doi: 10.1080/03014223.1994.9517992

Fan, J.J. and He, J.H. (1993) A new genus and species of Serphini (Hymenoptera: Serphidae) from China. *Entomotaxonomia* 15, 69–73.

He, J. and Chen X. (1991) *Xiphyropronia* gen. nov., a new genus of the Roproniidae (Hymenoptera: Proctotrupoidea) from China. *Canadian Journal of Zoology* 69, 1717–1719.

He, J.-H. and Xu, Z.-F. (2007) A new genus of Proctotrupinae (Hymenoptera: Proctotrupidae). *Entomotaxonomia* 29, 152–156.

Hedqvist, J.-J. (1959) A new species of *Ropronia* from Burma (Proctotrupoidea: Heloridae). *Entomologisk Tidskrift* 80, 137–139.

Hoebeke, R.E. (1978) Notes on the biology of *Codrus carolinensis* (Hymenoptera: Proctotrupidae), a parasite of *Platydracus violaceus* (Coleoptera: Staphylinidae). *Journal of the Kansas Entomological Society* 51, 507–511.

species that also occur in China. There are no recent keys to the genera of S.E. Asia but since many genera are cosmopolitan, it is advisable to try keys to the faunas of extra-limital areas such as Quadros and Brandão for part of Brazil.

Diapriinae

Three tribes are often recognised, viz. Diapriini, Spilomicrini and Psilini (Masner and García, 2002), all of which occur in S.E. Asia. The great majority of genera belong to the first of these. Male antennae are filiform, and often with long setae, and have the 2nd flagellomere sexually modified with a release-and-spread contact pheromone gland structure. Even in males of genera such as *Trichopria* (Fig. 21.2 b) with their very setose antennae, there is a sexual modification of the 2nd flagellomere; and Romani *et al.* (2008) described how during courtship the modifications are bought into contact with the two apical female flagellomeres. Female antennae are apically swollen and many have a distinct club (Fig. 21.2 c). Some genera have highly modified, almost opisthognathous heads (e.g. Fig. 21.3).

A few Old World species are termitophilous, but most are parasitoids of Diptera (e.g. Anthomyiidae, Calliphoridae, Muscidae, Sarcophagidae, Stratiomyiidae, Syrphidae, Tachinidae). Many of these have larvae that live in or on the ground and that is where most diapriids occur. A few are parasitoids of leaf-mining Diptera, and a few attack Coleoptera (e.g. Staphylinidae). Almost nothing is known about the host relationships of S.E. Asian species.

Simplified key to tribes of Diapriinae occurring in S.E. Asia

1	Flagellum with 11 segments in both sexes (Fig. 21.2 a) [but there are many variants] ....some **Spilomicrini**	
-	Flagellum with 10 segments in females (Fig. 21.2 c) and 12 in males (Fig. 21.2 d) 2	
2(1)	Notauli completely absent .... **Diapriini**	
-	Notauli present, at very least represented by distinct pit near trasnscutal articulation.. 3	
3(2)	Fore wing with submarginal vein present (Fig. 21.2 a), pigmented, reaching anterior wing margin distally where it becomes the marginal vein, often also with stigmal and other veins indicated; sometimes brachypterous or apterous .... some **Spilomicrini**	
-	Fore wing without submarginal vein **OR** if present then it is transparent (glassy) not reaching fore wing margin distally (Fig. 21.2 c), sometimes venation completely absent; always fully winged .... **Psilini**	

Baltazar (1966) recorded the genera *Acidopria, Ceratopria, Cologlyptus, Digalesus, Dilobopria, Hemigalesus, Lipoglyptus, Loxotropa, Oxypria, Phaenopria, Psilus, Scapopria, Spilomicrus, Stylopria* and *Trichopria* as occurring in the Philippines based on described species and based almost entirely on the much earlier work of Kieffer (1916). However, *Cologlyptus* and *Loxotropa* are now considered synonyms of *Spilomicrus*. For most of the rest of these names nothing has been published since their original descriptions, although some species of *Loxotropa, Spilomicrus* and *Trichopria* are parasitoids of economic pests in other parts of the world.

The widespread wasp *Synacra paupera* is a parasitoid of a sciarid pest (Diptera) in greenhouses in Europe and North America, but its origin is uncertain, though probably a warmer country, and it has been found in Malaysia (Notton, 1997). Notton (1991) confirmed that *Trichopria* and *Basalys* (both Diapriinae) are endoparasitoids of fly pupae within the puparium. Interestingly, he recorded the extralimital *T. clavatipes* as attacking members of two different dipteran suborders. Species of *Basalys*, a large, common

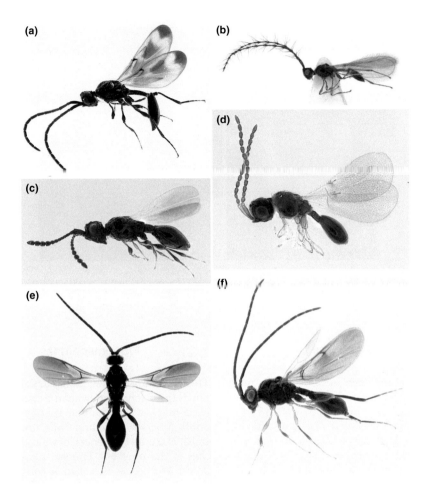

Fig. 21.2. Exemplar taxa of Diapriidae from Thailand: **(a)** *Spilomicrus* sp.; **(b)** male *Trichopria* sp.; **(c)** *Coptera* sp.; **(d)** *Entomacis* sp.; **(e)** Belytinae gen. sp., possibly *Belyta* sp.; **(f)** *Lyteba* sp.

and cosmopolitan genus, have a dense, woolly collar around the pronotum.

The cosmopolitan genus *Coptera* and related taxa have highly distinctive fore wings which are apically bilobed (incised) and in some genera have a glabrous, unpigmented longitudinal fold line (Fig. 21.2 c); the fold is present in a few other genera but the apical fore wing margin is not incised. Their biology is probably representative of many of those diapriines that have highly modified heads (Fig. 21.3). Their main hosts are Diptera puparia, buried shallowly in soils, which they locate using kairomones produced by the host before pupation (Granchietti *et al.*, 2012). Their highly modified heads are used to dig through the loose soil surrounding host puparium. They then drag the host puparium to the ground surface where they oviposit into it (Buckingham, 1975). Being endoparasitoids, *Coptera* and others diapriid parasitoids of Diptera probably face more physiological constraints on host species and so tend to have narrower host ranges than many ectoparasitoid pteromalids (Sivinski *et al.*, 1998). The North American *C. haywardi*, for example, appears to be largely or entirely restricted to tephritid fruit flies as hosts.

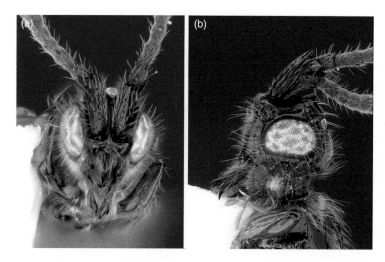

Fig. 21.3. Two views of the head of a Thai *Coptera* species showing the nearly opisthognathous shape, elongate mandibles that form a sort of beak, and the spiny ornamentation of the top of the head.

Ismaridae

These had long been recognised as constituting a separate and quite basal tribe in Diapriidae, and the group was raised to full family status as a result of Sharkey et al.'s (2012) large combined morphological and molecular analyses. Females have 15-segmented antennae, and males 14, with flagellomere 2 sexually modified (as in Diapriinae).

From the little that is known, ismarines are koinobiont endoparasitoids of dryinid (see Chrysidoidea, Chapter 16, this volume) larvae that are parasitising plant hoppers (Chambers, 1955, 1981; Jervis, 1979). The dryinid larva occupies a sack projecting from its host's abdomen and is thus easily accessible.

The *Ismarus* fauna of China was revised by Liu *et al.* (2011), who recognised eight species there. To date no species have been recorded from S.E. Asia, but Masner (1976) mentioned that he had seen three undescribed Oriental species in the Canadian National Insect Collection, Ottawa, but did not say which countries. We have recently collected a species in north Thailand as part of the Global Malaise Trap Programme.

Monomachidae

This rather uncommon family has a very disjunct (probably Gondwanan) distribution, occurring with numerous species both in South America, including Chile, and in Australia. There are two genera, of which *Monomachus* is the only one known from the Old World. Worldwide *Monomachus* comprises 21 described species. Fairly recently, two species have been described from New Guinea (Musetti and Johnson, 2000) and therefore this family may extend in to Irian Jaya.

Female members of the family have a very characteristic habitus (Fig. 21.4) which combined with their 'procto' type wing venation should make recognition easy. Johnson and Musetti (2012) revised all the world species and presented a phylogenetic analysis which showed that, within *Monomachus*, the Australian and Papuan species were basal.

The Australian species, *Monomachus antipodalis* (Fig. 21.4) has been reared from a *Boreoides* species (Diptera: Stratiomyidae: Chiromyzinae) (Naumann, 1985), which belongs to a subfamily of soldier flies with a Gondwanan origin. More recently, an unidentified Brazilian species was also reared from another chiromyzine stratiomyid (Lima *et al.*, 2001, cited in Perioto *et al.*, 2016).

Fig. 21.4. Female *Monomachus antipodialis* from Australia. (Source: photograph by and © Mollie-Rose Slater-Baker, reproduced with permission.)

References

Baltazar, C.R. (1966) A catalogue of Philippine Hymenoptera (with a bibliography, 1758–1963). *Pacific Insects Monographs* 8, 1–488.

Buckingham, G.R. (1975) The parasites of walnut husk flies (Diptera: Tephritidae: Rhagoletis) including comparative studies on the biology of *Biosteres juglandis* Mues. (Hymenoptera: Braconidae) and on the male tergal glands of the Braconidae (Hymenoptera). Unpublished PhD thesis, University of California, Berkeley.

Buhl, P.N. (1998) New or little known Oriental and Australasian Belytinae (Hymenoptera: Diapriidae), *Oriental Insects* 32, 41–58. doi: 10.1080/00305316.1998.10433766

Castro, L.R. and Dowton, M. (2006) Molecular analyses of Apocrita (Insecta: Hymenoptera) suggest that the Chalcidoidea are sister to the diaprioid complex. *Invertebrate Systematics* 20, 603–614.

Chambers, V.H. (1955) Some hosts of *Anteon* spp. (Hym. Dryinidae) and a hyperparasite *Ismarus* (Hym. Belytidae). *Entomologist's Monthly Magazine* 91, 114–115.

Chambers, V.H. (1971) Large populations of Belytinae (Hymenoptera, Diapriidae). *Entomologist's Monthly Magazine* 106, 149–154.

Chambers, V.H. (1981) A host for *Ismarus halidayi* Foerst (Hym, Diapriidae). *Entomologist's Monthly Magazine* 117, 29.

Dowton, M. and Austin, A.D. (2001) Simultaneous analysis of 16S, 28S, COI and morphology in the Hymenoptera: Apocrita – evolutionary transitions among parasitic wasps. *Biological Journal of the Linnean Society* 74, 87–111.

Early, J.W., Masner, L., Naumann, I.D. and Austin, A.D. (2001) Maamingidae, a new family of proctotrupoid wasp (Insecta: Hymenoptera) from New Zealand. *Invertebrate Taxonomy* 15, 341–352.

Granchietti, A., Sacchetti, P., Rosi, M.C. and Belcari, A. (2012) Fruit fly larval trail acts as a cue in the host location process of the pupal parasitoid *Coptera occidentalis*. *Biological Control* 61, 7–14. doi: 10.1016/j.biocontrol.2011.10.015

Hellqvist, S. (1994) Biology of *Synacra* sp. (Hym., Diapriidae), a parasitoid of *Bradysia paupera* (Dipt., Sciaridae) in Swedish greenhouses. *Journal of Applied Entomology* 117, 491–497. doi: 10.1111/j.1439-0418.1994.tb00766.x

Huggert, L. (1979) *Cryptoserphus* and Belytinae wasps (Hymenoptera, Proctotrupoidea) parasitizing fungus- and soil-inhabiting Diptera. *Notulae Entomologicae* 59, 139–144.

Huggert, L. (1982) New taxa of soil-inhabiting diapriids from India and Sri Lanka (Hymenoptera, Proctotrupoidea). *Revue Suisse de Zoologie* 89, 183–200.

Jervis, M.A. (1979) Parasitism of *Aphelopus* species (Hymenoptera: Dryinidae) by *Ismarus dorsiger* (Curtis) (Hymenoptera: Diapriidae). *Entomologist's Gazette* 30, 127–129.

Johnson, N.F. (1992) *Catalog of world species of Proctotrupoidea, exclusive of Platygastridae (Hymenoptera).* Memoirs of the American Entomological Institute 51, American Entomological Institute, Gainesville, Florida, 825 pp.

Johnson, N.F. and Musetti, L. (2012) Genera of the parasitoid wasp family Monomachidae (Hymenoptera: Diaprioidea). *Zootaxa* 3188(1), 31–41.

Kieffer, J.-J. (1916) *Diapriidae*. Das Tierreich, Vol. 44. Walter de Gruyter & Co., Berlin, 627 pp.

Liu, J., Chen, H. and Xu, Z. (2011) Notes on the genus *Ismarus* Haliday (Hymenoptera, Diapriidae) from China. *ZooKeys* 108, 49–60. doi: 10.3897/zookeys.108.768

Liu, J. and Xu, Z. (2012) Two new species of *Eccinetus* Muesebeck & Walkley, 1956 (Hymenoptera: Diapriidae) from China, with a key to the world species. *Entomological News* 122, 65–73.

Macek, J. (1990) Revision of European *Psilommina* (Hymenoptera, Diapriidae). I. *Psilomma* and *Acanosema* complex. *Acta Entomologica Musei Nationalis Pragae* 43, 335–360.

Macek, J. (1995) A taxonomic revision of European *Psilommina* (Hymenoptera: Diapriidae). Part 2. The *Synacra* complex. *European Journal of Entomology* 92, 469–482.

Masner, L. (1976) A revision of the Ismarinae of the New World (Hymenoptera, Proctotrupoidea, Diapriidae). *Canadian Entomologist* 108, 1243–1266. doi:10.4039/Ent1081243-11

Masner, L. (1993) Superfamily Proctotrupoidea. In: Goulet, H. and Huber, J.T. (eds) *Hymenoptera of the World: an Identification Guide to Families*. Agriculture Canada, Ottawa, pp. 537–557.

Masner, L. and García, J.L. (2002) The genera of Diapriinae (Hymenoptera: Diapriidae) in the New World. *Bulletin of the American Museum of Natural History* 268, 1–138.

Musetti, L. and Johnson, N.F. (2000) First documented record of Monomachidae (Hymenoptera: Proctotrupoidea) in New Guinea, and description of two new species. *Proceedings of the Entomological Society of Washington* 102, 957–963.

Naumann, I.D. (1982) Systematics of the Australian Ambositrinae (Hymenoptera: Diapriidae), with a synopsis on non-Australian genera of the subfamily. *Australian Journal of Zoology* 30, 1–239. doi: 10.1071/AJZS085

Naumann, I.D. (1985) The Australian species of Monomachidae (Hymenoptera: Proctotrupoidea), with a revised diagnosis of the family. *Australian Journal of Entomology* 24, 261–274. doi: 10.1111/j.1440-6055.1985.tb00241.x

Nixon, G.E.J. (1957a) Hymenoptera, Proctotrupoidea. Diapriidae subfamily Belytinae. *Handbooks for the Identification of British Insects* Vol. 8, Pt 3dii. Royal Entomological Society of London, 107 pp.

Nixon, G.E.J. (1957b) Hymenoptera, Proctotrupoidea. Diapriidae subfamily Diapriinae. *Handbooks for the Identification of British Insects* Vol. 8, Pt 3di. Royal Entomological Society of London, 55 pp.

Notton, D.G. (1991) Some Diptera host records for species of *Basalys* and *Trichopria* (Hym., Diapriidae). *Entomologist's Monthly Magazine* 127, 123–126.

Notton, D.G. (1997) *Synacra paupera* Macek (Hym., Duapriidae) new to Britain: a parasitoid of the greenhouse pest *Bradysia paupera* Tuomikoski (Dipt., Sciaridae). *Entomologist's Monthly Magazine* 133, 257–259.

Perioto, N.W., Lara, R.I.R., Fernandes, D.R.R., De Bortoli, C.P., Salas, C., Netto, J.C., Perez, L.A., Trevisan, M., Kubota, M.M., Pereira, N.A., Gil, O.J.A., Dos Santos, R.F., Jorge, S.J. and Laurentis, V.A. (2016) *Monomachus* (Hymenoptera, Monomachidae) from Atlantic rainforests in São Paulo State, Brazil. *Revista Colombiana de Entomologia* 42, 171–175.

Quadros, A.L. and Brandão, C.R.F. (2017) Genera of Belytinae (Hymenoptera: Diapriidae) recorded in the Atlantic dense ombrophilous forest from Paraíba to Santa Catarina, Brazil. *Papéis Avulsos de Zoologia* 57, 57–91.

Rajmohana, K. (2004) A key to the Oriental genera of Diapriidae (Hymenoptera: Proctotrupoidea: Diapriidae). In: Rajmohana, K. and Narendran, T.C. (eds) *Perspectives on Biosystematics and Biodiversity. Prof. T. C. Narendran commemoration volume*. Systematic Entomology Research Scholars Association, Kerala, pp. 519–526.

Romani, R., Rosi, M.C., Isidoro, N. and Bin, F. (2008) The role of the antennae during courtship behaviour in the parasitic wasp *Trichopria drosophilae*. *Journal of Experimental Biology* 211, 2486–2491.

Sharkey, M.J. (2007) Phylogeny and classification of Hymenoptera. *Zootaxa* 1668, 521–548.

Sharkey, M.J., Carpenter, J.M., Vilhelmsen, L., Heraty, J., Liljeblade, J., Dowling, A.P.G., Schulmeister, S., Murray, D., Deans, A.R., Ronquist, F., Krogmann, L. and Wheeler, W.C. (2012) Phylogenetic relationships among superfamilies of Hymenoptera. *Cladistics* 27, 1–33.

Sivinski, J., Vulinec, K., Menezes, E. and Aluja, M. (1998) The bionomics of *Coptera haywardi* (Ogloblin) (Hymenoptera: Diapriidae) and other pupal parasitoids of tephritid fruit flies (Diptera). *Biological Control* 11, 193–202. doi: 10.1006/bcon.1997.0597

22

Chalcidoidea and Mymarommatoidea

Abstract

The treatment of Chalcidoidea follows the changes made by Burks *et al.* (2022); however, since there is not yet a published key that includes all the 20 new families, every family that occurs in S.E. Asia is treated under the heading 'Former Pteromalidae'. A simplified key to the family occurring in the region is provided, all families are discussed, and most are illustrated photographically based on local fauna. References are given to the recent taxonomic literature. World and regional generic and species diversity are tabulated. Morphological characters peculiar to the superfamily are described and illustrated.

Although technically the Chalcidoidea should be referred to as chalcidoids, a lot of workers simply refer to them as chalcids, which is potentially confusing since 'chalcid' ought really to refer only to members of the Chalcididae. Monophyly of the Chalcidoidea + Mymarommatoidea has not been questioned for a considerable while (Gibson, 1986a). In the past the Mymarommatoidea were often included as a family of chalcidoids (e.g. Gibson, 1993), but they have been treated as a separate superfamily since 1989 because they differ in several important ways – see Gibson *et al.* (1999) and Gibson *et al.* (2007) for historical overviews. Nevertheless, all recent molecular phylogenies (See Figs 5.1, 5.2, 5.4) have recovered them as either the sister group to Chalcidoidea, or at least derived very close to its base, and therefore we treat them here as more or less together.

This is a large, extremely diverse and complex superfamily. Worldwide it comprises more than 23,000 described species and over 500,000 species are estimated to exist (Munro *et al.*, 2011; Janšta *et al.*, 2018). Chalcidoid wasps are typically small or very small, though there are a few larger species that are on a par with medium-sized ichneumonoids. The family composition has remained essentially unchanged: Gauld and Bolton (1988) in *The Hymenoptera* and Gibson (1993) in *Hymenoptera of the World* both recognised 20 families.

Molecular phylogenetic studies over the past few years (e.g. Munro *et al.*, 2011; Peters *et al.*, 2018; Zhang *et al.*, 2020; Cruaud *et al.*, 2022, 2023) have consistently shown that whilst about three-quarters of the traditional families are indeed monophyletic with a few diphyletic, two families (the Eupelmidae and Pteromalidae) are massively polyphyletic; therefore, in order to have a classification in which all families are monophyletic, more than 20 extra families need to be erected (Rasplus *et al.*, 2018). The process of splitting started rather recently in dribs and drabs, dealing with some of the smaller chalcidoid families. For example, Zhang et al. (2022) recently raised two subfamilies to family level, viz. Eutrichosomatidae and Chrysolampidae, and resurrected Perilampidae, raising the total number of families to 24.

The main explosion of families happened as we were nearly completing this book, with the appearance of Burks et al. (2022). They implemented many of the changes suggested by the above molecular studies and provided diagnoses of each of their new family-level taxa. Even so, several groups were left unplaced (*incertae sedis*) in the superfamily. The new classification recognises approximately double the previous number.

As we write, there is no published key to these, although one is in preparation and undergoing beta testing. Such is the nature of evolution that recognising some of the newly described families will require very good microscopes, as the characters may be rather tricky and small. Fortunately, many of the new families were previously recognised at subfamily or tribe level, and Zdenek Bouček's key to the Indo-Australian Chalcidoidea actually keyed many of them out separately (although not all with the same generic composition as now). Thus, in the interim, before the new comprehensive key is available, one can use Bouček's key (Bouček, 1988). Many tropical groups are absent from the Palaearctic and therefore Graham's (1969) key is likely to be of limited use for S.E. Asia, but may help in confirming some identifications. Many other keys to chalcidoid families struggle because they try, at least to some extent, to key out artificial, non-monophyletic units.

Because of the above, we have been forced to take a pragmatic approach that will reflect the family-level classification that material in museums and other collections is likely to be sorted to. Thus, we include the new families under the heading 'Former Pteromalidae' towards the end.

Approximate numbers of global and regional genera and species in all the chalcidoid families recognised by Burks et al. (2022) are given in Table 22.1, and the taxonomic changes they implemented, which mostly affect the former Pteromalidae, are summarised in Table 22.2.

Various other classificatory changes implemented by Burks et al. (2022) are: Chromeurytominae and Keiraninae transferred from Pteromalidae to Megastigmidae; Elatoidinae transferred from Pteromalidae to Neodiparidae; Nefoeninae transferred from Pteromalidae to Pelecinellidae; and Erotolepsiinae transferred from Pteromalidae to Spalangiidae. The subfamily Sycophaginae was transferred from Agaonidae to Pteromalidae. *Liepara* is transferred to Coelocybidae.

Chalcidoids collectively display an enormous range of biologies. Most are parasitoids attacking juvenile stages (eggs to pupae) of other insects, or sometimes arachnids. Phytophagy has evolved multiple times within Chalcidoidea (LaSalle, 2005) and includes seed-eaters (e.g., Eulophidae, Eurytomidae, Megastigmidae), stem-borers (Eurytomidae), gall-formers (e.g., Agaonidae, Tanaostigmatidae, Eulophidae, Pteromalidae), inquilinism of galls formed by other insects (e.g., Eulophidae, Eurytomidae, Megastigmidae) and entomophytophagy, in which the original gall-forming insect is consumed, but the chalcidoid then continues its development by consuming the plant gall tissue (Eurytomidae, Ormyridae). Lotfalizadeh (2012) surveyed knowledge of the host relationships of chalcidoid parasitoids of xylophagous beetles. For the whole of the Oriental Region the list comprised a total of 35 species in 28 genera, and these numbers include egg parasitoids such as Mymaridae and Trichogrammatidae.

At least in the case of one *Megastigmus* species (Megastigmidae), the wasp microbiome includes the bacterium *Ralstonia* sp. which is also present in the microenvironment of the host plant ovules (Paulson et al., 2014). It is speculated that the bacterium might be involved with nutrient recycling. The bacterium is present in the wasp adults as well as larvae and is presumed to be transmitted longitudinally.

Morphological Terminology

Chalcidoids display a number of distinctive features and some of these have specialised terminology, which we describe briefly below.

Table 22.1. Summary of families of Chalcidoidea as recognised prior to Burks *et al.*'s (2022) major re-organisation, and their approximate numbers of described genera and species (based largely on Noyes, 2021, and Ghahari *et al.*, 2021).

Family	Described genera	Described extant species	Described S.E. Asian genera	Described S.E. Asian species	Notes
Agaonidae	30	432	14	133	All fig pollinators belong here
Aphelinidae	42	1391	15	87	
Azotidae	1	94	2	7	
Baeomorphidae	2	3	0	0	Formerly called Rotoitidae; extralimital
Calesidae	1	14	0	0	Gondwanan, not yet recorded but introduced into N. America, Africa and S. Europe
Cerocephalidae	16	46	3	6	
Chalcedectidae	1	20	1	0	Several unidentified species in region
Chalcididae	85	1460	34	227	
Chrysolampidae	6	78	1	2	
Cleonymidae	6	85	3	3	Unidentified species of two more genera known in region
Coolooybidae	17	51	0	0	Known from New Guinea
Cynipencyrtidae	1	1	1	1	Vietnam
Diparidae	11	124	0	0	Many species present but none recorded
Encyrtidae	506	3710	64	236	
Epichrysomallidae	19	49	3	6	
Eucharitidae	68	467	14	47	
Eulophidae	328	c. 6050	69	241	(Including former Elasmidae)
Eunotidae	6	39	1	2	
Eupelmidae	45	854	12	38	
Eurytomidae	81	1424	19	64	
Eutrichosomatidae	4	5	0	0	Extralimital (New World, India, Australia)
Herbertiidae	2	8	1	1	
Hetreulophidae	3	6	0	0	Australia and New Zealand
Heydeniidae	1	19	1	2	
Idioporidae	1	1	0	0	
Leucospidae	4	151	1	21	
Lyciscidae	28	83	5	9	
Macromesidae	1	12	1	1	
Megastigmidae	12	211	4	10	
Melanosomellidae	31	83	1	2	
Metapelmatidae	1	39	1	5	
Moranilidae	15	49	1	1	
Mymaridae	117	1434	43	55	Approximate S.E. Asian figures (J. Huber, pers. comm.)
Neanastatidae	2	45	1	4	
Neodiparidae	2	7	0	0	
Ooderidae	1	22	1	8	
Ormyridae	5	153	2	31	
Pelecinellidae	3	10	1	1	
Perilampidae	17	277	5	33	
Pirenidae	25	232	6	9	

Continued

Table 22.1. Continued.

Family	Described genera	Described extant species	Described S.E. Asian genera	Described S.E. Asian species	Notes
Pteromalidae s.s.	443	3175	46	121	After removal by Burks et al. (2022) of 23 separate families
Signiphoridae	5	88	2	7	
Spalangiidae	6	74	1	9	
Systasidae	3	81	1	1	
Tanaostigmatidae	9	99	1	2	
Tetracampidae	16	50	1	1	
Torymidae	68	1100	11	29	
Trichogrammatidae	98	c. 800	14	47	

Antenna

The overall shape of the antennae is quite variable, but is most often geniculate (elbowed), that is there is a long scape, with the following pedicellus and flagellum at an angle to it. The basal part of the scapus is often demarked by a constriction, and is referred to as the radicle (not a separate segment but may look like one in some smaller-bodied taxa). The flagellum is often divided into zones (Fig. 22.1). Distally, one or more segments are often enlarged and may be fused to form a club which may be difficult to discern to extremely conspicuous. The flagellar segments (flagellomeres) between the pedicel and club are called the funicle. Sometimes the basal flagellomeres are extremely reduced, and lack elongate multiporous plate sensilla, in which cases these ring-like structures are called anelli.

Heraty et al. (2013) discussed the structure of the chalcidoid antenna and offered a somewhat modified interpretation than various previous workers.

The most distinctive feature of chalcidoid antennae is the presence of elongate placode sensilla (= multiporous plate sensilla) on the flagellar segments. These are superficially similar to those of ichneumonoids and various other apocritans, but the chalcidoid ones differ in their structure (Basibuyuk and Quicke, 1999). They lack an encircling groove (usually requires SEM to see) and, more conspicuously, their tips protrude above the flagellar cuticle, often extending a little beyond the apex of the flagellomere (Fig. 22.2). These are unique to chalcidoids. In contrast the placode sensilla of ichneumonoids and other groups lie flat apically and do not protrude distally. It has been hypothesised that the chalcidoid type sensilla evolved via a transformation from an erect sensory seta to a prostrate one with subsequent extension of the neural opening along its fused length.

Thoracic Morphology

Prepectus, Axillae and Axillulae

The chalcidoid mesosomal skeleton and musculature is extremely phylogenetically informative and has been extensively investigated (Gibson, 1986b). Especially at mid-level classification there are many structural terms that are hardly visited in work on other groups, and it is far too complicated for a detailed account in this book. Readers are referred to the more comprehensive illustrated descriptions of Gibson et al. (1997) and Heraty et al. (2013). One structure is of particular importance when it comes to both recognition of the superfamily and family-level identification and that is the prepectus or postspiracular sclerite, which is visible externally. The prepectus corresponds to the site of origin of the anterior thoracic spiracle occlusor muscle. In nearly all chalcidoids this is a separate sclerite which at least partly separates the

Table 22.2. Summary of major taxonomic changes at family group level implemented by Burks et al. 2022) with emphasis on those that affect the S.E. Asian fauna.

Taxon	Former status or current status if moved	Known distribution	Comments
Asaphinae	Subfamily of Pteromalidae	Australia	Chalcidoidea *incertae sedis*
Austrosystasinae	Subfamily of Pteromalidae	Neotropical	Chalcidoidea *incertae sedis*
Boucekiidae	Tribe of Pteromalidae: Cleonomynae	Neotropics, Australia and Taiwan	1 known species
Calesidae	Elevated to family from Aphelinidae	New World, Palaearctic, South Africa	1 genus
Ceidae	Subfamily of Pteromalidae	Cosmopolitan	3 genera
Cerocephalidae	Subfamily of Pteromalidae	Cosmopolitan	1 genera
Chalcedectidae	Tribe of Pteromalidae: Cleonomynae		1 genus
Chromeurytominae	Subfamily of Pteromalidae transferred to Megastigmidae		
Cleonymidae	Subfamily of Pteromalidae	Cosmopolitan	6 genera
Coelocybidae	Subfamily of Pteromalidae	Australia+New Zealand, one genus also in Papua New Guinea	17 genera
Ditropinotellinae	Subfamily of Pteromalidae	Principally Australian	Chalcidoidea *incertae sedis*; 1 genus
Diparidae	Subfamily of Pteromalidae	Cosmopolitan	11 genera
Elatoidinae	Subfamily of Pteromalidae transferred to Neodiparidae		
Eopelma	Genus of Eupelmidae	S.E. Asia and Nepal	Chalcidoidea *incertae sedis*; 6 species
Enoggerinae	Genera of Pteromalidae: Spalanginae	Australian	Chalcidoidea *incertae sedis*; 2 genera
Epichrysomallidae	Subfamily of Pteromalidae	Old World tropics	15 genera
Erotolepsiinae	Subfamily of Pteromalidae transferred to Spanalgiidae		
Eunotidae	Subfamily of Pteromalidae	Cosmopolitan	5 genera
Herbertiidae	Subfamily of Pteromalidae	Cosmopolitan	2 genera
Hetreulophidae	Subfamily of Pteromalidae	Australia and New Zealand	3 genera
Heydeniidae	Tribe of Pteromalidae: Cleonomynae	Nearly Cosmopolitan	1 extant genus
Idioporidae	Tribe of Pteromalidae	Meso-American	1 genus
Keiraninae	Subfamily of Pteromalidae transferred to Megastigmidae		
Keryinae	Tribe of Eulophidae	Australian	Chalcidoidea *incertae sedis*; 1 genus
Louriciinae	Subfamily of Pteromalidae	Known from Australia, Africa and S.E. Asia	Chalcidoidea *incertae sedis*; 1 genus
Lyciscidae	Tribe of Pteromalidae: Cleonomynae	Cosmopolitan	28 genera, two tribes
Macromesidae	Subfamily of Pteromalidae	Cosmopolitan except South America	1 genus
Melanosomellidae	Subfamily of Pteromalidae	Entirely Australian except for some African *Trichilogaster* species	31 genera

Continued

Table 22.2. Continued.

Taxon	Former status or current status if moved	Known distribution	Comments
Moranilidae	Tribe of Eunotinae of Pteromalidae	Mostly from Australian/New Zealand, 1 genus recorded from S.E. Asia	15 genera, 2 subfamilies
Neapterolelapinae	Genera of Pteromalidae: Diparini	Australian	Chalcidoidea *incertae sedis*; 3 genera
Nefoeninae	Subfamily of Pteromalidae transferred to Pelecinellidae		
Neodiparidae	Tribe of Pteromalidae	East Asia	2 genera
Ooderidae	Tribe of Pteromalidae: Cleonomynae	Cosmopolitan except South America	1 genus
Otitesellini	Comprises the former pteromalid subfamilies Otisellinae, Sycoecinae and Sycoryctinae	Cosmopilitan	Tribe of Pteromalinae
Parasaphodinae	Subfamily of Pteromalidae	Africa, Australia, Indo-Australian and Oriental regions	Chalcidoidea *incertae sedis*; 1 genus
Pelecinellidae	Subfamily of Pteromalidae as Leptofoeninae and Leptofoeninae	New World, S.E. Asia and Australia	3 genera
Pirenidae	Subfamily of Pteromalidae; Eriaporidae treated as synonym		25 genera, 5 subfamilies
Spalangiidae	Subfamily of Pteromalidae	Cosmopolitan	6 genera, 2 subfamilies
Storeyinae	Subfamily in Pteromalidae	Australia, Nepal, Thailand	Chalcidoidea *incertae sedis*; 1 genus
Sycophaginae	Transferred from Agaonidae to Pteromalidae		
Systasidae	Subfamily of Pteromalidae	Cosmopolitan except South America	3 genera, 2 subfamilies
Tomocerodinae	Raised from tribe and transferred to Moranilidae		

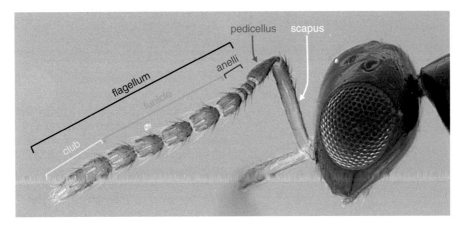

Fig. 22.1. Head and antenna of *Philotrypesis* sp. (Pteromalidae: Sycoryctinae) showing antenna structure terminology in Chalcidoidea. In this species, the club comprises three segments, the funicle five segments, and there are three very reduced flagellar segments called anelli (or ring segments).

Fig. 22.2. Flagellar segments of a pteromalid showing the elongate placoid sensilla on funicle and club segments, whose tips are elevated above the cuticle and often protrude a little beyond the apex of the flagellomere.

mesopleuron from the pronotum (Fig. 22.3 a, a'), although it is very occasionally fused to the pronotum in Perilampidae and some Eucharitidae (Fig. 22.3 c, c'). Its size varies greatly from minute in Chalcididae to very large, and its shape is also important. In the Eupelmidae it has a more 3-dimensional structure with distinct frontal and lateral parts (Fig. 22.3 b, b').

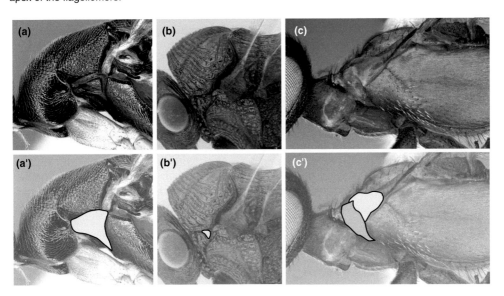

Fig. 22.3. Lateral views of mesosoma showing the prepectus, highlighted in lower member of each pair of micrographs, which is only visible externally, usually as a separate sclerite, in the Chalcidoidea, although it is reduced in some, rarely fused to the pronotum (c), or occasionally lost altogether: **(a, a')** torymid; **(b, b')** *Ancylotropus montanus* (Eucharitidae) showing fused state; **(c, c')** *Aprostocetus* sp. (Eulophidae), showing 3-dimensional prepectus with distinct frontal (orange) and lateral (yellow) parts.

Frenum and frenal area

In many members of the superfamily the posterior part of the scutellum is demarcated by a transverse furrow or raised line (the frenum), and the area behind this called the frenal area.

Wings and venation

The fore wing has greatly reduced venation and the terminology used is different from that employed with most other groups that have more extensive venation. In chalcidoids it usually comprises only four named veins: submarginal, marginal, postmarginal and stigmal (Fig. 22.4 a). A few taxa, such as some larger chalcidids, leucospids and perilampids, have various other veins indicated by pigmented wing membrane thickenings, although never as tubular veins. There have been several historical considerations of the homologies of chalcid veins, but the most sensible is that of Burks (1938) and we illustrate his interpretation of chalcidid venation in Fig. 22.4 b (using the terminology of Sharkey and Wharton (1997)). The stigma in chalcidoid wings is not homologous with the pterostigma, but is a swollen junction between vein r-rs and RS. Anterodistally the stigma may have a short spur-like projection called the uncus.

In many chalcids the distal part of the submarginal vein is differentiated, perhaps by a sharp bend or by being thickened and occasionally by a narrowing or break. This part is called the parastigma (Fig. 22.4 a) and is probably where vein RS exists fused to R. If there is a distinct spur or curved vein indication originating at the point of differentiation, it is called the basal vein and represents 1RS&1M.

Filum spinosum and linea calva

The *filum spinosum* is a row of erect setae that runs diagonally from near the stigmal vein towards the postero-basal part of the fore wing in some chalcidoids. Immediately adjacent to it on the side towards the base of the wing is a glabrous band called the *linea calva*.

Metasoma

Because many chalcidoids have a distinct small, tubular first metasomal segment, usually referred to as the petiole, the numbering of segments (e.g. by Bouček, 1988) starts after that, i.e. with the first segment of the gaster. The cerci (= pygostyles) are on the posterior margin of the 8th metasomal (= 7th gastral; = 9th abdominal) tergite, i.e., the penultimate tergite.

In many chalcidoids, the last visible dorsal structure of the female metasomal tergal series, beyond the cerci-bearing tergite, is generally called the epipygium. This is often an elongate structure and it is a syntergum composed of the fused metasomal tergites 7 and 8, the cerci being at the distal margin of tergum 7. These tergites are articulated in a few taxa, notably Torymidae and some Eulophidae.

Identification

There is no specific regional key to families, and identification of chalcidoids to family is not straightforward. Bouček (1988) and Gibson (1993) both provided keys to all families, and with practice it should be possible to use these to identify nearly all S.E. Asian specimens. Bouček's work, although principally on the Australian fauna, included all of the Island of New Guinea, East Timor, and southern Indonesia (Sulawesi) as well as offshore islands and New Zealand, and for these regions it covered all genera. Further, published keys to families have not been updated to take into account the relatively recent new additions (those from the formal splitting of some heterogeneous ones that have already been published). Subba Rao and Hayat (1985) provided a key to the families occurring in India and adjacent countries that may also be of use.

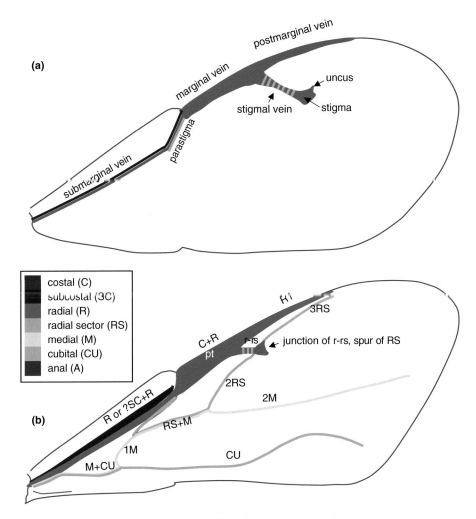

Fig. 22.4. Fore wing venation in Chalcidoidea: **(a)** the four named veins of most species and their usual terminology; **(b)** a chalcidid wing which has traces of more of the ground plan venation, and the presumed homologies of the veins to those of other groups.

Highly Simplified, Partial* Family Key to Female Fully Winged Species

(*does not include the Azotidae, Chrysolampidae, Cynipencyrtidae, Eriaporidae, Tanaostigmatidae, Tetracampidae or any of the recently erected families of Burks *et al.* (2022)).

1	Fore wing longitudinally folded; ovipositor curved over dorsal surface of metasoma lying in a groove (Fig. 22.23 c); (large bodied, often wasp-like mimics) .... **Leucospidae**
-	Fore wing not folded; ovipositor not curved over dorsal surface of metasoma 2
2(1)	Wing membrane glabrous (i.e. without setae) or with a single seta (Fig. 22.45 a, d); (rare; small bodied, less than 2 mm long) .... **Signiphoridae** (plus few **Trichogrammatidae**)
-	Wing membrane setose although setae may be arranged in rows 3
3(2)	Tarsi with 3 segments; setae of fore wing membrane usually arranged in four or more longitiudial rows (Fig. 22.49 a); (very small, usually 1 mm long or less .... **Trichogrammatidae**

-	Tarsi with 4 or 5 segments; setae of fore wing membrane not arranged in longitudinal rows 4
4(3)	Face with 'H'-shaped pattern of grooves (Fig. 22.26); hind wing stalked, i.e. wing membrane not extending all the way to its base (Fig. 22.27, b, e, g), occasionally hind wing membrane completely lost; (very common, usually less than 2 mm, but there are some larger species).. **Mymaridae**
-	Face without 'H'-shaped pattern of grooves; hind wing membrane extending all the way to base... 5
5(4)	Hind femur strongly enlarged **AND** ventrally dentate (Figs 22.10, 22.31, 22.48); often large wasps, body more than 2 mm long.. 6
-	Hind femur seldom enlarged **NEVER** with teeth; size variable................................ 8
6(5)	Tegula large, nearly 2× longer than its distance to pronotum (Fig. 22.10 a); prepectus small, narrow curved, not extending to base of fore coxa (Fig. 22.10 a); (quite common; not metallic) ....**Chalcididae**
-	Tegula smaller, approximately as long as or shorted than after as its distance to pronotum; (see Fig. 7.5 a) prepectus large, triangular and extending to base of fore coxa 7
7(6)	Eyes with inner margins diverging ventrally in anterior view; antennae inserted low on face (Fig. 22.31 a) .... a few former **Pteromalidae**
-	Eyes with inner margins parallel; antennae inserted near middle of face (Figs 22.47, 22.48) .... (some) **Torymidae**
8(5)	Head prognathous, highly elongate and flat in lateral view (Fig. 22.6); fore and usually also middle legs with their tibiae shorter (smaller) than their femurs, usually only half as long; stigmal vein often very long and basally nearly forming a right angle with fore wing margin (Fig. 22.6 a, f); eyes often very reduced; (common; not metallic, black, brown or yellow; head and body usually shiny and without sculpture)................**Agaonidae** (+ some fig-associated **Pteromalidae**: see **Agaonidae**)
-	Head usually orthogathous and not especially elongate and flattened **OR** fore and mid tibiae usually approximately as long as their femurs; stigmal vein much shorter and not at right angle with fore wing margin; eyes normal....................................... 9
9(8)	Cercal plates (indicated by conspicuous tufts of setae) located quite far anteriorly on the metasoma (Fig. 22.11 b-f), usually with the surrounding tergites narrowed and forming a 'V'-shaped arrangement (Figs 22.11 c, 22.12 f); anterior of middle coxa insertion at or anterior to middle of mesopleuron (Fig. 22.11 a); (common; seldom more than 2 mm long) .... **Encyrtidae**
-	Cerci usually short protruding rather than plate-like and situated very close to the posterior of the metasoma, and without conspicuous 'V'-shaped arrangement of tergites; anterior of middle coxa insertion clearly well behind midlength of mesopleuron... 10
10(9)	Fore wing with stigma much enlarged, forming a nearly rectangular blob (Fig. 22.24 c) (but beware some taxa that have a large dark pigmented spot there but it is in the wing membrane and not a swollen venous stigma)........................**Megastigmidae**
-	Fore wing with no or small stigma 11

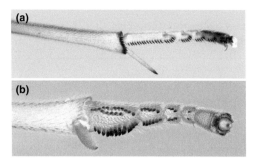

Fig. 22.5. Middle tibia of two eupelmids showing row of ventral pegs, as also found in Encyrtidae, Tanaostigmatidae and some Aphelinidae, though most eupelmids are larger.

11(10) At least first segment of mid tarsus with one or more rows of ventral pegs (Fig. 22.5) or dense, pad-like clusters of setae, these often contrastingly coloured; mesopleuron smoothly rounded without a diagonal division (Figs 22.8 d, 22.21 b–d) 12

- Mid tarsal segments without ventral rows of pegs; mesopleuron usually with an oblique or diagonal depression or groove (Figs 22.22 d, 22.29 d) ,,, ... 13

12(11) Antenna with 8–11 flagellomeres (Fig. 22.21 a, c, e); usually large bodied, > 3 mm ....,,, ... **Eupelmidae** (common) and **Tanaostigmatidae** (very rare)

- Antenna with 6 or fewer flagellomeres (rarely 7) (Fig. 22.8 c); body less than 2 mm long.. some **Aphelinidae**

13(11) All tarsi with only 4 segments (Figs 22.18 e, 22.19 b, d); fore tibial spur short and straight; antennae with 0–4 funicle segments between pedicellus and club .... **Eulophidae** and some **Aphelinidae**

- Tarsi with 5 segments (very rarely with 4, in which case fore tibial spur curved) and mid tarsus at least with 5; antennae usually with more funicle segments between pedicellus and club 14

14(13) Mesoscutum completely hiding vertical pronotum in dorsal view (Fig. 22.15); metasoma petiolate; mandibles sickle-shaped ....
....**Eucharitidae** and some **Perilampidae**

- Mesoscutum not completely hiding vertical pronotum in dorsal view (Fig. 22.29 c), pronotum at least visible laterally in dorsal view; metasoma variable; mandibles not sickle-shaped 15

16(15) Stigmal vein short, sometimes almost non-existent (Fig. 22.28 a); hind coxae much longer than fore coxae (Figs 22.28 a, 22.48); occipital carina present 17

- Stigmal vein variable but usually far more developed (Figs 22.29 b); hind coxae not particularly enlarged, approximately as long as fore coxae; occipital carina absent (Fig. 22.29 c) 18

17(16) Dorsal surface of metasomal tergite 3 onwards with transverse pattern of deep pits (Fig. 22.28); ovipositor not exserted.. **Ormyridae**

- Dorsal surface of metasomal tergite 3 onwards without deep pits; ovipositor exserted .... **Torymidae** (part)

18(16) Pronotum large, more or less rectangular in dorsal view, almost as wide as mesoscutum (Fig. 22.22 c); prepectus small; ovipositor never strongly exserted; never with metallic coloration.. **Eurytomidae**

- Pronotum smaller, not rectangular in dorsal view, often deeply emarginate posteriorly and considerably narrower than mesoscutum, prepectus usually larger; ovipositor variable; frequently with metallic coloration... 19

19(18) Prepectus fused with pronotum (Fig. 22.29 d); metasoma short (approximately as long as head and mesosoma combined), almost triangular in lateral view (Fig. 29.29 v); mesoscutum coarsely sculptured (Fig. 22.29 a, c), notauli complete; left mandible with two teeth, right with three large teeth........................ **Perilampidae**

- Prepectus separate from pronotum (Figs 22.32 a, b, 22.40 a, b, d, 22.44); metasoma usually relatively longer; mesoscutal sculpture variable; mandibles often different (Figs 22.33, 22.38, 22.39) .... former **Pteromalidae**

Agaonidae (Fig Wasps) *sensu lato*

This has been one of the most unstable families in the whole of the Hymenoptera. In the past it was often treated as including virtually all of the obligately fig-associated chalcidoids and thus comprising up to 7 subfamilies Agaoninae, Blastophaginae, Epichrysomallinae, Otitesellinae (= Sycoecinae; = Sycoryctinae) and Sycophaginae (now placed in Pteromalidae *sensu stricto*). The inclusion or exclusion of these from the Agaonidae has fluctuated. Bouček (1988) restricted the family based on biology (associated with figs) and two morphological characters. Earlier, the Apocryptinae and Sycophaginae (both as Sycophaginae) had been placed in the Torymidae, and the Epichrysomallinae and Otitesellinae in the Pteromalidae (Bouček *et al.*, 1981). Obviously, this was a mess, the culprit partly being convergent morphology associated with life within a fig syconium.

In an early molecular study, Rasplus *et al.* (1998) concluded that Agaonidae was not monophyletic and restricted it to just the pollinator clade, i.e. Agaoninae+Blastophaginae +Krabidiinae+ (extralimital, New World) Tetrapinae. The Sycoecinae, Otitesellinae and Sycoryctinae were assigned to the Pteromalidae, but they left the Sycophaginae and Epichrysomallinae unplaced within the superfamily. Wiebes (1967) had put the Sycophaginae in the Torymidae but Bouček (1988) and Heraty *et al.* (2013) treated them as Agaonidae; Cruaud *et al.* (2010) were not specific about their placement but treated them as an outgroup, and finally Cruaud *et al.* (2022) firmly placed them as the sister group to all remaining Pteromalidae s.s, and Epichrysomallidae its now raised to family status (Burks *et al.*, 2022).

Key to Separate True Agaonidae from Fig-associated Pteromalidae in S.E. Asia

1 Mandibles with conspicuous flat, transversely ridged appendages (Figs 22.6 b, e) on the underside of head; antennae usually placed in a broad median channel (Fig. 22.6 a, d), scapes always distinctly enlarged, third segment produced into a curved spine (Fig. 22.6 b, e); head usually prognathous, often longer than broad; fore tibia always much shorter than the femur .... **Agaonidae**
- Female mandibles normal, without appendages; third antennal segment normal, not produced into a spine (see Fig. 22.34 a); scape not enlarged; head usually orthognathous; fore tibia as long or a little longer than femur; ovipositor, or the narrowly tubular metasomal 'tail' often longer than body (Fig. 22.39 a, b); fore tibia at least as long as fore femur (Fig. 22.39 a, b).. **Pteromalidae**

The true fig wasps, the Agaonidae, have an obligate, mutualistic relationship with fig trees (genus *Ficus*, family Moraceae), whose flowers, which they pollinate, are inside the fig, which is actually an involuted flower head called a syconium. Agaonid larvae form galls on the fig ovules. These pollinating fig wasps are divided into three subfamilies: Agaoninae, Blastophaginae and Krabidiinae.

Members of the fig-associated Pteromalidae (see for example Otitesellinae) are particularly similar to the Agaonidae but their females lack mandibular appendages with rows of teeth or lamellae, which are characteristic of Agaonidae s.s. (Burks *et al.*, 2022). The mandibular appendages of Agaonini have many transverse combs of teeth, whereas those of Blastophagini only have transverse lamellae and usually rather fewer of them.

True fig wasps alone are responsible for fertilising the fig flowers in the unripe syconia, and thus have largely co-evolved along with the genus *Ficus*, giving rise to fairly tight association between fig wasp and fig tree species. Females of many of these (*Blastophaga*, *Ceratosolen* and *Pleistodontes*) have specially modified structures for carrying pollen to the next fig. The structures include 'pockets' on the thorax (Wiebes, 1979), corbicula (pollen baskets) on the fore leg (so called in reference to pollen-holding devices of bees), folds of the intersegmental membrane of the metasoma, parts of the antennae. Some *Blastophaga* simply get covered in fig pollen all over their body (passive pollination) or carry it inside their digestive tracts (Ramirez, 1969). Wiebes (1979) suggested that:

[T]he pre-agaonid is suggested to have been a gall-producing parasite of the pre-*Ficus* flower; the evolution of the fig syconium may be related to dispersal of the seeds rather than to the pollination of the flowers.

Female agaonids have exceptionally large and broad wings compared with their body size (Fig. 22.6 a), and this appears to be an adaptation to drifting in the slow air currents in the forest to reach host trees (Kjellberg et al., 2005).

There are some 850 species of fig worldwide and 440 described species of pollinating fig wasps. Molecular data reveal that some 35% of fig species have associations with more than one pollinator species.

Hill (1967) provided a key to females of world genera, but some of the nomenclature is out of date. Bouček's (1988) coverage obviously predates the limitation of Agaonidae to the two subfamilies considered here, so effectively deals with all the morphologically modified fig-associated groups, most of which are now in other families. Instead, the current Agaonidae now corresponds to Bouček's Agaoninae.

Those living in the western world are familiar with the widely cultivated edible fig, *Ficus carica*, which originated from the Mediterranean area. Of these figs, the persistent (or common) variety that arose during its long history of cultivation has all-female flowers and so reproduces parthenogenetically (parthenocarpic); thus it does not need pollination for fruiting, and so when you are eating these you are not consuming pollinating fig wasps. However, some varieties (i.e. Smyrna figs) do require agaonid pollinator wasps. The wasps do not affect the flavour much.

Eight genera of pollinating fig wasps occurring in S.E. Asia, five in the Agaoninae (*Deilagaon* 4 spp. (Fig. 22.6 a, b); *Eupristina*, 9 spp. (Fig. 22.6 c, d); *Platyscapa*, 3 spp.; *Pleistodontes*, 4 spp. and *Waterstoniella*, 20 spp.), two in Blastophaginae (*Blastophaga* (Fig. 22.6 f) and *Dolichoris*, with 3 spp. each), one in the Krabidiinae (*Ceratosolen*, 26 spp; Figs 22.6 e, 22.7).

Fertilised adult females enter figs of the appropriate *Ficus* species through the narrow ostiole, and then pollinate the female flowers with pollen they collected in the fig from which they had earlier emerged. During the process of entering the fig, the females almost always lose their wings (Fig. 22.7 a), and sometimes their antennae, and so are destined never to visit any more figs. Following entry, they attempt to oviposit on the flower ovules, which then become galled. The fig wasp larva consumes the gall tissue. Those pollinated flowers that are not oviposited in will go on to develop one seed each. Hatched, mated females exit the fig through holes chewed through the fig walls by males; indeed, males cooperate in this activity and figs with more males successfully release more females (Suleman et al., 2012).

Some male agaonids are wingless and highly modified (Fig. 22.7 c), often worm-like, and never leave the figs in which they completed their development. Male and female wasps mate within the fig after emerging from the galled ovules before the winged female wasps disperse, carrying the flower pollen with them. Much of the mating occurs between siblings and there is extreme local mate competition between males, which leads to very female-biased sex ratios overall. Males of *Ceratosolen* seek out and examine galls within the syconium to locate ones containing conspecific females at the correct stage and if so the male chews a hole through the gall and inserts his metasoma to try to inseminate her (Murray, 1990). In many species there are extremely morphologically divergent male morphs, with one being wingless and the

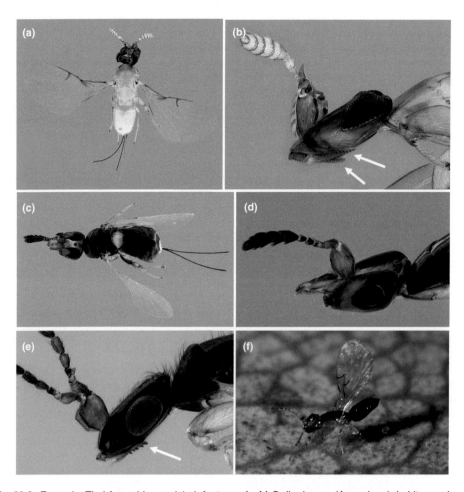

Fig. 22.6. Exemplar Thai Agaonidae and their features: **(a, b)** *Deilagion* sp. (Agaoninae), habitus and head lateral; **(c, d)** *Eupristina* sp. (Agaoninae), habitus and head lateral; **(e)** *Ceratosolen* sp. (Krabidiinae), head lateral; **(f)** live *Blastophaga* sp. (Blastophaginae). (Source: (f) photograph by and © Jean-Yves Rasplus, reproduced with permission.)

other fully winged (Greeff *et al.*, 2003). This difference has been termed a fighter-flyer dimorphism (Hamilton, 1979). It was long believed that, unlike non-pollinating fig wasps, agaonid males did not fight despite that in the vast majority of species, males show particular fighting adaptations. This was because of their highly female-bias sex ration, and that fighting would therefore mostly be between brothers. However, Greeff *et al.* (2003) showed that fighting between brothers had evolved independently in different lineages on between four and six occasions, and male dispersal had evolved at least twice.

Some fig species are dioecious; that is, they have separate 'male' and all-female trees (Weiblen *et al.*, 2001). 'Male' syconia contain male pollen-bearing flowers and female flowers with short styles that therefore have ovaries available for the female wasp to pollinate; female syconia contain only female flowers with long styles that cannot be oviposited in by the wasps. For these to be pollinated and figs to develop, the pollinator wasp must develop in a 'male' flower and then effectively commit evolutionary suicide by entering a female syconium in which she has no possibility of producing young. This system has been subject to much theoretical and increasingly experimental investigation. Apparently crucial is that the female wasps cannot distinguish the 'sexes' of the syconia before entering one.

Fig. 22.7. *Ceratosolen* (Krabidiinae) inside fig syconia of *Ficus cereicarpa*: **(a)** *Ceratosolen* sp. ovipositing into ovule of *Ficus ixoroides*; **(b)** freshly emerged females of *C. pilipes* inside *F. cereicarpa* in Sarawak; **(c)** highly modified male of *C. pilipes*. (Source: photographs by and © Jean-Yves Rasplus, reproduced with permission.)

Relationships between the pollinator taxa were investigated by Cruaud et al. (2010) based on multiple molecular markers. They found strong conflict between genes and morphology and considered that neither Agaoninae nor Blastophaginae were probably monophyletic but suggested maintaining them as useful informative groupings. Agaoninae and Blastophaginae are separated by the shape of the scapus (long vs short) and the structure of the 1st flagellomere (subdivided into three parts vs subdivided into two parts or not divided), respectively. The Blastophagines are included within the Agaoninae by some workers.

Wiebes (1963) revised the *Pleistodontes* species of the region but did not provide a key to species.

Aphelinidae (and Calesidae)

This is a tricky group of small to very small (0.5 – 2.5 mm) wasps. Their small size makes recognising them difficult and often they have to be keyed-out in multiple places (e.g. Gibson, 1993). The family is currently subdivided into four subfamilies: Aphelininae, Coccophaginae, Eretmocerinae and Eriaphytinae. The Aphelininae are further subdivided into three tribes: Aphelinini, Aphytini and Eutrichosomellini (Hayat, 1994; Polaszek et al., 2020).

Of the remaining Aphelinidae, 15 genera have been recorded from the region (Noyes, 2021): *Aphelinus*, *Aphytis*, *Botryoideclava*, *Centrodora* (Fig. 22.8 d), *Coccobius*, *Coccophagus*, *Encarsia* (Fig. 22.8 c), *Eretmocerus*, *Marietta* (Fig. 22.8 a), *Oenrobia*, *Paraphytis*, *Pteroptrix*, *Saengella*, *Timberlakiella* and *Wallaceaphytis*.

Most Aphelinidae are parasitoids of nymphs of Sterorrhyncha (Hemiptera), i.e. Aleyrodidae (whiteflies), Aphididae (greenflies) or Coccoidea (scale insects). A few genera (e.g. *Marietta*) are obligate hyperparasitoids of other chalcidoids (Hoddle et al., 2013). However, for some species, there is an added complication in that males and females have different developmental histories (ontogenies) – the females always developing as primary endoparasitoids of hemipterans (usually scale insects or whiteflies) but the males doing something quite different. This is sometimes referred to as heteronomy. There are four basic biologies exhibited by males: (i) ectoparasitism of the same host as that in which the female is an endoparasitoid; (ii) hyperparasitoids of their own conspecific females; (iii) facultative hyperparasitoids of their own or of other aphelinid female aphelinid species; and (iv) obligate hyperparasitism of other species of aphelinid (Williams and Polaszek, 1996).

The family includes a number of important biocontrol agents (Huang and Polaszek, 1998), particularly in the genus *Encarsia* (Coccophaginae) which comprises some 280 described species worldwide (Polaszek *et al.*, 1999; Schmidt and Polaszek, 2007). Most *Encarsia* species are primary endoparasitoids of whiteflies (Hemiptera: Aleyrodidae) or of armoured scale insects (Hemiptera: Diaspididae). *Eretmocerus* species are also endoparasitoids of whiteflies but, unlike *Encarsia*, the female has a blunt-tipped, soft ovipositor and oviposits between the ventral surface of the host and the leaf (Gerling *et al.*, 1998). Nevertheless, the *Eretmocerus* larva becomes an endoparasitoid because in a complex process it causes the host to engulf it into what is called a vital sac (Gerling *et al.*, 1991). Some aphelinids (approximately ten genera) display heteronomy; that is, males and females utilise different hosts (Walter, 1988). In some species, the males are obligate hyperparasitoids of juvenile conspecifics, and unfertilised haploid eggs cannot develop in the primary host (Quicke, 1997). Some *Encarsia* species are also facultatively heteronomous parasitoids of closely related conspecifics.

Egg parasitism occurs in a few genera (Polaszek, 1991), for example in *Centrodora* (Fig. 22.8). *Aphytis chionaspis* was introduced into South Africa from Thailand for the control of mango scale, *Aulacaspis tubercularis* (Hemiptera: Diaspididae) (Neser and Prinsloo, 2001). Eggs of Lepidoptera and Auchennorhycha are attacked by some *Encarsia* species (Polaszek and Luft-Albarracin, 2010).

Males of at least some aphelinids have exocrine glands in their scapes (Romani *et al.*, 1999; Shirley *et al.*, 2019).

Sugonyaev (1994, 1995, 2012) described two *Coccophagus* species, one from Vietnam and the other from Malaysia, that inhabit ant nests, where they parasitise soft scale insects. In both cases the wasps are somewhat brachypterous and it was proposed that this facilitated their movement within the confined spaces of the nest.

One extralimital and unusual group, the Calesinae, comprising the single genus Cales, was elevated to family rank by Burks *et al.* (2022) because molecular data recover it as the sister group to most of the remaining superfamily. Cales, is predominantly Neotropical but with two species in Australia and New Zealand, and one recently described from Taiwan (Polaszek *et al.*, 2015). These very tiny wasps (< 1 mm) are parasitoids of whiteflies (Hemiptera: Aleyrodidae) and are potentially easily overlooked. Given their distribution, it seems likely that they occur in S.E. Asia. One species, C. noacki, which attacks citrus scale, has been introduced as a biological control agent from South America into North America, the Mediterranean and Africa. All the legs of Cales have four tarsomeres so might be keyed mistakenly to Eulophidae.

Kim and Heraty (2012) provided a morphological phylogeny and a key to world genera of the Aphelininae. Hayat (1983) provided a well-illustrated key to the world genera. Geng and Li (2017) revised the *Encarsia flavoscutellum* species group, the nominal species of which parasitises several pest aphids in S.E. Asia (Suasa-ard, 2010).

Azotidae

This small family only contains the genus *Ablerus* (which is regarded as a senior synonym of *Azotus*, hence the family name); it comprises 94 described species from all over the world and is fairly common in S.E. Asia (Fig. 22.9). As a family it was erected by Nikol'skaya and Yasnosh (1966) but not widely adopted as separate until more recently, instead being treated as a member of the Aphelinidae. Compared with the aphelinids, the host relationships of azotids are far more diverse (Polaszek, 1991). *Ablerus clisiocampae* is a parasitoid of both diaspidid scale insects and lepidopteran eggs (Kattari

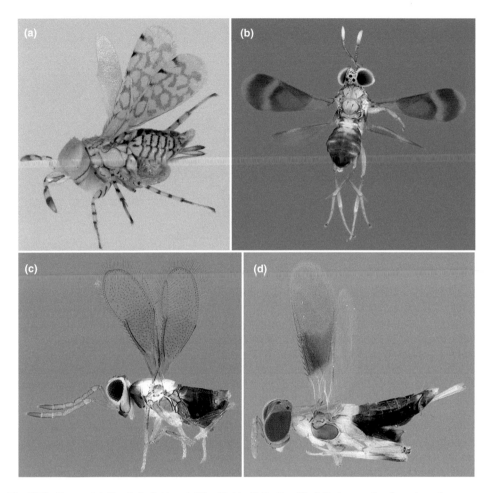

Fig. 22.8. Representative Aphelinidae: **(a)** the highly distinctive *Marietta leopardina* (specimen from South Africa but this species occurs in S.E. Asia); **(b)** Eutrichosomellini (Aphelininae) gen. sp. from Thailand; **(c)** *Encarsia* sp. nr *inquirenda*, from Thailand; **(d)** *Centrodora* sp., from Thailand. (Source: (a) photograph by and © Andrew Polaszek, reproduced with permission.)

et al., 1999). It has generally been presumed that the scale insect is a secondary host and that the *Ablerus* is actually a hyperparasitoid, but direct evidence for this hypothesis is lacking.

Chalcididae

Typically rather chunky wasps with swollen and often toothed hind femora (Fig. 22.10), they are classified into eight subfamilies and 14 tribes based on new molecular (ultraconserved element) data (Cruaud *et al.*, 2021), which is an increase from the six and 11, respectively, recognised in the Universal Chalcidoid Database (Noyes, 2021). All subfamilies are represented in S.E. Asia.

The new classification recognises Brachymeriinae as separate from the Chalcidinae, from which they differ in having a transverse petiole (Gibson, 1993). Species of 33 genera occur in our region (Gul *et al.*, 2020; Noyes, 2021), i.e. approximately 18% of the world total: *Antrocephalus* (Fig. 22.10 d), *Aplorhinus*, *Brachymeria* (Fig. 22.10 a), *Conura*, *Cratocentrus*, *Dirhinus* (Fig. 22.10 b, c), *Epitranus* (Fig. 22.10 f), *Haltichella* (Fig. 22.10 e), *Heydoniella*, *Hockeria*, *Kopinata*,

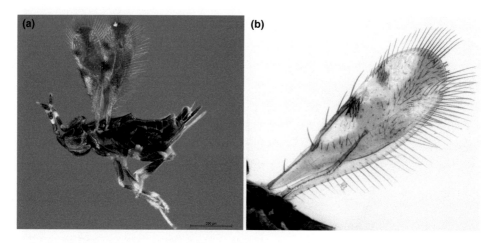

Fig. 22.9. *Ablerus* sp. (Source: photograph by and © Simon van Noort (Iziko Museums of South Africa), www.waspweb.org, reproduced with permission.)

Kriechbaumerella, Lasiochalcidia, Megachalcis, Megalocolus, Nearretocera, Neochalcis, Neohaltichella, Neokopinata, Notaspidiella, Notaspidium, Oxycoryphe, Phasgonophora (= *Muhabbetella,* = *Trigonura*), *Proconura, Psilochalcis, Smicromorpha, Steninvreia, Sthulapada, Tainaniella, Tanycoryphus, Thresiaella, Tropimeris* and *Uga*.

Cowan (1979) studied the function of the swollen hind femora in an extralimital species of *Chalcis* which oviposits into eggs of a stratiomyid fly that are lain in clusters; the study found that they are the only legs used for supporting the wasp during oviposition, and are also used in rearing aggressive/defensive interactions between pairs of females competing for access to the same egg mass. There are no published observations of their use in any S.E. Asian species.

Narendran and van Achterberg (2016) revised the Vietnamese fauna of Chalcididae, probably not including every Indo-Chinese species, and maybe not the majority of them, as it only includes 16 genera and therefore it misses much of the overall regional diversity.

Chalcidinae (Including Brachymeriinae)

One of the most frequently encountered genera is *Brachymeria* (Fig. 22.10 a, g), which has a worldwide distribution, especially in warmer climates. They are apparently all idiobiont endoparasitoids of holometabolous insect pupae, especially Lepidoptera, but also of Coleoptera, Hymenoptera and Diptera. Most are primary parasitoids but some are obligatory or facultatively hyperparasitoids of ichneumonids, braconids or tachinids within the pupa. They tend to have broad host ranges but there is some specialisation based on host group or niche.

Cratocentrinae

The subfamily Cratocentrinae is represented in the region by *Cratocentrus* and *Megachalcis*, the former just extending as far as Thailand, the latter with a more eastern distribution (Bouček, 1988). It is strongly suspected that cratocentrines are parasitoids of xylophagous beetles, because reared specimens all emerged from branches or twigs infested by such potential hosts, but detailed biological data are completely lacking (Abul-Sood et al., 2018). They might even be considered as an endangered group.

Dirhininae

Dirhininae are easily recognised by the shape of the head with a pair of protuberances (horns) projecting from the frons on either side of the antennal sockets (Fig. 22.10 b, c). They are represented in the region by *Dirhinus* (numerous species) and *Aplorhinus* (one species). They are idiobiont pupal parasitoids of Diptera, notably of synanthropic and corpse-associated species, but some are also able to attack fruit flies such as *Bactrocera* (Tephritidae) species and some extralimital species have been investigated as biocontrol agents against these too.

Epitraninae

The only genus in the region is *Epitranus* (Fig. 22.10 f), which was revised by Bouček (1982), who noted that they are not rare locally and that their hosts in the region are small moths of the genera *Corcyra* and *Tirathaba* (Pyralidae) and the genera *Tinea* and *Crypsithyris* (Tineidae). Twenty-four species are known from S.E. Asia.

Haltichellinae

The Haltichellinae (Fig. 22.10 e) is one of the largest tribes of the Chalcididae (Abhisree *et al.*, 2021) with 365 species described in 25 genera. The majority of the haltichelline genera are parasitoids of Lepidoptera (Delvare, 2017; Noyes, 2021) but species of two S.E. Asian genera have quite different host–parasitoid relationships. *Neochalcis* species are exclusively parasitoids of nesting Hymenoptera

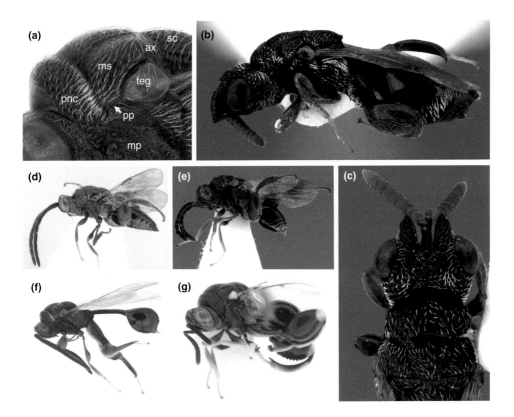

Fig. 22.10. Representatives of some of the chalcidid subfamilies and their mesosomal characters: **(a)** labelled anterior of a *Brachymeria* sp., showing the very tiny, elongate prepectus (pp); **(b, c)** *Dirhinus* sp. from Thailand; **(d)** *Antrocephalus* ?*decipiens*; **(e)** *Haltichella* sp.; **(f)** *Epitranis* sp.; **(g)** *Brachymeria* sp.

(Noyes, 2021; Abhisree et al., 2021) and *Proconura* species are primary parasitoids of bruchid beetles and hyperparasitoids of tachinids or braconids. Records of *Neochalcis* from other hosts are believed to be due to misidentifications of the parasitoid (Abhisree et al., 2021).

Smicromorphinae

One of the most easily recognised groups of chalcidids with the petiolate metasoma inserted very high on the propodeum well above hind coxae, somewhat reminiscent of evanioids. They are rare, predominantly Australian and Papuan wasps but with a few, mostly undescribed, species reported from Africa, India, Singapore, Indonesia and the Philippines (Ubaidillah and Kojima, 2004). The detailed biology is unknown for certain, *Smicromorpha* have been collected several times associated with nests of the weaver ant *Oecophylla smaragdina* (Naumann, 1986; Ubaidillah and Kojima, 2004), and finally reared from an *O. smaragdina* nest in Vietnam by Darling (2009). The distributions of the two genera are broadly congruent, suggesting that this could be the biology of all the species.

Darling (2009) presented a revised key to species to accommodate the new one he described from Vietnam.

Chrysolampidae

Recently elevated to family status by Zhang et al. (2022), and in other works such as Darling et al. (2021), regarded as a subfamily of Perilampidae. The Chrysolampidae comprises six genera, of which *Chrysolampus* itself occurs in S.E. Asia (Narendran and Sudheer, 2004). The life history of a North American species was described by Darling and Miller (1991). The larvae are hypermetamorphic, that is, they have a host-seeking planidial 1st instar larva (see Fig. 22.14), but, unlike eucharitids and perilampids, the hosts of chrysolampids are larvae of tychiine weevils (Curculionidae) feeding on seeds (Fabaceae), or nitidulid pollen beetles on various host plants (Darling et al., 2021).

Cynipencyrtidae

This family was erected by Heraty et al. (2013). It includes only the genus *Cynipencyrtus* from China and Japan, which was previously placed in the Tanaostigmatidae or Encyrtidae (LaSalle and Noyes, 1985). We mention it because it might just possibly occur in the mountains of northern Myanmar or Vietnam.

Encyrtidae

A common and diverse family of small (0.5–3.5 mm), often very beautiful parasitoids. Most are very easy to recognise based on the cercal plates (indicated by a tuft of setae) being located quite anteriorly on the metasoma, usually with the surrounding tergites narrowed and forming a 'V'-shaped arrangement (Fig. 22.11 c-f). Most species oviposit in direct contact with their hosts and consequently have ovipositors that hardly protrude past the end of the metasoma. However, a few have well exserted ovipositors, such as *Homalotylus* and *Pseudencyrtus* species (Fig. 22.12 c, d). Most have rather handsome colour patterns and a few are markedly metallic (e.g. Fig. 22.12 f).

All known host relationships up until 1980 were listed by Tachikawa (1981). Hayat et al. (2010) reported some encyrtids that are associated with lac insects in S.E. Asia. A few encyrtids have non-insect hosts. *Ixodiphagus hookeri* is an endoparasitoid of nymphal blood-sucking ticks (Araneae, Acari: *Amblyomma variegatum*) (Mwangi et al., 1994).

The Lepidoptera caterpillar parasitoids *Copidosoma* and *Ageniaspis* are definitely polyembryonic, and probably most other genera that are currently placed in the Copidosomatini such as *Parablastothrix* are too but in these other cases evidence is circumstantial. Polyembryony in *Copidosoma* has been extensively studied. A female wasp lays a

Fig. 22.11. Various representative Encyrtidae from Thailand. **(a)** *Encyrtus lecaniorum* (Encyrtinae), a widespread parasitoid of scale insects; **(b)** *Cheiloneurus* sp.; **(c)** *Ooencyrtus* sp., an important econoimic genus of egg parasitoids; **(d)** *Anicetus* sp.; **(e)** *Astymachus* sp.; **(f)** *Microterys* sp., probably *nietneri*.

single egg (sometimes a few) into a host egg, but the parasitoid egg divides to make two, these then divide again, and so on, such that inside large hosts, such as various noctuid caterpillars, a thousand or more parasitoid eggs are formed and develop, the resulting brood pupating within a mummified caterpillar skin. In the case of *Coelopencyrtus* several thousand individuals may emerge from a single *Xylocopa* host larva, but there is no proof that this is the result of polyembrony, and it is not certain that *Coelopencyrtus* belongs to the Copidosomatini, although it may do (J.S. Noyes, pers. comm.). It is not known whether *Coelopencyrtus* oviposits into the bee egg before the cells are sealed, or whether the *Coelopencyrtus* female is sealed in the chamber and then lays thousands of eggs into the bee larva host. All definite Copidosomatini are egg/larval parasitoids of Lepidoptera. The story of the polyembryonic *Copidosoma* development is more complicated, because a proportion of the divided eggs are never destined to develop into adult wasps; instead they develop into a morph called a precocious or killer larva which both helps prevent superparasitism and multiparasitism and preferentially kills male parasitoid polygerm and maybe larvae (see Quicke, 1997).

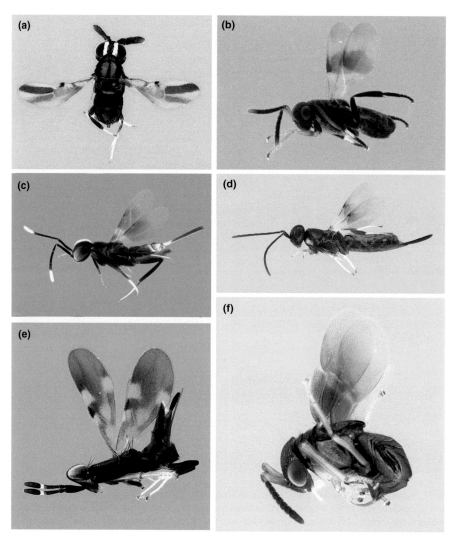

Fig. 22.12. More exemplar Encyrtidae: **(a)** *Comperiella* ?*bifasciata*; **(b)** *Carabunia* sp.; **(c)** *Homalotylus*, probably undescribed species; **(d)** near *Pseudencyrtus* sp.; **(e)** *Adelencyrtus* spp. (Encyrtinae); **(f)** *Gahania* sp.

Males of many, if not all, encyrtids have secretory glands in their scapes and these are associated with release and spread sites composed of 'scale-like' or 'peg-like' cuticular structures, which help spread the released secretions onto the female antenna during courtship (Guerrieri *et al.*, 2001).

Many species have been employed in biological control. Most species are endoparasitoids of scale insects (Hemiptera: Sternorrhymcha: Coccoidea). Probably one of the greatest success stories is that of *Anagyrus* (= *Apoanagyrus*; = *Epidinocarsis*) *lopezi* introduced into Africa to control the cassava mealybug, *Phenacoccus manihoti* (Neuenschwander, 2001). This wasp has also been introduced in recent years to S.E. Asia to control the same introduced pest. However, others are egg parasitoids with hosts belonging to numerous insect orders, e.g. *Ooencyrtus* (Fig. 22.11 c). Members of the Copidosomatini are egg-larval parasitoids of caterpillars and

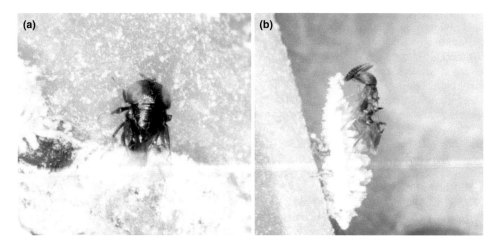

Fig. 22.13. Two species of encyrtid active in the community of parasitoids of the invasive pest mealybug *Phaenacoccus solenopsis*: **(a)** *Aenasius arizonensis* parasitizing *Phenacoccus solenopsis*; **(b)** *Cheiloneurus nankingensis* parasitising *P. solcopsis*, probably via *Aenasius arizonensis*. (Source: from Chen *et al.* (2021) reproduced under Creative Commons Attribution Licence CC-BY-4.0.)

of some aculeate Hymenoptera, and members of at least some genera are polyembryonic.

Noyes and Hayat (1984) reviewed the Indo-Pacific encyrtid genera and this is the most up-to-date work on the regional fauna, although various new genera have been described since then. Prior to that, Trjapitzin (1965) had made a study of S.E. Asian taxa with special reference to the Islands of Komodo and Padar (Indonesia). Noyes and Hayat (1984) recognised only two subfamilies, the Tetracneminae and the Encyrtinae. The most conspicuous difference is that, in the former, the *linea calva* of the fore wing does not have a differentiated margin, and the *filum spinosum* is (nearly) always absent, whereas in the Encyrtinae, the *linea calva* usually has stronger and longer setae on its basal edge and a *filum spinosum* is (nearly) always present. Most genera belong to the Encyrtinae.

Noyes (1991) described a new *Ooencyrtus* species from Malaysia that parasitises the prepupal stage of the cocoa pod borer, *Conopomorpha cramerella* (Lepidoptera; Gracillariidae), a rather unusual host relationship for the genus since most other species of *Ooencyrtus* are primary parasitoids of eggs, as the genus name suggests, although a few have also been recorded as primary parasitoids of syrphid larvae, aphid nymphs or hyperparasitoids of Dryinidae or Braconidae. *Homalotylus* species are parasitoids of coccinellid (ladybird) beetles.

Some extralimital species are egg parasitoids of lacewings (Neuroptera: Chrysopidae) and may have negative impacts on the latter's role in controlling hemipteran pests (Trjapitzin and Shuvakhina, 2019). For example, in southern India *Copidosomyia ambiguous* attacks eggs of *Mallada desjardinsi*, a significant predator of rugose spiralling whitefly *Aleurodicus rugioperculatus*, an invasive pest on coconut palm (Elango *et al.*, 2022). One Oriental species of *Cheiloneurus* is a parasitoid of a beneficial dryinid (Manickavasagam *et al.*, 2006) (Fig. 22.13).

Epichrysomallidae: see Former Pteromalidae (below)

Eriaporidae: see Former Pteromalidae: Pirenidae (below)

Eucharitidae and Perilampidae

Both Eucharitidae and traditional Perilampidae have always been recognised as being closely related, have similar planidial 1st instar larvae which appear to be homologous (Fig. 22.14) (Heraty and Darling, 1984), as well as a particularly complex ovipositor morphology (Heraty and Quicke, 2003). The shared planidial stage led to the group being collectively termed the 'planidial clade'.

However, Eucharitidae has also clearly been shown to be a derived clade within the former Perilampidae with the subfamilies Chrysolampinae, Philomidinae and Perilampinae forming a basal grade (Heraty et al., 2013). This might normally have led to synonymising Perilampidae with Eucharitidae but this would have meant the family name would have been less informative about biology. To maintain this nomenclatural stability, Zhang et al. (2022) raised two subfamilies to family level, i.e. Eutrichosomatidae and Chrysolampidae, which meant that Eucharitidae and remaining Perilampidae were both monophyletic. The two families are described separately below.

Fig. 22.14. Planidial larvae Eucharitidae and Perilampidae: **(a)** cleared microscopic preparation of *Schizaspidia nasua* planidium (Eucharitidae); **(b)** living larva of the extralimital *Perilampus auratus* on natural plant background (Perilampidae). (Source: (a) photograph by and © John Heraty, reproduced with permission; (b) photograph by and © Christopher Darling, reproduced with permission.)

Heraty and Murray (2013) surveyed the evolution of planidial feeding across the two families. In some groups there is no feeding; in some there is sustenance feeding but without growth; and in others they exhibit growth feeding.

Eucharitidae

A distinctive group of medium-sized to small chalcidoids; all species for which biology is known are parasitoids of juvenile stages of ants. Worldwide it comprises about 53 genera and slightly more than 400 species. Fifteen genera occur in our region (Noyes, 2021): *Ancylotropus* (Fig. 22.15 c), *Anorasema*, *Chalcura*, *Eucharis*, *Gollumiella*, *Hayatosema* (Fig. 22.15 f), *Losbanus*, *Neolosbanus* (Fig. 22.15 e), *Parapsilogastrus*, *Pogonocharis*, *Psilocharis* (Fig. 22.15 d), *Rhipipalloidea*, *Saccharissa*, *Schizaspidia* (Fig. 22.15 a, b) and *Stilbula* (see Heraty et al., 2009). Collectively these are represented by 47 S.E. Asian species.

Heraty (2019) catalogued the world species, and the world genera were revised and a key provided by Heraty (2002). Two (currently) extralimital groups have been transferred to other families: Philomidinae was removed to Perilampidae (see Heraty et al., 2019) and Echthrodapinae to Torymidae by Grissell (1995). The remaining eucharitids are divided among three subfamilies: Eucharitinae, Gollumiellinae and Oraseminae.

Eucharitinae

The biology of *Schizaspidia diacammae* was studied in Thailand by Heraty et al. (2015). The adults and immature stages can be obtained by collecting the cocooned pupae of its ant host, *Diacamma scalpratum* (Formicidae: Ponerinae: Ponerini). The biology of an African species of *Psilocharis* African species was described by Heraty et al. (2018).

Gollumiellinae

Heraty et al. (2004) erected this subfamily for *Gollumiella* which has been recorded

Fig. 22.15. Various Thai Eucharitidae: **(a, b)** *Schizaspidia aenea*, female and male respectively (Eucharitinae); **(c)** *Ancylotropus montanus* (Eucharitinae); **(d)** *Psilocharis hypaena* (Eucharitinae); **(e)** *Neolosbanus palgravei* (Eucharitinae); **(f)** *Hayatosema* sp. (Oraseminae).

from Malaysia, Singapore, Indonesia, Philippines, Thailand, Laos and Vietnam (Noyes, 2021), and also certainly occurs everywhere at least down to Wallace's line.

Oraseminae

Two genera of this subfamily, a *Hyatosema* (Fig. 22.15 f) and *Losbanus*, occur in S.E. Asia, *Orasema* itself being entirely from the New World (Burks *et al.*, 2017). Heraty (1994, 2000) revised the subfamily and the presented phylogenetic hypotheses, and this was followed by Burks *et al.* (2017) who revised the Old World species and described eight new genera. Most recently, a detailed molecular analysis was presented by Baker *et al.* (2020), whose results indicated that the Old World Oraseminae are a paraphyletic grade

-	Notauli complete and distinct, reaching or nearly reaching scutellum margin (Fig. 22.16 b-e) or curving towards and reaching axillae.. 5
4(3)	Scutellum with only 1 pair of setae and no other pilosity (Fig. 22.16 a); submarginal vein with only 2 (rarely only 1) dorsal bristles **AND** vein strongly tapering at apex ('broken' off from broad base of parastigma), not smoothly joining the parastigma some..**Entedoninae**
-	Scutellum with 2 (Fig. 22.16 b-e) or more pairs of setae and sometimes more generally setose; fore wing submarginal vein with at least 4 dorsal bristles and usually continuous with the parastigma, not or hardly barrowing apically ....some **Eulophinae**
5(3)	Axillae strongly angulately advanced along hind portion of the straight, groove-like notauli (Fig. 22.16 d) **OR** if only moderately advanced then anterior pair of scutellar setae situated near or behind middle (Fig. 22.16 d) **AND** scutum and scutellum shiny although maybe with fine microsculpture (Fig. 22.16 e); scutellum usually with submedian longitudinal grooves (Figs 22.16 d, c, 22.17); submarginal vein not smoothly continuous with parastigma .... **Tetrastichinae**
-	Axillae not angulately advanced (Figs 22.16, a–c, f) **OR** if somewhat approaching that condition, scutum and scutellum with strong, dense reticulate sculpture (Fig. 22.16 f) **AND** anterior scutellar setae in anterior third; submarginal vein often smoothly continuous with parastigma 6
6(5)	Fore wing submarginal vein dorsally with 2 (rarely only 1) bristles (Fig. 22.18 c) **AND** scutellum with only 1 pair of setae (Fig. 22.16 a) .... a few **Entedoninae**
-	Fore wing submarginal vein dorsally usually with more than 2 bristles (Fig. 22.16 b) **OR** scutellum with 2 pairs of setae.. 7
7(6)	Submarginal vein smoothly curving into parastigma; postmarginal vein mostly longer than, rarely only as long as, the stigmal vein; scutellum often with sublateral grooves separating the axillae; tergites 7 and 8 not separated dorsally....................... ....some **Eulophinae**

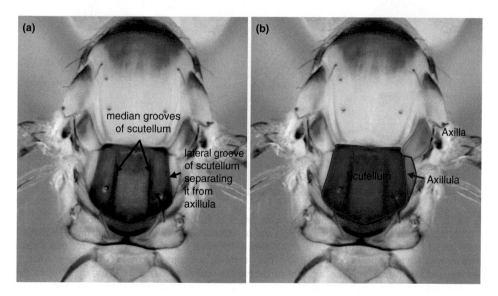

Fig. 22.17. Dorsal views of the mesosoma of a tetrastichine eulophid showing **(a)** submedial and sublateral longitudinal grooves on scutellum (= mesoscutellum) and **(b)** the relationships between the scutellum, axillae (far advanced anteriorly in this species) and axillulae.

Fig. 22.18. A selection of Thai Eulophidae: **(a)** *Omphale* sp. (Entedoninae); **(b)** gen. sp. Entedoninae, possibly *Pleurotropis* sp.; **(c)** *Closterocerus* sp. aff. *javanus* (Entedoninae); **(d)** *Euderus* sp. (Entiinae); **(e)** *Elasmus* sp. (Eulophinae: Eulophini).

	Submarginal vein tapering to apex, not continuous with parastigma **OR IF** smoothly curving into parastigma **THEN** scutellum without sublateral grooves (Fig. 22.18 d) **AND** female metasoma with tergite 8 dorsally separated from tergite 9 8
8(7)	Female with metasomal tergites 8 and 9 (the epypygium) weakly connected, there being a groove between the cerci at the posterior of the 8th tergite; postmarginal vein distinct, approximately as long as stigmal vein (Fig. 22.18 d); notauli usually incomplete posteriorly .... **Entiinae** (= **Euderinae**)
-	Female with metasomal tergites 8 and 9 (the epypygium) fused, without transverse groove at the level of the cerci at the posterior of the 8th tergite; fore wing postmarginal vein usually rudimentary or absent (Figs 22.19 b, e (inset)); notauli complete and straight.. **Tetrastichinae**

Fig. 22.19. A selection of Thai Eulophidae: **(a)** gen. sp. (Eulophinae: Cirrospilini); **(b)** *Melittobia* sp. (Tetrastichinae); **(c)** *Pediobius* (Entedoninae); **(d)** *Aprostocetus* sp. (Tetrastichinae); **(e)** *Henryana* sp. (Tetrastichinae), inset showing its peculiar hind wing base.

In several groups of chalcidoids, especially Eulophidae, various features of the scutellum, axillae and axillulae are important. The axillae are separated from the mesoscutum by the trans-scutal groove/articulation and are lateral to the scutellum. In many Testrastichinae the trans-scutal articulation is sometimes strongly deviated anteriorly lateral to the anterior margin of the scutellum and so they protrude forwards (Figs 22.16 d, e, 22.17).

Bouček (1988) provided a key to all the IndoAustralian genera including those occurring in S.E. Asia. Reina and LaSalle (2003) provided an online key to the world genera of Eulophidae that includes species that are parasitoids of leaf-mining Agromyzidae (Diptera).

Entedoninae

A large subfamily with more than 90 genera worldwide. Most entedonines are primary parasitoids or hyperparasitoids of weakly concealed insect larvae of various groups, although some attack host pupae or eggs. Hosts include Coleoptera, Diptera, Lepidoptera and Thysanoptera. Biology tends to be fairly conserved within genera. *Pediobius* species (Fig. 22.19 c) are often hyperparasitoids (Fig 22.18 a–c) and many important in biocontrol.

Entedonines are one of the most easily recognisable groups of eulophids (Schauff, 1991). Their unique characters include the presence of only one pair of scutellar setae, middle lobe of mesoscutum with two pairs of setae, submarginal vein with only two dorsal setae, and face with frontal sulcus distinctly removed from anterior ocellus.

Ubaidillah et al. (2000a) described an aberrant S.E. Asian entedonine from Peninsular Malaysia, Sulawesi and Papua New Guinea, *Ambocybe petiolata*, that in addition to its own peculiarities has five or six pairs of scutellar setae. and lacks a frontal sulcus. Since the number of scutellar setae is one of the best characters for recognising the Entedoninae, correct recognition of *Ambocybe* may be challenging. Yefremova et al. (2018) recorded the genera *Chrysonotomyia*, *Euderomphale*, *Neochrysocharis* and *Omphale* from Cambodia.

Entiinae (= Euderinae)

This subfamily is rather small with just 18 genera worldwide, of which *Acrias*, *Aoridus*, *Astichus* and *Euderus* have described species in S.E. Asia. They are parasitoids of eggs or larvae of Coleoptera (e.g. Buprestidae, Curculionidae, Cerambycidae and Cisidae in bracket fungi); some also attack wood-boring Lepidoptera. Narendran and Zubair (2013) revised the 10 world species of *Acrias*, of which two occur in the region. Yefremova et al. (2018) recorded *Acrias varicornis* from Cambodia.

Eulophinae

Most eulophines are gregarious idiobiont ectoparasitoids of weakly concealed caterpillars (typically leafminers, but also rollers and tiers). Gauthier et al. (2000) concluded that the subfamily comprises three tribes Eulophini (including Elasmini, Elachertini and Euplectrini), and Cirrospilini. However, following the results of Rasplus et al. (2020) UCE study, Burks et al. (2022) abolished the Elasmini as they appear to render the Eulophini paraphyletic.

The S.E. Asian fauna includes quite a few species of Euplectrini, including members of *Aroplectrus*, *Euplectrus*, *Metaplectrus* and *Platyplectrus*.

Some members of the closely related tribes Euplectrini and Elachertini, e.g., *Eulophus*, *Elachertus*, *Euplectrus* and some related genera (MacDonald and Caveney, 2004) are particularly interesting in that they are koinobiont ectoparasitoids of exposed lepidopteran caterpillars (Fig. 22.20). Some species of these are important parasitoids of economically and medically important hosts. For example, *Aroplectrus dimerus* has been used as a biological control agent of the nettle caterpillar, *Darna pallivitta* (Limacodidae).

Fig. 22.20. Ectoparasitoid Eulophinae (Euplectrini) parasitising small caterpillars in Thailand: **(a)** geometrid larva, possibly *Chiasmia* sp. based on DNA barcode; **(b)** young erebid caterpillar, probably *Dinumma* sp. based on DNA barcode. (Source: photograph by and © Pornthap Kerkig, reproduced with permission.)

Yao and Yang (2009) provided a key to world *Notanisomorphella* species which includes a single S.E. Asian species.

Opheliminae

This small and recently recognised (Burks et al., 2011) subfamily is largely Australian, but there are some species ranging from India through S.E. Asia, some endemic, some as accidental introductions. It comprises two tribes, Ophelimini (*Ophelimus*) and Anselmellini (*Aselmella, Perthiola*), which were treated and keyed out separately within Eulophinae by Bouček (1988). In addition to having more flagellar segments than all other eulophids (nine), both have strongly clavate (female) antennae and complete notauli. They differ from the Euplectrini in having both hind tibial spurs short.

The approximately 50 species of *Ophelimus* are all gall-formers on various *Eucalyptus* species (LaSalle et al., 2009; Borowiec et al., 2019) and some are potentially important economic pests, particularly *O. maskelli* and *O. eucalypti*. Although originally both come from Australia, they have been dispersed to many other countries where *Eucalyptus* is cultivated. *O. maskelli* has been recorded from Vietnam and Indonesia (Lawson et al., 2012). In Malaysia and Vietnam, *Anselmella malacia* has been reported as a pest of Java apple (= wax apple) fruit, *Syzygium samarangense* (Myrtaceae), on whose seeds its larvae feed (Xiao et al., 2006; Oanh et al., 2019). *Perthiola* is thus far only known from Australia and India (Reina and LaSalle, 2005), and therefore it might possibly occur in the region.

Tetrastichinae

By far the largest subfamily with more than 110 genera worldwide. They are enormously biologically diverse, collectively attacking members of more than 100 insect families, as well as including predators of spider eggs and gall-forming mites. A few are totally phytophagous.

The largest genus, *Aprostocetus* (Fig. 22.19 d), is also cosmopolitan and is represented in S.E. Asia by 19 described species. *Aprostocetus* species collectively attack a wide range of hosts, especially gall midges (Diptera: Cecidomyidae), but have even been investigated as biocontrol agents against the anthrophilic American cockroach *Periplaneta americana*. Other commoner hosts include various pest cerambycid beetles and stem-borer Lepidoptera. *Aprostocetus* species are endoparasitoids and they may be solitary or gregarious or facultatively hyperparasitic;

some are egg parasitoids, others attack host larval stages. Pest species attacked include the brown planthopper, *Nilaparvata lugens* (Vongpa *et al.*, 2016).

One genus, *Melittobia* (Fig. 22.19 b), even has an entire article in *Annual Review of Entomology* devoted to it, despite it having only 12 known species (Matthews *et al.*, 2009). *Melittobia* wasps are gregarious ectoparasitoids of solitary and social bees and wasps, including the honey bee. In addition they will parasitise various cohabitants of their main hosts' nests, such as Coleoptera, Lepidoptera and Diptera. They are particularly interesting because of their intra- and intersexual dimorphism There are two female morphotypes (castes), one being fully winged the other brachypterous, the development of which particular form being believed to result from differences in nutrition. Their males are flightless and lack eyes, being adapted for fighting and killing their male siblings, which are probably determined by nutrition. The surviving male mates with its sisters, so obviously these do not have single locus sex determination (see Chapter 2, this volume).

Members of one of the most distinctive genera, *Zagrammosoma*, which have been called tattooed wasps, are niche-specific, primary ectoparasitoids mainly of leaf-rolling and leaf-mining Lepidoptera and Diptera but sometimes also Coleoptera and Hymenoptera with similar biology (Bouček, 1988; LaSalle, 1989). They often attack leaf miners on both agricultural and ornamental plants. Ubaidillah *et al.* (2000b) described a new species of *Zagrammosoma* from Indonesia and Australia which is a parasitoid of the leaf-mining fly, *Liriomyza huidobrensis* (Agromyzidae), an introduced pest in S.E. Asia. Some tetrastichines possess anchored eggs which function in a similar way to those of tryphonine ichneumonids (q.v.) (Zinna, 1955).

Separation of *Zagrammosoma* from *Cirrospilus* is tricky and readers should consult Zhu *et al.* (2002) and other specialist works in addition to more general works such as Bouček (1988). *Zagrammosoma* was revised recently by Perry and Heraty (2021). Narendran (2007) revised the Indian species of Tetrastichinae and this may also be a useful aid for identifying S.E. Asian representatives.

Eupelmidae and *Eopelma*

The Eupelmidae is a cosmopolitan family but primarily occurring in the tropics. Body size ranges from a little over 1 mm up to nearly 2 cm excluding ovipositor. Most commonly these wasps are observed walking over tree trunks or dead wood underneath which their hosts live, although they are also frequent in Malaise-trap samples. The family is divided into three subfamilies, Calosotinae, Eupelminae and Eusandalinae, with the former subfamilies Neanstatinae and Metapelmatinae having been removed and elevated to full family status by Burks *et al.* (2022). Eusandalinae was raised to subfamily status by Burks *et al.* (2022), having been treated previously as belonging to the Calosotinae. A morphological phylogeny of the family was presented by Gibson (1989), who included a key to the subfamilies then included as well as differentiating them from other potentially confusing chalcidoids.

All species appear to be idiobionts but including egg endoparasitoids as well as larval ectoparasitoids.

Key to S.E. Asian Subfamilies of Eupelmidae

1 Anterior profile of mesoscutum in dorsal view evenly rounded........**Eupelmatinae**
- Anterior profile of mesoscutum in dorsal view trisinuate, with obvious 'shoulders' .... **Calosotinae**

Calosotinae

The subfamily includes eight genera of which two occur in S.E. Asia: *Balcha* (ten species) and *Tanythorax* (one species). These are usually rather large chalcidoids with at least some metallic coloration. Females are often seen searching for hosts walking over dead wood. Gibson (1989) revised the world genera of Calosotinae (including Eusandalinae).

Sexual dimorphism is slight. They are mostly ectoparasitoids of xylophagous beetle larvae, although some other host associations have been found for extralimital taxa.

Gibson (2005) revised the world species of *Balcha*.

Eupelminae

Globally the two largest genera are *Eupelmus* and *Anastatus* (Fig. 22.21 b, c). *Eupelmus* species are mainly ectoparasitoids of insect larvae or pupae, whereas species of *Anastatus* are almost always endoparasites of insect eggs belonging to a range of hosts, including Lepidoptera and Heteroptera with a few species known from Orthoptera, cockroaches and praying mantises (Bouček, 1978). The type species of the genus, *A. mantoidae*, has been reared from mantid eggs in Sabah (Malaysia). Some more host-specialised species of *Anastatus* have been involved in biological pest control, for example of the moth *Dendrolimus* and beetle *Tessarotoma* in China (Bouček, 1978).

There is strong sexual dimorphism, especially in their mesosomal characters. If not approached cautiously, the females rapidly escape by jumping. This is particularly

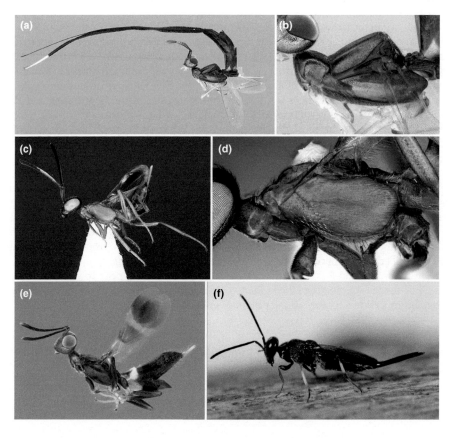

Fig. 22.21. Exemplar Eupelmidae from Thailand: **(a, b)** *Merostenus* (*Hirticauda*) sp., habitus in common death pose of family and detail of mesosoma showing undivided mesopleuron and flat, non-protruding prepectus; **(c, d)** *Anastatus* sp. (Eupelminae), habitus and detail of mesosoma; **(e)** *Zaischnopsis* sp. (Eupelminae); **(f)** *Balcha* sp. (Calosotinae), a somewhat atypical eupelmid, living female. (Source: (f) photograph by and © Jean-Yves Rasplus, reproduced with permission.)

true of members of the subfamily Eupelminae. The enlarged muscles enabling this will contract on death in alcohol, meaning that the dead wasps assume a fairly characteristic death posture (Fig. 22.21 a). Indeed, it has been questioned whether female eupelmines can fly at all, because of a trade-off between jumping and flight capabilities (Gibson, 1986b).

Eopelma

Eopelma includes several species in S.E. Asia, but the genus was left *incertae sedis* in the Chalcidoidea (i.e. not placed to family) by Burks *et al.* (2022). Gibson had described *Eopelma* based on a species from the Philippines; Fusu and Polaszek (2017) then described another from Sabah (Malaysia). Gibson (2017) revised the genus, including

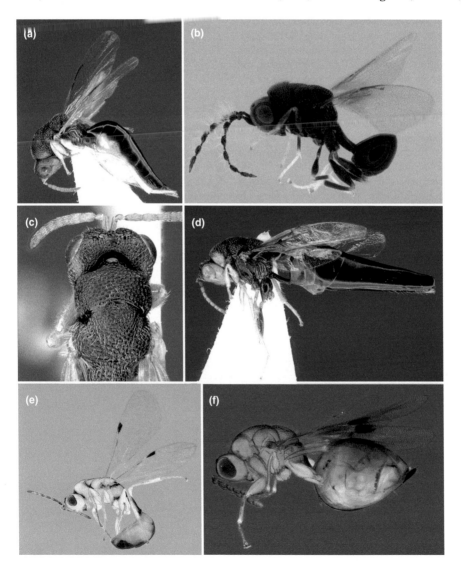

Fig. 22.22. Exemplar Eurytomidae and characters: **(a)** *Bephrata* sp., from New Guinea; **(b)** male *Axemopsis* sp. from Thailand; **(c, d)** female of an undescribed genus of Eurytominae from New Guinea; **(e)** *Sycophila* sp., from Thailand; **(f)** *Ficomila* sp., from Thailand.

descriptions of three new species in S.E. Asia, and a sixth in Nepal.

Eurytomidae

This is a rather large, cosmopolitan family of mostly medium-sized wasps and comprises some 90 genera and 1400 species, of which 19 and 64, respectively, have been recorded from S.E. Asia (Fig. 22.22).

In keys and books, eurytomids have usually been characterised by having a rather quadrate pronotum (in dorsal view), being densely and rather coarse punctate on the head and mesosoma, and by having propodeum medially depressed or grooved. However, this combination is not totally reliable as not all eurytomids have all these characters, and some are found in members of various other families.

Lotfalizadeh *et al.* (2007) presented a morphological phylogenetic analysis of the Eurytominae but found deep relationships within the family to be unstable. The family is nowadays divided into three subfamilies, Eurytominae, Rileyinae, and Heimbrinae, the latter being entirely restricted to New World and Australia (Gates, 2008).

Many eurytomids have a largely internalised ovipositor involving coiling of the 2nd gonocoxae which allows it to be protruded quite a long way whilst being protected within the metasoma when not in use (Copland and King, 1972).

They have diverse biologies. Most developed inside plant tissues, and many are phytophagous in seeds or stems, others are typical parasitoids, and some are associated with figs.

Leucospidae

These are some of the largest, most conspicuous and most easily recognisable chalcidoids. Apart from their characteristic general appearance, note that their fore wings in life are longitudinally folded when not in flight (Fig. 22.23). Several are unambiguously mimics of vespid wasps in terms of coloration.

Leucospis is by far the largest genus, cosmopolitan with 131 species worldwide,

Fig. 22.23. Living and mounted *Leucospis* spp.: **(a)** species from Singapore; **(b)** species from Sarawak; **(c, d)** species from Thailand. (Source: (a) photograph by and © Jean-Yves Rasplus, reproduced with permission; (b) photograph by and © Zestin Soh, reproduced with permission.)

and 20 species recorded from S.E. Asia; the other three recognised genera are all extralimital (Gibson, 1993). They are nearly all ectoparasitoids of solitary aculeate Hymenoptera, mostly of bees but also of eumenine vespids and some Sphecidae s.l. More recently, in Iran the W. Palaearctic *L. dorsigera* (Hesami et al., 2005) has been reported to be a hyperparasitoid of a cerambycid beetle via its xoridine ichneumonid primary parasitoid. Most species are solitary but one extralimital species is known to be gregarious (Grissell and Cameron, 2002).

Bouček (1974) revised the world Leucospidae. Since that work no new species have been described from S.E. Asia (Noyes, 2021). Further, Ye et al. (2017) provided a well-illustrated key to the 12 species known from China, of which four are also known from S.E. Asia.

Megastigmidae

These wasps were for a long while treated as members of the Torymidae, which they closely resemble in their general appearance. However, there have been molecular studies over recent years (Campbell et al., 2000; Munro et al., 2011) and they were re-elevated to family status by Janšta et al. (2018) based on molecular evidence. Burks et al. (2022) transferred Keiraninae and Chromeurytominae, with an undescribed species from Myanmar, from the Pteromalidae to the Megastigmidae.

As their name suggests, they are largely characterised by having a large stigma at the apex of the stigmal vein. The body is typically rather yellowish with little or no metallic lustre, together with a large, rectangular pronotum reminiscent of Eurytomidae (Fig. 22.24). They differ from Torymidae in wing venation but both have a complete occipital carina, long exserted ovipositor and digitiform cerci.

Milliron (1949) described the developmental biology of the extralimital *Megastigmus nigrovariegatus*.

The family includes approximately 180 species in 12 genera. Collectively the genera display a wide range of juvenile feeding strategies (Bouček, 1988; Grissell, 1995; Janšta et al., 2018). Of the genus, *Megastigmus*, approximately one-third of the described species are inquilines of gall-forming insects, another one-third are phytophagous within seeds; and the biology of the rest is unknown (Grissell and Prinsloo, 2001). Recently one species in Indonesia has been found to parasitise galls on *Eucalyptus* caused by the eulophid *Leptocybe invasa* (de Souza Tavares et al., 2023). Some genera include only purely phytophagous species (Böhmová et al., 2022), for example, *Bootania leucospoides* (Myanmar), *B. fascia* (Malayan Cubah), *D. orba* (Malaysia) and *B. piliformis* (Malaysia: Sarawak), the first three all reared from seeds of *Pandanus* (Grissell and Desjardins, 2002). *Mangostigmus* are parasitoids of gall-forming Diptera (mainly Cecidomyiidae), and *Bootanomyia* species are parasitoids of gall-forming cynipoids and chalcidoids.

The larval developmental strategy of at least some seed-feeding Megastigmidae is very similar to that of koinobionts, in that the 'host' seed is attacked early in its development when it is relatively soft and penetrable but offers few resources. The larva at that stage affects the seed little and waits until the seed has accumulated sufficient nutritive resources before completing development (Jansen-González et al., 2020).

In S.E. Asia various species of *Megastigmus* are pests. In Thailand, together with a *Eurytoma* species (see Eurytomidae), up to 76% of seeds of the edible and widely cultivated *Sesbania grandiflora* (Fabaceae) may be attacked (Hellum and Sullivan, 1990). Some of seed samples tested in 1984 had insect exit holes. However, even though partially eaten, approximately half could still germinate, especially if allowed to absorb water.

Metapelmatidae

Burks et al. (2022) restored this monotypic family from synonymy with Neanastatidae (q.v.) which they also removed from Eupelmidae. *Metapelma* (Fig. 22.25) is a distinctive cosmopolitan genus comprising about 40 described species (Noyes, 2021). Gibson (1989) revised the world genera of Metapelmatinae which he included in Eupelmidae at the time. Cao et al. (2020a) described a new Chinese species which is a parasitoid of the buprestid

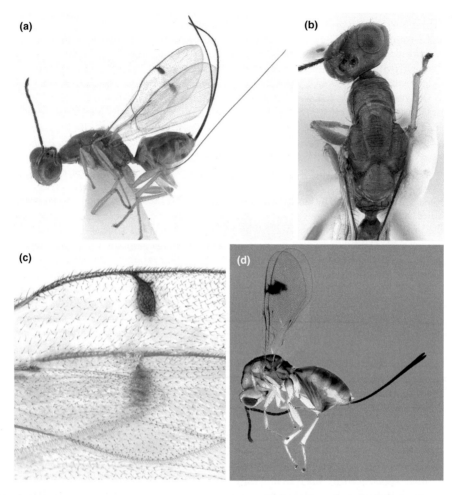

Fig. 22.24. Exemplar Thai Megastigmidae species: **(a-c)** *Megastigma* sp., habitus; **(b)** dorsal view of head and mesosoma showing rather large pronotum; **(c)** fore wing detail showing large 'stigma', short stigmal vein and long postmarginal vein; **(d)** *Bootanomyia* sp.

Fig. 22.25. *Metapelma* species from Thailand.

beetle *Coraebus cavifrons*, a pest of the landscape tree *Symplocos stellaris* (Symplocaceae), and provided a key to the East Palaearctic, Oriental and S.E. Asian species.

Mymaridae (Fairy Flies)

A well-defined, cosmopolitan and abundant family of mostly small to very small egg parasitoids whose hosts include members of several insect orders, notably Orthoptera, Hemiptera, Psocoptera, Coleoptera and Diptera, but a few are also known to attack Odonata and Thysanoptera. The smallest species are about 0.3 mm long but a few can be as large as 5.0 mm. Some have even lost or have reduced adult brain cell nuclei (Polilov et al., 2023).

The majority are solitary but a few are gregarious, and their hosts are mostly concealed, hence some mymarids have relatively long ovipositors. This is probably one reason why they do not attack Lepidoptera, because their eggs are seldom concealed. Most attack Hemiptera eggs but some parasitise members of various other orders, including beetles, barklice (Psocoptera), Orthoptera, Odonata and Diptera (Gibson, 1993; Huber, 1995).

Mymarids may be recognised by the presence of an 'H'-shaped pattern of (usually) dark bars on the face and frons (Fig. 22.26). In addition, when the hind wings are fully developed, they are almost always stalked, that is, the wing membrane does not extend to the base, leaving a naked stem. However, in some genera, the hind wings are reduced to just the stalk (Fig. 22.27 c). The fore wings nearly always have a border of very long setae (Fig. 22.27 b, c, d, g).

Anagrus nilaparvatae and *A. optibalis* are important parasitoids of various plant hoppers, including rice pests such as *Nilaparvata*, *Laodelphax*, *Sogatella* and *Toya* spp., throughout the Oriental and S.E. Asian region (Zhu et al., 2013).

The cosmopolitan genus *Gonatocerus* (in the past often called *Lymaenon* but now regarded as two separate genera) is the largest in the family with, until recently, 410 nominal species. Huber (2015) reclassified these into 14 genera, all within the Gonatocerini, which is largely defined by a distinct arrangement of fore wing macrosetae. Fairly certain host relationships are known for only approximately half of the genera: *Cosmocomopsis* attacks Orthoptera; *Cosmocomoidea* (Fig. 22.27 e, f), *Gonatocerus* and *Lymaenon* mostly attack Cicadellidae; *Gahanopsis* parasitises Membracidae and Cicadellidae; *Gastrogonatocerus* attacks Membracidae.

Subba Rao and Hayat (1983) provided a key to genera from the Oriental Region (comprising India, Pakistan, Sri Lanka, Nepal, Myanmar, Cambodia, Malaysia, Indonesia and the Philippines) and Lin et al. (2007) provided a key to the Australian genera. Although somewhat out of date, these two works should enable most S.E. Asian specimens to be placed to genus. Triapitsyn and Berezovskiy (2007) reviewed the Oriental and Australasian species of *Acmopolynema* which are often common and rather large (Fig. 22.27 d). Anwar et al. (2019) treated the species of *Anophes* (Fig. 22.27 g) from India, of which one also occurs in Indonesia. Sankararaman et al. (2020) provided a key to the world species of *Camptopteroides* and described a new species from Malaysia and Thailand. Triapitsyn et al. (2020) recorded *Tanyxiphium harriet* in the Oriental part of China, Sulawesi Island in Indonesia, Peninsular Malaysia, and Thailand. Narendran and Anjana (2013) described a new *Australomymar* species from Vietnam, partly filling the gap between India, China and Australia where other species occur (Jin and Li, 2015).

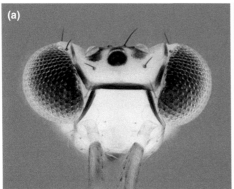 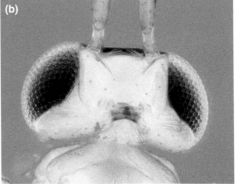

Fig. 22.26. Two views of the head of a *Cosmocomoidea* sp. showing the 'H'-shaped pattern of cuticular bars diagnostic of Mymaridae: **(a)** dorsal; **(b)** ventral.

Fig. 22.27. Exemplar mymarids from Thailand: **(a)** *Ooctonus* sp., probably *O. sinensis*, the largest species in the genus and one of the largest mymarids; **(b)** *Allanagrus* sp.; **(c)** *Mymar* sp.; **(d)** *Acmopolynema* sp., another large species; **(e, f)** *Cosmocomoidea* spp., female and male respectively; **(g)** *Anaphes* sp., probably *A. kailashchandrai*.

Neanastatidae

This was formerly included within Eupelmidae (e.g. Bouček, 1988; Gibson, 2009) but was elevated to family status by Burks *et al.* (2022). It comprises only two extant genera (*Lambdobregma* and *Neanastatus*). Bouček (1988) additionally included *Eopelma* and *Metapelma*, commenting that the Metapelmatinae comprised only *Metapelma* and *Neanastatus* but that they were very different from one another, drawing into question their monophyly, and, based on Cruaud *et al.*'s next-generation sequencing work, Burks *et al.* (2022) removed *Metapelma* to a separate family, Metapelmatidae (q.v.). *Neanastatus* occurs throughout the Old World but *Lambdobregma* is Neotropical. *Eopelma* was left *incertae sedis* by Burks *et al.* (2022) and we treat it together with Eupelmidae above.

S.E. Asian species are parasitoids of wood-boring beetle larvae but species of the extralimital genus *Lambdobregma* are believed to be endoparasitoids of cricket (Gryllidae) eggs. *Neanastatus grallarius* is an occasionally important parasitoid of the rice gall midge, *Orseolia oryzae* (Cecidomyiidae), in S.E. Asia (Hidaka *et al.*, 1988).

Gibson (2009) described three genera from Baltic amber and as a result concluded that the Neanastatinae were not definable in terms of any single synapomorphy.

Ormyridae

This is an easily recognised family that turns up infrequently in Malaise-trap samples and rearings from fruit. For a long while it was considered to include only three genera (the cosmopolitan *Ormyrus*, extralimital *Ormyrulus* and the Malaysian *Eubeckerella*) (Zerova *et al.*, 2021) but Burks *et al.* (2022) included two more extralimital genera: *Asparagobius* and *Hemadas*. The characteristic feature is the transverse row of very, very deep, setiferous pits on metasomal tergites 3–5 (Fig. 22.28) and two robust curved hind tibial spurs. They are nearly all entomophytophagous parasitoids of various gall-forming insects (LaSalle, 2005; Gómez *et al.*, 2017) including cynipids, chalcids and Diptera. However, some may be parasitoids of Eurytomidae developing in seeds. Some are associated with galls in figs.

Perilampidae

Wasps with a generally chunky appearance (c. 3–5 mm); this is a relatively small family. The metasoma is rather triangular in lateral view (Fig. 22.29), the antennae have 13 segments with one anellus and seven funicular segments.

With the removal of the Chrysolampinae into a separate family, Chrysolampidae, only two subfamilies are recognised: Perilampinae and Philomidinae. Nevertheheless, Bouček's (1988) comment that the family 'is difficult to define by simple tangible characters' is still quite appropriate. Only four genera, viz. *Euperilampus*, *Krombeinius*, *Monacon* and *Perilampus* (Fig. 22.29), and 31 species have been recorded from S.E. Asia.

Hosts of *Perilampus* are extremely diverse and include (but are not limited to) beetle, sawfly, sphecid wasps and moth larvae. Many species are primary parasitoids but some are hyperparasitoids via ichneumonoids or tachinid flies. *Euperilampus* appear to be hyperparasitoids of ichnemonids. *Krombeinius* are parasitoids of eumenine vespid wasps. Darling (1995) described two new species of *Krombeinius* from Indonesia and also illustrated the planidial 1st instar

Fig. 22.28. *Ormyrus* sp., Thailand, showing the diagnostic transverse rows of deep pits (creulation) on metasomal tergites, characteristic metallic coloration and short, 13-segmented antennae.

Fig. 22.29. Exemplar Perilampidae: **(a)** living extralimital *Perilampus* species from North America; **(b)** *Perilampus singaporensis* from the Philippines; **(c)** *Monacon gawai*; **(d)** *Monacon* sp., from Indonesia. (Source: (a) cropped from photograph by Christina Butler (Georgia, United States) via Wikimedia Commons under terms of Creative Commons Attribution licence 2.0.; (b-d) photographs by and © Christopher Darling, reproduced with permission.)

larva (see Fig. 22.14 b). Darling and Roberts (1999) described the life history of *Monacon*, parasitoids of ambrosia beetles (Curculionidae: Platypodinae), based on fieldwork in Papua New Guinea. Darling and Tatarnic (2020) documented the life history of another *Monacon* based on a new species from Sarawak (Malaysia).

Heraty *et al.* (2019) reviewed the aberrant subfamily Philomidinae. This is an entirely Old World and principally Afrotropical subfamily. It is a rarely collected group, despite being widespread, displaying a number of morphological peculiarities such that it has even been placed within the Eucharitidae. Three genera are recognised, *Philomides*, *Aperilampus* and *Vidlinus*, but only the former occurs in S.E. Asia, with one species, *P. frater*, being widespread in the region (Myanmar, Philippines and Indonesia).

Finally, we note that Argaman (1990) reviewed the genus *Perilampus* and described a total of 26 new genera. Qabir Argaman was a 'splitter' and it should come as no surprise that these have all been synonymised with *Perilampus* (see Darling, 1996). Darling commented appropriately that Argaman's paper '*threatens the stability of nomenclature and the predictability of the classification of Perilampidae*'.

Former Pteromalidae

Various published phylogenetic papers over the past 15 years (Desjardins *et al.*, 2007; Heraty *et al.*, 2013) have shown that the Pteromalidae needed to be split into quite a lot of new families (mostly by elevating the status of various previously recognised subfamilies). The traditional family comprised more than 4000 species in more than 800 genera. The Pteromalidae had long been a dustbin for a lot of genera that do not fit into other subfamilies and it is not surprising that many of its members have now been split into a considerable number of new family-level taxa. The family certainly could not be defined previously by any simple combination of characters, and therefore identification usually mostly proceeded by a process of elimination. Even some subfamilies appeared to be polyphyletic.

Reflecting the above, a large number of classificatory changes were implemented with the publication of Burks et al. (2022) which appeared in December of that year with a wonderful title based on a quote from Herman Melville's famous novel "Moby Dick". These changes were based on both published and unpublished molecular and morphological studies. The new classification splits the Pteromalidae into 23 monophyletic families some of which at least are relatively easily diagnosable. The main thing is that many of the new families created resulted from elevating the rank various pteromalid subfamilies and some tribes. The Pteromalidae s.s. is nevertheless still a very large family in its own right.

The resulting families that occur in S.E. Asia are the Cerocephalidae, Chalcedectidae, Cleonymidae, Colotrechninae, Diparidae, Eunotidae, Herbertiidae, Pelecinellidae (= Leptofoeninae), Louriciinae, Macromesinae, Miscogasterinae, Ormocerinae, Parasaphodinae, Pirenidae, Pteromalidae, Spalangiidae, Storeyinae and Sycoryctinae. In addition, the Erotolepsiidae are known from Papua New Guinea and so might also occur. The Chromeurytominae was transferred to the Megastigmidae. Seven previous pteromalid subfamilies and genera were transferred out of that family and left unplaced (incertae sedis) within the superfamily, viz. Asaphinae, Austrosystasinae, Ditropinotellinae, Keryinae, Louriciinae, Micradelinae, Rivasia, and Storeyinae. Of these, Ditropinotellinae (Papua New Guinea), Louriciinae and Storeyinae occur in S.E. Asia. The Storeyinae was previously only known from Australia and Nepal, but we have recently collected one in Thailand so probably they may occur rarely throughout the region. Collectively, the former Pteromalidae are represented in the region by 77 genera (Noyes, 2021) although this is certainly a marked under estimate.

For simplicity, because there is no new key to the families that were classified in the Pteromalidae we treat them here under the umbrella term "Former Pteromalidae" because as we write that is the family to which they will key in existing publications. Whilst this is not ideal, a new key to the families is in preparation and will be published, we understand, in a book edited by John Heraty and James [Jim] Woolley. Graham (1969) and Bouček (1988) provide keys to subfamilies of the then Pteromalidae. Graham's is somewhat antiquated now and only deals with the Western European fauna.

Here we deal with these following splits from former Pteromalidae in the following order:
- Asaphinae
- Cerocephalidae
- Chalcedectidae
- Cleonymidae
- Diparidae
- Ditropinotellinae
- Epichrysomallidae
- Eunotidae
- Herbertiidae
- Heydenidae
- Louriciinae
- Lyciscidae
- Macromesidae
- Moranilidae
- Ooderidae
- Parasaphodinae
- Pelecinellidae (= Leptofoeninae)
- Pirenidae (= Eriaporidae)
- Pteromalidae *sensu stricto*
    - Colotrechninae
    - Miscogasterinae
    - Pachyneurinae
    - Pteromalinae
        - Otitesellini (= former Otitisellinae; = Sycoecinae; = Sycoryctinae)
        - Pteromalini
    - Sycophaginae
    - Trigonoderinae
- Spalangiidae
    - Storeyinae
- Systasidae

Asaphesidae (Chalcidoidea *incertae sedis*)

A small group with three genera, of which *Hyperimerus* has been recorded from Thailand (Schender *et al.*, 2014). Their biology is very diverse (Gauld and Bolton, 1988) with *Asaphes* being hyperparasitoids of aphids via braconid and chalcidoid primary hosts, and *Hyperimenus* are parasitoids of Neuroptera. At least one extralimital brachypterous, egg-parasitic species is phoretic on the adult female of its host species, waiting for her to start laying eggs (Naumann and Reid, 1990). *Bairamlia* was removed by Burks *et al.* (2022) to the Sphegigastrinae of the real Pteromalidae.

Cerocephalidae

A small family (Fig. 22.30) with only 16 genera worldwide, of which three have species in the region. Several species of *Cerocephala* have been recorded from S.E. Asia, and one species each of *Neocalosoter* and *Theocolax*. *Theocolax elegans* is an important cosmopolitan parasitoid of pest stored grain beetles, especially the bostrichid, *Rhyzopertha dominica*, but also the curculionids *Sitophilus oryzae*, *S. granarius* and *S. zeamais*, all of whose larvae feed inside the grain kernel. There is circumstantial evidence that some extralimital species of *Cerocephala* and *Neocalosoter* are parasitoids of disease-carrying scolytine weevils (Cooke-McEwen, 2020) on trees.

Chalcedectidae

A monotypic family that are fairly easy to recognise based on their swollen and ventrally dentate hind femurs (Fig. 22.31). Extralimintal species are parasitoids of bark-associated buprestid beetles (Doğanlar and Laz, 2023). An extralimital species of *Chalcedectus* is a parasitoid of the Rosaceae branch borer *Osphranteria coerulescens* (Coleoptera: Cerambycidae) (Steffan, 1968; Sharifi and Javadi, 1971).

Cleonymidae

Monophyly of the former subfamily based on published molecular analyses seemed uncertain so it comes as no surprise that it has now been split. Gibson (2003) revised the world cleonymine fauna and provided a key to the six tribes that he recognised, all of which are now treated as separate families

Fig. 22.30. Unidentified Thai genus of Cerocephalidae.

Fig. 22.31. *Chalcedectus* sp. from Thailand.

following Burks *et al.* (2022), viz. Ooderidae, Heydeniidae, Cleonymidae Chalcedectidae, Boucekiidae (extralimital) and Lyciscidae. Gibson (2003) provided keys to the genera of each of these.

Members of five genera occur in the region: *Callocleonymus*, *Cleonymus*, *Dasycleonymus*, *Notanisus* and *Zolotarewskya* (Fig. 22.32).

Cleonymids are mostly parasitoids of wood boring beetles but some attack stem or mud nestling aculeates.

Diparidae

The subfamily is well-supported as monophyletic based on morphology: all members have a cercal brush (Desjardins, 2007). The subfamily is predominantly Afrotropical, Australian and Oceanic. None of the 11 world genera appear to have been recorded from S.E Asia (Desjardins, 2007; Noyes, 2021) but we have collected a few in Thailand (Fig. 22.33). *Australolaelaps* and *Malinka* are known from Papua New Guinea (Bouček, 1988). Bouček also included *Neapterolaelaps* in this group, also known from Papua New Guinea, but Burks *et al.* (2022) treated Neapterolaelapinae as Chalcidoidea *incertae sedis*. Many show marked sexual dimorphism, often with apterous or brachypterous females.

Host include mantid egg cases and curculionid beetles.

Ditropinotellinae (Chalcidoidea *incertae sedis*)

Burks *et al.* (2022) left this as a subfamily *incertae sedis* in the Chalcidoidea. The sole genus, *Ditropinotella*, occurs in Australia and

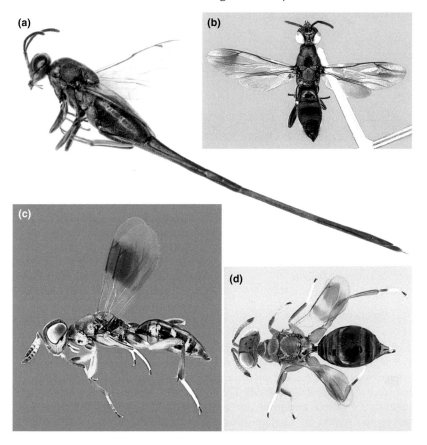

Fig. 22.32. Representatives of selected Lyciscidae and Cleonymidae subfamilies: **(a)** female *Solenura ania* (Lyciscidae), one of the largest chalcidoides, which is widespread in S.E. Asia and adjacent countries; **(b)** *Cleonomus* sp., (Cleonymidae); **(c)** *Callocleonomus* sp., (Cleonymidae); **(d)** *Notanisus* sp., (Cleonymidae). (Source: (a) modified from Sun *et al.* (2021), reproduced under Creative Commons Attribution Licence CC-BY-4.0.)

Fig. 22.33. Four exemplar Thai Diparidae.

New Guinea (two species) (Bouček, 1988). They probably occur at least in Irian Jaya. They have been reared from galls on various trees (*Acacia*, *Casuarina* and *Eucalyptus*) and one from a gall known to be induced by the scale insect *Apiomorpha frenchi*.

Epichrysomallidae

Members of this group are associated with figs and were accordingly included in the Agaonidae in former times. Some bear a considerable superficial resemblance to various megastigmids. The group comprises 19 genera worldwide and is represented in the region by *Camarothorax*, *Leeuweniella* (Fig. 22.34 b), *Neosycophila*, *Odontofroggatia* and *Sycobia* (Fig. 22.34 a). Although *Odontofroggatia* is not a member of the Agaonidae and is not a fig pollinator, they are 'primary' fig wasps in that figs into which they have oviposited do not fall and actually continue to ripen (Galil and Copland, 1981). Mating occurs outside the fig and the males are not massively modified as in the pollinator species or in various inquilines. Interestingly, with a dataset comprising mitochondrial and transcriptome data but very few taxa, and very few outgroups, Epichrysomallinae and Sycophaginae were recovered closest to Agaonidae (Zhao *et al.*, 2021). However, molecular studies with very extensive taxon sampling recover both of them quite far removed from Agaonidae.

Eunotidae

Bouček (1988) divides the Eunotinae into Eunotini, Moranilinae and Tomocerodini but only the former comprises the Eunotidae of Burks *et al.* (2022). This small family of seven genera is represented in the region

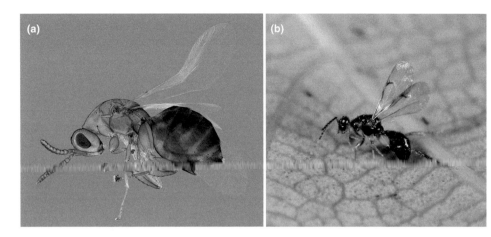

Fig. 22.34. (a) Female *Sycobia* species from Thailand; **(b)** *Leeuweniella* sp. emerged from *Ficus araneosa* in Thailand. (Source: (b) photograph by and © Jean-Yves Rasplus, reproduced with permission.)

by two described species of *Cephaleta*. Their main host association is with various Heteroptera, for example Aleyrodidae, Aphididae, Coccidae, Diaspididae, Eriococcidae, Pseudococcidae and others, mainly as egg predators, but some as parasitoids on the eggs or adults, or are larva–pupal parasitoids. They may also develop as hyperparasitoids, for example, via Encyrtidae and Coccinellidae (Coleoptera). In addition to Bouček (1988), Xiao and Huang (2001) provided a key that separated the Oriental Asian genera but this also includes *Moranila* and *Ophelosia*, which are now in Moranilidae.

Herbertiidae

Cosmopolitan parasitoids of leaf-mining Diptera, including various Agromyzidae, and found in all warmer parts of the world (Bouček, 1988). Of the two known extant genera, the widespread *Herbertia indica* has been found in Malaysia.

Heydeniidae

Heydenia (not illustrated) females have the fore femurs highly enlarged, as also in Ooderidae. The genus is nearly cosmopolitan and two species of *Heydenia* have been recorded from S.E. Asia. They are parasitoids of various wood-boring beetles belonging to the Buprestidae, Cerambycidae and Curculionidae (including Scolytinae) (Sureshan, 2009). One extralimital species is a parasitoid of a pest bark beetle and appears to be attracted to the host's sex pheromones (Camors and Payne, 1972) as well as to volatiles emanating from the host's fungal associates (Adams and Six, 2008). Sureshan (2009) provided a key to the three species known from the Indian subcontinent.

Louricinae (Chalcidoidea *incertae sedis*)

The group includes only the genus *Callimomoides*, members of which are egg parasitoids of cerambycid beetles. *C. ovivorus* occurs in Malaysia and other species occur in Australia and Papua New Guinea. Burks *et al.* (2022) were unable to assign it satisfactorily to family and so left it as Chalcidoidea *incertae sedis*.

Lyciscidae

Burks *et al.* (2022) recognised two subfamilies, Lycisinae with most genera, and Solenurinae with two genera (*Grooca* and *Solenura*), both of which occur in S.E. Asia. *Solenura* includes some of the largest chalcidoids. The stunning metallic blue female of *Solenura ania* (Lyciscidae) reaches approximately 28 mm in length (see Fig. 22.32 a); males only reach about 6 mm. This species is an ectoparasitoid of several species of Cerambycidae and

Buprestidae such as the cerambycids *Trichoferus campestris* (a pest of many live trees and wood furniture and a quarantine pest in many countries), *Clytocera chinospila* and *Olenecamptus bilobus* (Cao et al., 2020b), as well as the buprestid *Chrysobothris succedanea*. It also attacks hosts infesting already decaying wood, for example of *Ficus* (Sureshan, 2005).

Macromesidae

The single recognised genus *Macromesus* contains 12 species, of which only *M. javensis* is recorded in the region but a few more are known from China and India. *Macromesus* have a greatly reduced prepectus. An Indian species has recently been reared from Scolytidae (Khanday et al., 2019).

Moranilidae

A single species of this principally Australian family has been recorded from S.E. Asia, viz. *Ophelosia crawfordi* from Indonesia (Java) (Bouček, 1988). However, *Ismaya* and *Tomicobomorpha* have been recorded from Papua New Guinea, so might occur. Extralimital species have been recorded as parasitoids of mealy bugs (Pseudococcidae) and cottony scale insects including the pest *Icerya purchasi* (Monophlebidae).

Ooderidae

Oodera females have highly enlarged fore femurs (Fig. 22.35). It is widespread, with S.E. Asian species. *Oodera* are principally parasitoids of Scolytinae and Buprestidae (Lotfalizadeh, 2012). They are often found on trunks of dying trees. However, most host records are unverified (Bouček, 1988). The species of *Oodera* were revised by Werner and Peters (2018).

Parasaphodinae

A rare monotypic subfamily (Fig. 22.36) with only four described species, widely distributed in the warm temperate and tropical Old World. *Parasaphodes townsendi* is recorded from the Philippines, and unidentified species are known from Malaysia, Myanmar (Bouček, 1988) and Thailand (this work). The subfamily was left *incertae sedis* in the Chalcidoidea (i.e. not placed to family) by Burks et al. (2022).

Fig. 22.35. Female *Oodera longicollis* (Ooderidae) from Thailand: **(a)** lateral habitus; **(b)** dorsal habitus.

Pelecinellidae (= Leptofoeninae)

This group comprises large to very large species including the largest known chalcidoid, *Doddifoenus wallacei*, which reaches nearly 50 mm in length (including ovipositor) and occurs in S.E. Asia (Thailand and Laos) (Krogmann and Burks, 2009) (Fig. 22.37). Other species of *Doddifoenus* are known from Australia and Papua New Guinea. The other genera in the subfamily are *Leptofoenus*, which is restricted to the New World from southern USA to Argentina, and *Nefoenus*, placed in Noefoeninae by Burks et al. (2022), from Australia. They are thought to be parasitoids of wood-boring beetle larvae, probably in tree branches, and share morphological characters typical of several groups with that biology; for example, the head has rows of crests or denticles on the frons and vertex (see Ichneumonidae: Poemeniinae) and the mesoscutum has transverse ridges (see Ichneumonidae: Rhyssinae).

Fig. 22.36. Female *Parasaphodes* sp. from Thailand.

Pheniidae (= Eriaporidae)

These are small wasps with body length generally less than 2.0 mm, and are not particularly

Fig. 22.37. Female *Doddifoenus wallacei* female from Thailand. These wasps reach up to 42 mm (including ovipositor).

easily recognised. Members of the genus *Zebe* are atypical in having only four tarsal segments and so could easily key to Eulophidae.

The Eriaporidae was synonymised with the Pirenidae by Burks *et al.* (2022). LaSalle *et al.* (1997) had questioned the placement of the Eriaporinae within the Chalcidoidea. It had mainly been placed in the Aphelinidae but showed some features in common with the eulophid genus *Elasmus* and also with the former pteromalid subfamily Eunotinae (see Eunotidae). Heraty *et al.* (2013) elevated Eriaporinae from being a subfamily of Aphelinidae to a full family, and recognised in it two subfamilies: Eriaporinae and Euryischinae, although LaSalle *et al.* (1997) had questioned the strength of the evidence for their monophyly.

Burks *et al.* (2022) recognised five subfamilies (Cecidellinae, Eriaporinae, Euryischiinae, Pireninae, Tridyminae), of which Eriaporinae, Euryischiinae and Pireninae are present in the region (Mitroiu, 2011; Noyes, 2021), represented by six genera: *Keesia*, *Myiocnema*, *Promuscidea* and *Watshamia* each represented by a single species, *Macroglenes* by two and *Zebe* by three. Hayat and Verma (1980) provided a key to the tribes and genera of Eriaporinae (Fig. 22.38) plus Euryischiinae. *Promuscidea unafasciaventris* is sometimes an important parasitoid of various pest mealy bugs and scale insects, including the cotton mealy bug *Phenacoccus solenopsis* (Chen *et al.*, 2021).

Fig. 22.38. *Eunotiscus* sp. (Source: photograph by, and © Simon van Noort (Iziko Museums of South Africa), www.waspweb.org, reproduced with permission.)

Pteromalidae *sensu stricto*

Even after the removal of many subfamilies and tribes by Burks *et al.* (2022), the remaining group is still a very large cosmopolitan subfamily with 417 genera worldwide. Burks *et al.* recognised eight subfamilies, some further divided into tribes. Essentially, the Pteromalidae comprises the former Pteromalinae together with the Colotrechninae, Miscogastrinae, Sycophaginae and Trigonoderinae, which are each considered separate subfamilies, and in addition the extralimital Erixestinae (monotypic, Neotropical), Ormocerinae (worldwide but two genera occur in Australia) and Pachyneurinae.

Several pteromalines are important in the natural or controlled regulation of pests in S.E. Asia and elsewhere. *Pachycrepoideus vindemmiae* (Pachyneurinae) is a widespread generalist ectoparasitoid of Diptera pupae within the puparium, and may play a role in the control of filth flies, fruit flies and other pests in S.E. Asia and elsewhere (Sulaiman *et al.*, 1990). *Propicroscytus* (Pteromalini) species are important parasitoids of rice gall midge. *Pycnetron* species are parasitoids of weevils (Curculionidae).

Mitroiu and van Achterberg (2013) revised Oriental *Apsilocera* (Pteromalini) and described several species from S.E. Asia.

Colotrechninae

Of the 18 genera in this subfamily, only *Colotrechnus agromyzae* has been found in S.E. Asia (Indonesia). As its specific name suggests, it has been reared from agryomyzid leaf-mining flies. In Indonesia (Sumatra) it has been reared from the soybean stem fly, *Melanagromyza sojae* (Diptera: Agromyzidae) (Van Den Berg *et al.*, 1995). The world species of *Colotrechnus* were revised by Li *et al.* (2014).

Miscogasterinae

A large cosmopolitan subfamily with 50 recognised genera, of small chalcidoids. They are parasitoids of Diptera boring in plant soft tissue, including leaf-miners. *Cryptoprymna*, *Cyrtogaster*, *Miscogaster*, *Notoglyptus*,

Sphegigaster and *Syntomopus* are recorded from the region (Noyes, 2021). *Sphegigaster* species are ecto- or endoparasitic on gall-forming Cecidomyiidae or on mining larvae of Agromyzidae in leaves, fruit, stems or flowers. The extralimital *Bairamlia* are of particular interest because they are ectoparasitoids of fleas (Siphonaptera), emerging from the host cocoon.

Pachyneurinae

Of the 22 genera of Pachyneurinae, only *Acroclisoides* has so far been recorded from S.E. Asia: *A. luzonensis* from the Philippines and *A. indicus* from Myanmar. Because of its importance as an egg parasitoid of the pest brown marmorated stinkbug, the biology of the currently extralimital *Acroclisoides sinicus* (Pachyneurinae) has been described in some detail (Giovannini *et al.*, 2021). It and other *Acroclisoides* species are specialist hyperparasitoids of scelionids that attack the eggs of various stink bugs and relatives (Clarke and Seymour, 1992).

Pteromalinae

The largest subfamily with only two tribes, the entirely fig-associated Otitisellini (= Sycoecinae; = Sycoryctinae) with 28 genera, and the remaining 267 genera worldwide which are placed in Pteromalini.

OTITESELLINI (= FORMER OTITISELLINAE; = SYCOECINAE; = SYCORYCTINAE). One of the more conspicuous changes made is that most members of three long-standing, fig-associated former subfamilies (Otitisellinae, Sycoecinae and Sycoryctinae) were demoted or synonymised into the tribe Otitisellini, by Burks *et al.* (2022), which is transferred to Pteromalinae. For a long while these fig wasps were classified in the Agaonidae because of their association with figs and in some cases considerable morphological convergence, but they are 'mostly' non-pollinators and were later transferred to the Pteromalidae (Rasplus *et al.*, 1998). Burks *et al.* (2022) noted an 'amazing disparity' within this group, which underlies their previous classification into three subfamilies. As recently as ten years ago chalcidoidologists were still recognising these groups as probably monophyletic entities (e.g. Cruaud *et al.*, 2013). The newly prescribed tribe Otitisellini is cosmopolitan and comprises 28 genera worldwide, and in S.E. Asia is represented by 16 genera and just over 60 species (Table 22.3).

Most Otitisellini are non-pollinators, with females ovipositing into the fig from the outside during the early stage of its development, and are believed to be phytophagous, gall formers (Beardsley, 1998) although some, perhaps many, are kleptoparasites of gall-forming Agaonidae (Jousselin *et al.*, 2001). However, a few species can enter figs to oviposit, e.g. *Diaziella* and *Lipothymus* species (van Noort and Compton, 1996), and some of these are efficient pollinators in their own right (Zhang *et al.*, 2008). These internal ovipositing wasps show several similar morphological adaptations to those of the pollinating species such as flattened heads (Fig. 22.39 d), smooth bodies and spurs on the legs, but they lack morphological adaptations for carrying out pollination, instead being passive transferrers of pollen. Concordant with this, the fig species pollinated by these wasps have a higher ratio of male anther-bearing flowers in their syconia. Since they are obligate kleptoparasites, they cannot substitute for the agaonid pollinator.

The former 'Sycoryctinae' is a cosmopolitan group, the females having exceedingly long ovipositors (Fig. 22.39 a). It was divided into four tribes (Wong *et al.*, 2022), three with members in in S.E. Asia, viz. Apocryptini, Philotrypesini and Sycoryctini, but these are no longer recognised. Collectively they were associated with all six subgenera of *Ficus*. Some species have quite complicated mechanisms for approximating the ovipositor tip to the fig surface for oviposition; for example, Wenquan *et al.* (2004) described how this was achieved by female *Apocrypta* which have especially modified anterior metasomal segments (Fig. 22.39 c, d).

Males of the former Otitisellinae and Sycoryctinae are mainly apterous, whereas those of former Sycoecinae are fully winged. The apterous males are highly modified males that spend their entire lives within

Table 22.3. Representation of Otitisellini in S.E. Asia grouped according to previous classificatory units.

Former group prior to Burks et al. (2022)	Genera known from S.E. Asia with number of recorded species
'Otitisellinae'	*Aepocerus* (1 sp.), *Eujacobsonia* (2 spp.), *Grandiana* (1 sp.), *Grasseiana* (3 spp.), *Lipothymus* (4 spp.), *Micranisa* (2 spp.), *Micrognathophora* (1 sp.), *Otitesella* (1 sp.) and *Walkerella* (3 spp.).
'Sycoecinae'	*Diaziella* (11 spp.) and *Robertsia* (0, but occurs in Papua New Guinea so may occur)
'Sycoryctinae'	*Apocrypta* (Apocryptini) (7 spp.), *Arachonia* (Sycoryctini) (1 sp.), *Philotrypesis* (Philotrypesini) (16 spp.), *Sycoscapter* (Sycoryctini) (9 spp.), and *Watshamiella* (Philotrypesini) (1 sp.).

the figs and often show adaptations for fighting; in the genus *Sycoscapter*, up to 25% of fights result in the death of one of the individuals (Bean and Cook, 2001). Most species of *Philotrypesis* have two distinct male morphs, normal small ones and far larger soldier males that are adapted for fighting (Murray, 1990), and some species have three male morphs (Jousselin et al., 2004) including also winged dispersers. The soldier morph males, although also mating, spend a large amount of time in often injurious and sometimes fatal combat with conspecifics. The small ones are thought to be specialised sneaky maters. Males of *Apocrypta* are exceedingly elongate and tubular with an extremely extendable abdomen and this appears to enable them to move easily between tightly packed galls. Brood sizes of fig wasps fit theoretical predictions, with those of species that only have wingless males being far larger than those of species with winged or dimorphic ones (Cook et al., 1997).

Of the former Sycoecinae, four genera are entirely Afrotropical but *Diaziella* (Fig. 22.39 d) and *Robertsia* occur in greater S.E. Asia, the former mostly from Indonesia and the Philippines, but also recorded from Thailand and China (van Noort et al., 2006). Some *Diaziella* species can have winged and flightless male morphs but even the former seem mostly to mate in their natal patch (Zhang and Yang, 2010). The former Otitisellinae genus *Lipothymus* species are also internally ovipositing species which follow the primary pollinating agaonid, *Eupristina* species, into the fig (Zhang et al., 2008).

Not surprisingly, the species associated with domesticated figs (*Ficus carica*) are best studied. (Joseph, 1958; Kjellberg et al., 2022). *Philotrypesis caricae* is a phytophagous kleptoparasitoid of its 'host' *Blastophaga psenes* (Agaonidae). It oviposits through the fig wall into ovules that contain a pollinator larva and its young larva initially feeds on the seed endosperm that the *Blastophaga* larva had induced. The *Philotrypesis* larva develops quickly and by the second larval instar it outcompetes the *Blastophaga*, resulting in the latter's death.

Even though the fig ecosystem has been very well studied, many species of Otitisellini are yet to be described (e.g. Segar et al., 2012). Wiebes (1974) provided keys to the Philippines species of *Grasseiana* and *Lipothymus*. Wong et al. (2022) provided a key to female former sycoryctines associated with *Ficus hirta*, a fig species which also occurs in the northern part of S.E. Asia. The Papuan species of *Robersia* were revised by van Noort and Rasplus (2005). A key to species of *Diaziella* is provided by van Noort et al. (2006).

PTEROMALINI. Only 19 genera of Pteromalini have species recorded from S.E. Asia, viz. *Aepocerus*, *Agiommatus*, *Anisopteromalus*, *Apsilocera*, *Dinarmus*, *Eurydinotomorpha*, *Frena*, *Kumarella*, *Lariophagus*, *Mokrzeckia*, *Norbanus*, *Oxysychus*, *Propicroscytus*, *Psilocera*, *Pteromalus*, *Pterosemopsis*, *Pycnetron*, *Trichomalopsis* and *Uniclypea*, so obviously the group is very understudied in the region (Fig. 22.40). *Anisopteromalus calandrae* is, or

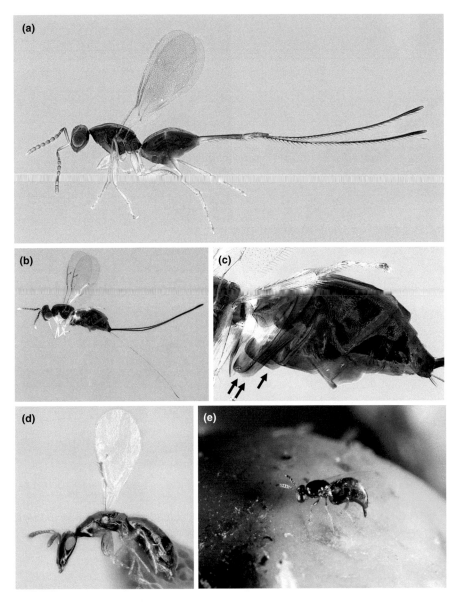

Fig. 22.39. Exemplar Otitisellini: **(a)** *Philotrypesis* sp.; **(b, c)** *Apocrypta* sp., arrows in (c) pointing to the modified anterior segments that allow the metasoma to flex at its front during oviposition; **(d)** *Diaziella bizzarea*, an extralimital (Chinese) species of a genus widespread in S.E. Asia; **(e)** *Micranisa* sp. on *Ficus crassiramea*, Sulawesi. (Source: (d, e) photographs by and © Jean-Yves Rasplus, reproduced with permission.)

used to be, an important laboratory insect parasitising several stored-product (grain and beans) pest beetles, and has been involved in many types of experimental study (see Baur et al., 2014). Larval development was recently described in detail by Oanh (2022).

Sycophaginae

Members of this subfamily of non-pollinating fig wasps are gall-inducers. Molecular phylogenetics have indicated that this group originated in Gondwanaland (Australia) and

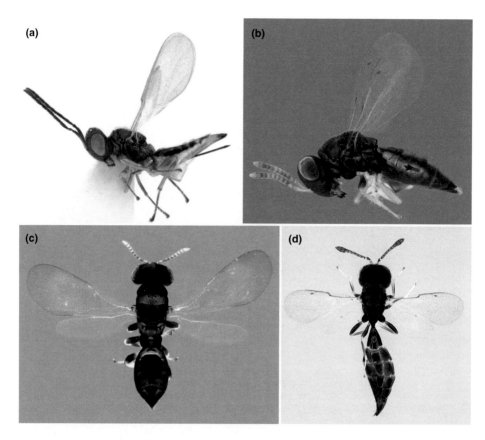

Fig. 22.40. Exemplar Pteromalini from Thailand: **(a)** *Propicroscytus* sp., common natural enemy *Orseolia oryzae*; **(b)** gen. sp. indet.; **(c)** probably *Homoporus* sp; **(d)** gen. sp. indet.

from there spread into Asia and the Americas (Cruaud *et al.*, 2011). Worldwide there are nine genera, of which *Conidarnes*, *Eukoebelea* and *Sycophaga* (Fig. 22.41) have described species in the region.

Trigonoderinae

This subfamily is represented in the region by three species of *Miscogasteriella* (Fig. 22.42) and two of *Trigonoderus*. They appear to be parasitoids of wood-boring beetles, especially of Anobiidae and Bostrychidae (Bouček, 1988).

Fig. 22.41. Highly modified male (long and thin) *Sycophaga* sp. inside syconium of *Ficus cereicarpa*. (Source: photograph by and © Jean-Yves Rasplus, reproduced with permission.)

Fig. 22.42. *Miscogasteriella* sp. from Thailand. (Source: photograph by and © Jean-Yves Rasplus.)

Spalangiidae

This is a small but very important family, whose members are fairly easy to recognise because of their distinctive habitus (Fig. 22.43). There are two subfamilies and six genera (Burks et al., 2022), of which the cosmopolitan genus *Spalangia* occurs in S.E. Asia. The Erotolepsiinae includes *Papuopsia* from Papua New Guinea, which might also occur in the region. All species are solitary or gregarious ectoparasitoids within the puparial stage of cyclorrhaphous Diptera, including many synanthropic species, and therefore they have been used in biological control of these pests. Several *Spalangia* species occur in the region and attack filth flies (Calliphoridae and Muscidae) in general refuse situations and especially in poultry farms (Sulaiman et al., 1990). Their bodies are mostly black or dark brown and they lack metallic coloration. The 10-segmented antennae have no anelli and are inserted widely separated just above the base of each mandible.

Spalangia was revised by Bouček (1963).

Storeyinae (Chalcidoidea: *incertae sedis*)

A rare subfamily known from only two species of *Storeya*, which were previously known collectively only from Australia, India and Nepal, although we have recently discovered a third species in Thailand (Fig. 22.44). They seem to be adapted for jumping and the head is almost prognathous, with the antennae inserted very close to the mouth. The notauli are virtually absent and the wings have long marginal setae, a tuft of thickened setae on the parastigma but the surface of the membrane is essentially hairless with only dots marking where setae would normally be. Their biology is unknown, as are their phylogenetic affinities (Cruaud et al., 2021).

Systasidae

Of the three genera, only *Systasis* has been recorded from S.E. Asia; however, the closely related *Semiotellus* is known from Australia, south Asia and the Palaearctic and it might well occur in the region. Bouček (1988) noted some uncertainty about their biology, with some reported as being reared from seeds and others from cecidomyiid galls, and suggested that in the latter case they may be entomophytophagous. Records of both genera from Bruchinae (Coleoptera: Chrysomelidae) should be discounted and the true hosts are believed to be Cecidomyiidae (Diptera) (Bellifa and Chapelin-Viscardi, 2021).

Fig. 22.43. *Spalangia cameroni* female. (Source: photograph by and © Mircea-Dan Mitroiu (Universitatea Alexandru Ioan Cuza, Romania), reproduced with permission.)

Signiphoridae

Fig. 22.44. An undescribed *Storeya* species from Thailand.

Small wasps (c. 0.5 – 2.0 mm) that are characterised by having the bulk of the fore wing membrane (disc) without setae or with a single seta, antennae with a large unsegmented clava and 2–4 anelli (Fig. 22.45 c), the presence of a triangular area on the propodeum, and the metasoma being broadly joined to the mesosoma without a wasp waist (Fig. 22.45 b). Woolley (1988), as a result of a morphological cladistic analysis, recognised four extant genera: *Chartocerus*, *Clytina*, *Thysanus* and *Signiphora*. Of these only *Chartocerus* and *Signiphora* are known from the region; *Clytina* is entirely Palaearctic and unlikely to be found. Woolley (1988) also presented a key to the genera of the family as well as to the species-groups of *Signiphora*.

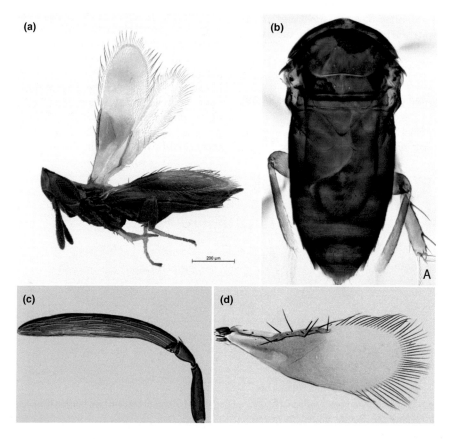

Fig. 22.45. Exemplar Signiphoridae: **(a)** Extralimital (South African) *Signiphora* sp. habitus; **(b-d)** slide-mounted parts of *Chartocerus javensis*, body, antenna and fore wing, respectively. (Source: (a) photograph by and © Simon van Noort (Iziko Museums of South Africa), www.waspweb.org, reproduced with permission; (b–d) from Schmidt *et al.*, 2019, reproduced under terms of Creative Commons Attribution licence CC-BY 4.0).

Signiphorids comprise a cosmopolitan group of 80 described species and are particularly diverse in the Neotropics, with several species with extremely wide ranges incorporating the Neotropics and Australia. In the past they were frequently placed in either the Aphelinidae, Encyrtidae or Eulophidae (Woolley, 1988).

The Signiphoridae of Indonesia were revised by Schmidt et al. (2019), the subfamily having only been discovered in that country about eight years previously (Muniappan et al., 2012). Muniappan et al. had reared *Signiphora bifasciata* from the introduced cycad aulacaspis scale, *Aulacaspis yasumatsui*, (Hemiptera, Diaspididae), in the Bogor Botanic Garden. Amazingly, the three species of *Chartocerus*, and four species of *Signiphora*, reported on by Schmidt et al. (2019), are the only records of Signiphoridae from the whole of S.E. Asia. Undoubtedly there are species of both genera in every country of the region. Hayat (2009) reviewed the Indian species, which might also be useful, especially as subsequent authors have tended to use his terminology.

Tanaostigmatidae

Within our region to date this small family has only been recorded from Indonesia, where two species of *Protanaostigma* are noted. However, it undoubtedly must be present in every country. The two *Protanaostigma* species were reared from galls on *Derris elliptica* (Fabaceae), and from fruit galls on *Milletia sericea*, both members of the Fabaceae (LaSalle, 2005). Tanaostimatids may be distinguished from Eupelmidae largely by having the prepectus very bulbous, protruding anteriorly and overlapping the pronotum in dorsal view.

Tetracampidae

This small family is represented in the region by the widespread *Cassidocida aspidomorphae*, a specialist egg parasitoid of the papery oothecae (egg masses) of the cassidine chrysomelid (tortoise beetle) genus *Aspidimorpha* (Fig. 22.46 b). The hosts are associated mainly with Convolvulaceae. Although only recorded to date from Indonesia, Malaysia and Myanmar, it must undoubtedly occur in most other countries. In addition, there are unpublished records of other genera including *Foersterella* (Fig. 22.46 a), which is widespread from Europe to Papua New Guinea. Some tetracampids have four tarsal segments so would key here to Eulophidae, but they usually have the central part of the propodeum setose rather than glabrous, and they usually have six flagellomeres.

Fig. 22.46. Exemplar Tetracampidae: **(a)** female *Foersterella* specimen from Thailand; **(b)** *Cassidocida* sp. female ovipositing into the ootheca (egg mass) of the cassidine beetle, *Aspidimorpha miliarus* (Chrysomelidae) on *Ipomea batatus*. (Source: (a) photograph by and © Alex Gumovsky, reproduced with permission; (b) from Flickr via Wikimedia, reproduced under terms of Creative Commons Attribution-Share Alike 2.0 Generic.)

Torymidae

A large and cosmopolitan family of mostly medium-sized wasps. The main diagnostic features of torymids are: ovipositor clearly exserted (Figs 22.47, 22.48), sometimes much longer than abdomen, fore wing stigmal vein very short so that apex of the uncus almost touches the anterior margin of fore wing (Fig. 22.48 a), petiole strongly transverse and indistinct, and cercal plates slightly raised (papilliform) not flat. Many species have metallic coloration, especially in the subfamilies Podagrioninae, and Toryminae and some have the hind femur swollen and with ventral teeth and/or serrations.

In torymids, the occipital carina is present, the marginal vein is at least twice as long as the short stigmal vein, and often far longer, whereas postmarginal vein is short. The hind coxa is far larger (at least 2.5× longer) than middle coxa (Fig. 22.48). The cerci are protruding, digitiform.

Grissell (1995) recognised just two subfamilies, the Megastigminae and the Toryminae, but subdivided the Toryminae into seven tribes. With the elevation of the Megastigminae to family level (Janšta et al., 2018), the remaining torymids were then classified into six subfamilies: Chalcimerinae (= Thaumatotoryminae), Erimerinae (= Microdontomerinae), Glyphomerinae, Monodontomerinae, Podagrioninae and Toryminae (Janšta et al., 2019; Zerova et al., 2021). Of these, Monodontomerinae, Podagrioninae and Toryminae occur in S.E. Asia; Chalcimerinae and Erimerinae are restricted to the West Palaearctic and Near East; Glyphomerinae is Palaearctic.

Monodontomerinae

In monodontomerines, the hind femur is greatly enlarged and ventrally serrate or dentate, and correspondingly, the hind tibia is strongly curved and usually produced into an apical spine.

Four genera of the subfamily Monodontomerinae have been recorded from S.E. Asia: *Anneckeida* (four species, Laos, Malaysia and Thailand), *Chrysochalcissa* (two species, Malaysia and Myanmar), *Monodontomerus* (Fig. 22.47) and *Rhynchoticida* Thailand (Janšta et al., 2018; Noyes, 2021).

Monodontomerus dentipes plays a role in

Fig. 22.47. *Monodontomerus* sp. (Source: photograph by and © Jean-Yves Rasplus, reproduced with permission.)

the control of the pest pine sawfly, *Neodiprion biremis*, in northern Thailand (Beaver and Laosunthorn, 1974, 1975), although in addition to being a primary parasitoid of the sawfly cocoon it can also act as a pseudohyperparasitoid of ichneumonid and tachinid primaries. It is a gregarious ectoparasitoid and from each cocoon emerge one (rarely two) males and between two and 11 females. *Chrysochalcissa* spp. appear to be specialist egg parasitoids of coreid bug eggs. The genus is widely distributed from India through to Australia and species are also known from Africa. The Vietnamese species of *Rhynchoticida* were revised by Narendran and van Achterberg (2012), who provided a key to world species. There is one host record for this genus from eggs of a heteropteran (Bouček, 1978).

Podagrioninae

Podagrionines display a number of distinctive features (Grissell, 1995). The subfamily is divided into two tribes: Podagrionini (Fig. 22.48 a) and Palachiini (Fig. 22.48 b). The generic classification of Podagrionini is still in need of further work (Janšta et al. 2018). The world species until approximately 1994 were catalogued by Grissell

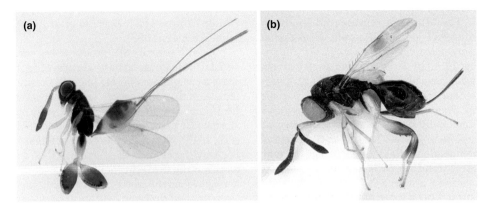

Fig. 22.48. Exemplar Podagrioninae. **(a)** *Podagrion* sp., Thailand (Podagrionini); **(b)** *Propalachia* sp., possibly *P. beaveri*, Thailand (Palachiini).

(1995). The largest genus is *Podagrion* (Fig. 22.48 a) with more than 100 species worldwide and many more undescribed; but only one described species, *P. ahlonei*, has thus far been reported from S.E. Asia (Myanmar) (Narendran and Sheela, 2013). Material included in Janšta et al.'s (2018) study also included Malaysian *Podagrion*, *Podagriomicron* from Laos and Thailand, and *Propalachia* from Laos.

The great majority of Podagrionini for which there are reliable published host associations are parasitoids of praying mantis (Mantodea) eggs within their hardened oothecae.

Regarding the tribe Palachiini, two species of *Palachia* and two species of *Propalachia* (Fig. 22.48 b) have been described from S.E. Asia by Bouček (1998), who commented that he was surprised that there was only one host record for this tribe (from a mantid ootheca in India) given how many entomologists had collected and reared mantid eggs; perhaps they mainly utilised a different host group, and the published record might represent a host misidentification.

Toryminae

Torymines are nearly always at least partly metallic. Out of the 21 world genera, only a few species of three genera (*Odopoia*, *Torymus* and *Torymoides*) are recorded from S.E. Asia.

The sole S.E. Asian described species of *Torymus*, *T. kovaci* from Malaysia, is phytophagous on culms of the bamboo, *Gigantochloa scortechinii* (Narendran and Kumar, 2005). An extralimital *Diomorus* species, *D. aiolomorphi*, is an inquiline of bamboo galls induced by the eurytomid *Aiolomorphus rhopaloides* in Japan (Shibata, 2002). The ecology of *Diomorus aiolomorphi* has been subject to quite extensive investigation.

Odopoia was revised by Xiao *et al.* (2012).

Trichogrammatidae

A large, cosmopolitan and important family of tiny wasps with many economically important species. In total it comprises approximately 839 species and 83 genera. Almost all species are idiobiont primary endoparasitoids of the eggs of other insects. The family includes species with solitary and gregarious biology and most of the hosts belong to Lepidoptera, Hemiptera, Coleoptera, Thysanoptera, Hymenoptera, Diptera and Neuroptera. They may be diagnosed as follows: (i) small to minute insects 0.3–1.2 mm in length; (ii) never metallic; (iii) funicle with no more than two segments; and (iv) three-segmented tarsi.

The family is represented in the region by 11 genera: *Doirania*, *Hispidophila*, *Lathromeromyia*, *Megaphragma*, *Oligosita* (Fig. 22.49 b), *Paracentrobia*, *Pseudobrachysticha*, *Pseudoligosita* (Fig. 22.49 a), *Trichogramma* (Fig. 22.50), *Trichogrammatoidea* and *Ufens*. One species of *Trichogramma* occasionally develops as a facultative hyperparasitoid via a *Telenomus* sp. (Scelionidae) inside its lepidopteran egg primary host (Strand and Vinson, 1984). One *Lathromeris* sp. and one *Oligosita* sp. have been recorded as larval parasitoids of cecidomyiids (Diptera) (Viggiani and Laudonia, 1994). *Paracentrobia* species attack various rice leaf hoppers throughout S.E. Asia (Vungsilabutr, 1981; Barrion *et al.*, 1989; Ishii-Eiteman, 1993).

As with several other groups of egg parasitoids (those whose hosts lay clusters of eggs), some trichogrammatids are phoretic on adult females of their host species which they somehow locate, and disembark from her when she starts to lay a batch of eggs (Pinto, 1994; Shih *et al.*, 2013). They may quite often be found on the wing bases or bodies of butterflies and moths in the field, but they are so small that it requires careful observation to spot them (Fatouros and Huigens, 2012).

Much of the higher systematics of trichogrammatids has relied upon features of male genitalia because so many other of their

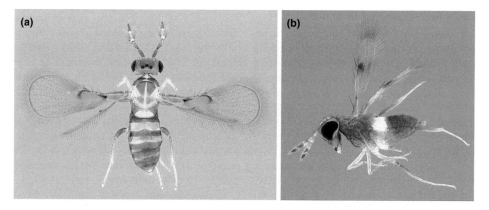

Fig. 22.49. Exemplar Thai Trichogrammatidae: **(a)** *Pseudoligosita* sp., showing tracts of setae in fore wing; **(b)** *Oligosita* sp., a genus which has lost the wing setal tracts.

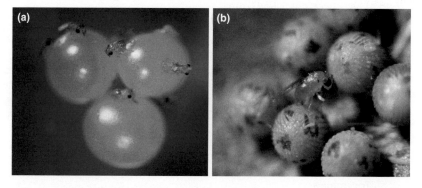

Fig. 22.50. Two examples of living Trichogrammatidae: **(a)** *Trichogramma dendrolimi* females on eggs of a hawk moth (Sphingidae); **(b)** *Trichogramma dendroliti* female ovipositing into egg of armyworm moth (Noctuidae). (Source: photographs by Victor Fursov reproduced under terms of Creative Commons Attribution-Share Alike 4.0 International.)

features are so reduced as a consequence of small body size. Surprisingly their adaptations to small size go much deeper and it has been shown that some of their adult brain neurones lack nuclei (Polilov, 2012, 2023).

Yousuf and Shafee (1987) provided a key to the subfamilies, tribes and genera of Trichogrammatidae occurring on the Indian subcontinent, but it does not include quite a few S.E. Asian genera, so it is likely only useful for higher level categories. Polaszek et al. (2022) revised the world species of *Megaphragma*, including many species from S.E. Asia.

References

Abhisree, P., Ranjith, M.D., Nasser, M. and Delvare, G. (2021) A review of the biology of *Neochalcis* Kirby (Hymenoptera, Chalcididae) with a new host and biological data for *N. breviceps* (Masi). *Journal of Natural History* 55, 1769–1780. doi: 10.1080/00222933.2021.1965237

Abul-Sood, M.I., Gadallah, N.S., Hossni, M.T. and Delvare, G. (2018) The subfamily Cratocentrinae (Hymenoptera: Chalcididae): reappraisal of their morphological characters and review of the West Palaearctic species, with the description of two new species. *Zootaxa* 4377, 490–516.

Adams, A.S. and Six, D.L. (2008) Detection of host habitat by parasitoids using cues associated with mycangial fungi of the mountain pine beetle, *Dendroctonus ponderosae*. *The Canadian Entomologist* 110, 124–127.

Anwar, F.I., Zeya, S.B. and Huber, J.T. (2019) Two new species of *Anaphes* Haliday (Hymenoptera: Mymaridae) from India and Indonesia. *Zootaxa* 4623, 26–40. doi: 10.11646/zootaxa.4623.1.2

Argaman, Q. (1990) A synopsis of *Perilampus* Latreille with descriptions of new genera and species (Hymenoptera: Perilampidae), I. *Acta Zoologica Hungarica* 36, 189–263.

Baker, A.J., Heraty, J.M., Mottern, J., Zhang, J., Hines, H.M., Lemmon, A.R. and Moriarty Lemmon, E. (2020) Inverse dispersal patterns in a group of ant parasitoids (Hymenoptera: Eucharitidae: Oraseminae) and their ant hosts. *Systematic Entomology* 45, 1–19. doi: 10.1111/syen.12371

Barrion, A.T., Litsinger, J.A. and Morrill, W.L. (1989) *Egg parasitoids of rice leafhoppers in the Philippines: a review and assessment of their biocontrol potential.* Paper presented at the 20th Annual Convention of the Pest Control Council of the Philippines, Baguio City (Philippines), 9–12 May 1989 (Biological Control, Parasite Predator, Cultural Control, Planting Density, Philippines).

Basibuyuk, H.H. and Quicke, D.L.J. (1999) Gross morphology of multiporous plate sensilla in the Hymenoptera (Insecta). *Zoologica Scripta* 28, 51–67.

Baur, H., Kranz-Baltensperger, Y., Cruaud, A., Rasplus, J.-Y., Timokhov, A.V. and Gokhman, V.E. (2014) Morphometric analysis and taxonomic revision of *Anisopteromalus* Ruschka (Hymenoptera: Chalcidoidea: Pteromalidae) – an integrative approach. *Systematic Entomology* 39, 691–709.

Bean, D. and Cook, J.M. (2001) Male mating tactics and lethal combat in the nonpollinating fig wasp *Sycoscapter australis*. *Animal Behaviour* 62, 535–542.

Beardsley, W.J. (1998) Chalcid wasps (Hymenoptera: Chalcidoidea) associated with fruit of *Ficus microcarpa* in Hawaii. *Proceedings of the Hawaiian Entomological Society* 33, 19–34.

Beaver, R.A. and Laosunthorn, D. (1974) Pine sawflies in northern Thailand. *Natural History Bulletin of the Siam Society* 25, 199–202.

Beaver, R.A. and Laosunthorn, D. (1975) The biology and control of the pine sawfly, *Nesodiprion biremis* (Konow) (Hymenoptera, Diprionidae), in northern Thailand. *Bulletin of Entomological Research* 65, 117–128.

Bellifa, M. and Chapelin-Viscardi, J.-D. (2021) Synthesis of the interactions between the European species of the genus *Bruchus* (Coleoptera: Chrysomelidae: Bruchinae) and their natural enemies. *Annales de la Société Entomologique de France (N.S.)* doi: 10.1080/00379271.2021.1927180

Böhmová, J., Rasplus, J.-Y., Taylor, G.S. and Janšta, P. (2022) Description of two new Australian genera of Megastigmidae (Hymenoptera: Chalcidoidea) with notes on the biology of the genus *Bortesia*. *Journal of Hymenoptera Research* 90, 75–99. doi: 10.3897/jhr.90.82585

Borowiec, N., LaSalle, J., Brancaccio, L., Thaon, M., Warot, S., Branco, M., Ris, N. Malausa, J.-C. and Burks, R. (2019) *Ophelimus mediterraneus* sp. n. (Hymenoptera, Eulophidae): a new *Eucalyptus* gall wasp in the Mediterranean region. *Bulletin of Entomological Research* 109, 678–694.

Bouček, Z. (1963) A taxonomic study in *Spalangia* Latr. (Hymenoptera, Chalcidoidea). *Acta Entomologica Musei Nationalis Pragae* 35, 429–512.

Bouček, Z. (1974) A revision of the Leucospidae (Hymenoptera: Chalcidoidea) of the world. *Bulletin of the British Museum (Natural History), Entomology* Supplement 23, 1–241.

Bouček, Z. (1978) A study of the non-podagrionine Torymidae with enlarged hind femora, with a key to the African genera (Hymenoptera). *Journal of the Entomological Society of Southern Africa* 41, 91–134.

Bouček, Z. (1982) Oriental chalcid wasps of the genus *Epitranus*. *Journal of Natural History* 16, 577–622. doi: 10.1080/00222938200770451

Bouček, Z. (1988) *Australasian Chalcidoidea (Hymenoptera): A Biosystematic Revision of Genera of Fourteen Families, with a Reclassification of Species*. CAB International, Wallingford, UK.

Bouček, Z. (1998) A taxonomic revision of the species of Palachiini (Hymenoptera; Torymidae). *Journal of Natural History* 32, 217–262. doi: 10.1080/00222939800770121

Bouček, Z., Watsham, A. and Wiebes, J.T. (1981) The fig wasp fauna of the receptacles of *Ficus thonningii* (Hymenoptera Chalcidoidea). *Tijdschrift voor Entomologie* 124, 149–233.

Burks, B.D. (1938) A study of chalcidoid wings (Hymenoptera). *Annals of the Entomological Society of America* 31, 157–161.

Burks, R.A., Heraty, J.M., Gebiola, M. and Hansson, C. (2011) Combined molecular and morphological phylogeny of Eulophidae (Hymenoptera: Chalcidoidea), with focus on the subfamily Entedoninae. *Cladistics* 27, 581–605.

Burks, R.A., Heraty, J.M., Mottern, J., Dominguez, C. and Heacox, S. (2017) Biting the bullet: revisionary notes on the Oraseminae of the Old World (Hymenoptera, Chalcidoidea, Eucharitidae). *Journal of Hymenoptera Research* 55, 139–188. doi: 10.3897/jhr.55.11482

Burks, R., Mitroiu, M.-D., Fusu, L., Heraty, J.M., Janšta, P., Heydon, S., Papilloud, N.D.-S., Peters, R.S., Tselikh, E.V., Woolley, J.B., van Noort, S., Baur, H., Cruaud, A., Darling, C., Haas, M., Hanson, P., Krogmann, L. and Rasplus, J.-Y. (2022) From hell's heart I stab at thee! A determined approach towards a monophyletic Pteromalidae and reclassification of Chalcidoidea (Hymenoptera). *Journal of Hymenoptera Research* 94, 13–88.

Camors, F.B. Jr and Payne, T.L. (1972) Response of *Heydenia unica* (Hymenoptera: Pteromalidae) to *Dendroctonus frontalis* (Coleoptera: Scolytidae) pheromones and a host-tree terpene. *Annals of the Entomological Society of America* 65, 31–33.

Campbell, B.C., Heraty, J.M., Rasplus, J.Y., Chan, K., Steffen-Campbell, J.D. and Babcock, C.S. (2000) Molecular systematics of the Chalcidoidea using 28S-D2 rDNA. In: Austin A.D. and Dowton, M. (eds) *Hymenoptera, Evolution, Biodiversity and Biological Control*, CSIRO Publishing, Collingwood, Australia, pp. 59–73.

Cao, L.M., van Achterberg, C., Tang, Y.L., Yang, Z.Q., Wang, X.Y. and Cao, T.W. (2020a) Redescriptions of two parasitoids, *Metapelma beijingense* Yang (Hymenoptera, Eupelmidae) and *Spathius ochus* Nixon (Hymenoptera, Braconidae), parasitizing *Coraebus cavifrons* Descarpentries & Villiers (Coleoptera, Buprestidae) in China with keys to genera or species groups. *ZooKeys* 926, 53–72.

Cao, L.M., Cui, J., Wang, X., Wang, G. and Yang, Z. (2020b) First description of the male of *Solenura ania* (Walker) (Hymenoptera: Pteromalidae), a giant pteromalid parasitoid of *Trichoferus campestris* (Faldermann), with special reference to its sexual dimorphism. *Biodiversity Data Journal* 8, e54961. doi: 10.3897/BDJ.8.e54961

Chen, H.-Y., Li, H.-L., Pang, H., Zhu, C.-D. and Zhang, Y.-Z. (2021) Investigating the parasitoid community associated with the invasive mealybug *Phenacoccus solenopsis* in Southern China. *MDPI Insects* 12, 290. doi: 10.3390/insects12040290

Clarke, A.R. and Seymour, J.E. (1992) Two species of *Acroclisoides* Girault and Dodd (Hymenoptera: Pteromalidae) parasitic on *Trissolcus basalis* (Wollaston) (Hymenoptera: Scelionidae), a parasitoid of *Nezara viridula* (L.) (Hemiptera: Pentatomidae). *Australian Journal of Entomology* 31, 299–300.

Cook, J.M., Compton, S.A., Herre, E.A. and West, S.A. (1997) Alternative mating tactics and extreme male dimorphism in fig wasps. *Proceedings of the Royal Society of London, Series B* 264, 747–754.

Cooke-McEwen, C.L. (2020) Morphological, taxonomic, and molecular contributions to *Cerocephala* Westwood (Hymenoptera: Pteromalidae: Cerocephalinae). *Proceedings of the Entomological Society of Washington* 122, 42–54.

Copland, M.J.W. and King, P.E. (1972) The structure of the female reproductive system in the Eurytomidae (Chalcidoidea: Hymenoptera). *Journal of Zoology (London)* 166, 185–212.

Cowan, D.P. (1979) The function of enlarged hindlegs in oviposition and aggression by *Chalcis canadensis*. *Great Lakes Entomologist* 12, 133–136.

Cruaud, A., Jabbour-Zahab, R., Genson, G., Cruaud, C., Couloux, A., Kjellberg, F., Van Noort, S. and Rasplus, J.-Y. (2010) Laying the foundations for a new classification of Agaonidae (Hymenoptera: Chalcidoidea), a multilocus phylogenetic approach. *Cladistics* 26, 359–387.

Cruaud, A., Jabbour-Zahab, R., Genson, G., Couloux, A., Yan-Qiong, P., da Rong, Y., Ubaidillah, R., Pereira, R.A.S., Kjellberg, F., van Noort, S., Kerdelhué, C. and Rasplus, J.-Y. (2011) Out of Australia and back again: the world-wide historical biogeography of non-pollinating fig wasps (Hymenoptera: Sycophaginae). *Journal of Biogeography* 38, 209–225.

Cruaud, A., Underhill, J.G., Huguin, M., Genson, G., Jabbour-Zahab, R., Tolley, K.A., Rasplus, J.-Y. and van Noort, S. (2013) A multilocus phylogeny of the world Sycoecinae fig wasps (Chalcidoidea: Pteromalidae). *PLoS ONE* 8(11), e79291.

Cruaud, A., Delvare, G., Nidelet, S., Sauné, L., Ratnasingham, S., Chartois, M., Blaimer, B.B., Gates, M., Brady, S.G., Faure, S., van Noort, S., Rossi, J.-P. and Rasplus, J.-Y. (2021) Ultra-conserved elements and morphology reciprocally illuminate conflicting phylogenetic hypotheses in Chalcididae (Hymenoptera, Chalcidoidea). *Cladistics* 37, 1–35.

Cruaud, A., Rasplus, J.-Y., Zhang, J., Burks, R., Delvare, G., Fusu, L., Gumovsky, A., Huber, J.T., Janšta, P., Mitroiu, M.-D., Noyes, J.S., van Noort, S., Baker, A., Böhmová, J., Baur, H., Blaimer, B.B., Brady, S.G., Bubeníková, K., Chartois, M., Copeland, R.S., Dale-Skey Papilloud, N., Dal Molin, A., Dominguez, C., Gebiola, M., Guerrieri, E., Kresslein, R.L., Krogmann, L., Moriarty Lemmon, E., Murray, E.A., Nidelet, S., Nieves-Aldrey, J.L., Perry, R.K., Peters, R.S., Polaszek, A., Sauné, L., Torréns, J., Triapitsyn, S., Tselikh, E.V., Yoder, M., Lemmon, A.R., Woolley, J.B. and Heraty, J.M. (2022) The Chalcidoidea bush of life – a massive radiation blurred by mutational saturation. *bioRxiv* 2022.09.11.507458. doi: 10.1101/2022.09.11.507458

Cruaud, A., Rasplus, J.-Y., Zhang, J., Burks, R., Delvare, G., Fusu, L., Gumovsky, A., Huber, J.T., Janšta, P., Mitroiu, M.-D., Noyes, J.S., van Noort, S., Baker, A., Böhmová, J., Baur, H., Blaimer, B.B., Brady, S.G., twenty three other authors, Woolley, J.D. and Heraty, J.M. (2023) The Chalcidoidea bush of life: evolutionary history of a massive radiation of minute wasps. *Cladistics*. https://doi.org/10.1111/cla.12561

Darling, D.C. (1995) New species of *Krombeinius* (Hymenoptera: Chalcidoidea: Perilampidae) from Indonesia, and the first description of first-instar larva for the genus. *Zoologische Mededelingen* 69, 209–229.

Darling, D.C. (1996) Generic concepts in the Perilampidae (Hymenoptera: Chalcidoidea): an assessment of recently proposed genera. *Journal of Hymenoptera Research* 5, 100–130.

Darling, D.C. (2009) A new species of *Smicromorpha* (Hymenoptera, Chalcididae) from Vietnam, with notes on the host association of the genus. *ZooKeys* 20, 155–163.

Darling, D.C. and Miller, T.D. (1991) Life history and larval morphology of *Chrysolampus* (Hymenoptera: Chalcidoidea: Chrysolampinae) in western North America. *Canadian Journal of Zoology* 69, 2168–2177. doi: 10.1139/z91-303

Darling, D.C. and Roberts, H. (1999) Life history and larval morphology of *Monacon* (Hymenoptera: Perilampidae), parasitoids of ambrosia beetles (Coleoptera: Platypodidae). *Canadian Journal of Zoology* 77, 1768–1782.

Darling, D.C. and Tatarnic, N.J. (2020) On the horns of a dilemma: toward a better understanding of the *Monacon* species (Hymenoptera: Perilampidae) of Borneo. *Journal of Natural History* 54, 723–734. doi: 10.1080/00222933.2020.1776906

Darling, D.C., Gharari, H. and Gibson, G.A.P. (2021) Family Perilampidae Förster, 1856. In: Gharari, H., Gibson, G.A.P. and Viggiani, G. (eds) *Chalcidoidea of Iran (Insecta: Hymenoptera)*. CABI International, Wallingford, UK, pp. 289–293.

Das, G.M. (1963) Preliminary studies on the biology of *Orasema assectator* Kerrich (Hymenoptera: Eucharitidae) parasitic on *Pheidole* and causing damage to leaves of tea in Assam. *Bulletin of Entomological Research* 54, 393–398. doi: 10.1017/S0007485300048884

de Souza Tavares, W., Sinulingga, N.G.H.B., Saha, M.A., Sunardi, K., Sihombing, I.F.L., Tarigan, M., Kkadan, S.K. and Duran, A. (2023) *Leptocybe invasa* (Hymenoptera: Eulophidae) galls parasitized by *Megastigmus* sp. (Hymenoptera: Torymidae): first record in Indonesia on a new host plant, *Eucalyptus brassiana* (Myrtaceae). *Journal of Plant Diseases and Protection* 130, 1149–1153.

Delvare, G. (2017) Order Hymenoptera, family Chalcididae. *Arthropod fauna of the UAE* 6, 225–274.

Desjardins, C.A. (2007) Phylogenetics and classification of the world genera of Diparinae (Hymenoptera: Pteromalidae). *Zootaxa* 1647, 1–88. doi: 10.11646/zootaxa.1647.1.1

Desjardins, C.A., Regier, J.C. and Mitter, C. (2007) Phylogeny of pteromalid parasitic wasps (Hymenoptera: Pteromalidae): initial evidence from four protein-coding nuclear genes. *Molecular Phylogenetics and Evolution* 45, 454–469.

Doğanlar, M. and Laz, B. (2023) *Chalcedectus balachowskyi* Steffan, 1968 (Hymenoptera: Pteromalidae), collected from bark of *Robinia pseudoacacia* (Umbraculifera), in Kahramanmaraş, Turkey. *Munis Entomology & Zoology* 18, 178–180.

Elango, K., Aravind, A., Nelson, S.J. and Ayyamperumal, M. (2022) First report of the encyrtid parasitoid, *Copidosomyia ambiguous* (Subba Rao) (Hymenoptera: Chalcidoidea: Encyrtidae) on *Mallada desjardinsi* (Navas) an indigenous predator of the rugose spiralling whitefly, *Aleurodicus rugioperculatus* Martin in India. *International Journal of Tropical Insect Science* 42, 2019–2021. doi: 10.1007/s42690-021-00667-5

Fatouros, N.E. and Huigens, M.E. (2012) Phoresy in the field: natural occurrence of *Trichogramma* egg parasitoids on butterflies and moths. *BioControl* 57, 493–502. doi: 10.1007/s10526-011-9427-x

Fusu, L. and Polaszek, A. (2017) Description, DNA barcoding and phylogenetic placement of a remarkable new species of *Eopelma* (Hymenoptera: Eupelmidae) from Borneo. *Zootaxa* 4263, 557–566.

Galil, J. and Copland, J.W. (1981) *Odontofroggatia galili* Wiebes in Israel, a primary fig wasp of *Ficus microcarpa* L. with a unique ovipositor mechanism (Epichrysomallinae, Chalcidoidea). *Proceedings of the Koninklijke Nederlandse Akademie van Wetenschappen, series C: Biological and Medical Sciences* 84, 183–195.

Gates, M.W. (2008) *Species Revision and Generic Systematics of World Rileyinae (Hymenoptera: Eurytomidae)*. UC Publications in Entomology, 127. University of California Press, Berkeley, 332 pp. Available at http://escholarship.org/uc/item/6d0851rn

Gauld, I.D. and Bolton, B. (1988) *The Hymenoptera*. British Museum (Natural History), London, and Oxford University Press, Oxford, 332 pp.

Gauthier, N., LaSalle, J., Quicke, D.L.J. and Godfray, H.C.J. (2000) The phylogeny and classification of the Eulophidae (Hymenoptera, Chalcidoidea) and the recognition that the Elasmidae are derived eulophids. *Systematic Entomology* 25, 521–539.

Geng, H. and Li, C.-D. (2017) The *Encarsia flavoscutellum*-group key to world species including two new species from China (Hymenoptera, Aphelinidae). *ZooKeys* 662, 127–136. doi: 10.3897/zookeys.662.11809

Gerling, D., Tremblay, E. and Orion, T. (1991) Initial stages of the vital capsule formation in the *Eretmocerus-Bemisia tabaci* association. *Redia* 74, 411–415.

Gerling, D., Quicke, D.L.J. and Orion, T. (1998) Oviposition mechanisms in the whitefly parasitoids *Encarsia transvena* and *Eretmocerus mundus*. *Biocontrol* 43, 289–297.

Ghahari, H., Gibson, D. and Viggiani, G. (2021). *Chalcidoidea of Iran (Insecta-Hymenoptera)*. CAB International, Wallingford, UK.

Gibson, G.A.P. (1986a) Evidence for monophyly and relationships of Chalcidoidea, Mymaridae, and Mymarommatidae (Hymenoptera: Terebrantes). *Canadian Entomologist* 118, 205–240.

Gibson, G.A.P. (1986b) Mesothoracic skeletomusculature and mechanics of flight and jumping in Eupelminae (Hymenoptera, Chalcidoidea: Eupelmidae). *Canadian Entomologist* 118, 691–728. doi: 10.4039/Ent118691-7

Gibson, G.A.P. (1989) Phylogeny and classification of Eupelmidae, with a revision of the world genera of Calosotinae and Metapelmatinae (Hymenoptera: Chalcidoidea). *Memoirs of the Entomological Society of Canada* 121, Supplement S149, 3–121. doi: 10.4039/entm121149fv

Gibson, G.A.P. (1993) Superfamilies Mymarommatoidea and Chalcidoidea. In: Goulet, H. and Huber, J. (eds) *Hymenoptera of the World: An Identification Guide to Families*. Research Branch, Agriculture Canada, Ottawa, Canada, pp. 570–655.

Gibson, G.A.P. (2003) Phylogenetics and classification of Cleonyminae (Hymenoptera: Chalcidoidea). *Memoirs on Entomology, International* 16, 1–339.

Gibson, G.A.P. (2005) The world species of *Balcha* Walker (Hymenoptera: Chalcidoidea: Eupelmidae), parasitoids of wood-boring beetles. *Zootaxa* 1033, 1–62.

Gibson, G.A.P. (2009) Description of three new genera and four new species of Neanastatinae (Hymenoptera: Eupelmidae), from Baltic amber, with discussion of their relationships to extant taxa. *ZooKeys* 20, 175–214.

Gibson, G.A.P. (2017) Revision of *Eopelma* Gibson (Hymenoptera: Chalcidoidea: Eupelmidae: Neanastatinae). *Proceedings of the Entomological Society of Washington* 119, 741–777.

Gibson, G.A.P., Huber, J.T. and Woolley, J.B. (1997) *Annotated Keys to the Genera of the Nearctic Chalcidoidea (Hymenoptera)*. NRC Research Press, Ottawa, Ontario.

Gibson, G.A.P., Heraty J.M. and Woolley J.B. (1999) Phylogenetics and classification of Chalcidoidea and Mymarommatoidea – a review of current concepts (Hymenoptera, Apocripta). *Zoologica Scripta* 28, 87–124. doi: 10.1046/j.1463-6409.1999.00016.x

Gibson, G.A.P., Read, J. and Huber, J.T. (2007) Diversity, classification and higher relationships of Mymarommatoidea (Hymenoptera). *Journal of Hymenoptera Research* 16(1), 51–146.

Giovannini, L., Sabbatini-Peverieri, G., Tillman, P.G., Hoelmer, K.A. and Roversi, P.F. (2021) Reproductive and developmental biology of *Acroclisoides sinicus*, a hyperparasitoid of scelionid parasitoids. *Biology* 2021, 10, 229. doi: 10.3390/biology10030229

Gómez, J.F., Hernández Nieves, M., Gayubo, S.F. and Nieves-Aldrey, J.L. (2017) Terminal-instar larval systematics and biology of west European species of Ormyridae associated with insect galls (Hymenoptera, Chalcidoidea). *ZooKeys* 644, 51–88. doi: 10.3897/zookeys.644.10035

Graham, M.W.R. de V. (1969) Pteromalidae of North-Western Europe (Hymenoptera: Chalcidoidea). *Bulletin of the British Museum (Natural History), Entomology* Supplement 16, 1–908.

Greeff, J.M., van Noort, S., Rasplus, J.-Y. and Kjellberg, F. (2003) Dispersal and fighting in male pollinating fig wasps. *Comptes Rendus Biologies* 326, 121–130.

Grissell, E.E. (1995) Toryminae (Hymenoptera: Chalcidoidea: Torymidae): a redefinition, generic classification and annotated world catalogue of species. *Memoirs on Entomology International* 2, 1–470.

Grissell, E.E. and Cameron, S.A. (2002) A new *Leucospis* Fabricius (Hymenoptera: Leucospidae), the first gregarious species. *Journal of Hymenoptera Research* 11, 271–278.

Grissell, E.E. and Desjardins, C.A. (2002) Revision of *Bootania* Dalla Torre and recognition of *Macrodasyceras* Kamijo (Hymenoptera: Torymidae). *Journal of Hymenoptera Research* 11, 279–311.

Grissell, E.E. and Prinsloo, G.L. (2001) Seed-feeding species of *Megastigmus* (Hymenoptera: Torymidae) associated with Anacardiaceae. *Journal of Hymenoptera Research* 10, 271–279.

Guerrieri, E., Pedata, P.A., Romani, R., Isidoro, N. and Bin, F. (2001) Functional anatomy of male antennal glands in three species of Encyrtidae (Hymenoptera: Chalcidoidea). *Journal of Natural History* 35, 41–54. doi: 10.1080/00222930014477980

Gul, M.A., Soliman, A.M., Gadallah, N.S., Al Dhafer, H.M. and Delvare, G. (2020) The genus *Phasgonophora* Westwood, 1832 (Hymenoptera, Chalcididae) in Saudi Arabia: re-evaluation of its limits and description of three new species. *Journal of Hymenoptera Research* 76, 1–38.

Hamilton, W.D. (1979) Wingless and fighting males in fig wasps and other insects. In: Blum, M.S. and Blum, N.A. (eds) *Sexual Selection and Reproductive Competition in Insects*. Academic Press, London, pp. 167–220.

Hansson, C. and Straka, J. (2009) The name Euderinae (Hymenoptera: Eulophidae) is a junior homonym. *Proceedings of the Entomological Society of Washington* 111, 272–273. doi: 10.4289/0013-8797-111.1.272

Hayat, M. (1983) The genera of Aphelinidae (Hymenoptera) of the world. *Systematic Entomology* 8(1), 63–102.

Hayat, M. (1994) Notes on some genera of the Aphelinidae (Hymenoptera: Chalcidoidea), with comments on the classification of the family. *Oriental Insects* 28, 81–96. doi: 10.1080/00305316.1994.10432297

Hayat, M. (2009) A review of the Indian Signiphoridae (Hymenoptera: Chalcidoidea). *Biosystematica* 3, 5–27.

Hayat, M. and Verma, M. (1980). The aphelinid subfamily Eriaporinae (Hym.: Chalcidoidea). *Oriental Insects* 14, 29–40. doi: 10.1080/00305316.1980.10434580

Hayat, M., Schroer, S. and Pemberton, R.W. (2010) On some Encyrtidae (Hymenoptera: Chalcidoidea) on lac insects (Hemiptera: Kerriidae) from Indonesia, Malaysia and Thailand. *Oriental Insects* 44, 23–33. doi: 10.1080/00305316.2010.10417603

Hellum, K. and Sullivan, F. (1990) Symbiosis between insects and the seeds of *Sesbania grandiflora* Desv. near Muak Iek, Thailand. *Embryon* 3, 37–39.

Heraty, J.M. (1994) Classification and evolution of the Oraseminae in the Old World, with revisions of two closely related genera of Eucharitinae (Hymenoptera: Eucharitidae). *Life Sciences Contributions, Royal Ontario Museum* 157, 1–174.

Heraty, J.M. (2000) Phylogenetic relationships of Oraseminae (Hymenoptera: Eucharitidae). *Annals of the Entomological Society of America* 93, 374–390.

Heraty, J.M. (2002) A revision of the genera of Eucharitidae (Hymenoptera: Chalcidoidea) of the World. *Memoirs of the American Entomological Institute* 68, 1–359.

Heraty, J.M. (2017) World catalog of Eucharitidae. Available at http://www.hymenoptera.ucr.edu/eucharitidae (accessed August, 2022)

Heraty, J.M. (2019) Catalog of World Eucharitidae, 2019. Available at https://www.antwiki.org/wiki/images/7/7e/EucharitidaeCatalog2017.pdf (accessed 20 July 2023)

Heraty, J.M. and Darling, D.C. (1984) Comparative morphology of the planidial larvae of Eucharitidae and Perilampidae (Hymenoptera: Chalcidoidea). *Systematic Entomology* 9, 309–328.

Heraty, J.M. and Murray, E. (2013) The life history of *Pseudometagea schwarzii*, with a discussion of the evolution of endoparasitism and koinobiosis in Eucharitidae and Perilampidae (Chalcidoidea). *Journal of Hymenoptera Research* 35, 1–15.

Heraty, J.M. and Quicke, D.L.J. (2003) Phylogenetic implications of ovipositor structure in Eucharitidae and Perilampidae (Hymenoptera: Chalcidoidea). *Journal of Natural History* 37, 1751–1764.

Heraty, J.M., Hawks, D., Kostecki, J.S. and Carmichael, A. (2004) Phylogeny and behaviour of the Gollumiellinae, a new subfamily of the ant-parasitic Eucharitidae (Hymenoptera: Chalcidoidea). *Systematic Entomology* 29, 544–559.

Heraty, J.M., Heraty, J.M. and Torréns, J. (2009) A new species of *Pseudochalcura* (Hymenoptera, Eucharitidae), with a review of antennal morphology from a phylogenetic perspective. *ZooKeys* 20, 215–231.

Heraty, J.M., Burks, R.A., Cruaud, A., Gibson, G.A.P., Liljeblad, J., Munro, J., Rasplus, J.-Y., Delvare, G., Janšta, P., Gumovsky, A., Huber, J., Woolley, J.B., Krogmann, L., Heydon, S., Polaszek, A., Schmidt, S., Darling, D.C., Gates, M.W., Mottern, J., Murray, E., Dal Molin, A., Triapitsyn, S., Baur, H., Pinto, J.D., van Noort, S., George, J. and Yoder, M. (2013) A phylogenetic analysis of the megadiverse Chalcidoidea (Hymenoptera). *Cladistics* 29, 466–542.

Heraty, J.M., Mottern, J. and Peeters, C. (2015) A new species of *Schizaspidia*, with discussion of the phylogenetic utility of immature stages for assessing relationships among eucharitid parasitoids of ants. *Annals of the Entomological Society of America* 108, 865–874.

Heraty, J.M., Burks, R.A., Mbanyana, N. and van Noort, S. (2018) Morphology and life history of an ant parasitoid, *Psilocharis afra* (Hymenoptera: Eucharitidae). *Zootaxa* 4482, 491–510. doi:10.11646/zootaxa.4482.3.3

Heraty, J.M., Derafsha, H.A. and Ghafouri Moghaddam, M. (2019) Review of the Philomidinae Ruschka (Hymenoptera: Chalcidoidea: Perilampidae), with description of three new species. *Arthropod Systematics & Phylogeny* 77, 39–56.

Herreid, J.S. and Heraty, J.M. (2017) Hitchhikers at the dinner table: a revisionary study of a group of ant parasitoids (Hymenoptera: Eucharitidae) specializing in the use of extrafloral nectaries for host access. *Systematic Entomology* 42, 204–229.

Hesami, S., Akrami, M.A. and Baur, H. (2005) *Leucospis dorsigera* Fabricius (Hymenoptera, Leucospidae) as a hyperparasitoid of Cerambycidae (Coleoptera) through Xoridinae (Hymenoptera: Ichneumonidae) in Iran. *Journal of Hymenoptera Research* 14, 66–68.

Hidaka, T., Budyanto, E., Ya-Klai, V. and Joshi, R.C. (1988) Recent studies on natural enemies of the rice gall midge, *Orseolia oryzae* (Wood-Mason). *Japan Agricultural Research Quarterly* 22, 175–180.

Hill, D.S. (1967) Fig-wasps (Chalcidoidea) of Hong Kong: I. Agaonidae. *Zoologische Verhandelingen* 89, 1–54.

Hoddle, C., Hoddle, M. and Triapitsyn, S. 2013. *Marietta leopardina* (Hymenoptera: Aphelinidae) and *Aprostocetus* (*Aprostocetus*) sp. (Hymenoptera: Eulophidae) are obligate hyperparasitoids of *Tamarixia radiata* (Eulophidae) and *Diaphorencyrtus aligarhensis* (Hymenoptera: Encyrtidae). *Florida Entomologist* 96, 643–646. doi: 10.1653/024.096.0236

Huang, J. and Polaszek, A. (1998) A revision of the Chinese species of *Encarsia* Förster (Hymenoptera: Aphelinidae): parasitoids of whiteflies, scale insects and aphids (Hemiptera: Aleyrodidae, Diaspididae, Aphidoidea). *Journal of Natural History* 32, 1825–1966.

Huber, J.T. (1995) Mymaridae. In: P.E. Hanson, P.E. and Gauld, I.D. (eds) *The Hymenoptera of Costa Rica*. Oxford University Press, New York, pp. 344–349.

Huber, J.T. (2015) World reclassification of the *Gonatocerus* group of genera (Hymenoptera: Mymaridae). *Zootaxa* 3967, 1–184.

Ishii-Eiteman, M.J. (1993) Egg parasitism of green rice leafhoppers in rice cultivation systems in northern Thailand. Dissertation. Cornell University, Ithaca, New York.

Jansen-González, S., Teixeira, S.P. and Pereira, R.A.S. (2020) Larval strategy of two species of seed-feeding Chalcidoidea parallels that of parasitoid koinobionts. *Oecologia Australis* 24, 903–916. doi: 10.4257/oeco.2020.2404.13

Janšta, P., Cruaud, A., Delvare, G., Genson, G., Heraty, J., Křížkoá, B. and Rasplus, J.Y. (2018) Torymidae (Hymenoptera, Chalcidoidea) revised, molecular phylogeny, circumscription and reclassification of the family with discussion of its biogeography and evolution of life-history traits. *Cladistics* 34, 627–651.

Jin, X. and Li, C. (2015) A new species of *Australomymar* (Hymenoptera: Mymaridae), with a key to species. *Turkish Journal of Zoology* 39, 421–424.

Joseph, K.J. (1958) Recherches sur les chalcidiens *Blastophaga psenes* (L.) et *Philotrypesis caricae* (L.) du figuier (*Ficus carica* L.). *Annales des Sciences Naturelles, Zoologie, 11ème série* 20, 197–260.

Jousselin, E., Rasplus, J.Y. and Kjellberg, F. (2001) Shift to mutualism in parasitic lineages of the fig/fig wasp interaction. *Oikos* 94, 287–294.

Jousselin, E., Van Noort, S. and Greeff, J.M. (2004) Labile male morphology and intraspecific male polymorphism in the *Philotrypesis* fig wasps. *Molecular Phylogenetics and Evolution* 33, 706–718.

Kattari, D., Heimpel, G.E., Ode, P.J. and Rosenheim, J.A. (1999) Hyperparasitism by *Ablerus clisiocampae* Ashmead (Hymenoptera: Aphelinidae). *Proceedings of the Entomological Society of Washington* 101, 640–644.

Khanday, A.L., Sureshan, P.M., Buhroo, A.A., Ranjith, A.P. and Tselikh, E. (2019) Pteromalid wasps (Hymenoptera: Chalcidoidea) associated with bark beetles, with the description of a new species from Kashmir, India. *Journal of Asia-Pacific Biodiversity* 12, 262–272. doi: 10.1016/j.japb.2019.01.014

Kim, J.W. and Heraty, J. (2012) A phylogenetic analysis of the genera of Aphelininae (Hymenoptera: Aphelinidae), with a generic key and descriptions of new taxa. *Systematic Entomology* 37, 497–549. doi: 10.1111/j.1365-3113.2012.00625.x

Kjellberg, F., Jousselin, E., Hossaert-Mckey, M. and Rasplus, J.-Y. (2005) Biology, ecology, and evolution of fig-pollinating wasps (Chalcidoidea, Agaonidae). *Biology, Ecology and Evolution of Gall-Inducing Arthropods* 2, 539–572.

Kjellberg, F., van Noort, S. and Rasplus, J.Y. (2022) Fig wasps and pollination. In: Sarkhosh, A., Yavari, A. and Ferguson, L. (eds) *The Fig. Botany, Production and Uses*. CAB International, Wallingford, UK, pp. 231–254.

Krogmann, L. and Burks, R.A. (2009) *Doddifoenus wallacei*, a new giant parasitoid wasp of the subfamily Leptofoeninae (Chalcidoidea: Pteromalidae), with a description of its mesosomal skeletal anatomy and a molecular characterization. *Zootaxa* 2194, 21–36.

LaSalle, J. (1989) Notes on *Zagrammosoma* (Hymenoptera: Eulophidae) with description of a new species. *Proceeding Entomological Society of Washington* 91, 230–236.

LaSalle, J. (2005) Biology of gall inducers and evolution of gall induction in Chalcidoidea (Hymenoptera: Eulophidae, Eurytomidae, Pteromalidae, Tanaostigmatidae, Torymidae). In: Raman, A., Schaefer, C.W. and Withers, T.M. (eds) *Biology, Ecology, and Evolution of Gall-inducing Arthropods*. Science Publishers, Enfield, New Hampshire, pp. 507–537.

LaSalle, J. and Noyes, J.S. (1985) New family placement for the genus *Cynipencyrtus* (Hymenoptera: Chalcidoidea: Tanaostigmatidae). *Journal of the New York Entomological Society* 93, 1261–1264.

LaSalle, J. and Schauff, M.E. (1994) Systematics of the tribe Euderomphalini (Hymenoptera: Eulophidae): parasitoids of whiteflies (Homoptera: Aleyrodididae). *Systematic Entomology* 19, 235–258.

LaSalle, J., Polaszek, A., Noyes, J.S. and Zolnerowich, G. (1997) A new whitefly parasitoid (Hymenoptera: Pteromalidae: Eunotinae) with comments on its placement, and implications for classification of Chalcidoidea with particular reference to the Eriaporinae (Hymenoptera: Aphelinidae). *Systematic Entomology* 22, 131–150.

LaSalle, J., Arakelian, G., Garrison, R.W. and Gates, M.W. (2009) A new species of invasive gall wasp (Hymenoptera: Eulophidae: Tetrastichinae) on blue gum (*Eucalyptus globulus*) in California. *Zootaxa* 2121, 35–43.

Lawson, S., Griffiths, M., Nahrung, H., Noack, A., Wingfield, M., Wilcken, C., Silippers, B., Lo, N., Pham, Q.T., Lee, S.-S., Lelana, N.E. Ketphanh, H., Zhou, X. and Eungwijarnpanya, S. (2012) *Biological Control of Eucalypt Pests Overseas and in Australia*. Final report number FST/2011/028.

Le, K.H. and Tran, D.H. (2022) Biology of *Brachymeria kamijoi* Habu (Hymenoptera: Chalcididae), a pupal parasitoid of the coconut black-headed caterpillar, *Opisina arenosella* Walker (Lepidoptera: Xyloryctidae). *Research on Crops* 23, 866–869.

Li, Q. Hu, H.-Y., Li, Z. and Xiao, H. (2014) First record of *Colotrechnus* Thomson (Hymenoptera: Chalcidoidea: Pteromalidae) from China, with description of one newly-recorded species and a key to known species. *Entomotaxonomia* 36, 134–140.

Lin, N.-Q., Huber, J.T. and LaSalle, J. (2007) The Australian genera of Mymaridae (Hymenoptera: Chalcidoidea). *Zootaxa* 1596, 1–111. doi: 10.11646/zootaxa.1596.1.1

Lotfalizadeh, H. (2012) Review of chalcidoid parasitoids (Hymenoptera: Chalcidoidea) of xylophagous beetles. *Munis Entomology & Zoology Journal* 7, 309–333.

Lotfalizadeh, H., Delvare, G. and Rasplus, J.Y. (2007) Phylogenetic analysis of Eurytominae (Chalcidoidea: Eurytomidae) based on morphological characters. *Zoological Journal of the Linnean Society* 151, 441–510. doi: 10.1111/j.1096-3642.2007.00308.x

MacDonald, K.E. and Caveney, S. (2004) Unusual life history characteristics of *Elachertus scutellatus* Howard (Hymenoptera: Eulophidae), a koinobionic ectoparasitoid. *Environmental Entomology* 33, 227–233.

Manickavasagam, S., Prabhu, A. and Kanagarajan, R. (2006) Record of a hyperparasitoid on *Pseudogonatopus nudus* Perkins (Dryinidae: Chrysidoidea) parasitizing *Nilaparvata lugens* (Stål) from Asia. *International Rice Research Notes* 31, 24–25.

Matthews, R.W., González, J.M., Matthews, J.R. and Deyrup, L.D. (2009) Biology of the parasitoid *Melittobia* (Hymenoptera: Eulophidae). *Annual Review of Entomology* 54, 251–266.

Milliron, H.E. (1949) Taxonomic and biological investigations in the genus *Megastigmus* with particular reference to the taxonomy of the Nearctic species (Hymenoptera: Chalcidoidea; Callimomidae). *The American Midland Naturalist* 41, 257–420.

Mitroiu, M.D. (2011) New Pireninae (Hymenoptera: Pteromalidae) from South-East Asia. *Zootaxa* 3065, 1–13.

Mitroiu, M.-D. and van Achterberg, C. (2013) Revision of the Oriental species of *Apsilocera* Bouček (Hymenoptera, Pteromalidae), with description of twelve new species. *Zootaxa* 3717, 448–468.

Muniappan, R., Watson, G.W., Evans, G.A., Rauf, A. and Von Ellenrieder, N. (2012) Cycad *Aulacaspis* Scale, a newly introduced insect pest in Indonesia. *HAYATI Journal of Biosciences* 19, 110–114. doi: 10.4308/hjb.19.3.110

Munro, J.B., Heraty, J.M., Burks, R.A., Hawks, D., Mottern, J., Cruaud, A., Rasplus, J.-Y. and Jansta, P. (2011) A molecular phylogeny of the Chalcidoidea (Hymenoptera). *PLoS ONE* 6, e27023. doi: 10.1371/journal.pone.0027023

Murray, M.G. (1990) Comparative morphology and mate competition of flightless male fig wasps. *Animal Behaviour* 39, 434–443. doi: 10.1016/S0003-3472(05)80406-3

Mwangi, E.N., Kaaya, G.P., Essuman, S. and Kimondo, M.G. (1994) Parasitism of *Amblyomma variegatum* by a hymenopteran parasitoid in the laboratory, and some aspects of its basic biology. *Biological Control* 4, 101–104.

Narendran, T.C. (2007) *Indian Chalcidoid Parasitoids of the Tetrastichinae (Hymenoptera: Eulophidae)*. Occasional Paper No. 272, Records of Zoological Survey of India, 386 pp.

Narendran, T.C. and Anjana, M. (2013) A new record and a new species of *Australomymar* Girault (Hymenoptera: Mymaridae) from Vientam with key to species. *Journal of Experimental Zoology India* 16, 438–439.

Narendran, T.C. and Kumar, P.G. (2005) A new species of *Torymus* Dahlman (Hymenoptera: Torymidae) from West Malaysia. *Zoos' Print Journal* 20, 1805–1806.

Narendran, T.C. and Sheela, S. (2013) A new species of *Podagrion* Spinola (Hymenoptera: Torymidae) from India with a checklist of species of India and adjacent countries. *Records of the Zoological Survey of India* 113, 35–40.

Narendran, T.C. and Sudheer, K. (2004) A new species of *Chrysolampus* Spinola (Hymenoptera: Perilampidae) from Vietnam (South-East Asia). *Ecobios* 2, 5–7.

Narendran, T.C. and van Achterberg, C. (2012) A taxonomic study on the genus *Rhynchoticida* Bouček (Hymenoptera: Chalcidoidea: Torymidae) of Vietnam. *Zoologische Mededelingen Leiden* 86, 485–496.

Narendran, T.C. and van Achterberg, C. (2016) Revision of the family Chalcididae (Hymenoptera, Chalcidoidea) from Vietnam, with the description of 13 new species. *ZooKeys* 576, 1–202. doi: 10.3897/zookeys.576.8177

Narendran, T.C. and Zubair, A. (2013) A review of *Acrias* Walker (Hymenoptera: Eulophidae: Entiinae) with description of a new genus from Saudi Arabia. *Prommalia* I, 1–16.

Naumann, I.D. (1986) A revision of the Indo-Australian Smicromorphinae (Hymenoptera: Chalcididae). *Memoirs of the Queensland Museum* 22, 169–187.

Naumann, I.D. and Reid, C.A.M. (1990) *Ausasaphes shiralee* sp. n. (Hymenoptera, Pteromalidae, Asaphinae), a brachypterous wasp phoretic on a flightless chrysomelid beetle (Coleoptera, Chrysomalidae). *Journal of the Australian Entomological Society* 29, 319–325.

Neser, O.C. and Prinsloo, G. (2001) *Aphytis chionaspis* (Hymenoptera: Aphelinidae), a parasitoid introduced to South Africa from Thailand for the control of mango scale, *Aulacaspis tubercularis* (Hemiptera: Diaspididae). *African Entomology* 9, 199–201. doi: 10.10520/EJC32946

Neuenschwander, P. (2001) Biological control of the cassava mealybug in Africa: a review. *Biological Control* 21, 214–229.

Nikol'skaya, M.N. and Yasnosh, A. (1966) Aphelinids of the European part of the USSR and the Caucasus (Hymenoptera, Aphelinidae). No. 91 in series *Opredeleteli po Faune SSSR*. Akademii Nauk SSSR, Moscow and Leningrad, 296 pp. [in Russian]

Noyes, J.S. (1991) A new species of *Ooencyrtus* (Hymenoptera; Encyrtidae) from Malaysia, a prepupal parasitoid of the cocoa pod borer, *Conopomorpha cramerella* (Snellen) (Lepidoptera; Gracillariidae). *Journal of Natural History* 25; 1617–1622. doi: 10.1080/00222939100771011

Noyes, J.S. (2021) Universal Chalcidoidea Database. World Wide Web electronic publication. Available at http://www.nhm.ac.uk/chalcidoids (accessed 20 July, 2023)

Noyes, J.S. and Hayat, M. (1984) A review of the genera of Indo-Pacific Encyrtidae (Hymenoptera: Chalcidoidea). *Bulletin of the British Museum (Natural History) (Entomology)* 48, 131–139.

Oanh, N.T. (2022) Morphology and development of *Anisopteromalus calandrae* (Howard) (Hymenoptera: Pteromalidae) parasitizing *Lasioderma serricorne* (F.) (Coleoptera: Anobiidae). *International Journal of Tropical Insect Science* 42, 3033–3043.

Oanh, N.T., Bup, N.K. and Long, K.D. (2019) First record of *Anselmella malacia* Xiao & Huang, 2006 (Hymenoptera: Eulophidae), a new insect pest of water apple (*Syzygium samarangense*) and its associated parasitoid in Dong Thap province, Vietnam. *Tap Chi Sinh Hoc* 41, 7–14.

Polilov, A.A., Hakimi, K.D. and Makarova, A.A. (2023) Extremely small wasps independently lost the nuclei in the brain neurons of at least two lineages. *Scientific Reports* 13, 4320. doi.org/10.1038/s41598-023-31529-4

Paulson, A.R., von Aderkas, P. and Perlman, S.J. (2014) Bacterial associates of seed-parasitic wasps (Hymenoptera: *Megastigmus*). *BMC Microbiology* 14(1), 1–16.

Perry, R.K. and Heraty, J.M. (2021) Read between the lineata: a revision of the tattooed wasps, *Zagrammosoma* Ashmead (Hymenoptera: Eulophidae), with descriptions of eleven new species. *Zootaxa* 4010, 1–108.

Peters, R.S., Niehuis, O., Gunkel, S., Bläser, M., Mayer, C., Podsiadlowski, L. *et al.* (2018) Transcriptome sequence-based phylogeny of chalcidoid wasps (Hymenoptera: Chalcidoidea) reveals a history of rapid radiations, convergence, and evolutionary success. *Molecular Phylogenetics and Evolution* 120, 286–196.

Pinto, J.D. (1994) A taxonomic study of *Brachista* (Hymenoptera: Trichogrammatidae) with the description of the two new species phoretic on robberflies of the genus *Efferia* (Diptera: Asilidae). *Proceedings of the Entomological Society of Washington* 96, 120–132.

Polaszek, A. (1991) Egg parasitism in Aphelinidae with special reference to *Centrodora* and *Encarsia* species. *Bulletin of Entomological Research* 81, 97–106.

Polaszek, A. and Luft-Albarracin, E. (2010) Two new *Encarsia* species (Hymenoptera: Aphelinidae) reared from eggs of Cicadellidae (Hemiptera: Auchenorrhyncha) in Argentina: an unusual new host association. *Journal of Natural History* 45, 55–64.

Polaszek, A., Abd-Rabou, S. and Huang, J. (1999) The Egyptian species of *Encarsia* (Hymenoptera: Aphelinidae): a preliminary review. *Zoologische Mededelingen* 73, 131–163.

Polaszek, A., Shih, Y. and Ward, S. (2015) A new species of *Cales* (Hymenoptera: Aphelinidae) parasitizing *Bemisia pongamiae* (Takahashi) (Hemiptera: Aleyrodidae) in Taiwan, with a key to world species of the *Cales spenceri*-group. *Biodiversity Data Journal* 3, e6352. doi: 10.3897/BDJ.3.e6352

Polaszek, A., Lahey, Z. and Woolley, J.B. (2020) *Noyesaphytis* (Chalcidoidea: Aphelinidae) – an unusual new genus from Madagascar, and a reassessment of Aphelininae classification based on morphology. *Journal of Natural History* 54, 9–12, 647–664, doi: 10.1080/00222933.2020.1773559

Polaszek, A., Fusu, L., Viggiani, G., Hall, A., Hanson, P. and Polilov, A.A. (2022) Revision of the World Species of *Megaphragma* Timberlake (Hymenoptera: Trichogrammatidae). *Insects* 13(6), 561. doi: 10.3390/insects13060561

Polilov, A.A. (2012) The smallest insects evolve anucleate neurons. *Arthropod Structure & Development* 41, 29–34. doi: 10.1016/j.asd.2011.09.001.

Quicke, D.L.J. (1997) *Parasitic Wasps*. Chapman & Hall, London, 470 pp.

Ramirez, B.W. (1969) Fig wasps: mechanism of pollen transfer. *Science* 163, 580–581.

Rasplus, J.-Y., Kerdelhue, C., Le Clainche, I. and Mondor, G. (1998) Molecular phylogeny of fig wasps: Agaonidae are not monophyletic. *Comptes Rendus de l'Academie des Sciences (III), Paris. Sciences de la vie* 321, 517–527.

Rasplus, J.-Y., Blaimer, B.B., Brady, S.G., Burks, R.A., Copeland, R.S., Dale-Skey Papilloud, N., Delvare, G., Fisher, N., Fusu, L., Gate, M., Gibson, G.A.P., Gumovsky, A.V., Hanson, P., Heraty, J., Hube, J.T., Jansta, P., LaSalle, J., Mitroiu, M.-D., Nidele, S., Nieves-Aldrey, J.L., Polaszek, A., Saune, L., Triapitsyn, S., Tselikh, E., van Noort, S., Woolley, J.B. and Cruaud, A. (2018) *Phylogenomics of the Chalcid wasps: When UCEs take Pteros away from the garbage can*. Conference paper, 9th International Conference of the International Society of Hymenopterists, Matsuyama City, Japan. Available at https://www.researchgate.net/publication/327137649_Phylogenomics_of_the_Chalcid_wasps_When_UCEs_take_Pteros_away_from_the_garbage_can

Rasplus, J.-Y., Blaimer, B.B., Brady, S.G., Burks, R.A., Delvare, G., Fisher, N., Gates, M., Gauthier, N.A., Gumovsky, A.V., Hansson, C., Heraty, J.M., Fusu, L., Nidelet, S., Pereira, R.A.S., Sauné, L., Ubaidillah, R. and Cruaud, A. (2020) A first phylogenomic hypothesis for Eulophidae (Hymenoptera, Chalcidoidea). *Journal of Natural History* 54, 597–609. doi: 10.1080/002229 33.2020.1762941

Reina P. and La Salle J. (2003) Key to the world genera of Eulophidae parasitoids (Hymenoptera) of leaf-mining Agromyzidae (Diptera). https://keys.lucidcentral.org/keys/v3/eulophidae_parasitoids/ [accessed 10 August 2022]

Reina, P. and LaSalle, J. (2005) Revision of the genus *Perthiola* (Hymenoptera: Eulophidae: Anselmellini) with the description of a new species. *Acta Societatia Zoologicae Bohemicae* 69, 219–224.

Romani, R., Isidoro, N. and Bin, F. (1999) Further evidence of male antennal glands in Aphelinidae: the case of *Aphytis melinus* DeBach (Hymenoptera: Aphelinidae). *Journal of Hymenoptera Research* 8, 109–115.

Sankararaman, H., Manickavasagam, S., Triapitsyn, S.V., Huber, J.T. and Kharbisnop. B. (2020) Two new species of *Camptopteroides* (*Camptopteroides*) (Hymenoptera: Mymaridae) from the Oriental region with a key to Old World species. *Zootaxa* 4868, 243–256. doi: 10.11646/zootaxa.4868.2.4

Schauff, M.E. (1991) The Holarctic genera of Entedoninae (Hymenoptera: Eulophidae): *Contributions of the American Entomological Institute* 26, 1–109.

Schender, D., Katz, K. and Gates, M.W. (2014) Review of *Hyperimerus* (Pteromalidae: Asaphinae) in North America, with redescription of *Hyperimerus corvus* (Girault). *Proceedings of the Entomological Society of Washington* 116, 408–420

Schmidt, S. and Polaszek, A. (2007) *Encarsia* or *Encarsiella*? – redefining generic limits based on morphological and molecular evidence (Hymenoptera, Aphelinidae). *Systematic Entomology* 32, 81–94. doi: 10.1111/j.1365-3113.2006.00364.x

Schmidt, S., Hamid, H., Ubaidillah, R., Ward, S. and Polaszek, A. (2019) A review of the Indonesian species of the family Signiphoridae (Hymenoptera, Chalcidoidea), with description of three new species. *ZooKeys* 897, 29–47. doi: 10.3897/zookeys.897.38148

Segar, S.T., Lopez-Vaamonde, C., Rasplus, J.Y. and Cook, J.M. (2012) The global phylogeny of the subfamily Sycoryctinae (Pteromalidae): parasites of an obligate mutualism. *Molecular Phylogenetics and Evolution* 65, 116–125.

Sharifi S. and Javadi, I. (1971) Control of Rosaceae branch borer in Iran. *Journal of Economic Entomology* 64, 484–486.

Sharkey, M.J. and Wharton, R.A. (1997) Morphology and terminology. In: Wharton, R.A., Marsh, P.M. and Sharkey, M.J. (eds) *Identification Manual to the New World Genera of Braconidae, Special Publication of the International Society of Hymenopterists*, Vol. 1. International Society of Hymenopterists, Washington, DC, pp. 19–37.

Shibata, E. (2002) Potential fecundity of the bamboo gall maker, *Aiolomorphus rhopaloides* (Hymenoptera: Eurytomidae), and its inquiline, *Diomorus aiolomorphi* (Hymenoptera: Torymidae), in relation to gall size and body size. *Journal of Forest Research* 7, 117–120. doi: 10.1007/BF02762517

Shih, Y.T., Ko, C.C., Pan, K.T., Lin, S.C. and Polaszek, A. (2013) *Hydrophylita* (*Lutzimicron*) *emporos* Shih & Polaszek (Hymenoptera: Trichogrammatidae) from Taiwan, parasitising eggs, and phoretic on adults, of the damselfly *Psolodesmus mandarinus mandarinus* (Zygoptera: Calopterygidae). *PLoS ONE* 8(7), e69331.

Shirley, X.A., Woolley, J.B., Hopper, K.R., Isidoro, N. and Romani, R. (2019) Evolution of glandular structures on the scape of males in the genus *Aphelinus* Dalman (Hymenoptera, Aphelinidae). *Journal of Hymenoptera Research* 72, 27–43. doi: 10.3897/jhr.72.36356

Steffan, J.R. (1968) Observations sur *Chalcedectus sinaiticus* (MS.) et descriptions de *C. balachowskyi* sp.n. (Hym. Chalcedectidae) et d'*Oopristus safavii* gen.n., sp.n. (Hym.: Torymidae), deux parasites d'importance économique en Iran. *Entomophaga* 13, 209–216.

Strand, M.R. and Vinson, S.B. (1984) Facultative hyperparasitism by the egg parasitoid *Trichogramma pretiosum* (Hymenoptera: Trichogrammatidae). *Annals of the Entomological Society of America* 77, 679–686. doi: 10.1093/aesa/77.6.679

Suasa-ard, W. (2010) Natural enemies of important insect pests of field crops and utilization as biological control agents in Thailand. Paper presented at the International Seminar on Enhancement of Functional Biodiversity Relevant to Sustainable Food Production in ASPAC (Asia Pacific Network of Science and Technology Centres), 9–11 November, Tsukuba, Japan, 15 pp.

Subba Rao, B.R. and Hayat, M. (1983) Key to the genera of Oriental Mymaridae, with a preliminary catalog (Hymenoptera: Chalcidoidea). *Contributions of the American Entomological Institute* 20, 125–150.

Subba Rao, B.R. and Hayat, M. (eds) (1985) The Chalcidoidea (Insecta: Hymenoptera) of India and the adjacent countries. Part I: reviews of families and keys to families and genera. *Oriental Insects* 19, 163–310.

Sugonyaev, E.S. (1994) Chalcid wasps (Hymenoptera, Chalcidoidea) parasites on soft scales (Coccinea, Coccidae) in Vietnam. Two new unusual species of the aphelinid genus *Coccophagus* Westw. found in the nests of ants. I. *Éntomologicheskoe Obozrenie* 73, 427–432.

Sugonyaev, E.S. (1995) Chalcidoid Wasps (Hymenoptera, Chalcidoidea) parasitizing on soft scales (Homoptera, Coccidae) in Vietnam: II. New species of the genus *Coccophagus* Westwood—inhabitants of ant nests. *Entomologicheskoe Obozrenie* 74, 884–888.

Sugonyaev, E.S. (2012) A new species of the genus *Coccophagus* Westwood (Hymenoptera, Chalcidoidea, Aphelinidae) inhabiting ants' nests (Hymenoptera, Formicidae) in Malaysia. *Entomological Review*, 92, 101–102 [Original Russian text: *Entomologicheskoe Obozrenie* 2011, 90, 439–441].

Sulaiman, S., Omar, B., Omar, S., Jeffery, J., Ghauth, I. and Busparani V. (1990) Survey of microhymenoptera (Hymenoptera: Chalcidoidea) parasitizing filth flies (Diptera: Muscidae, Calliphoridae) breeding in refuse and poultry farms in Peninsular Malaysia. *Journal of Medical Entomology* 27, 851–855. doi: 10.1093/jmedent/27.5.851

Suleman, N., Raja, S. and Compton, S.G. (2012) Only pollinator fig wasps have males that collaborate to release their females from figs of an Asian fig tree. *Biology Letters* 8, 344–346.

Sun, J.-W., Hu, H.-Y., Nkunika, P.O.Y., Dai, P., Xu, W., Bao, H.-P., Desneux, N. and Zang, L.-S. (2021) Performance of two trichogrammatid species from Zambia on Fall Armyworm, *Spodoptera frugiperda* (J. E. Smith) (Lepidoptera: Noctuidae). *MDPI Insects* 12, 859. doi: 10.3390/insects12100859

Sureshan, P.M. (2005) New host and distributional records for *Solenura ania* (Walker) from India and redescription of *Trichilogaster femtrium* (Walker) (Hymenoptera: Chalcidoidea: Pteromalidae). *Records of the Zoological Survey of India* 105, 111–116.

Sureshan, P.M. (2009) A new species of *Heydenia* Förster (Hymenoptera: Chalcidoidea: Pteromalidae) from Sri Lanka, with a key to species of the Indian subcontinent. *Journal of Threatened Taxa* 1, 114–116.

Tachikawa, T. (1981) Hosts of encyrtid genera in the World (Hymenoptera: Chalcidoidea). *Memoirs of the College of Agriculture, Ehime University* 25(2), 85–110.

Triapitsyn, S.V. and Berezovskij, V.V. (2007) Review of the Oriental and Australasian species of *Acmopolynema*, with taxonomic notes on *Palaeoneura* and *Xenopolynema* stat. rev. and description of a new genus (Hymenoptera: Mymaridae). *Zootaxa* 1455, 1–68.

Triapitsyn, S.V., Aishan, Z. and Huber, J.T. (2020) Description of the male of *Tanyxiphium harriet* (Hymenoptera: Mymaridae), with new distribution records and synonymy. *Zootaxa* 4896, 105–112. doi: 10.11646/zootaxa.4896.1.5

Trjapitzin, V.I. (1965) Contribution to the knowledge of the encyrtid fauna of the Comodo and Padar Islands with a catalogue of Indonesian species (Hymenoptera, Encyrtidae). *Treubia* 26, 309–327.

Trjapitzin V.A. and Shuvakhina, E.Ya. (2019) Contribution to the knowledge of the encyrtid-wasp genus *Isodromus* Howard, 1887 (Hymenoptera, Encyrtidae: Homalotylini), parasitoids of lacewings (Neuroptera, Chrysopidae) in the Palaearctic. *Entomological Review* 99, 1382–1388.

Ubaidillah, R. and Kojima, J.I. (2004) Record of *Smicromorpha*, (Hymenoptera: Chalcididae: Smicromorphinae) possible parasitoids of weaver ants, from Halmahera, the North Moluccas. *Treubia* 33, 199–201.

Ubaidillah, R., LaSalle, J. and Quicke, D.L.J. (2000a) A peculiar new genus and species of Entedoninae (Chalcidoidea: Eulophidae) from Southeast Asia. *Journal of Hymenoptera Research* 9, 170–175.

Ubaidillah, R., LaSalle, J. and Rauf, A. (2000b) A new species of *Zagrammosoma* (Hymenoptera: Eulophidae) from the Indo-Australian Region, a parasitoid of the invasive pest species *Liriomyza huidobrensis* (Diptera: Agromyzidae), *Oriental Insects* 34, 221–228. doi: 10.1080/00305316.2000.10417260

Van Den Berg, H., Ankasah, D., Hassan, K., Muhammad, A., Widayanto, H.A., Wirasto, H. B. and Yully, I. (1995) Soybean stem fly, *Melanagromyza sojae* (Diptera: Agromyzidae), on Sumatra: seasonal incidence and the role of parasitism. *International Journal of Pest Management* 41, 127–133.

van Noort, S. and Compton, S.G. (1996) Convergent evolution of agaonine and sycoecine (Agaonidae, Chalcidoidea) head shape in response to the constraints of host fig morphology. *Journal of Biogeography* 23, 415–424.

van Noort, S. and Rasplus, J.-Y. (2005) Revision of the Papua New Guinean fig wasp genus *Robertsia* Boucek (Hymenoptera: Chalcidoidea: Pteromalidae: Sycoecinae). *Zootaxa* 929, 1–35.

van Noort, S., Peng, Y.-Q. and Rasplus, J.Y. (2006) First record of the fig wasp genus *Diaziella* Grandi (Hymenoptera: Chalcidoidea: Pteromalidae: Sycoecinae) from the Asian mainland with description of two new species from China. *Zootaxa* 1337, 39–59.

Viggiani, G. and Laudonia, S. 1994. Description of a new species of *Lathromeris* Förster (Hymenoptera: Trichogrammatidae) larval parasitoid of *Lasioptera* sp. (Diptera: Cecidomyiidae). *Bollettino del Laboratorio di Entomologia Agraria 'Filippo Silvestri'* 49, 169–172.

Vongpa, V., Amornsak, W. and Gordh, G. (2016) Development, reproduction and longevity of *Aprostocetus* sp. (Hymenoptera: Eulophidae), an egg parasitoid of the brown planthopper, *Nilaparvata lugens* (Stål) (Hemiptera: Delphacidae). *Agriculture and Natural Resources* 50, 291–294.

Vungsilabutr, P. (1981) Relative composition of egg-parasite species of *Nilaparvata lugens*, *Sogatella furcifera*, *Nephotettix virescens* and *N. nigropictus* in paddy fields in Thailand. *Tropical Pest Management* 27, 313–317.

Walter, G.H. (1988) Heteronomous host relationships in aphelinids - Evolutionary pathways and adaptive significance (Hymenoptera: Chalcidoidea). In Gupta, V. K. (ed.) *Advances in Parasitic Hymenoptera Research*. E.J. Brill, Leiden, pp. 313–326.

Weiblen, G.D., Yu, D.W. and West, S.A. (2001) Pollination and parasitism in functionally dioecious figs. *Proceedings of the Royal Society of London B* 268, 651–659.

Wenquan, Z., Dawei, H., Darong, Y. and Chongdong, Z. (2004) Oviposition behavior of *Apocrypta westwoodi*. *Kun Chong Zhi Shi* 4, 446–448.

Werner, J. and Peters, R.S. (2018) Taxonomic revision of the genus *Oodera* Westwood, 1874 (Hymenoptera, Chalcidoidea, Pteromalidae, Cleonyminae), with description of ten new species. *Journal of Hymenoptera Research* 63, 73–123.

Wiebes, J.T. (1963) Indo-Malayan and Papuan fig wasps (Hymenoptera: Chalcidoidea) 2. The genus *Pleistodontes* Saunders (Agaonidae). *Zoologische Mededelingen* 38, 303–321.

Wiebes, J.T. (1967) Redescription of Sycophaginae from Ceylon and India, with designation of lectotypes, and a world catalogue of the Otitesellini (Hymenoptera, Chalcidoidea, Torymidae). *Tijdschrift voor Entomologie* 110, 399–452.

Wiebes, J.T. (1974) Philippines fig wasps 1. Records and descriptions of Otitesellini (Hymenoptera Chalcidoidea, Torymidae). *Zoologische Mededelingen* 48, 145–161.

Wiebes, J.T. (1979) Co-evolution of figs and their insect pollinators. *Annual Review of Entomology* 10, 1–12.

Williams, T. and Polaszek, A. (1996) A re-examination of host relations in the Aphelinidae (Hymenoptera: Chalcidoidea). *Biological Journal of the Linnean Society* 57, 35–45.

Wong, D.-M., Fan, S. and Yu, H. (2022) Seven sycoryctine fig wasp species (Chalcidoidea: Pteromalidae) associated with dioecious *Ficus hirta* inhabiting South China and Southeast Asia. *MDPI Biology* 11, 2079–7737.

Woolley, J.B. (1988) Phylogeny and classification of the Signiphoridae (Hymenoptera: Chalcidoidea). *Systematic Entomology* 13, 465–501. doi: 10.1111/j.1365-3113.1988. tb00256.x

Xiao, H. and Huang, D.-W. (2001) A review of Eunotinae (Hymenoptera: Chalcidoidea: Pteromalidae) from China. *Journal of Natural History* 35, 1587–1605.

Xiao, H., Xu, L.-N., Huang, D.W. and Zhao, Z.-Y. (2006) *Anselmella malacia*, a new pest wasp (Hymenoptera: Chalcidoidea: Eulophidae) reared from *Syzygium samarangense* in Malaysia. *Phytoparasitica* 34, 261–264.

Xiao, H., Jiao, T. and Hu, T. (2012) Description of two new species of *Odopoia* Walker, 1871 (Hymenoptera: Chalcidoidea: Torymidae) from China, with a key to known species. *Zootaxa* 3239, 35–42.

Yao, Y.-X. and Yang, Z.-Q. (2009) Key to world species of *Notanisomorphella* Girault (Hymenoptera: Eulophidae), and description of a new species parasitizing the three-striped pyralid *Dichocrocis chlorophanta* Butler (Lepidoptera: Pyralidae) on Chinese silkvine in China. *Entomologica Fennica* 20, 105–110.

Ye, X.-H., van Achterberg, C., Yue, Q. and Xu, Z.-F. (2017) Review of the Chinese Leucospidae (Hymenoptera, Chalcidoidea). *ZooKeys* 651, 107–157. doi: 10.3897/zookeys.651.11235

Yefremova, Z.A., Yegorenkova, E. and Dekoninck, W. (2018) First records of Eulophidae from Cambodia (Hymenoptera: Chalcidoidea). *Belgian Journal of Entomology* 75, 1–13.

Yefremova, Z.A., Viggiani, G., Ghahari, H., Gibson, G.A.P. and Doganlar, M. (2021) Family Eulophidae Westwood, 1829. In: Gharari, H., Gibson, G.A.P. and Viggiani, G. (eds) *Chalcidoidea of Iran (Insecta: Hymenoptera)*. CABI International, Wallingford, UK, pp. 161–209.

Yousuf, M. and Shafee, S.A. (1987) Taxonomy of Indian Trichogrammatidae (Hymenoptera: Chalcidoidea). *Indian Journal of Systematic Entomology* 4, 55–200.

Zerova, M.D., Nieves-Aldrey, J.L., Ghahari, H., Gibson, G.A.P. and Fursov, V.N. (2021) Family Ormyridae Förster, 1856. In: Gharari, H., Gibson, G.A.P. and Viggiani, G. (eds) *Chalcidoidea of Iran (Insecta: Hymenoptera)*. CABI International, Wallingford, UK, pp. 281–286.

Zhang, F. and Yang, D. (2010) Study on mating ecology and sex ratio of three internally ovipositing fig wasps of *Ficus curtipes*. *Bulletin of Entomological Research* 100, 241–245.

Zhang, F.-P., Peng, Y.-Q., Guan J.-M. and Yang, D.-R. (2008) A species of fig tree and three unrelated fig wasp pollinators. *Evolutionary Ecology Research* 10, 611–620.

Zhang, J., Lindsey, A.R.I., Peters, R.S., Heraty, J.M., Hopper, K.R., Werren, J.H., Martinson, E.O., Woolley, J.B., Yoder, M.J. and Krogmann, L. (2020) Conflicting signal in transcriptomic markers leads to a poorly resolved backbone phylogeny of chalcidoid wasps. *Systematic Entomology* 45, 783–802.

Zhang, J., Heraty, J.M., Kresslein, R.L., Rasplus, J.Y., Darling, C., Baker, A.J., Torréns, J., Lemmon, A. and Lemmon, E.M. (2022) Anchored phylogenomics and a revised classification of the planidial larva clade of jewel wasps (Hymenoptera: Chalcidoidea). *Systematic Entomology* 47, 329–353. 10.1111/syen.12533

Zhao, D., Xin, Z., Hou, H., Zhou, Y., Wang, J., Xiao, J. and Huang, D. (2021) Inferring the phylogenetic positions of two fig wasp subfamilies of Epichrysomallinae and Sycophaginae using transcriptomes and mitochondrial data. *Life* 11, 40. doi: 10.3390/life11010040

Zhu, C.-D., La Salle, J. and Huang, D.-W. (2002) A study of Chinese *Cirrospilus* Westwood (Hymenoptera: Eulophidae). *Zoological Studies* 41, 23–46.

Zhu, P., Gurr, G.M., Lu, Z., Heong, K., Chen, G., Zheng, X., Xu, H. and Yang, Y. (2013) Laboratory screening supports the selection of sesame (*Sesamum indicum*) to enhance *Anagrus* spp. parasitoids (Hymenoptera: Mymaridae) of rice planthoppers. *Biological Control* 64, 83–89. doi: 10.1016/j.biocontrol.2012.09.014.

Zinna, G. (1955) Un nuovo parassita della *Dioryctria splendidella* H.S., *Crataepoides russoi* n.sp., rappresentante di un nuovo genere. *Bollettino del Laboratorio di entomologia agraria "Filippo Silvestri" Portici* 14, 65–82.

Mymarommatoidea (False Fairy Wasps)

These are very small and seldom collected wasps, less than 1 mm long (Fig. 22.51). Historically, they were treated variously within the Mymaridae (sometimes as a subfamily) or as a separate family within Chalcidoidea. Gibson (1986) demonstrated, based on detailed morphological study, that they constituted a sister group to the remaining Chalcidoidea but left them unplaced within

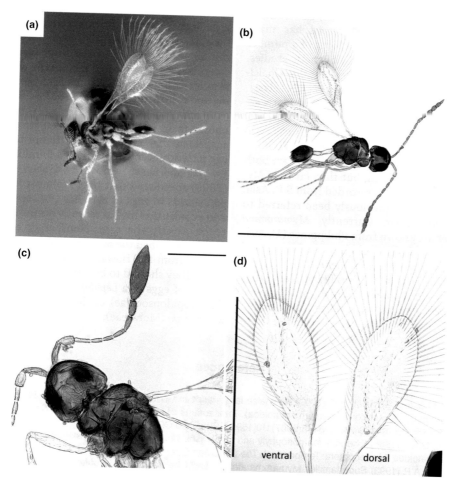

Fig. 22.51. The extralimital *Mymaromma* species, *M. menehune*: **(a)** specimen mounted on a dark card point; **(b)** slide-mounted male showing 2-segmented petiole and virtual absence of hind wings; **(c)** detail of head of female showing antennal club and 6-segmented funicle; **(d)** wings showing long marginal fringe and different dorsal and ventral patterns of microtrichia. Scale bars in **(b)** and **(d)** are 1 mm. (Source: from Honsberger *et al.* (2022) reproduced under terms of Creative Commons Attribution License CC-BY-4.0.)

23

Collection, Preservation and Rearing

Abstract

Three major topics are dealt with, including methods of collecting parasitoid wasps such as light trapping, Malaise traps and yellow pan traps. Specimen preparation methods (for example sorting from the Malaise trap samples, drying fragile specimens, direct pinning and card mounting) are described, with notes particularly aimed at beginners to this crucial process. Finally, some guidelines are given for those wishing to rear parasitoid wasps successfully, and the need to consider those circumstances that the parasitoid would experience in the wild.

The number of times we get asked to identify a parasitoid wasp from a photograph taken in the wild is enormous. If the inquirer is lucky, they may get a family name, rarely a genus and essentially never a species. That is because identifying parasitoid wasps is difficult, and we always tell them that they must collect the specimen. This then leads on to the next problem: most of the photos sent are of specimens in protected areas such as national parks, where permits are needed to collect, and such permits are not readily available to the lay public (see *The Collecting Permit Problem*, Chapter 4, this volume). Indeed, even for university researchers, obtaining permits can be a very long and laborious task – it is not usually sufficient to say that you want to get an identification name for a specimen you have photographed. However, in the tropics, there are actually a lot of interesting insects, including parasitoid wasps, to be found in non-protected areas, even in private gardens.

To be able to estimate insect diversity, sampling methods are crucial because they must show samples representative of the community or taxon selected for the investigation. There are many ways to collect parasitoid wasps, including sweep nets, Malaise traps, yellow pan traps, flight intercept traps, beating trays and vacuum samplers (Shweta and Rajmohana, 2016). To decide which sampling method is best for collecting the parasitoids, many factors need to be considered such as the design and costs of the respective sampling equipment, as well as the ecological traits and habitat conditions of the target taxa (Gullan and Cranston, 2010). For example, to collect Chalcidoidea effectively, the sweep net is the best way, but the Malaise trap is suitable for collecting Ichneumonoidea and Chalcidoidea and very effective in forest edges, while the yellow pan trap was effective in more open habitats with increased visibility of the traps (Noyes, 1989).

The most important question when it comes to collecting is: what is the aim of the study? Is it ecological and therefore needing comparable replicates? Is it to try to build an inventory of all the species of interest in an area, in which case a diversity of methods is desirable? If the former, it must be emphasised that it is unimportant how many insects there are in a single sample, such as a Malaise trap bottle after two weeks, because statistically that is just a single sample. Therefore, to make comparisons, or to estimate variance, many independent traps are necessary.

Having collected a specimen, for example using a butterfly net or a jam jar, there are several stages to go through. The specimen needs to be killed, as painlessly as possible and without causing obvious visible damage.

It is also important to recognise that when traps collect large numbers of individuals, these may not include many potentially interesting rare ones. Setting traps in a sunny open habitat with many flowers may collect a lot of butterflies, flies, bees, etc. but will almost never catch parasitoids of wood-boring beetles. The high biodiversity of the tropics is far more to do with forested areas.

Depending on one's interests, it can be very profitable to seek out particular host substrates where parasitoids may also be found. This is particularly true for parasitoids of xylophagous insects. Dead wood, either still standing or fallen to the ground, can yield many interesting wasps. When walking trails or in the forest, look out for dead wood with signs of beetle activity (fresh-looking sawdust spilling from holes). However, many of the larger wasps that might be present are very flighty; and therefore, starting several metres away, you are advised to approach stealthily, perhaps even with chameleon-type behaviour. A pair of binoculars can be useful.

Collecting Methods

Aerial or Butterfly Nets

The butterfly net is the archetypal entomological tool. It is a light fine-mesh net affixed to a circular or triangular, solid or slightly flexible frame, with a handle of desired length (Fig. 23.1). Traditionally it is used by butterfly collectors and so most on the market have a large diameter (c. 40 cm), ideal for scooping a butterfly off some flower head. However, parasitoid Hymenoptera are often seen roaming over and among vegetation, and to catch these it is often preferable to have a net with an opening of 15–20 cm, which is more manoeuvrable among the vegetation. The principle is simple, but efficient use requires some practice. The swipe should be quick, followed by a deft flick of the wrist that throws the end of the net bag around the edge of the frame, which hopefully traps the sought-after specimen at the end of the net bag.

Although many such nets are sold with white netting, this causes problems in brightly lit places because the reflected light makes it hard to see what is in the net and where it is. Therefore, black or green netting is to be preferred. Insects cannot see red, so some people use that colour.

Obviously, female aculeate hymenopterans can and will sting if you grab them through the net, but with practice one can develop skills to minimise the risk. Most non-aculeate parasitoid wasps can be handled with impunity but female ophionine ichneumonids and female *Netelia* (Ichneumonidae:

Fig. 23.1. Standard kite-shaped butterfly net, easily snagged on thorny/prickly vegetation.

Tryphoninae) can also deliver a sharp sting, though the pain abates fairly quickly.

When your net bag contains several small insects, it is often possible to tell Hymenoptera from flies by their hardness – flies tend to be far more easily squashed.

Sweep Nets and Insect Separators

A sweep net is a heavyweight version of a butterfly net. The fabric is thick and durable, and the frame more robust. It is designed to sweep into vegetation hard, causing creatures that are sitting or walking on the hit vegetation to let go and fall into the net. It is used by sweeping it to-and-fro in a continuous series of swipes, typically ten or 20, and then examining the catch (Fig. 23.2). Avoid the common sweep-net design that has a central bar. The best way to extract insects of interest is to insert your head and arm into the sweep net bag, with an aspirator to suck up all the insects of interest. Big stinging wasps will normally quickly fly out when you finish a cycle of sweeping, but the smaller parasitoids will stay in the net far longer. If you hold the bag end up towards the light with your free hand, they will almost all fly up to the bag wall and then are easily collected using a pooter.

If the sweep samples contain a lot of plant debris, it can take a long while for all the parasitoids and other small insects to make their way out to the side of the net. Rather than waiting for ages, there are various designs of insect separators into which the whole pile of debris is placed for wasps to emerge, thus allowing the entomologist to carry on sweeping. The basic features of a standard separator are a large box into which the material can be inserted and removed easily, with an opening near the top that leads to a collecting jar. Insects will escape from the debris and fly to the light and hence into the collecting container (Fig. 23.3 right side of image).

Fig. 23.2. Collecting specimens from a sweep net by inserting one's hand and using a pooter.

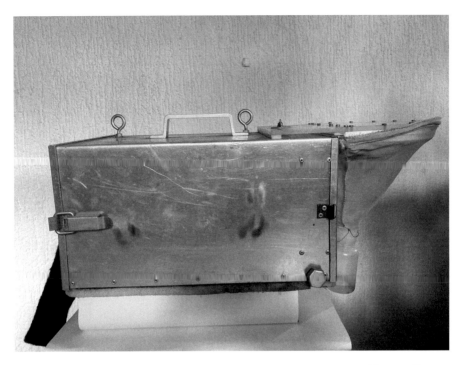

Fig. 23.3. One particularly robust type of insect separator, designed and built by P. Thomas. (Source: photograph by Kees (C.) van Achterberg, reproduced with permission.)

Inevitably when the net encounters many robust thorny plants it can get entangled, which is a nuisance. However, sweep netting is by far the most efficient collecting method, especially for Chalcidoidea, Proctotrupoidea, Diaprioidea, and Ceraphronoidea. Also available are tough protectively screened sweep nets, the outer material being very resilient to thorns and spines.

A modified sweep net especially suited for collecting 'microhymenoptera' was designed by Dr John Noyes of the Natural History Museum, London, and it is particularly useful for sweeping low-growing vegetation (Fig. 23.4). There are three important modifications: (i) the net frame is triangular, giving the net an arrow-like silhouette; (ii) there is an additional, removable, robust wiremesh screen that is attached over the net bag opening; and (iii) the handle is rather long. What are these modifications for? The shape of the net means that a long side will be close to the ground or surface of the low vegetation, because a lot of the 'microhymenoptera' spend most of their time close to the ground, where their hosts are found. The removable screen prevents most pieces of vegetation from entering the net bag and then bashing the small wasps to pieces. The long handle means that each sweep covers a large arc.

Malaise Traps

Malaise traps are designed to collect a wide range of flying insects, including Hymenoptera. Although in terms of specimens per hour their efficiency may be rather low, they operate all day and all night for as long as you service them. The flying insects get intercepted by chance and crawl towards a collecting bottle, located in the upper part of the trap, often filled with a killing agent (e.g. 70–95% ethanol) (Campos *et al.*, 2000; Yi *et al.*, 2012). It is a passive method; the trap is installed at a fixed place for a period of time, normally one year, and it is expensive compared with other traps. The Malaise trap

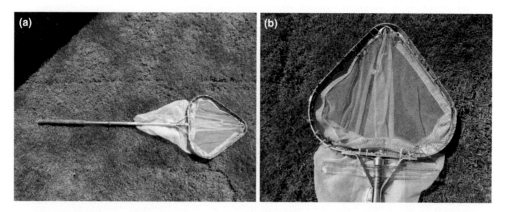

Fig. 23.4. Noyes-design sweep net ideal for collecting 'microhymenoptera' from low vegetation. Note the shape enabling it to be swept close to the ground, the robust construction, and the detachable wire mesh screen (b) that prevents the net becoming filled with large pieces of plant debris. (Source: photograph by Andrew Polaszek, reproduced with permission.)

was invented by the Swedish dipterist, René Malaise, in 1934. It is a tent-like trap made from fine mesh material. Normally it has one central wall, back and front walls and a roof with one end peaked up where the collecting bottle is located (Fig. 23.5). The wall is normally black while the roof is generally white, with the collecting bottle at the highest point of the trap. Malaise traps are easy to operate and a very efficient way to collect parasitoid wasps. There are numerous variants of the design (van Achterberg, 2009). The colour of the roof has little effect on the catch of most insect groups – and once you are under the roof, plenty of sky light is visible and the roof will always be the brightest place. However, Uhler et al. (2022) found that traps with white roofs did collect a greater species richness of pollinators.

Darling and Packer (1988) investigated the importance of various aspects of Malaise traps in relation to collecting all groups of Hymenoptera, and also the importance of using them in conjunction with pan traps (see below) placed under or next to the central curtain. Mesh size had little effect on the catch of the generally larger Ichneumonoidea, but fine mesh was important for the 'microhymenoptera' catch. Uhler et al. (2022) compared two basic design types, Townes versus shorter Bartak type, and concluded that the Townes type (Fig. 23.5 a) was better in terms of total insect biomass collected and greater taxonomic richness of highly mobile taxa.

Our own observations indicate that Townes-design traps with a higher roof and pointed collecting head end collect a considerably larger number of insects compared with the BugDorm and similarly designed traps whose roofs are very shallow in the mid-region part of the trap (Fig. 23.5 b).

Setting up the trap takes some skill and experience. These are the key principles.

- Try to find an area of fairly level ground – that makes it far easier to set up well.
- It is often good to place the rear of the trap among vegetation and have the front sticking into the open, or into a forest trail.
- Choose the orientation so that the collecting head will be pointing in the direction of most light. In the open this generally means towards the equator, but in forest it might be towards a tree gap or trail.
- Clear away any vegetation that happens to be where the central curtain will go.
- Adjust all the support cords so that there are as few creases in the netting, especially that of the roof, and if perfect smoothness cannot be obtained, then manipulate the cords so that any creases

Fig. 23.5. Standard-design Malaise traps: (a) Townes-type Malaise trap set up on Kranji Nature Trail, Singapore; (b) Bugdorm design of Malaise trap with interlocking, easy-to-assemble aluminium alloy tube supports but low roof, especially in the centre. (Source: (a) from Pauly (2012), reproduced under terms and conditions of Creative Commons Attribution 3.0 Unported license.)

run towards the collecting head, not away from it.
- Using tent pegs or bits of vegetation, make sure that there are no big gaps between the ground and the central curtain, otherwise insects that initially fall when bumping into the curtain may simply fly under it and continue on their way.

Although you can use poles such as canes or plastic water-piping at all the corners, this is not usually necessary if you are setting the trap up among taller vegetation, because you can use some small tree or shrub to secure the back cords, and then a vertical support is only needed for the centre front where the collecting bottle is attached. The support cords are usually secured in place using metal tent pegs. Obviously, exactly what is used will depend on the condition of the ground and longer pegs may be needed if the soil is very soft. In the tropics, wood is not recommended for long-term traps as it rapidly rots or gets eaten by termites.

The collecting bottle should be filled with at least 75% ethanol but if specimens are required for molecular work, it is much better to use 95% (Quicke, 2015), as the quicker the DNAase enzymes in the insects are inactivated, the better will be the quantity and quality of the DNA. There is a trade-off, of course, because the stronger the ethanol, the more brittle the specimens become, and therefore morphologists prefer 75%.

Malaise traps do not last for ever, and may be damaged by large mammals such as elephants, wild pigs or bison, or more often in the tropics by monkeys. Sharp-eyed insectivorous birds sometimes peck through the netting near the collecting head to get an easy meal, and so on. Therefore, when servicing the traps, it is wise to carry with you a repair kit of needle and thread, duct tape and string and spare tent pegs. Normally, if enough alcohol is put in the collecting jar, the traps can be left operating for about two weeks between emptying the catch, even in a hot climate, but for best results (quality of specimens) weekly servicing is recommended.

There are many studies on the efficiency of the Malaise traps compared with the other collecting methods. Shweta and Rajmohana (2016) studied efficiencies of sweep nets, yellow pan traps and Malaise traps in sampling the Platygastridae in Kozhikode district, India, over a 6-month period from December 2013 to May 2014. They recommended that, for Platygastridae collections, the Malaise trap is the most effective method when compared with sweep nets and yellow pan traps. They also concluded that sweep netting and yellow pan traps provided better quantitative estimates but that Malaise traps were better qualitatively, and that a combination of methods was important for obtaining a comprehensive taxonomic coverage.

With some ingenuity, it is possible to construct Malaise traps that can be hoisted into the forest canopy, where the composition of flying insect species may vary markedly from that at ground level. These are called aerial or canopy Malaise traps (Fig. 23.6) (Skvarla *et al.*, 2021).

Yellow Pan Traps

Yellow pan traps are a passive method for collecting insects and it is widely used for 'microhymenopteran' groups that tend to be active near the ground layer. The pans can be set out in the field by someone who has little training in insect collecting. They are very cost-effective and collect diverse group of insects, including parasitoids for research purposes (Buffington *et al.*, 2021). Nevertheless, there is quite a lot of effort involved, such as carrying all the water. A pan trap is normally circular in shape, usually cheap yellow plastic bowls or dishes (Fig. 23.7), but you can use alternatives such as aluminium foil dishes and then spray-paint them yellow. The pans should be moderately deep, because it is often quite difficult to set them perfectly level in the field and water will easily slop out of a shallow one. Carrying the water is the hardest work. Into each trap pour some water to which has been added a small drop of a mild detergent such as washing-up liquid – this reduces the surface tension and so insects landing on the water, especially small ones, will sink and drown. If the pans are to be left out for a few days, then bacteria will multiply and rot or coat the specimens, so a tiny amount of a safe antibacterial agent should be added to the water. The ideal one is sodium benzoate, approximately 0.1% w/v. This chemical is used widely in the food

Fig. 23.6. Aerial Malaise traps hoisted high into trees in Khao Yai National Park, Thailand. (Source: photographs by and © Pornthap Kerkig, reproduced with permission.)

Fig. 23.7. Yellow pan traps in action at Danum Valley Field Centre, Sabah, Malaysia: **(a)** some traps set out on reasonably flat ground; **(b)** catch after a few hours. Traps contain water with a little detergent to break the surface tension so that insects sink. (Source: Photograph by and © Andrew Polaszek, Natural History Museum London, reproduced with permission.)

industry and is found in many drinks as well as cosmetics and wet-wipes, so it is safe. Table salt also works, but if it crystallises at a much higher concentration it will damage specimens.

Obvious problems, particularly in the tropics, are that heavy rain storms will wash out the samples and that in hot drier situations the liquid in the pan will quickly evaporate. Some workers have replaced water with very harmful low-evaporation media such as ethylene glycol, but in such cases they have often returned to find many dead and dying mammals and birds around their traps who had come to take a drink from them. Please stick with water.

Samples from each pan may be separated by filtering through a sieve such as a fine tea-strainer or muslin cloth, then washed off into alcohol. The now insect-free liquid is then returned to the pan and topped up as necessary.

Daniel et al. (2018) studied variability and effectiveness of sampling methods in the rice fields during 2015–2016 in the Paddy Breeding Station, Tamil Nadu Agricultural University, Coimbatore. They used six different methods: sweep net, yellow pan trap kept at ground level, yellow pan trap erected at canopy levels, Malaise trap, sanction trap and light trap. They found that the yellow pan trap at ground level was the most effective method in trapping the parasitoids, followed by the sweep net method, confirming that the yellow pan trap is another easy and convenient method to collect parasitoid wasps – depending on which group of parasitoids one is interested in and what type of habitats. Used in conjunction with Malaise traps, pan traps help greatly with collecting aculeates and 'microhymenoptera' (Darling and Packer, 1988).

Buffington et al. (2021) found that the diversity, richness and abundance of hymenopterans in pan trapping projects were significantly impacted by the colour of the pan trap set up. Overall, yellow pans did collect more species of Hymenoptera, but they were followed closely by green and by white. Yellow was particularly effective for Platygastroidea and Braconidae, and yellow and green collectively best for Ichneumonidae, Chalcidoidea and Cynipoidea. They also found that some individual species show significant preferences not only for yellow pan traps but also for white, fluorescent yellow, blue and fluorescent blue pans. Therefore, more research should be done on the effect of colour of the pan traps for use as a collecting method in studying species diversity. Particularly if the aim is to inventory the 'microhymenoptera' species of a habitat, it might be best to employ traps of additional colours. Aguiar and Sharkov (1997) reported that blue pans were quite effective at collecting Stephanidae.

Light Traps

Many insects, not only moths, are attracted to lights at night, especially when the source also emits a small amount of UV. Among the parasitoid Hymenoptera, lights are very effective at attracting those groups that search for their hosts between dusk and dawn, such as ophionine Ichneumonidae, rogadine Braconidae and, quite often, fig wasps. Winged male ants can sometimes cause confusion and be a nuisance.

The normal set-up used by hymenopterists comprises a white sheet strung on a cord between suitable supports, and some combination of bright white + UV light bulbs (such as a 250 W high-pressure mercury vapour lamp) (Fig. 23.8), and sometimes some additional black light tubes. The latter can be used by themselves but are not very effective alone. Lepidopterists also use white sheets but may use Robinson or Heath traps into which moths (and some parasitoids) fly and are then retained until the trap is opened in the morning. Not many (even very enthusiastic) entomologists want to stay up all night. The lights of these traps are left running all night and then their catch is examined in the morning.

High-pressure mercury vapour lamps require mains electricity and a choke in the circuit, otherwise they will blow up. They are easy to operate near to habitation, but in the wild it necessitates a portable generator. Make sure that the electric plugs have been safely protected, in case of rain. More recently, expensive designs that use visible plus UV light-emitting diodes which operate off a simple battery have started to appear and circuit diagrams are readily available so that you can construct them yourself.

Light traps should be set in place and the lights switched on just before sunset. Parasitoid wasps tend to arrive at the trap early in the night. One should be very careful when collecting the wasps near the traps, as insects may fly into your eyes, nose and ears. Protective UV-blocking goggles are recommended and certainly one should not look directly at the bright mercury vapour bulb.

It should be noted that mercury vapour bulbs get very hot and if they get hit by large drops of rain the outer protective envelope will shatter, but the inner vial with the mercury arc

Fig. 23.8. The junior author (some years ago) collecting parasitoids from among all the moths at a light sheet in Taiwan.

may continue to work. It gives out really quite dangerous UV light, so it is important to turn that off and not to look at it. To avoid these risks and expenses, it is a good idea to contrive a rain hat to suspend above the bulb.

Canopy Fogging

In a Siberian winter, military tanks would often become frozen into the ice and getting them moving was a problem. Therefore someone invented a small, portable two-stroke engine and exhaust pipe that would shoot out a jet of very hot exhaust to melt the ice. Move on some years and some entomologist got the idea of attaching a vial of volatilisable insecticide just after the engine-exhaust junction, and so a cloud of insecticide can be sprayed through vegetation and the affected insects will drop towards the ground, where they can be collected on sheets or fabric funnels of known area. Nowadays the insecticide most often used is a natural pyrethroid which gets degraded rapidly and so causes no long-term environmental harm.

A handheld version used to sample insects from the canopy of oil palms (*Elaeis* spp.) near ground level is shown in Fig. 23.9, but for natural forests a strong line is shot over the bough of a tall forest tree and attached to a fogger, which is then started up and hauled way up into the canopy and allowed to swing about, spraying the insecticide mist through the canopy for about 6–9 minutes. Obviously it will not work if there is any wind, but in general there is almost no wind in forest canopies – especially in the early hours of the morning, before the sun has started heating things up. Therefore, canopy fogging is usually conducted as soon after dawn as possible. Also, whilst the air is still cool, larger more active species may be overcome before they have a chance to fly away.

Fig. 23.9. A handheld canopy fogger being used in Indonesia. (Source: photograph by and © Jochen Drescher, University of Goettingen), reproduced with permissions.)

Since this method has been employed mostly in quantitative ecological studies, it has become the practice to use special, elaborate collecting funnels each with a 1 m² surface area. If quantification is not required we suspect that simply laying large white bed sheets on the ground and searching them for specimens would be just as effective and take far less time to set up.

Vacuum Cleaners (Suction Traps)

Mostly used in agro-entomological scenarios, the D-Vac has been an important sampling machine, well suited for extracting tiny insects from vegetation close to the ground. It is a sort of vacuum cleaner carried as a backpack, with an extra-wide input nozzle. It has the potential advantage of also sampling other groups such as spiders, as well as rather immobile taxa, even seeds. It is most suited to fairly consistent vegetation types such as short grassland, vineyards or tree trunks, and can be used at least semi-quantitatively.

Lesser-used Methods

Several collecting techniques more often used by non-hymenopterists still often yield interesting parasitoid wasp specimens. Many insects dwell on tree trunks or among leaf litter. Coleopterists and arachnologists frequently collect large samples of leaf-litter and place them in Burlese extraction funnels. Among the arthropods that emerge are often small parasitoid wasps, including members of seldom-collected groups. Essentially a dustpan and brush (ideally with a curved flexible lip) can be used to brush insects off tree trunks, among which there are often many apterous parasitoids that search for hosts in the bark crevices, such as various (often apterous) cryptine ichneumonids and bethylids.

Conclusions

In places like Europe, North America and Australia, it is becoming increasingly common to involve interested members of the general public and schools in insect research projects – 'citizen science'. After all, it is usually the public's taxation that has funded university research. Malaise traps are often used for this public interaction as it does not take much training to run them. The scientists can demonstrate the value of the research by showing the amazing diversity that can be collected, even in the middle of cities. Sometimes by, say, naming new taxa after particular schools it may be possible to recruit enthusiastic volunteer labour for servicing traps.

Specimen Preparation

Here we briefly describe some methods of specimen preparation of particular relevance to parasitoid wasps. Readers who are not already very experienced entomologists might also find it worthwhile consulting some more general texts such as Gibb and Oseto (2000) and Krogmann and Holstein (2010).

A Note on Essential Equipment and Consumables

Exactly what is needed by an entomologist depends to some extent on the group of insects being studied, and perhaps particularly on their body size. Parasitoid wasps are generally small, and sometimes very tiny indeed. Apart from the obvious pins, card, paper for labels and store-boxes/drawers, there are some other things. An adequate range of commodities is shown in Fig. 23.10.

Fine forceps

Fine watchmaker's forceps, minimally size 4, ideally size 5 with far sharper points. Good ones are not cheap but they are vital. The second author (DQ) always told his PhD students to take great care of them and not to drop them, because they will probably land on their points and they will break. This is not good, but should it happen (and DQ has

Fig. 23.10. A suggested general set of equipment and materials that the average parasitoid hymenopterist might need: **(a)** postal box or unit tray with plastazote bottom; **(b)** archival-quality insect glue that can be soaked off if needed; **(c)** fine permanent ink pens (Pigma® Micron pens are almost universally used); **(d)** extra plastazote, can be used to make right-angle stages for specimen examination; **(e)** entomological card rectangles, various sizes; **(f)** entomological card points; **(g)** gelatin capsules, can store broken parts, host remains on same pin as specimen; **(h)** dense silicone rubber foam or plastazote secondary staging material; **(i)** cork for pinning – the end of a wine bottle cork is nearly perfect as the quality of the cork is high; **(j)** standard entomological pins of various sizes; **(k)** minuten pins of various sizes; **(l)** fine forceps, watchmaker's No. 5s are ideal; **(m)** pipettes (glass Pasteur ones are also good); **(n)** artist's fine paint brushes for handling specimens and sometimes for cleaning off debris; **(o)** specimen manipulator, allows specimens to be viewed easily from a large range of angles (this one invented by Jochen Drescher).

certainly dropped many over his 45-year career) all is not necessarily lost. With an engineer's fine carborundum block, pair of small tin-snips, pliers and a bit of effort, it is usually possible to restore the pair to a reasonably good working condition.

FINE FORCEPS – DEXTERITY IS A GODSEND. If you are not particularly dextrous, do not try to handle tiny wasps with hard metal instruments – it will end badly.

Do not try to pick up tiny wasps by closing the forceps tips on them – the force will surely badly damage the specimen. Rather, using fine forceps to remove small insects from, say, a Malaise trap soup (Fig. 23.11) does not involve pinching the wasp, or part of it, with the forceps. Rather, one should learn to cradle the insect between the tips whilst only touching it lightly.

If you cannot learn this easily practising on unimportant material, a slightly less precise way is to use a fine artist's paintbrush which will only touch the specimen gently, or with very small specimens you can suck them up with a pipette (Fig. 23.11).

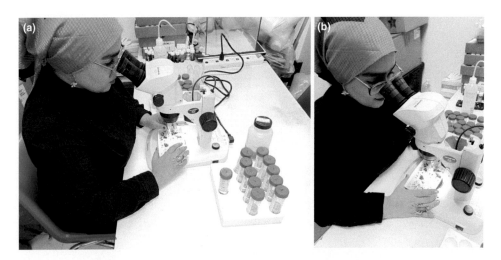

Fig. 23.11. Sorting a Malaise trap sample. Notes: only a small portion should be sorted at one time because small insects are often hidden underneath larger ones; a sorting dish with pointed corner(s) makes pouring tray residue back into container much easier as it minimises spillage – Petri dishes are not good.

Sorting Wet Samples

When one is presented with a large full Malaise trap bottle there are a few tips for extracting relevant hymenopterans for your work. Firstly, it takes time and if you want to extract/count all individuals of some groups, you must only process small amounts at a time. Do not pour the whole sample bottle into a big tray but take a pinch and put it in a smaller dish. We recommend a white light plastic dish with at least one pointed corner (Fig. 23.11).

If you are only interested in generally relatively small-bodied Hymenoptera, essentially 0.2–5 mm in body size, it is possible to use a fractionator set of sieves that will extract virtually all of the small insects away from the large Lepidoptera, Coleoptera, etc. (Buffington and Gates, 2008) and Haas-Renninger et al. (2023) have shown that with a 2 mm mesh it is good for separating 71% of families. This still leaves the problem of transferring the small fraction to a vial, a time-consuming and difficult procedure. An easy and effective method was introduced by Schweizer et al. (2020) that involves straining the fraction through a tea filter bag, and these can then be stored with their enclosed specimens in a larger, alcohol-filled storage container.

Sieving into different insect size categories is also very important for metabarcoding projects, because the sheer biomass of the large-bodied insects in a Malaise sample means that their DNA will swamp the miniscule amount coming from the smallest wasps and flies (Elbrecht et al., 2021; Kirse et al., 2023).

Drying Small, Fragile Specimens

Very early on after starting to study parasitoids, you will find that many of the smaller species, particularly of Chalcidoidea, collapse when dried direct from alcohol or a killing jar. This is because their cuticle is thin and rather soft and as the liquid within them slowly evaporates it contracts, whilst still adhering strongly to the cuticle, and so it pulls the cuticle in, following the surface of the liquid because of its surface tension. The result is not only ugly but can also hide critical features. Methods must therefore be employed that reduce or completely eliminate this effect.

Chalcidoidologists, as well as those working on various other delicate, small-bodied specimens such as tiny Diptera, have essentially employed two types of method: (i) critical point drying (CPD); and (ii) chemical drying. Both add a bit of extra time but the specimens obtained are vastly better.

CPD, introduced for small insects such as chalcidoids by Gordh and Hall (1979), was the method of choice for a long time, but requires rather expensive apparatus – the sort of kit that might be available in a university or museum but not in the home of your average amateur entomologist. It involves taking the specimens through a graded ethanol series until they are in 100% (not 96%) dry alcohol, and then, under high pressure in a specialised pressure vessel, gradually replacing the ethanol with liquid carbon dioxide. Once the ethanol has been replaced by the CO_2, the pressure is gradually released (the CO_2 slowly vented off) and in this process the CO_2 changes directly from liquid to a space-filling gas-like state that gently diffuses from the specimen. Finally, back to room temperature and pressure and you end up with very pristine specimens.

Importantly for all the methods described below, Austin and Dillon (1997) found that the procedures had little or no negative effect on the quantity and quality of DNA that could be extracted and amplified from the specimens afterwards.

Chemical drying – the AXA method

CPD is clearly not for the beginner and soon a far simpler procedure called the AXA method (standing for Alcohol-Xylene-Amyl acetate) was described by van Achterberg (2009) which involved replacing the ethanol with a mixture of 6 parts 96% ethanol and 4 parts xylene for approximately 2 days, followed by transfer to amyl acetate for another 1–2 days, from which they are finally dried. The original version was invented for drying delicate hoverflies (Vockeroth, 1966) at the Canadian National Collection of Insects. A not inconsiderable advantage of this method compared with the next is the low cost of the reagents. As with many organic reagents, perform these tasks in a well-ventilated space, ideally under a fume hood, and keep away from naked flames.

Just amyl acetate

Recently, Pérez-Benavides *et al.* (2023) showed that, at least for small soft-bodied chalcidoid wasps, a procedure involving only amyl acetate can be used for drying specimens from 70% ethanol in a way that yields good, non-collapsed specimens for morphological study. Their protocol involves transferring the specimens to a 50:50 amyl acetate and ethanol mixture for 1 hour, followed by another hour in 100% amyl acetate, from which they are then dried.

Hexamethyldisilazane (HMDS)

Use of this compound for dehydrating soft specimens for electron microscopy was introduced by Nation (1983), who recommended that small specimens should be dehydrated through a graded series of ethanol, transferred to HMDS for 5 minutes and then air-dried. It was introduced into entomology by the chalcidoidologists John Heraty and David Hawks (Heraty and Hawks, 1998). Small insects having a more impermeable cuticle require longer in the HMDS. Unfortunately, whilst excellent for the purpose, this chemical is rather expensive and also rather toxic. To quote Heraty and Hawks more than 20 years ago, 'at $30 U.S. per 400 ml of HMDS and 5 ml per large lot of about 100 chalcidoids, we estimate a cost of about 37.5 cents per run, or 0.4 cents per specimen'.

Mounting

There are two main aims and one subsidiary aim of mounting specimens. For Lepidoptera and a few other groups, there is usually a fourth, which is beauty, but for parasitoid wasps that is not normally a high priority: the sheer numbers of specimens largely precludes it, because mounting specimens beautifully takes a lot more time.

So, what are the principles to abide by?

- Making clearly visible as many characters as possible that are important for identification in the group.
- Protecting the specimen from damage, especially for very small species.
- Ease of remounting should the first criterion not be met as much as needed.

Finally, a note on pushing the entomological pin through the card point/rectangle in the first place: the card should be placed on a very firm but easily penetrated substrate. If the substrate is not firm enough, the card will bend at the place where the pin is being pushed into it. We highly recommend good-quality natural cork, such as the end of a wine-bottle cork (Fig. 23.12), which in our experience with multiple daily use will last well over a year or even five years. This is because the dimensions of the cork cells are so small. Do not try to exert the pressure needed from the top of the pin, because it will buckle and, in any case, you will hurt your fingers. Instead, grasp the pin firmly near its point with forceps and apply the pressure with that hand (Fig. 23.12 a). Once the tip of the pin has gone through the card, use the forceps to grasp the card along its whole length and slide it up the pin, keeping them at right angles all of the way (Fig. 23.12 b).

There are commercial pinning blocks but avoid any with a metal bottom, because the very sharp point of the entomological pin will become bent.

Store Boxes

Historically, mounted specimens were kept in cabinet drawers lined with cork, which is a useful and long-lasting material, although it tends to become harder with age. Nowadays this has been supplanted by a high-density, fine-grained polyethylene foam called Plastazote®. This is highly recommended but buy the best, finest quality, because, unlike cork, each time you push a pin into it you inevitably do break some of the bubbles. Nevertheless, it will last and function well for a long time.

Never ever use expanded polystyrene, especially not for boxes that are going to be posted. This is because polystyrene is not plastic in the materials sense, and when a pin is inserted into it, it permanently crushes the bubbles and can very quickly work loose and fall out.

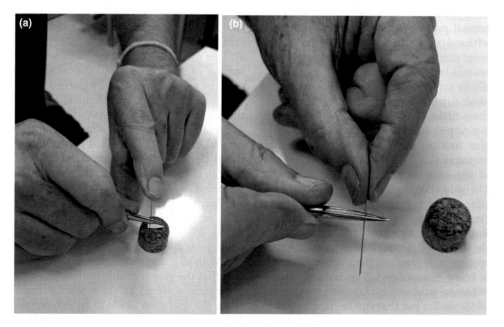

Fig. 23.12. Pushing an entomological pin through a card point without bending it: **(a)** making the initial hole – a firm substrate (in this case wine-bottle cork) behind the pin prevents the card from folding, the finger at the top is for keeping the pin vertical, the force is applied by the forceps holding the pin low down; **(b)** the card is grasped along its full length in the forceps, and slid up the pin to the desired height.

In the humid tropics, pests are particularly problematic. These include fungi, book lice and dermestid beetles. Obviously, one ideally needs to keep a collection in a made-for-purpose, humidity- and temperature-controlled room, in proper entomological cabinets or in unit trays within glass-topped entomological drawers in a compactor system. Only a few well-financed institutions can afford such 'luxury'. We have found that a cheap and effective solution is to keep specimens in postal boxes and then, when not in use, store these in zip-locked plastic bags together with a good amount of anhydrous silica gel (which we enclose in easily obtained tea-bags). Changing the silica gel when its colour has changed significantly will keep the humidity below that which is required for pests. Lower temperature also helps.

When specimens are received by major museums, they are first placed in a −60°C freezer for approximately two weeks, which is sufficient to kill any pest insects that might inadvertently have come with them.

Microscope Slide Preparations

Small chalcidoids and similar wasps are often unsuitable for dry mounting, even when dried in ways that avoid their collapse. The smaller the wasp, the truer this is. Therefore, chalcidologists have developed special techniques for making permanent microscope-slide preparations.

A very good aspect of modern DNA extraction procedures is that they can be applied to entire specimens of small insects (e.g. Qiagen DNeasy Blood and Tissue Kit) without damaging the specimen (e.g. Cruaud et al., 2019; Polaszek et al., 2022). Further, since the lysis buffer contains proteinase K, the specimens do not need clearing with chemicals such as chloral hydrate.

Genitalia Preparations

For larger-bodied, robust parasitoid wasps, the standard technique used is to cut off the apex of the male metasoma (even using scissors) and to soak it in approximately 5–10% aqueous KOH (potassium hydroxide) (Quicke, 1988).

For small insects such as tiny bethylids and embolemids, the metasomas need to be softened before removing the genitalia. Trying to use a small pair of scissors to cut off the end of the abdomen of a dry specimen is certain to lead to destruction of the specimen.

For small delicate insects, potassium hydroxide treatment is too harsh and often damages structures. Options are to use a less alkaline solution such as dilute ammonia. Shop-bought general-purpose ammonia cleaner is typically approximately 27%.

Martinelli et al. (2017) presented an alternative that used proteinase K (a broad-spectrum serine proteinase) to remove soft tissue from small wasp (such as bethylids) male genitalia, with very nice results (Fig. 23.13).

Rearing Parasitoids from Wild-collected Hosts

Of course, understanding the host relationships of parasitoids and their hosts is of great scientific importance. From one-off host records, research has developed to the construction of complex host–parasitoid food webs. Such data are vital aspects of insect community ecology. It is widely appreciated among parasitoid wasp specialists that many species that are collected in traps seldom appear in host rearing and conversely many reared species hardly ever turn up in Malaise trap samples (Sharkey et al., submitted). When parasitoid taxonomists are presented with a reared member of their group, they often metaphorically scratch their heads and have been known to question whether the rearing method may have caused some developmental differences. In truth it is much more likely that the reared specimen represents a previously unknown species.

There is not a huge amount of literature on how to go about rearing parasitoids. Of course, lepidopterists often have a lot of experience in rearing caterpillars, but sadly, often have destroyed any emerging parasitoids as if they were criminals preventing them from getting perfect butterfly or moth speci-

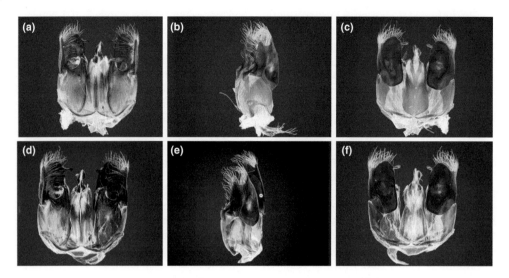

Fig. 23.13. Male genitalia dissected from a *Pristocera* sp. (Bethylidae): **(a–c)** dorsal, lateral and ventral views respectively, before treatment; **(d–f)** as **(a–c)** but after proteinase K treatment. (Source: modified from Martinelli *et al*. (2017) reproduced under terms of Creative Commons Attribution licence CC-BY 4.0.)

men. Thus, historically incalculable amounts of biological data have been lost for ever.

There are a few very crucial key points when it comes to rearing, but the most important scientifically is:

- Be certain that the parasitoid has actually emerged from the host you think it did.

It may come as a surprise that a very large proportion of published parasitoid host records are wrong – maybe as large a proportion as 50%. This horrible error rate has a number of basic causes:

- The parasitoid may have been misidentified.
- The host may have been misidentified.
- Both the parasitoid and host might have been misidentified.
- The parasitoid did not actually emerge from the purported host.

Because of all of the above, keeping the remains of reared specimens and their host inexorably linked together is really important, if at all possible. So often in the past, host remains have been discarded or, if they were retained, then they were mounted on a separate pin or kept in a separate vial. Inevitably things get separated and future scientists are deprived of reliable host–parasitoid records. When possible, host remains, no matter how small, incomplete or damaged, should be mounted on the same pin as the parasitoid specimen, either glued to a card rectangle (Fig. 23.10 e) or placed (after drying) in a gelatin capsule (Fig. 23.10 g) and the insect pin with the specimen passed through this.

In the past, with relatively less understanding of taxonomy and not so many readily available identification resources as there are now, it is possible that the first three of these error-rate causes were more common. There are publications suggesting that microgastrines attack beetles that aphidiines attack caterpillars. They do not. Full stop! However, the fourth cause is still very common. It is particularly common with substrate rearings; that is, a chunk of substrate, such as a piece of dead wood, is placed in a rearing container, and eventually some parasitoids and some beetles or Diptera emerge. It is so easy to suspect, and then publish, that the parasitoid had been attacking the non-parasitoid.

A particularly interesting parasitoid clade from an evolutionary point of view is the *Polysphincta* group of pimpline ichneumonids which are koinobiont ectoparasitoids of spiders. Mark Shaw wrote a note specifically on raising these (Shaw, 1990).

Culturing Parasitoids

This adds another level of complexity for the research: sufficiently large populations of both the host and parasitoid need to be maintained. And do not forget that hosts require food, and if they specialise on a given plant species, this may be a further issue. Further, even when successfully managed, cultures may become infected with pathogens of various degrees of harmfulness, such as microsporidians. In the case of wasps with single-locus sex determination, sex ratio (proportion of males) will rapidly increase and so either allele diversity will have to be 'topped up' by introducing fresh individuals from the wild, or multiple separate genetic lines maintained so that they can from time to time be mixed, thus restoring a sufficient level of allelic diversity.

The basic principle, at least when developing a new culture system, is to maintain wasps in as close to outside (natural) conditions as possible. Obviously, wasps that need to undergo a period diapause at some stage in their life cycle are going to be vastly harder to work with and any experimental studies will have to be spread over very long time periods.

Mating in most species involves a combination of chemical and tactile or vibratory intersexual signalling. If the chemicals are volatile pheromones that can build up in the container, this may confuse members of both sexes; for example, a male surrounded by female perfume from all directions may be unable to actually locate the female. Therefore, it is usually best to introduce the female into the male's container (Shaw, 1997). Alternatively, keep everything in a large, airy, well-ventilated system.

When it comes to parasitoids of concealed hosts, there may be other problems. Some parasitoids will happily attack their host species whether or not it is in its normal concealed situation (e.g. Ueno, 2000), but this is not always the case. Researchers at Texas A&M University some years ago wanted to culture a braconid wasp that was a potential biocontrol agent against the stem-borer pest *Eoreuma loftini* (Crambidae) (Wharton *et al.*, 1989). They found that female *Digonogastra* would not oviposit onto naked exposed host caterpillars, so they invented a special clear Plexiglass container into which multiple host larvae could be place (Fig. 23.14) which had very fine (ovipositor-sized) access holes drilled into it.

The take-home messages are:

- Think like the wasp.
- Be inventive.

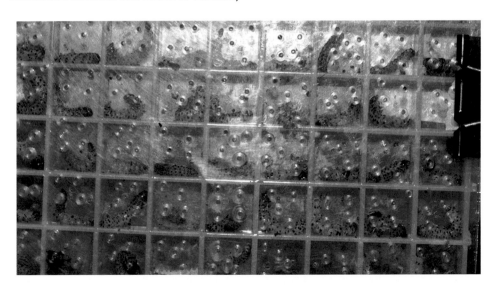

Fig. 23.14. Container for getting a braconid parasitoid of naturally concealed hosts to oviposit successfully in captivity, invented and developed Jim Smith, Pat Gillogly and Brad Hawkins at Texas A&M University.

References

Aguiar, P.A. and Sharkov, A. (1997) Blue pan traps as a potential method for collecting Stephanidae (Hymenoptera). *Journal of Hymenoptera Research* 6, 422–423.

Austin, A.D. and Dillon, N. (1997) Extraction and PCR of DNA from parasitoid wasps that have been chemically dried. *Australian Journal of Entomology* 36, 241–244.

Buffington, M.L. and Gates, M.W. (2008) The Fractionator: a simple tool for mining 'Black Gold'. *Skaphion* 2 (24), 1–4.

Buffington, M.L., Garretson, A., Kula, R.R., Gates, M.W., Carpenter, R., Smith, D.R. and Kula, A.A.R. (2021) Pan trap color preference across Hymenoptera in a forest clearing. *Entomologia Experimentalis et Applicata* 169, 298–311.

Campos, W.G., Pereira, D.B.S. and Schoereder, J.H. (2000) Comparison of the efficiency of flight-interception trap models for sampling Hymenoptera and other insects. *Annals of Entomological Society of Brazil* 29, 381–389.

Cruaud, A., Nidelet, S., Arnal, P., Weber, A., Fusu, L., Gumovsky, A., Huber, J., Polaszek, A. and Rasplus, J.-Y. (2019) Optimized DNA extraction and library preparation for minute arthropods: application to target enrichment in chalcid wasps used for biocontrol. *Molecular Ecology Resources* 19, 702–710.

Daniel, J.A., Ramaraju, K., Mohan Kumar, S., Jeyaprakash, P. and Chitra, N. (2018) A study on five sampling methods of parasitic hymenopterans in rice ecosystem. *Journal of Biological Control* 32, 187–192.

Darling, D.C. and Packer, L. (1988) Effectiveness of malaise traps in collecting Hymenoptera: the influence of trap design, mesh size, and location. *Canadian Entomologist* 120, 787–796.

Deans, A.R. (2018) A review of adhesives for entomotaxy. *PeerJ Preprints* 6, e27184v1.

Elbrecht, V., Bourlat, S. J., Hörren, T., Lindner, A., Mordente, A., Noll, N.W., Schäffler, L., Sorg, M. and Zizka, V.M.A. (2021) Pooling size sorted malaise trap fractions to maximize taxon recovery with metabarcoding. *PeerJ* 9, e12177.

Gibb, T. and Oseto, C. (2000) *Insect Collection and Identification: Techniques for the Field and Laboratory (Second Edition)*. Academic Press/ Elsevier, London, 339 pp.

Gordh, G. and Hall, J. (1979) A critical point drier used as a method of mounting insects from alcohol. *Entomological News* 90, 57–59.

Gullan, P.G. and Cranston, P.S. (2010). *The Insects: An Outline of Entomology*, 4th edition. Wiley-Blackwell, Oxford, 565 pp.

Haas-Renninger, M., Schwabe, N., Moser, M. and Krogmann, L. (2023) Black gold rush - Evaluating the efficiency of the Fractionator in separating Hymenoptera families in a meadow ecosystem over a two week period. *Biodiversity Data Journal* 11, p. e107051.

Heraty, J. and Hawks, D. (1998) Hexamethyldisilazane – a chemical alternative for drying insects. *Entomological News* 109, 369–374.

Kirse, A., Bourlat, S.J., Langen, K., Zapke, B. and Zizka, V.M.A. (2023). Comparison of destructive and nondestructive DNA extraction methods for the metabarcoding of arthropod bulk samples. *Molecular Ecology Resources* 23, 92–105.

Krogmann, L. and Holstein, J. (2010) Preserving and specimen handling: insects and other invertebrates. In: Eymann, J., Degreef, J., Häuser, C., Monje, J. C., Samyn, Y. and van den Spiegel, D. (eds) *Manual on Field Recording Techniques and Protocols for All Taxa in Large-Scale Biodiversity Surveys*. ABC Taxa, The Belgian Development Cooperation, Brussels, pp. 463–481.

Martinelli, A.B., Waichert, C., Barbosa, D.N., Fagundes, V. and Azevedo, C.O. (2017) The use of Proteinase K to access genitalia morphology, vouchering and DNA extraction in minute wasps. *Anais da Academia Brasileira de Ciências* 89, 1629–1633.

Masner, L. and Goulet, H. (1981) A new model of flight interception trap for some hymenopterous insects. *Entomological News* 92, 199–202.

Nation, J.L. (1983) A new method using hexamethyldisilazane for preparation of soft insect tissues for scanning electron microscopy. *Stain Technology* 58, 347–351.

Noyes, J.S. (1982) Collecting and preserving chalcid wasps (Hymenoptera: Chalcidoidea). *Journal of Natural History* 16, 315–334.

Noyes, J.S. (1989) A study of five methods of sampling Hymenoptera (Insecta) in a tropical rainforest, with special reference to the Parasitica. *Journal of Natural History* 23, 285–298.

Pauly, A. (2012) Three new species of *Eupetersia* Blüthgen, 1928 (Hymenoptera, Halictidae) from the Oriental Region. *European Journal of Taxonomy* 14, 1–12.

Pérez-Benavides, A.L., Ospina-Peñuela, E., Gamboa, J. and Duran-Bautista, E.H. (2023) Amyl acetate: an alternative technique to dry mount Chalcidoidea (Hymenoptera) from alcohol, faster and inexpensively. *Journal of Insect Science* 23(2), 4.

Polaszek, A., Fusu, L., Viggiani, G., Hall, A., Hanson, P. and Polilov, A.A. (2022) Revision of the world species of *Megaphragma* Timberlake (Hymenoptera: Trichogrammatidae). *Insects* 13, 561.

Quicke, D.L.J. (1988) Inter-generic variation in the male genitalia of the Braconinae (Insecta, Hymenoptera, Braconidae). *Zoologica Scripta* 17, 399–409.

Quicke, D.L.J. (2015) *The Braconid and Ichneumonid Parasitic Wasps: Biology, Systematics, Evolution and Ecology*. Wiley Blackwell, Oxford, 688 pp.

Schmidbaur, T., Wanke, D., Haas, M. and Krogmann, L. (2020) Biodiversi-TEA—A quick and easy handling method for arthropod trap material in ethanol. *Zootaxa* 4731(2), 275–278.

Shaw, M.R. (1990) Rearing parasitic wasps from spiders and their egg sacs. In: *British Arachnological Society – Members' Handbook*, Section 2.8, 2 pp.

Shaw, M.R. (1997) *Rearing Parasitic Hymenoptera*. Amateur Entomologist's Society, Orpington, UK, 46 pp.

Shweta, M. and Rajmohana, K. (2016) A comparison of efficiencies of sweep net, yellow pan trap and Malaise trap in sampling Platygastridae (Hymenoptera: Insecta). *Journal of Experimental Zoology India* 9, 393–396.

Shweta, M. and Rajmohana, K. (2018) A comparison of sweep net, yellow pan trap and Malaise trap for sampling parasitic Hymenoptera in a backyard habitat in Kerala. *Entomon* 43, 33–44.

Skvarla, M.J., Larson, J.L., Fisher, J.R. and Dowling, A.P.G. (2021) A review of terrestrial and canopy Malaise traps. *Annals of the Entomological Society of America* 114, 27–47.

Ueno, T. (2000) Host concealment: a determinant for host acceptance and feeding in an ectoparasitic wasp. *Oikos* 89, 223–230.

Uhler, J., Haase, P., Hoffmann, L., Hothorn, T., Schmidl, J., Stoll, S., Welti, E.A.R., Buse, J. and Müller, J. (2022) A comparison of different Malaise trap types. *Insect Conservation and Diversity* 15, 666–672.

van Achterberg, C. (2009) Can Townes type Malaise traps be improved? Some recent developments. *Entomologische Berichten* 69, 129–135.

Vockeroth, J.R. (1966) A method of mounting insects from alcohol. *Canadian Entomologist* 98, 69–70.

Wharton, R.A., Smith, J.W. Jr, Quicke, D.L.J. and Browning, H.W. (1989) Two new species of *Digonogastra* Viereck (Hymenoptera: Braconidae) parasitic on Neotropical pyralid borers (Lepidoptera) in maize, sorghum and sugarcane. *Bulletin of Entomological Research* 79, 401–410.

Yi, Z., Jinchao, F., Dayuan, X., Weiguo, S. and Axmacher, J.C. (2012) A comparison of terrestrial arthropod sampling methods. *Journal of Resources and Ecology* 3, 174–182.

Appendix 1. Taxa, Authors and Dates

Ablerus Howard, 1894
    Ablerus clisiocampae (Ashmead, 1894)
Abzariini Santos & Wahl, 2021
Acacia Gerrit Smith Miller [Plantae]
Acaenitinae Förster, 1869
Acampsohelcon Tobias, 1987
Acampsohelconinae Tobias, 1987
Acerataspis Uchida, 1934
Acaenitellus Morley, 1913
Acaenitus Latreille, 1809
    Acaenitus dubitator (Panzer, 1800)
Acampsis Wesmael, 1835
Acanthaegilips Ashmead, 1897
Acanthaegilopsis Mata-Casanova and Pujade-Villar, 2013
Acanthocoris Amyot & Serville, 1843 [Hemiptera]
    Acanthocoris scaber (Linnaeus, 1763)
Acanthormius Ashmead, 1906
Acanthoserphinae Townes and Townes, 1981
Acanthoplus Stål, 1873 (Orthoptera)
    Acanthoplus discoidalis (Walker, 1869)
Acerataspis Uchida, 1934
Acerophagus papayae Noyes and Schauff, 2003
Acidopria Kieffer, 1913
Acidopria Masner, 1964
Aclista Förster, 1856
    Aclista albohirta (Dodd, 1920)
    Aclista philippinensis (Kieffer, 1913)
    Aclista soror (Kieffer, 1909)
    Aclista zenobia Buhl, 1998
Acmopolynema Ogloblin, 1946
Acrepyris Kieffer, 1905
Acrias Walker, 1847

Acrias varicornis (Girault, 1922)
Acrisis Förster, 1862
Acroclisoides Girault and Dodd, 1915
    Acroclisoides indicus Ferrière, 1931
    Acroclisoides sinicus (Huang and Liao, 1988)
Acrodactyla Haliday, 1839
Acropimpla Townes, 1960
Adelencyrtus Ashmead, 1900
Adeliini Viereck, 1918
Adelius Haliday, 1833
Adesha Cameron, 1912
Aegilips Haliday, 1835
Aenasius Walker, 1846
    Aenasius arizonensis (Girault, 1915)
Aepocerus Mayr, 1885
Afrocampsis van Achterberg & Quicke, 1990
Afromegischus van Achterberg, 2002
    Afromegischus gigas (Schletterer, 1889)
Afrostilba Benoit, 1956
Aganaspis Lin, 1987
    Aganaspis daci (Weld, 1951)
Agaonidae Walker, 1848
Agathacrista Sharkey, 2013
Agathidinae Haliday 1833
Agathirsini Sharkey, 2017
Agathis Latreille, 1804
Ageniaspis Dahlbom, 1857
Agiommatus Crawford, 1911
Agriotypinae Haliday, 1838
Agriotypus Curtis, 1832
    Agriotypus chaoi Bennett, 2001
    Agriotypus masneri Bennett, 2001
Agrothereutes Förster, 1850

Agupta Fernandez-Triana, 2018
Aiolomorphus Walker, 1871
    Aiolomorphus rhopaloides Walker, 1871
Alcidodes Marshall, 1939 [Coleoptera]
    Alcidodes dipterocarpi (Marshall, 1921)
    Alcidodes humeralis Heller, 1940
Aleiodes Wesmael, 1838
    Aleiodes concoronarius Butcher, Smith, Sharkey & Quicke, 2012
Aleiodini Muesebeck, 1928
Aleurodicus Douglas, 1892 [Hemiptera]
    Aleurodicus rugioperculatus Martin, 2004
Aleyroctonus Masner & Huggert, 1989
Aleyrodidae Westwood, 1840 [Hemiptera]
Allanagrus Noyes & Valentine, 1989
Allobethylus Kieffer, 1905
Allophrys Förster, 1869
Alloplasta Förster, 1869
Alomya Panzer, 1806
Alomyini Förster, 1869
Alloxysta Foerster, 1869
    Alloxysta asiatica Ferrer-Suay and Pujade-Villar, 2013
    Alloxysta arcuata (Kieffer, 1902)
    Alloxysta areeluckator Ferrer-Suay, Selfa and Pujade-Villar, 2014
    Alloxysta brevis (Thomson, 1862)
    Alloxysta castanea (Hartig, 1841)
    Alloxysta melanogaster (Hartig, 1840)
    Alloxysta paretasmartinezi, Ferrer-Suay and Pujade-Villar, 2013
    Alloxysta petchabunensis Ferrer-Suay, Selfa and Pujade-Villar, 2014
    Alloxysta pilosa Ferrer-Suay and Pujade-Villar, 2013
    Alloxysta pusilla (Kieffer, 1902)
    Alloxysta sawoniewiczi (Kierych, 1988)
Alydidae Amyot and Serville, 1843 [Hemiptera]
Alysiinae Leach, 1815
Amauromorpha Ashmead, 1905
Amblyomma Koch, 1844 [Araneae]
    Amblyomma variegatum Fabricius, 1794
Amblyoponinae Forel, 1893
Ambositrinae Masner, 1961
Amiseginae Krombein, 1957
Amitus Haldeman, 1850
Amphirhachis Townes, 1970
Ampulicomorpha Ashmead, 1893
    Ampulicomorpha collinsi Olmi, 1996
Amylostereum (Pers.) Boidin (1958) [Fungi]
Amyosoma Viereck, 1913
Anacharis Dalman, 1823

Anacharoides Cameron, 1904
Anagrus Haliday, 1833
    Anagrus nilaparvartae Pang & Wang, 1985
    Anagrus optabilis (Perkins, 1905)
Anagyrus Haward, 1896
    Anagyrus lopezi (De Santis, 1964)
Anamalysia van Achterberg, 2022
Anaphes Haliday, 1833
    Anaphes kailashchandrai Anwar & Zeya, 2019
Anastatus Motschulsky, 1859
    Anastatus mantoidae Motschulsky, 1859
Anastrepha Schiner, 1868 [Diptera]
Ancylotropus Cameron, 1909
    Ancylotropus montanus (Girault, 1928)
Andreimyrme Lelej, 1995
Andricus Hartig, 1840
    Andricus mukaigawae (Mukaigawa, 1913)
Aneuclis Förster, 1869
Aneurobracon Brues, 1930
Angustibracon Quicke, 1987
Anicetus Howard, 1896
Anisopteromalus Ruschka, 1912
    Anisopteromalus calandrae (Howard, 1881)
Anneckeida Bouček, 1978
Anomalon Panzer, 1804
Anomaloninae Viereck, 1918
Anorasema Bouček, 1988
Anselmella Girault, 1926
    Anselmella malacia Xiao & Huang, 2006
Anselmellini Bouček, 1988
Anteon Jurine, 1807
    Anteon abatanense Olmi, 1991
    Anteon acre Olmi, 1991
    Anteon atrum Olmi, 1998
    Anteon austini Olmi, 1989
    Anteon autumnale Olmi, 1991
    Anteon borneanum Olmi, 1984
    Anteon cerberum Olmi, 1992
    Anteon chui Xu & He, 1998
    Anteon clavatum Olmi & Currado, 1979
    Anteon debile Olmi, 1984
    Anteon devriesi Olmi, 1998
    Anteon diaoluoshanense Xu, Olmi, Guglielmino & Chen, 2011
    Anteon dignum Olmi, 1989
    Anteon doiense Olmi, 2008
    Anteon expolitum Olmi, 1984
    Anteon fidum Olmi, 1991
    Anteon fyanense Olmi, 1984
    Anteon hilare Olmi, 1984
    Anteon hirashimai Olmi, 1993

Anteon huettingeri Olmi, Xu & Guglielmino, 2015
Anteon ingenuum Olmi, 1984
Anteon insertum Olmi, 1991
Anteon khaokhoense Olmi, 2008
Anteon kresli Olmi, 2008
Anteon krombeini Olmi, 1984
Anteon laminatum Olmi, 1987
Anteon luzonense Olmi, 1991
Anteon malaise Olmi, 1993
Anteon meifenganum Olmi, 1991
Anteon motuoidum Olmi, 1987
Anteon muiri Olmi, 1984
Anteon munitum Olmi, 1984
Anteon naduense Olmi, 1987
Anteon nanlingense Xu, Olmi & He 2011
Anteon mysorense Olmi, 1984
Anteon pahanganum Olmi, 1991
Anteon papillatum Olmi, 2000
Anteon parupiscum Olmi, 1991
Anteon peterseni Olmi, 1984
Anteon phetchabunense Olmi, 2008
Anteon philippinum Olmi, 1984
Anteon phuphayonense Olmi, 2008
Anteon priscum Olmi, 1991
Anteon provinciale Olmi, 1991
Anteon quatei Olmi, 1991
Anteon raptor Olmi, 1998
Anteon sabahnum Olmi, 1998
Anteon sarawaki Olmi, 1984
Anteon semipolitum Olmi, 2008
Anteon silvicolum Olmi, 1984
Anteon spenceri Olmi, 1991
Anteon striolaforceps Xu & He 1997
Anteon subdignum Olmi, 1992
Anteon sulawesianum Olmi, 1991
Anteon superbum Dodd, 1913
Anteon thai Olmi, 1984
Anteon viraktamathi Olmi, 1987
Anteon yangi Xu, He & Olmi, 1998
Anteon yasiri Olmi, 1998
Anteon yasumatsui Olmi, 1984
Anteoninae Krombein, 1979
Anthoboscinae Turner, 1912
Anthomyiidae Robineau-Desvoidy, 1830 [Diptera]
Antrocephalus Kirby, 1883
Aoridus Yoshimoto, 1971
Apanteles Förster, 1862
Apsilocera Bouček, 1988
Apechthis Förster, 1869
Aperilampus Walker, 1871

Aperileptus Förster, 1869
Aphanogmus Thomson, 1858
   Aphanogmus hakonensis Ashmead, 1904
Aphanomerus Perkins, 1905
Aphelinidae Thomson, 1876
Aphelinus Dalman, 1820
Aphelopinae Perkins, 1912
Aphelopus Dalman, 1823
   Aphelopus albiclypeus Xu, He & Olmi, 1999
   Aphelopus alebroides Xu, Olmi, Guglielmino & Dong, 2011
   Aphelopus birmanus Olmi, 1984
   Aphelopus borneanus Olmi, 1984
   Aphelopus fuscoflavus Guglielmino & Olmi & Marletta & Xu, 2017
   Aphelopus maculiceps Bergman, 1957
   Aphelopus malayanus Olmi, 1984
   Aphelopus ochreus Olmi, 1984
   Aphelopus orientalis Olmi, 1984
   Aphelopus penanganus Olmi, 1984
   Aphelopus philippinus Olmi, 1984
   Aphelopus sabahnus Olmi, 1991
   Aphelopus spadiceus Xu & He, 1997
   Aphelopus taiwanensis Olmi, 1991
   Aphelopus wushensis Olmi, 2010
Aphidiinae Haliday, 1833
Aphis Linnaeus, 1758 [Hemiptera]
Aphytis Howard, 1900
   Aphytis chionaspis Ren, 1988
Apinae Latreille, 1802
Aplomerus Provancher, 1886
Aplorhinus Masi, 1924
Apoanagyrus Compere, 1947
Apocharips Fergusson, 1986
Apocrypta Coquerel, 1855
Apoderus Olivier, 1807 [Coleoptera]
Apophua Morley, 1913
Aporus Spinola, 1808
Apozyx Mason, 1978
Aprostocetus Westwood, 1833
Apsilocera Bouček, 1956
Apterogyninae André, 1899
Aptesini Smith & Shenefelt, 1955
Aptesis Foerster, 1850
Arachnospila Kincaid, 1900
Arachonia Joseph, 1957
Aradoidea Brullé, 1836 [Hemiptera]
Archaphidius Starý et Schlinger, 1967
   Archaphidius greenideae Starý and Schlinger, 1967
Archaphidius Starý et Schlinger, 1967
Archaeromma Yoshimoto, 1975

Arhytis, Townes 1969
Ariphron Erichson, 1842
Arkaditilla Okayasu, 2021
Aroplectrus Lin, 1963
Arthula Cameron, 1900
Asaphes Walker, 1834
Asaphinae Ashmead, 1904
Asparagobius Mayr, 1905
Aspicera Dahlbom, 1842
Aspidimorpha Hope, 1840
    Aspidimorpha miliaris (Fabricius, 1775)
Aspidobraconina van Achterberg, 1984
Aspidomorpha Hope, 1984 [Coleoptera]
Astichus Förster, 1856
Astiphromma Förster, 1869
Ateleute Förster, 1869
Ateleutinae Townes, 1970
Atoposega Krombein, 1957
    Atoposega lineata (Krombein, 1957)
Atopotrophos Cushman, 1940
Atrophini Seyrig, 1932
Atree Ranjith, van Achterberg and Priyadarsanan, 2022
Aulacocentrum philippinense (Ashmead, 1904)
Aulacaspis Cockerell, 1893
    Aulacaspis tubercularis (Newstead, 1906)
    Aulacaspis yasumatsui Takagi, 1977
Aulacidae Shuckard, 1842
Aulacocentrum Brues, 1922
Aulacus Jurine, 1807
Aulosaphes Muesebeck, 1935
Aulosaphoides van Achterberg, 1995
Austinicotesia Fernández-Triana, 2018
Australolaelaps Girault, 1925
Australomymar Girault, 1929
Austrocynipidae Riek, 1971
Austroniidae Kozlov, 1975
Austroserphinae Kozlov, 1970
Austrosystasinae Bouček, 1988
Austrozele Roman, 1910
Azotidae Nikol'skaya & Yasnosh, 1966
Azotus Howard, 1898
Bactrocera Macquart, 1835 (Diptera)
    Bactrocera correcta (Bezzi, 1916)
    Bactrocera dorsalis (Hendel, 1912)
Baeomorphidae Yoshimoto, 1975
Baeus Haliday, 1863
Bairamlia Waterston, 1929
Bakeronymus Rohwer, 1922
Balcha Walker, 1862
Balna Cameron, 1883
Banchinae Wesmael, 1845

Banchus Fabricius, 1798
Bareogonalos Schulz, 1907
Barronia Gauld & Wahl, 2002
Barycnemis Förster, 1869
Basalys Westwood, 1833
Belisana Thorell, 1898 [Araneae]
    Belisana khaosok Huber, 2005
Bellimeris Betrem, 1972
Belyta Jurine, 1870
Belytinae Kieffer, 1916
Bephrata Cameron, 1884
Bethylidae Forster, 1856
Batrocera Macquart, 1835 [Diptera]
    Batrocera correcta (Bezzi, 1916)
    Batrocera dorsalis (Hendel, 1912)
Betylobraconini Tobias, 1979
Billmasonius Fernández-Triana & Boudreault, 2018
Binodoxys Mackauer, 1960
    Binodoxys nr *indicus* (Subba Rao & Sharma, 1958)
Biosteres Förster, 1869
Biroia Szépligeti, 1900
    Biroia fuscicornis (Cameron, 1903)
Bischoffitilla Lelej, 2002
    Bischoffitilla lamellata (Mickel, 1933)
Blacini Förster, 1863
Blastophaga Gravenhorst, 1829
    Blastophaga psenes (Linnaeus, 1758)
Blastophaginae Kirchner, 1867
Bobekia Niezabitowski, 1910
Bocchinae Richards, 1939
Bocchus Ashmead, 1893
    Bocchus achterbergi Olmi, 1991
    Bocchus banianus Olmi, 1993
    Bocchus beckeri Olmi, 1991
    Bocchus laotianus Olmi, 1984
    Bocchus levis Olmi, 1991
    Bocchus muluensis Olmi, 1984
    Bocchus pedunculatus Nagy, 1969
    Bocchus rubricus Olmi, 1984
Bootania Dalla Torre, 1897
    Bootania fascia Grissell & Desjardins, 2002
    Bootania leucospoides (Walker, 1862)
    Bootania orba Desjardins & Grissell, 2002
    Bootania pilicornis (Cameron 1909)
Bootanomyia Girault, 1915
Boreoides Hardy, 1920 [Diptera]
Bostrychidae Latreille, 1802 [Coleoptera]
Botryoideclava Subba Rao, 1980
Boucekelimini Kim and La Salle, 2005

Boucekiidae Gibson, 2003
Brachistinae Förster, 1863
Brachycistidinae Kimsey, 1991
Brachycyrtinae Viereck, 1919
Brachycyrtus Kriechbaumer, 1880
    Brachycyrtus nawaii (Ashmead, 1906)
Brachygaster Leach, 1815
Brachymeria Westwood, 1829
Brachyscleroma Cushman, 1940
Brachyscleromatinae Townes, 1961
Brachyserphus Hellén, 1941
Brachyzapus Gauld and Dubois, 2006
Bracon Fabricius, 1805
    Bracon apoderi Watanabe, 1933
Braconinae Nees, 1811
Bradynobaeninae Saussure, 1892
Braunsia Kriechbaumer, 1894
    Braunsia sumatrana Enderlein, 1906
    Braunsia wallacei Turner, 1918
Bruchinae Latreille, 1802 [Coleoptera]
Brulleia Szépligeti, 1904
Buprestidae Leach, 1815 [Coleoptera]
Caenosclerogibba Yasumatsu, 1958
    Caenosclerogibba longiceps (Richards, 1958)
    Caenosclerogibba rossi Olmi, 2005
Cales Howard, 1987
Calesidae Mercet, 1929
    Cales noacki Howard, 1907
Callajoppa Cameron, 1903
Callaspidia Dahlbom, 1842
Callimomoides Girault, 1926
    Callimomoides ovivorus (Ferriere, 1936)
Calliphoridae Brauer & Bergenstamm, 1889 [Diptera]
Calliscelio Ashmead, 1893
    Calliscelio elegans Perkins, 1910
Callocleonymus Masi, 1940
Caloapenesia Terayama, 1995
Calofigites Kieffer, 1909
Calosotinae Bouček, 1958
Calyoza Westwood 1837
Camarothorax Mayr, 1906
Campanotus Mayr, 1861
Campoletis Förster, 1869
    Campoletis chlorideae Uchida, 1957
Campopleginae Förster, 1869
Campsomerini Bradley, 1957
Camptopteroides Viggiani, 1974
Camptotypus Kriechbaumer, 1889
Canalicephalus Gibson, 1977
Carabidae Latreille, 1802 [Coleoptera]
Carabunia Waterston, 1928

Cardiochiles Nees von Esenbeck, 1818
    Cardiochiles philippinensis Ashmead, 1905
Cardiochilinae Ashmead, 1900
Carinopius Tan & van Achterberg, 2016
Carminator Shaw, 1988
    Carminator affinis Shaw, 1988
    Carminator ater Shaw, 1988
    Carminator coronatus Mita & Konishi, 2010
    Carminator gracilis Mita & Konishi, 2010
Cassidocida Crawford, 1913
Cassidocida aspidomorphae Crawford, 1913
Castania Mill. [Plantae]
Castanopsis (D. Don) Spach, 1841 [Plantae]
Casuarina L. [Plantae]
Cecidomyiidae Newman, 1835 [Diptera]
Cercobarconinae Tobias, 1979
Cedria Wilkinson, 1934
    Cedria paradoxa Wilkinson, 1934
Ceidae Bouček, 1961
Cenocoeliinae Szépligeti, 1901
Centistidea Rohwer, 1914
Centistini Čapek, 1970
Centrodora Förster, 1878
Cephaleta Motschulsky, 1859
Cephalonomia Westwood, 1833
    Cephalonomia tarsalis (Ashmead, 1893)
    Cephalonomia waterstoni Gahan, 1932)
Cerambycidae Latreille, 1802 [Coleoptera]
Ceraphron Jurine, 1807
    Ceraphron manilae Ashmead, 1904
Ceraphronoidea Haliday, 1833
Ceratobaeus Ashmead, 1893
Ceraspidia Belzin, 1852
Ceratomansa Cushman, 1922
Ceratopria Ashmead, 1893
Ceratosolen Mayr, 1885
    Ceratosolen pilipes Wiebes, 1963
Cerocephala Westwood, 1832
Cerocephalidae (Gahan, 1946)
Cerrena Gray, 1821 [Fungi]
Certonotus Kriechbaumer, 1889
Chablisea Gauld and Dubois, 2006
Chalcedectidae Ashmead, 1904
Chalcedectus Walker, 1852
Chalcididae Latreille, 1817
Chalcimerinae Bouček, 1978
Chalcura Kirby, 1886
Chamaemyiidae [Diptera]
Chrionotini Uchida, 1957
Charipinae Dalla Torre & Kieffer, 1910
Charmon Haliday, 1833
    Charmon extensor Linnaeus, 1758

Charmontia van Achterberg, 1979
Charmontinae van Achterberg, 1979
Chartocerus Motschulsky, 1859
Cheiloneurus Westwood 1833
    Cheiloneurus nankingensis Li & Xu, 2020
Cheloninae Nees, 1926
Chelonus Panzer, 1806
    Chelonus formosanus Sonan, 1932
Chiasmia Hübner, 1823 [Lepidoptera]
Chilo Zincken, 1817 [Lepidoptera]
Chiromyzinae Brauer, 1880 [Diptera]
Chloridea Duncan & Westwood, 1841 [Lepidoptera]
    Chloridea virescens (Fabricius, 1777)
Chlorocryptus Cameron, 1903
    Chlorocryptus purpuratus (Smith, 1852)
Choreutidae Stainton, 1858 [Lepidoptera]
Chremylus Haliday, 1833
Chriodes Förster, 1868
Chromeurytominae Bouček, 1988
Chrysididae Latreille, 1802
Chrysobothris Eschscholtz, 1829 [Coleoptera]
    Chrysobothris succedanea Saunders, 1873
Chrysochalcissa Girault, 1915
Chrysolampidae Torre, 1898
Chrysolampus Spinola, Spinola, 1811
Chrysomelidae Latreille, 1802 [Coleoptera]
Chrysonotomyia Ashmead, 1904
Chrysopidae Esben-Petersen, 1918 [Neuroptera]
Chyphotidae Ashmead, 1896
Cibdela Konow, 1899 (Hymenoptera: 'Symphyta')
    Cibdela janthina (Klug, 1834)
Cicadellidae Latreille, 1802 [Hemiptera]
Cidaphus Förster, 1869
Cirrospilini LaSalle, 2000
Cirrospilus Westwood, 1832
Cisaris Townes, 1970
Cladobethylus Kieffer, 1922
Clatha Cameron, 1905
Cleonyminae Walker, 1905
Cleonymus Latreille, 1809
Cleptes Latreille, 1802
    Cleptes asianus Kimsey, 1987
    Cleptes humeralis Geoffroy, 1785
    Cleptes thaiensis Tsuneki, 1982
    Cleptes townesi Kimsey, 1987
Cleptidea Mocsáry, 1904
Cleptinae Latreille, 1802
Clinocentrini van Achterberg, 1991
Clinocentrus Haliday, 1833

Clistopyga Gravenhorst, 1829
Closterocerus Westwood, 1833
    Closterocerus javanus Perkins, 1912
Clypeodromus Tereshkin, 1992
Clytina Erdos, 1957
Clytocera Gahan, 1906 [Coleoptera]
    Clytocera chionospila Gahan, 1906
Cnaphalocrocis Lederer, 1863 [Lepidoptera]
    Cnaphalocrocis medinalis (Guenée, 1854)
Cnastis Townes, 1957
Cobunus Uchida, 1926
Coccinellidae Latreille, 1807 [Coleoptera]
Coccobius Ratzeburg, 1852
Coccoidea Handlirsch, 1903 [Hemiptera]
Coccophaginae Förster, 1878
Coccophagus Westwood, 1833
Coccygomimus Saussure, 1892
Coeliniaspis Fischer, 2010
Coelinius Nees, 1819
Coelocybidae Bouček, 1988
Coelopencyrtus Timberlake, 1919
Colastes Haliday, 1833
Colastinus Belokobylskij, 1984
Colastomion Baker, 1917
Coleophoridae Hübner, 1825 [Lepidoptera]
Colletinae Lepeletier, 1841
Collyriinae Cushman, 1924
Cologlyptus Crawford, 1910
Colotrechninae Thomson, 1876
Colotrechnus Thomson, 1878
    Colotrechnus agromyzae Subba Rao., 1981
Colpa Dufour, 1841
Colpotrochia Holmgren, 1856
Compsophorini Heinrich, 1967
Conganteoninae Olmi, 1984
Conidarnes Farache and Rasplus, 2015
Conobregma van Achterberg, 1995
Conopomorpha Meyrick, 1885 [Lepidoptera]
    Conopomorpha cramerella (Snellen, 1904)
Conostigmus Dahlbom, 1858
Conura Spinola, 1837
Copidosoma Ratzeburg, 1844
Copidosomyia Girault, 1915 AND move species
    Copidosoma floridanum Ashmead, 1900
    Copidosomyia ambiguous (Subba Rao, 1979)
Copidosomatini Hoffer, 1955
Coptera Say, 1836
    Coptera haywardi Loiácono, 2003
Coraebus Gory & Laporte, 1839 [Coleoptera]
    Coraebus cavifrons Descarpentries & Villiers, 1967
Corcyra Ragonot, 1885 [Lepidoptera]

Coreidae Leach, 1815 [Hemiptera]
Cornutorogas Chen, Belokobylskij, van Achterberg & Whitfield, 2004
Coronagathis van Achterberg & Long, 2010
Cosmocomoidea Howard, 1908
Cosmocomopsis Huber, 2015
Cotesia Cameron, 1891
    Cotesia glomerata (Linnaeus, 1758)
Crambidae Linnaeus, 1758 [Lepidoptera]
Cratobracon Cameron, 1901
Cratocentrus Cameron, 1907
Cratolabus Heinrich, 1974
Cremastinae Förster, 1869
Cremastobaeus Ashmead, 1893
Cremnoptini Sharkey, 1992
Cremnoptoides van Achterberg & Chen, 2004
    Cremnoptoides yui Sharkey, 2011
Cricula Walker, 1855 [Lepidoptera]
Crinibracon Quicke, 1988
Crovettia Olmi, 1984
Crypsithyris Meyrick, 1907 [Lepidoptera]
Cryptinae Kirby, 1837
Cryptopimpla Taschenberg, 1863
Cryptoprymna Förster, 1856 (fly)
Cryptoserphus Kieffer, 1907
Cryptus Fabricius, 1804
Ctenopelmatinae Förster, 1869
Curculionidae Latreille, 1802 [Coleoptera]
Cyanotiphia Cameron, 1907
Cyanoxorides Cameron, 1903
Cybocephalus Erichson, 1844 [Coleoptera]
    Cybocephalus nipponicus Endrödy-Younga, 1971
Cylloceria Schiodte, 1838
    Cylloceria melancholica (Gravenhorst, 1820)
Cylloceriinae Wahl, 1990
Cynipencyrtidae Trjapitzin, 1973
Cynipencyrtus Ishii, 1928
Cynipoidea Latreille, 1802
Cyoceraphron Dessart, 1975
Cyphacolus Priesner, 1951
Cyrtogaster Walker, 1833
Cyrtorhyssa Baltazar, 1961
    Cyrtorhyssa moellerii Bingham, 1898
Dacnusa Haliday, 1833
Dacnusini Förster, 1863
Dallatorrella Kieffer, 1911
Dallatorrellinae Kieffer, 1911
Darna Walker, 1862 [Lepidoptera]
    Darna pallivitta (Moore, 1877)
Dasycleonymus Gibson, 2003
Dasylabrinae Skorikov, 1935
Deilagaon Wiebes, 1977

Deinodryinus Perkins, 1907
    Deinodryinus asiaticus Olmi, 1984
    Deinodryinus constrictus (Olmi, 1998)
    Deinodryinus malaisei Olmi, 1993
    Deinodryinus philippinus Olmi, 1984
Delias Hübner, [1819] [Lepidoptera]
Delomerista Förster, 1869
Delomeristini Hellén, 1915
Dendrocerus Ratzeburg, 1852
    Dendrocerus carpenteri (Curtis, 1829)
Dendrolimus Germar, 1812 [Lepidoptera]
Dermestidae Latreille, 1804 [Coleoptera]
Derris Lour. [Plantae]
    Derris elliptica (Wall.) Benth
Diacamma Mayr, 1862
    Diacamma scalpratum (Smith, 1858)
Diachasmimorpha Viereck, 1913
    Diachasmimorpha longicaudata (Ashmead)
Diacritinae Townes, 1965
Diadegma Förster, 1869
Diamminae Ashmead, 1903
Diaparsis Förster, 1869
Diapetimorpha Viereck, 1913
Diaphania Hübner, 1818 [Lepidoptera]
    Diaphania hyalinata (Linnaeus, 1767)
Diapriidae Haliday, 1833
Diaspididae Maskell, 1879 [Hemiptera]
Diplazon visayensis Baltazar, 1955
Diaziella Grandi, 1928
Dicamptus Szépligeti, 1905
Dicroscelio Kieffer, 1913
Dictyonotus Kriechbaumer, 1894
Digalesus Kieffer, 1914
Digonogastra Viereck, 1912
Dilobopria Kieffer, 1914
Dilyta Förster, 1869
    Dilyta orientalis (Ferrer-Suay & Paretas-Martínez, 2011)
Dimaetha Cameron, 1901
Dinapsini Waterston, 1922
Dinarmus Thomson, 1878
Dinocampus Förster, 1862
    Dinocampus coccinellae (Schrank, 1802)
Dinocryptus Cameron, 1905
    Dinocryptus niger Cameron, 1905
Dinumma Walker, 1858 [Lepidoptera]
Diolcogaster Ashmead, 1900
Diomorus Walker, 1834
    Diomorus aiolomorphi Kamijo, 1964
Diospilini Forster, 1863
Diparidae Thomson, 1876
Diplazon Nees von Esenbeck, 1818
    Diplazon laetatorius (Fabricius, 1781)
    Diplazon orientalis (Cameron, 1905)

Diplazontinae Viereck, 1918
Dirhinus Dalman, 1818
Dirrhopinae van Achterberg, 1984
Disophrini Sharkey, 1992
Dissomphalus Ashmead 1893
Distilirella van Achterberg, 1979
Ditropinotella Girault, 1915
Ditropinotellinae Bouček, 1988
Doddifoenus Bouček, 1988
    Doddifoenus wallacei Burks and Krogmann, 2009
Doirania Waterston, 1928
Dolichoris Hill, 1967
Dominibythus Prentice et Poinar in Prentice et al., 1996
Doryctinae Foerster, 1863
Doryctobracon Enderlein, 1920
Drepanoctonus Pfankuch, 1911
Drosophila Fallén, 1823
Drosophilidae Rondani, 1856 [Diptera]
Dryinidae Haliday, 1833
Dryinus Latreille, 1804
    Dryinus achterbergi Olmi, 1991
    Dryinus alboniger Olmi, 1992
    Dryinus asiaticus Olmi, 1984
    Dryinus bellicus Olmi, 1987
    Dryinus browni Ashmead, 1905
    Dryinus bruneianus Olmi, 1987
    Dryinus flavus Xu & He 1994
    Dryinus gibbosoides Olmi, 1993
    Dryinus indianus Olmi, 1984
    Dryinus indicus (Kieffer, 1914)
    Dryinus irregularis Olmi, 1984
    Dryinus kiefferi (Fouts, 1922)
    Dryinus krombeini Ponomarenko, 1981
    Dryinus latus Olmi, 1984
    Dryinus lini Olmi, 2007
    Dryinus longipes Olmi, 1984
    Dryinus lucens Olmi, 1984
    Dryinus mansus Olmi, 1992
    Dryinus parvulus Olmi, 1987
    Dryinus peterseni Olmi, 1984
    Dryinus praeclarus Olmi, 1984
    Dryinus pyrillae (Kieffer, 1911)
    Dryinus pyrillivorus Olmi, 1986
    Dryinus sinicus Olmi, 1987
    Dryinus stantoni Ashmead, 1904
    Dryinus trifasciatus Kieffer, 1906
    Dryinus viet Olmi
Dryocosmus Giraud, 1859
    Dryocosmus okajimai Abe, Ide, Konishi & Ueno, 2014
Dvivarnus Rajmohana & Veenakumari, 2011

Dvivarnus agamades (Kozlov & Le, 1986)
Earinini Sharkey, 1992
Eccinetus Muesebeck & Walkley, 1956
Echthrodelphax Perkins, 1903
    Echthrodelphax fairchildii Perkins, 1903
    Echthrodelphax laotianus Olmi, 1984
    Echthrodelphax rufus Olmi, 1984
Echthromorpha Holmgren, 1868
Echthrus Gravenhorst, 1829
Ecnomiini van Achterberg, 1985
Ecnomius Mason, 1979
Eileeniella Fergusson, 1992
Elaeis Jacquin [Plantae]
    Elaeis guineensis Jacq.
Elampini Dahlbom, 1854
Elasmus Westwood, 1833
Elateridae Leach, 1815 [Coleoptera]
Eleonoria Braet and van Achterberg, 2000
Embidobiini Kozlov
Embolemidae Förster, 1856
Embolemus Westwood, 1833
    Embolemus krombeini Olmi, 1996
Encarsia Foerster, 1878
    Encarsia flavoscutellum Zehntner, 1900
Encyrtidae Walker, 1837
Encyrtoscelio Dodd, 1914
Encyrtus Latreille, 1809
    Encyrtus lecaniorum (Mayr, 1876)
Enicospilini Townes, 1971
Enicospilus Stephens, 1835
Ennogerinae Burks, 2022
Entedoninae Förster, 1856
Entiinae Hedqvist, 1974
Entomacis Foerster, 1856
Eopelma Gibson, 1989
Eoreuma Ely, 1910
    Eoreuma loftini (Dyar, 1917)
Eotrogaspidia Lelej, 1996
Epactiothynnus Turner, 1910
Ephedrini Mackauer, 1961
Ephialtes Gravenhorst, 1829
Ephialtini Hellén, 1992
Ephutinae, Ashmead 1903
Epichrysomallidae Hill & Riek, 1967
Epidinocarsis Girault, 1913
    Epidinocarsis lopezi (De Santis, 1964)
Epipompilus Kohl, 1884
Epirhyssa Cresson, 1865
Epitranus Walker, 1834
Epyrinae Kieffer, 1914
Eretmocerinae Shafee and Khan, 1978
Eretmocerus Haldeman, 1850
Eriaphytinae Hayat, 1978

Eriaporinae Ghesquiere, 1955
Eriborus Förster, 1869
Erimerinae Crawford, 1914
Erimyrmosa Lelej, 1984
    Erimyrmosa burmanensis Lelej, 1984
Erionota Mabille, 1878 [Lepidoptera]
Erotolepsiinae Bouček, 1988
Erotylidae Latreille, 1802 [Coleoptera]
Etha Cameron, 1903
Ettchellsia Cameron, 1909
Ettchellsia ignita Mita and Shaw, 2012
    Ettchellsia nigripes Mita & Shaw, 2012
    Ettchellsia philippinensis Mita & Shaw, 2012
    Ettchellsia reidi Mita & Shaw, 2012
Euagathis Szépligeti, 1900
Eubeckerella Narendran, 1999
Eucalyptus L'Hér [Plantae]
Euceros Gravenhorst, 1829
    Euceros albitarsus Curtis, 1837
    Euceros frigidus Cresson, 1869
    Euceros pruinosus Gravenhorst 1829
    Euceros unifasciatus Vollenhoven, 1878
Eucerotinae Viereck, 1919
Eucharis Clausen, 1940
Eucharitidae Latreille, 1809
Eucnemidae Eschscholtz, 1829
Eucoilinae Thomson, 1862
Euderinae Erdos, 1956
Euderomphale Girault, 1916
Euderus Haliday, 1844
Eugalta Cameron, 1899
Eujacobsonia Grandi, 1923
Eukoebelea Ashmead, 1904
Eulophidae Westwood, 1829
Eulophus Geoffroy, 1762
Eumenes Latreille, 1802
    Eumenes arcuata (Fabricius, 1775)
Eumeninae (Leach, 1815)
Eunotidae Ashmead, 1904
Eunotiscus Compere, 1928
Eupelmidae Walker, 1833
Eupelmus Dalman, 1820
Euperilampus Walker, 1871
Euphorinae Foerster, 1863
Euplectrus Westwood, 1832
Eupristina Saunders, 1882
Eurema Hübner, [1819]
Eurydinotomorpha Girault, 1915
Eurylabini Heinrich, 1934
Euryproctini Thomson, 1883
Eurytoma Illiger, 1807

Eurytomidae Walker, 1832
Eusandalinae Fusu, 2022
Eusterinx Förster, 1869
Eutrichosomatidae Peck, 1951
Eutrichosomellini (Trjapitzin, 1973)
Euurobracon Ashmead, 1900
Evania Fabricius, 1775
    Evania appendigaster (Linnaeus, 1758)
Evaniidae Latreille, 1802
Exetastes Gravenhorst, 1829
Exochus Gravenhorst, 1829
Exoryza Mason, 1981
Exothecinae Foerster, 1863
Facitorini Van Achterberg, 1995
Facydes Cameron, 1901
Ficobracon van Achterberg & Weiblen, 2000
Ficomila Bouček, 1981
Ficus Linnaeus [Plantae]
    Ficus araneosa King
    Ficus carica L.
    Ficus cereicarpa Corner, 1960
    Ficus crassiramea (Miq.) Miq.
    Ficus ixoroides Corner, 1960
Figitidae Forster, 1869
Fissicaudus Starý and Schlinger, 1967
    Fissicaudus thailandicus Starý & Rakhshani, 2010
Flavopimpla Betrem, 1932
Foenatopus Smith, 1860
    Foenatopus turcomanorum (Semenov, 1891)
Foenobethylus Kieffer, 1913
Fopius Wharton, 1987
Formicoidea Latreille, 1802
Frena Bouček, 1988
Friona Cameron, 1902
Fulgoroidea Latreille, 1807 [Hemiptera]
Fusicornia Risbec, 1950
Fustiserphus Townes, 1981
    Fustiserphus intrudens (Smith, 1878)
Gabunia Kriechbaumer, 1895
Gabuniina Townes, 1970
Gahanopsis Ogloblin, 1946
Galodoxa Nagy, 1974
Gammabracon Quicke, 1894
Gastraspis Lin, 1988
Gasteruptiidae Ashmead, 1900
Gasteruption Latreille, 1796
Gelis Thunberg, 1827
Geoscelionidae Engel & Huang, 2017
Gerridae Leach, 1815
Gigantochloa Kurz ex Munro
    Gigantochloa scortechinii Gamble

Gilen Reshchikov & Achterberg, 2018
    Gilen orientalis Reshchikov & Achterberg, 2018
Glodianus Cameron, 1902
Glyphomerinae Janšta, 2018
Glyptini Cushman & Rohwer, 1920
Glyptomorpha Holmgren, 1868
Glyptopimpla Morley, 1913
Gnamptodontini Fischer, 1970
Gnathochorisis Förster, 1869
Gollumiella Hedqvist, 1978
Gonatocerus Nees von Esenbeck, 1834
Gonatopodinae Kieffer, 1906
Gonatopus Ljungh, 1810
    Gonatopus achterbergi Olmi, 1991
    Gonatopus asiae Olmi, 1993
    Gonatopus asiaticus (Olmi, 1984)
    Gonatopus attenuatus Olmi, 1984
    Gonatopus beaveri Olmi, 1998
    Gonatopus bicuspis (Olmi, 1993)
    Gonatopus bonaerensis Virla, 1997
    Gonatopus borneanus Xu & Olmi & He 2013
    Gonatopus cantonensis Olmi, 1987
    Gonatopus cristatus (Kieffer, 1922)
    Gonatopus flavifemur (Esaki & Hashimoto, 1932)
    Gonatopus iarensis Olmi, 2005
    Gonatopus insulae Olmi, 1993
    Gonatopus javanus (Perkins, 1912)
    Gonatopus lucens (Olmi, 1984)
    Gonatopus maurus Kieffer, 1906
    Gonatopus malesiae (Olmi, 1984)
    Gonatopus medius Olmi, 1984
    Gonatopus muiri (Olmi, 1984)
    Gonatopus nearcticus (Fenton, 1927)
    Gonatopus nigricans Perkins, 1905
    Gonatopus nudus (Perkins, 1912)
    Gonotapus pajanensis (Olmi, 1989)
    Gonatopus perpolitus Perkins, 1984
    Gonatopus philippinus Olmi, 1984
    Gonatopus plebeius (Perkins, 1912)
    Gonatopus rufoniger Olmi, 1993
    Gonatopus sarawakensis (Olmi, 1984)
    Gonatopus validus (Olmi, 1984)
    Gonatopus yasumatsui Olmi, 1984
Goryphus Holmgren, 1868
Gotra Cameron, 1902
Gracillariidae Stainton, 1854 [Lepidoptera]
Grandiana Wiebes, 1961
Grasseiana Abdurahiman & Joseph 1967
Gronaulax Cameron, 1910
Grooca Sureshan and Narendran, 1997

Gryon Haliday, 1833
    Gryon ancinla Kozlov & Lê, 1996
    Gryon flavipes (Dodd, 1914)
Gryonini Kozlov, 1970
Gyragathis van Achterberg & Long, 2010
Gyrolasomyiini Bouček, 1988
Habrobracon Ashmead, 1895
    Habrobracon brevicornis (Wesmael, 1838)
    Habrobracon hebetor (Say, 1836)
Habroteleia Kieffer, 1905
Halictinae Thomson, 1869
Haltichella Spinola, 1811
Haltichellinae Ashmead, 1904
Halyomorpha Mayr, 1864 [Hemiptera]
    Halyomorpha halys Stål, 1855
Haplogonatopus Perkins, 1905
    Haplogonatopus apicalis Perkins, 1905
Hartemita Cameron, 1910
Hayatosema Heraty and Burks, 2017
    Hayatosema initiator (Kerrich, 1963)
Hebichneutes Sharkey & Wharton, 1994
Heimbrinae Burks, 1971
Helcon Nees, 1814
Helconinae Foerster, 1862
Helicoverpa Hardwick, 1965 [Lepidoptera]
    Helicoverpa armigera (Hübner, [1808])
Helictes Haliday, 1837
Helictinae Gill, 1872
Heliothis Ochsenheimer, 1816 [Lepidoptera]
    Heliothis virescens (Fabricius, 1777)
Heliozelidae Heinemann & Wocke, 1876 [Lepidoptera]
Hellwigia Gravenhorst, 1829
Heloridae Förster, 1856
Helorimorphini Schmiedeknecht, 1907
Helorus Latreille, 1802
Hemadas Crawford, 1909
Hemerobiidae Latreille, 1802 [Neuroptera]
Hemigalesus Kieffer, 1913
Hemiphanes Förster, 1869
Henosepilachna Li & Cook, 1961 [Coleoptera]
    Henosepilachna vigintioctopunctata (Fabricius, 1775)
Hepialidae Stephens, 1829
Heptascelio Kieffer 1916
Herbertia Howard, 1894
    Herbertia indica Burks, 1959
Herbertiidae Bouček, 1988
Heresiarchini Ashmead, 1900
Heresiarches Wesmael, 1859
Hesperiidae Latreille, 1809 [Lepidoptera]
Heteribalia Sakagami, 1949
    Heteribalia confluens (Maa, 1949)

Heterogamus Wesmael, 1838
Heterospilus Haliday, 1836
Hetreulophidae Girault, 1915
Hetrodinae Brunner von Wattenwyl, 1878
Hevea Aubl. [Plantae]
    Hevea brasiliensis Muell. Arg.
Heydenia Förster, 1856
Heydeniidae Hedqvist, 1961
Heydoniella Narendran, 2003
Hexacharis Kieffer, 1907
Himertosoma Schmiedeknecht, 1900
Hidari Distant, 1886 [Lepidoptera]
Hispidophila Viggiani, 1968
Histeromerus Wesmael, 1838
    Histeromerus orientalis Chou & Chou, 1991
Hockeria Walker, 1834
Holcojoppa Cameron, 1902
Holophris Mocsáry, 1890
Homalotylus Mayr, 1876
Homolobinae van Achterberg, 1979
Homolobus Förster, 1862
Homonotus Dahlbom, 1843
Homotropus Förster, 1869
Hormiinae Förster, 1863
Hormiini Förster, 1863
Hormius Nees, 1818
Hormoserphus Townes, 1981
Hyalonema Özdikmen, 2009
Hyblaea Fabricius, 1793
    Hyblaea puera (Cramer, 1777)
Hybomischos Baltaza, 1961
Hylcalosia Fischer, 1967
Hylomesa Krombein, 1968
    Hylomesa bakeri Krombein, 1968
    Hylomesa longiceps (Turner, 1918)
Hyperbius Förster, 1878
Hyperimerus Girault, 1917
Hypsicera Latreille, 1829
Hyptiogaster Kieffer, 1903
Hyptiogastrinae Crosskey, 1953
Ibalia Latreille, 1802
    Ibalia kalimantanica Liu, 1998
Ibaliidae Forster, 1869
Icerya Signoret, 1875 [Hemiptera]
    Icerya purchase Maskell, 1878
Ichneumonidae Latreille, 1802
Ichneutinae Foerster, 1863
Idioporidae LaSalle, Polaszek & Noyes, 1997
Idris Förster, 1856
Inbioia Gauld & Ugalde, 2002
    Indaphidius curvicaudatus Starý, 2010
Inostemmatinae Ashmead, 1903

Ischnobracon Baltazar, 1963
Ischnojoppini Heinrich, 1938
Ischnus Gravenhorst, 1829
Ischyracis Förster, 1868
Ishtarella Martens, 2021
    Ishtarella thailandica Martens, 2021
Ismaridae (Thomson, 1858)
Ismarus Haliday, 1835
Ismaya Bouček, 1988
Isotima Förster, 1868
Itoplectis Förster, 1869
Ixodiphagus Howard, 1907
    Ixodiphagus hookeri (Howard, 1908)
Janzenellidae Johnson & Austin, 2021
Jezarotes Uchida, 1928
    Jezarotes yanensis Sheng & Sun 2013
Jimwhitfieldius Fernandez-Triana, 2018
Joppocryptini Viereck, 1918
Karlidia Lelej, 1999
    Karlidia peterseni Lelej & Krombein, 1999
Kaeeia Mitroiu, 2011
Keiraninae Bouček, 1988
Kerevata Belokobylskij, 1999
Keryinae Bouček, 1988
Khoikhoinae Mason, 1983
Klutiana Betrem, 1932
Kopinata Bouček, 1988
Kriechbaumerella Dalla Torre, 1897
Kristotomus Mason, 1962
Krombeinella Pate, 1947
Krombeinius Bouček, 1978
Kudakrumia Krombein, 1979
Kumarella Sureshan, 1999
Kurzenkotilla Lelej, 2005
Labeninae Ashmead, 1900
Lagynodes Förster, 1840
Lagynodinae Masner and Dessart, 1967
Lambdobregma Gibson, 1989
Laodelphax Fennah, 1963 [Hemiptera]
Lariophagus Crawford, 1909
Lasiochalcidia Masi, 1916
Lathrolestes Förster, 1869
Lathromeromyia Girault, 1914
Laxiareola Sheng & Sun, 2011
    Laxiareola ochracea Sheng and Sun, 2011
Leeuweniella Ferrière, 1929
Leiophron Nees, 1819
Lentocerus Dong Naito, 1999
Lepidopsocus Enderlein, 1903 [Psocodea]
Leptacis Förster, 1856
Leptobatopsis Ashmead, 1900
Leptocorisa Latreille, 1829 [Hemiptera]

Leptocorisa oratoria (Fabricius, 1764)
Leptocybe Fisher and La Salle, 2004
    Leptocybe invasa Fisher & La Salle, 2004
Leptofoeninae Handlirsch, 1924
Leptofoenus Smith, 1862
Leptophion Cameron, 1901
Leptopilina Förster, 1862
    Leptopilina boulardi (Barbotin, Carton & Kelner-Pillault, 1979)
    Leptopilina heterotoma Thomson, 1862
Leptopimpla Townes, 1961
Leucospidae Fabricius, 1775
Leucospis Fabricius, 1775
    Leucospis dorsigera Fabricius, 1775
Liepara Bouček, 1988
Limacodidae Duponchel, 1845 [Lepidoptera]
Linshcosteus Distant, 1904
Liopteridae Ashmead, 1895
Liosphex Townes, 1977
    Liosphex trichopleurum Townes, 1977
Liotryphon Ashmead, 1900
Lipoglyptus Crawford, 1910
Lipolexis Förster, 1862
    Lipolexis oregmae (Gahan, 1932)
Lipothymus Grandi, 1922
Liriomyza Mik, 1894
    Liriomyza huidobrensis (Blanchard, 1926)
Lissonota Gravenhorst, 1829
Lissopimpla Kriechbaumer, 1889
Listrodomini Förster, 1869
Listrodomus Wesmael, 1845
Lithocarpus Blume [Plantae]
Lithosaphonecrus Tang, Melika & Bozsó, 2013
    Lithosaphonecrus mindatus Ide, Aung & Tanaka 2020
    Lithosaphonecrus papuanus Schweger et al. 2015
    Lithosaphonecrus vietnamensis (Abe, Ide, Konishi & Ueno, 2014)
Loboscelidia Westwood, 1874
Loboscelidiinae Maa and Yoshimoto, 1961
Lonchodryinus Kieffer, 1905
    Lonchodryinus sinensis Olmi, 1984
Lorio Cheesman, 1936
Losbanus Ishii, 1932
Louriciinae Hedqvist, 1961
Loxotropa Förster, 1856
Lucanidae Latreille, 1804 [Coleoptera]
Lustrina Kurian, 1955
    Lustrina assamensis Kurian, 1955
Lyciscidae Bouček, 1958
Lycogaster Shuckard, 1841

Lycogaster flavonigrata Chen, Achterberg & He, 2014
Lycogaster rufiventris (Magretti, 1897)
Lycorina Holmgren, 1859
Lycorininae Cushman & Rohwer, 1920
Lyctidae Billberg, 1820 [Coleoptera]
Lygaeoidea Schilling, 1829 [Hemiptera]
Lymaenon Walker, 1846
Lymeon Forster, 1869
Lyonetiidae Stainton, 1854 [Lepidoptera]
Lysitermini Tobias, 1968
Lysitermus Förster, 1862
Lyteba Thomson, 1859
Lytopylini Sharkey, 2017
Lytoxysta Kieffer, 1909
Maamingidae Early, Manser, Naumann, & Austin, 2001
Macrobracon Szépligeti, 1902
    Macrobracon flavus Chishti & Quicke, 1994
Macrocentrinae Foerster, 1863
Macrocentrus Curtis, 1833
    Macrocentrus cingulum Brischke, 1882
    Macrocentrus philippinensis Ashmead, 1904
Macroglenes Westwood, 1832
Macromesidae Graham, 1959
Macromesus Walker, 1848
    Macromesus javensis Hedqvist, 1968
Macrostomion Szépligeti, 1900
Macroteleia Westwood, 1835
Mahinda bo Kimsey, Mita & Pham, 2016
Mahinda Krombein, 1983
Malinka Bouček, 1988
Mallada Navás, 1925 [Neuroptera]
    Mallada desjardinsi (Navás, 1911)
Mangostigmus Bouček, 1986
Mansa Tosquinet, 1896
Mantibaria Kirby, 1900
Marietta Motschulsky, 1863
Markshawius Fernandez-Triana, 2018
Masner Mikó & Deans, 2009
Masona van Achterberg, 1995
Masonbeckia Sharkey & Wharton, 1994
Maxfischeriinae Papp, 1994
Mecotes Townes, 1971
Megachalcis Cameron, 1903
Megachilinae Latreille, 1802
Megalocolus Kirby, 1883
Megalomya Uchida, 1940
Megalyra Westwood, 1832
    Megalyra longiseta Szépligeti, 1902

Megalyra rufiventris Szépligeti, 1902
Megalyra sedlaceki Shaw, 1990
Megalyra spectabilis Shaw, 1990
Megalyra tawiensis Petersen, 1966
Megalyridae Schletterer, 1890
Megaphragma Timberlake, 1924
Megarhyssa Ashmead, 1900
Megascogaster Baker, 1926
Megascolia Betrem, 1928
   Megascolia azurea (Christ, 1791)
Megaspilidae Ashmead, 1893
Megaspilinae Ashmead, 1903
Megastigmidae Thomson, 1876
Megastigmus Dalman, 1820
   Megastigmus nigrovariegatus Ashmead, 1890
Megastylus Schiodte, 1838
Megischus Brullé, 1846
   Megischus insularis Smith, 1857
   Megischus saussurei (Schulz, 1907)
Melanagromyza Hendel, 1920 [Diptera]
   Melanagromyza sojae (Zehntner, 1900)
Melandryidae Leach, 1815
Melanips Haliday (in Walker, 1835)
Melanosomellidae Girault, 1913
Melittobia Westwood, 1848
Membracidae Rafinesque, 1815 [Hemiptera]
Merostenus Walker, 1837
Mesitiinae Barbosa & Azevedo, 2012
Mesochorinae Förster, 1869
Mesochorus Gravenhorst, 1829
   Mesochorus curvulus Thomson, 1886
Mesocoelini van Achterberg, 1990
Mesocynips Cameron, 1903
Mesostenus Gravenhorst, 1829
Metapelma Westwood, 1835
Metapelmatidae Bouček, 1988
Metaplectrus Ferrière, 1941
Meteoridea Ashmead, 1900
   Meteoridea hutsoni (Nixon, 1941)
Meteorideinae Capek, 1970
Meteorini Cresson, 1887
Meteorus Haliday, 1835
   Meteorus pulchricornis (Wesmael, 1835)
Methocha Latrielle, 1804
   Methocha malayana Pagden, 1949
   Methocha granulosa Narita & Mita, 2018
Methocinae André, 1903
Metisa Walker, 1855 [Lepidoptera]
   Metisa plana Walker, 1855
Metopheltes Uchinda, 1932
Metopiinae Förster, 1869

Metopius Panzer, 1806
Mickelomyrme Lelej, 1995
   Mickelomyrme puttasoki Williams, 2019
Microtypinae Szépligeti, 1908
Micranisa Walker, 1875
Microgastrinae Förster, 1863
Micrognathophora Grandi, 1922
   Miscogastrinae Walker, 1833
Microleptes Gravenhorst, 1829
Microleptinae Townes, 1958
Microterys Thomson, 1876
Microthoron Masner, 1972
Millettia Wight & Arn. [Plantae]
   Millettia sericea (Vent.) Benth
Mimagathidini Enderlein, 1905
Minagenia Banks, 1934
Miracinae Viereck, 1918
Mirax Haliday, 1833
Mirobaeoides Dodd, 1914
Miscogaster Walker, 1833
Miscogasteriella Girault, 1915
Miscogastrinae Walker, 1833
Mocsarya Konow, 1897
   Mocsarya metallica Mocsáry, 1896
Mokrzeckia Mokrzecki, 1934
Monacon Waterston, 1922
Monacon gawai Darling, 2020
Monema Walker, 1855 [Lepidoptera]
   Monema flavescens Walker, 1855
Monoblastus Hartig, 1837
Monodontomerinae Ashmead, 1899
Monodontomerus Westwood, 1833
   Monodontomerus dentipes (Dalman, 1820)
Monomachidae Ashmead, 1902
Monomachus Klug, 1841
   Monomachus antipodalis Westwood, 1874
Monophlebidae Signoret, 1875 [Hemiptera]
Moranila Cameron, 1883
Moranilidae Bouček, 1988
Mordellidae Latreille, 1802 [Coleoptera]
Muesebeckiini Mason, 1969
Muhabbetella Ko ak and Kemal, 2008
Muscidae Latreille, 1802 [Diptera]
Mutillidae Latreille, 1802
Mycetophilidae Newman, 1834 [Diptera]
Myiocephalini Chen & van Achterberg, 1997
Myiocnema Ashmead, 1900
Myllenyxis Baltazar, 1961
Mymaridae Haliday, 1833
Mymaromma Girault, 1920
   Mymaromma menehune Honsberger and Huber, 2022

Mymaromella Girault, 1931
Mymarommatoidea Debauche, 1948
Myrmosidae Fox, 1894
Neanastatidae Kalina, 1984
Neanastatus Girault, 1913
    Neanastatus grallarius (Masi, 1926)
Neapterolelapinae Rasplus, Burks & Mitroiu, 2022
Nearretocera Girault, 1913
Nefoeninae Bouček, 1988
Neoapenesia Terayama, 1995
Neocalosoter Girault and Dodd, 1915
Neochalcis Kirby, 1883
Neochrysocharis Kurdjumov, 1912
Neodiparidae Erixestinae Burks & Rasplus, 2022
Neodiprion Rohwer, 1918
    Neodiprion sertifer (Geoffroy, 1785)
Neodryinus Perkins, 1905
    Neodryinus beaveri Olmi, 1998
    Neodryinus chelatus Olmi, 1984
    Neodryinus diffusus Olmi, 1984
    Neodryinus javanus (Roepke, 1916)
    Neodryinus leptopus Richards, 1953
    Neodryinus malayanus (Olmi, 1991)
    Neodryinus phuphayonensis Olmi, 2008
    Neodryinus pseudodiffusus Olmi, 1984
    Neodryinus reticulatus Baltazar, 1966
    Neodryinus sumatranus Enderlein, 1907
Neofacydes Heinrich, 1960
Neohaltichella Narendran, 1989
Neokopinata Narendran, 2003
Neolosbanus Heraty, 1994
    Neolosbanus palgravei (Girault, 1922)
Neoneurini Bengtsson, 1918
Neosycophila Grandi, 1923
Neoxorides Clément, 1938
Nephotettix Matsumura, 1902 [Hemiptera]
Nepidae Latreille, 1802 [Hemiptera]
Nepticulidae Braun, 1917 [Lepidoptera]
Neralsia Cameron, 1883
Nesaulax Roman, 1913
Nesodiprion biremis (Konow, 1899)
Nesomesochorinae Ashmead, 1905
Nesomesochorus Ashmead, 1905
Neurocrassus Šnoflak, 1945
Neurogenia Roman, 1910
Neurolarthra Fischer, 1976
Neuroscelio Dodd, 1913
    Neuroscelio doddi Galloway, Austin & Masner, 1992
Neuroscelionidae Johnson & Austin 2021
Nezara Amyot & Serville, 1843 [Hemiptera]
    Nezara viridula (Linnaeus, 1758)
Nilaparvata Distant, 1906
    Nilaparvata lugens (Stål, 1854)
Nipisa Huber, 2018
    Nipisa phyllicola Huber, Eberle & Dimitrov, 2018
Nipponosega Kurzenko & Lelej, 1994
    Nipponosega lineata Mita, 2021
Nitidulidae Latreille, 1802 [Coleoptera]
Nixonia Masner, 1976
    Nixonia krombeini Johnson & Masner, 2006
    Nixonia watshami Johnson & Masner, 2006
Nixoniidae Masner, 1976
Noctuidae Latreille, 1809 [Lepidoptera]
Nomosphecia Gupta, 1962
    Nomosphecia zebroides (Krieger, 1906)
Nonnus Cresson, 1874
Norbanus Walker, 1843
Nordeniella Lelej, 2005
Notanisomorphella Girault, 1913
Notanisus Walker, 1837
Notaspidiella Bouček, 1988
Notaspidium Dalla Torre, 1897
Nothofagus Blume. [Plantae]
Nothoserphus Brues, 1940
    Nothoserphus afissae (Watanabe, 1954)
Notocyphus Smith, 1855
Notoglyptus Masi, 1917
Notosemini Townes, Towns & Gupta, 1961
Notosemus Förster, 1869
Nyleta Dodd, 1926
Oberthuerellinae Hedicke and Kerrich, 1940
Odontacolus Kieffer, 1910
Odontepyris Kieffer, 1904
Odontocolon Cushman, 1942
Odontofroggatia Ishii, 1934
Odontosphaeropygini Zettel, 1990
Odontosphaeropyx Cameron, 1910
Odopoia Walker, 1871
Oecophylla Smith, 1860 [Hymenoptera]
    Oecophylla smaragdina Fabricius, 1775
Oedemopsini Woldstedt, 1877
Oedicephalini Heinrich, 1934
Oenrobia Hayat, 1995
Olenecamptus Chevrolat, 1835 [Coleoptera]
    Olenecamptus bilobus (Fabricius, 1801)
Olethrodotini Townes, 1970
Oligoneurus Szépligeti, 1902
Oligosita Haliday in Walker, 1851
Olixon Lohrmann, Volker & Ohl, Michael, 2007

Omalaspis Giraud, 1860
Omphale Haliday, 1833
Oncophanes Förster, 1862
Ooctonus Haliday, 1833
    Ooctonus sinensis Subba Rao, 1969
Oodera Westwood, 1874
    Oodera longicollis (Cameron, 1903)
Ooderidae Bouček, 1958
Ooencyrtus Ashmead, 1990
Opheliminae Ashmead, 1904
Ophelimus Haliday, 1844
    Ophelimus eucalypti (Gahan)
    Ophelimus maskelli (Ashmead, 1900)
Ophelosia Riley, 1890
    Ophelosia crawfordi Riley, 1890
Ophion Fabricius, 1798
Ophioninae Shuckard, 1840
Ophiusa Ochsenheimer, 1816 [Lepidoptera]
    Ophiusa simillima Guenée, 1852
Ophrynopinae Benson, 1935
Ophrynopus Konow, 1897
    Ophrynopus kuhlii (Konow, 1897)
    Ophrynopus maculipennis (F. Smith, 1859)
Opiinae Blanchard, 1845
Opisina Walker, 1864 [Lepidoptera]
    Opisina arenosella Walker, 1864
Opius Wesmael, 1835
Orasema Cameron, 1884
Oraseminae Burks, 1979
Orgilinae Ashmead, 1900
Orgilini Ashmead, 1900
Orgilonia van Achterberg, 1987
Orgilus Haliday, 1833
Orientilla Lelej, 1979
    Orientilla vietnamica Lelej, 1979
Orientocardiochiles Kang & Long, 2020
Orientocolastes Belokobylskij, 1999
Ormyridae Förster, 1856
Ormyrulus Bouček, 1986
Ormyrus Westwood, 1832
Orseolia Kieffer & Massalongo, 1902 [Diptera]
    Orseolia oryzae (Wood-Mason, 1889)
Orthocentrinae Förster, 1869
Orthocentrus Gravenhorst, 1829
Orthostigma Ratzeburg, 1844
Orthotrichia Eaton, 1873 [Neuroptera]
    Orthotrichia muscari Wells, 1983
Orseolia Kieffer & Massalongo, 1902 [Diptera]
    Orseolia oryzae (Wood-Mason, 1889)
Orussidae Newman, 1834
Orussinae Newman, 1834
Orussus Latreille, 1796

Orussus bensoni Guiglia, 1937
Orussus decoomani Maa, 1950
Orussus loriae Mantero, 1899
Orussus punctulatissimus Blank & Vilhelmsen, 2014
Orussus striatus Maa, 1950
Osphranteria Redtenbacher, 1849 [Coleoptera]
    Osphranteria coerulescens Redtenbacher, 1850
Osphrynchotus Spinola, 1841
Otitesella Westwood, *1883*
Otitesellini Joseph, 1964
Oxycoryphe Kriechbaumer, 1894
Oxypria Kieffer, 1908
Oxyrrhexis Förster, 1869
Oxyserphus Masner, 1961
    Oxyserphus rossica (Kolyada, 1996)
Oxysychus Delucchi, 1956
Oxytorinae Thomson, 1883
Oxytorus Förster, 1869
    Oxytorus carinatus Riedel, 2021
    Oxytorus rufopronodealis Riedel, 2021
Pachycrepoideus Ashmead, 1904
    Pachycrepoideus vindemmiae (Rondani, 1875)
Pachyneurinae Schiner, 1864
Palachia Bouček, 1969
Palachiini Bouček, 1976
Palaomymar Meunier, 1901
Palpostilpnus Aubert, 1961
Pambolinae Marshall, 1885
Pambolus Haliday, 1836
Pandanus Parkinson [Plantae]
Paniscomima Enderlein, 1904
Pantisarthrus Förster, 1871
Pantoclis Förster, 1856
    Pantoclis javensis (Dodd, 1920)
    Pantoclis convexa Buhl, 1998
    Pantoclis fuscicorpa Buhl, 1998
Papatuka Deans, 2002
Papuopsia Bouček, 1988
Parabioxys Shi & Chen, 2001
    Parabioxys songbaiensis Shi & Chen, 2001
Parablastothrix Mercet, 1917
Parabrulleia van Achterberg, 1983
Paracentrobia Howard, 1897
Paracoccus marginatus Williams and Granara de Willink, 1992 (Hemiptera)
Paracyphononyx Gribodo, 1884
Parahormius Nixon, 1940
Paramblynotus Cameron, 1908
Paraphytis Compere, 1925

Paraplitis Mason, 1981
Parapsilogastrus Ghesquière, 1946
Pararhabdepyris Gorbatvsky, 1995
Parasaphodes Schulz, 1906
 Parasaphodes townsendi (Ashmead, 1905)
Parasaphodinae Bouček, 1988
Parasapyga Turner, 1910
 Parasapyga boschi van Achterberg, 2014
 Parasapyga yvonnae van Achterberg, 2014
Paraspicera Kieffer, 1907
Parasteatoda Archer, 1946
 Parasteatoda merapiensis Yoshida & Takasuka, 2011
Parastephanellus Enderlein, 1906
Paratelenomus Dodd, 1914
Pareumenes Saussure, 1855
 Pareumenes quadrispinosus (de Saussure, 1855)
Parevania Kieffer, 1907
Paridris Kieffer, 1908
 Paridris nephta (Kozlov, 1976)
Parnara Moore, [1881] [Lepidoptera]
 Parnara guttatus (Bremer & Grey, 1853)
Paroligoneurus Muesebeck, 1931
Peckidium Masner & Garcia 2002
 Peckidium enigmaticum Masner & Garcia 2002
Pediobius Walker, 1846
Pelecinellidae Ashmead, 1895
Pelecinidae Haliday, 1840
Pentatermini Belokobylskij, 1990
Pentatermus Hedqvist, 1963
Pentatomidae Leach, 1815 [Hemiptera]
Peradeniidae Naumann & Masner, 1985
Perilampidae Latreille, 1809
Perilampus McClelland, 1838
 Perilampus auratus (Panzer, 1798)
 Perilampus singaporensis Rohwer, 1923
Perilissini Thomson, 1883
Periplaneta Burmeister, 1838 [Blattodea]
 Periplaneta americana (Linnaeus, 1758)
Peristenus Foerster, 1862
Perithous Holmgren, 1859
 Perithous changaishanus (He, 1996)
Perthiola Bouček, 1988
Petersenidia Lelej, 1992
Phenococcus Cockerell, 1902 [Hemiptera]
 Phaenococcus solenopsis (Tinsley, 1898)
Phaenodus Foerster, 1863
Phaenoglyphis Förster, 1869

Phaenoglyphis chiangmaiensis Ferrer-Suay and Pujade-Villar, 2014
Phaenoglyphis wongchaiensis Ferrer-Suay and Pujade-Villar, 2014
Phaenopria Ashmead, 1893
Phaenospila van Achterberg & Yao, 2022
Phaeogenini Förster, 1869
Phalgea Cameron, Cameron, 1905
Phaneroserphus Pschorn-Walcher, 1958
 Phaneroserphus calcar (Haliday, 1839)
Phalacridae Leach, 1815
Phanerotoma Wesmael, 1838
Phanerotomini Baker, 1926
Phanuromyia Dodd, 1914
Phasgonophora Westwood, 1832
Phenacoccus Cockerell, 1902 [Hemiptera]
 Phenacoccus manihoti Matile-Ferrero, 1977
 Phenacoccus solenopsis Tinsley, 1898
Pherotilla Brothers, 2015
 Pherotilla rufitincta (Hammer, 1957)
Philomides Haliday, 1862
 Philomides frater Masi, 1927
Philomidinae Ruschka, 1924
Philoplitis Nixon, 1965
Philotrypesini Wiebes, 1966
Philotrypesis Förster, 1878
 Philotrypesis caricae (Linnaeus, 1762)
Phoenoteleia Kieffer, 1916
Phygadeuontinae Förster, 1869
Physaraia Shenefelt, 1978
Phytodietini Hellen, 1915
Phytodietus Gravenhorst, 1829
Pieris Schrank, 1801 [Lepidoptera]
 Pieris brassicae (Linnaeus, 1758)
Pilocutis Vargas, Colombo & Azevedo, 2020
Pimpla Fabricius, 1804
Pimplinae Wesmael, 1845
Pionini Smith & Shenefelt, 1955
Pirenidae Haliday, 1844
Plagiotrochus Mayr, 1881
 Plagiotrochus indochinensis Abe, Ide, Konishi & Ueno 2014
Planitorini van Achterberg, 1995
Plastanoxus Kieffer, 1905
Platygaster Förster, 1856
 Platygaster oryzae Cameron, 1891
 Platygaster foersteri (Gahan, 1919)
Platygastridae Haliday, 1833
Platygastroidea *sensu* Sharkey, 2007
Platylabini Berthoumieu, 1904
Platyplectrus Ferrière, 1941

Platypodinae Shuckard, 1840 [Coleoptera]
Platyscapa Motschoulsky, 1863
Platytetracampini Bouček, 1988
Playaspalangia Yoshimoto, 1976
Platyscelio Kieffer, 1905
Plectiscidea Viereck, 1914
Plectiscus Gravenhorst, 1829
Plectochorus Uchida, 1933
Pleistodontes Saunders, 1882
Plesiocoelus van Achterberg, 1990
Plumariidae Bischoff, 1914
Plutella Schrank, 1802 [Lepidoptera]
    Plutella xylostella (Linnaeus, 1758)
Pochazia Amyot & Audinet-Serville, 1843 [Hemiptera]
    Pochazia shantungensis (Chou & Lu, 1977)
Podagriomicron Narendran and Merry, 2010
Podagrion Spinola, 1811
Podagrioninae Ashmead, 1904
Podoschistus Townes, 1957
Poecilogonalos Schulz, 1906
    Poecilogonalos thwaitesi var. *gestroi* Schulz, 1908
Poemenia Holmgren, 1859
Poemeniinae Smith & Shenefelt, 1955
Pogonocharis Heraty, 2002
Polydegmon Foerster, 1862
Polysphincta Gravenhorst, 1829
Pompilidae Latreille, 1804
Ponerinae Lepeletier, 1835
Popillia Serville, 1825 [Coleoptera]
    Popillia japonica Newman, 1841
Praestochrysis Linsenmaier, 1959
    Praestochrysis shanghaiensis (Smith, 1874)
Praini Mackauer, 1961
Praon Haliday, 1833
Pristapenesia Brues, 1933
    Pristapenesia asiatica Azevedo, Xu and Beaver, 2011
Pristaulacus Kieffer, 1900
    Pristaulacus comptipennis Enderlein, 1912
Pristepyris Kieffer, 1905
Pristocerinae Mocsary, 1881
Pristomerus Curtis, 1836
Proanacharis Kovalev, 1996
Prochazia Amyot & Audinet-Serville, 1843 [Hemiptera]
    Prochazia shantungensis Chou & Lu, 1977
Proclitus Förster, 1869
Proconura Dodd, 1915

Proctotrupoidea Latreille, 1802
Proctotrupes Latreille, 1796
Promecidia Lelej, 1996
Promethes Forster, 1869
    Promethes philippinensis Baltazar, 1955
Promuscidea Girault, 1917
    Promuscidea unafasciaventris Girault, 1917
Pronkia van Achterberg, 1990
Propalachia Bouček, 1978
    Propalachia beaveri Bouček, 1978
Propicroscytus Szelényi, 1941
Prorops Waterston, 1923
Prosaspicera Kieffer, 1907
Proscoliinae Rasnitsyn, 1977
Prosevania Kieffer, 1911
Protanaostigma Ferriere, 1929
Proteropinae van Achterberg, 1976
Proterops Wesmael, 1835
    Proterops borneoensis Szépligeti, 1902
    Proterops fumosus Belokobylskij, 1993
Protichneumonini Heinrich, 1934
Psammocharidae Banks, 1947
Pselaphanus Szépligeti, 1902
Pseudencyrtus Ashmead, 1900
Pseudeugalta Ashmead, 1990
Pseudobrachysticha Girault, 1915
Pseudococcidae Heymons, 1915
Pseudococcus Westwood, 1840 [Hemiptera]
    Pseudococcus jackbeardsleyi Gimpel & Miller, 1996
Pseudodryinus Olmi, 1991
    Pseudodryinus beckeri Olmi, 1991
    Pseudodryinus piceus Olmi, 2011
    Pseudodryinus sinensis Olmi, 1993
Pseudofoenus Kieffer, 1902
Pseudognamptodon Fischer, 1964
Pseudoligosita Girault, 1913
Pseudomegischus van Achterberg, 2002
Pseudonomadina Yamane & Kojima, 1982
Pseudophanerotomini Zettel, 1990
Pseudophanomeris Belokobylskij, 1984
Pseudophotopsidinae Bischoff, 1920
Pseudorhyssa Merrill, 1915
Pseudoshirakia van Achterberg, 1983
Psilini Hellén, 1963
Psilocera Walker, 1833
Psilochalcis Kieffer, 1905
Psilocharis Heraty, 1994
    Psilocharis hypena Heraty, 1994
Psilini Hellén, 1963
Psilus Panzer, 1801

Psix Kozlov & Lê, 1976
Psychidae Boisduval, 1828 [Lepidoptera]
Psyllidae Latreille, 1807 [Hemiptera]
Psyttalia Walker, 1860
Pteromalidae Dalman, 1820
Pteromalus Swederus, 1795
Pteroptrix Westwood, 1833
Pterosemopsis Girault, 1917
Pycnetron Gahan, 1925
Pyralidae Latreille, 1809 [Lepidoptera]
Pyrrhocoroidea Amyot & Serville, 1843 [Hemiptera]
Queequeg Wahl & Sime in Sime & Wahl, 2002
Quercus (*Cyclobalanopsis*) (Oersted) Schneider [Plantae]
    Quercus griffithii Hook.f. & Thomson ex Miq. [Plantae]
Ralstonia Yabuuchi *et al.* 1996 [Bacteria]
Rhaconotini Fahringer, 1928
Rhadinoscelidia Kimsey, 1988
    Rhadinoscelidia lixa Mita, 2020
Rhagigaster Guérin-Méneville, 1838
Rhipipalloidea Girault, 1934
Rhopalomutillinae Schuster, 1949
Rhopalosomatidae Brues, 1922
Rhorus Förster, 1869
Rhynchoticida Bouček, 1978
Rhysipolinae Belokobylskij, 1984
Rhysipolis Foerster, 1862
    Rhysipolis parnarae Belokobylskij & Vu, 1988
    Rhysiolis taiwanicus Belokobylskij, 1988
Rhyssa Gravenhorst, 1829
    Rhyssa persuasoria (Linnaeus, 1758)
Rhyssalinae Forster, 1862
Rhyssalus Haliday, 1833
Rhyssinae Morley, 1913
Rhytidaphora. Reshchikov & Quicke, 2022
Rhyzopertha Stephens, 1830 [Coleoptera]
    Rhyzopertha dominica (F., 1792)
Ricaniidae Amyot & Audinet-Serville, 1843 [Hemiptera]
Rileyinae Ashmead, 1904
Rivasia Askew and Nieves-Aldrey, 2005
Robertsia Bouček, 1988
Rogadinae Foerster, 1862
Rogadini Förster, 1862
Ropronia Provancher, 1886
    Ropronia malaisei Heqvist, 1959
Roproniidae Viereck, 1916
Rothneyia Cameron, 1897
Saccharissa Kirby, 1886

Sachtlebenia Townes, 1963
Saengella Kim and Heraty, 2012
Saphonecrus Dalla Torre & Kieffer 1910
Sapyga Latreille, 1796
    Sapyga luteomaculata Pic, 1920
Sapygidae Latreille, 1809
Sarcophagidae Macquart, 1834 [Diptera]
Scaphepyris Kieffer, 1904
Scapopria Kieffer, 1913
Scarabaeidae Latreille, 1802 [Coleoptera]
Scelio Latreille, 1805
    Scelio pembertoni Timberlake, 1932
Scelionidae Haliday, 1839
Sceliotrachelinae Brues, 1908
Schlettererius Ashmead, 1900
    Schlettererius cinctipes (Cresson, 1880)
Schizaspidia Westwood, 1835
    Schizaspidia aenea (Girault, 1913)
    Schizaspidia diacamma Heraty, Mottern & Peters, 2015
    Schizaspidia nasua Walker, 1846
Schizoprymnus Förster, 1862
Schizopyga Gravenhorst, 1829
Schoenlandella Cameron, 1905
Schoenobius Duponchel, 1836 [Lepidoptera]
Scirpophaga Treitschke, 1832
Scleroderminae Kieffer, 1914 [Lepidoptera]
Sclerogibba Riggio & De Stefani-Perez, 1888
    Sclerogibba africana (Kieffer, 1904)
    Sclerogibba berlandi Benoit, 1963
    Sclerogibba impressa Olmi, 2005
    Sclerogibba madegassa Benoit, 1952
    Sclerogibba rapax Olmi, 2005
    Sclerogibba rossi Olmi, 2005
    Sclerogibba talpiformis Benoit, 1950
Sclerogibbidae Ashmead, 1902
Scolebythidae Evans, 1963
Scolobatini Schmiedeknecht, 1911
Scolebythus Evans, 1963
Scolia Fabricius, 1775
Scoliidae Latreille, 1802
Scudderopsis Bennett, 2012
Semiotellus Westwood, 1839
Senwot Wharton, 1983
Separatatus Chen & Wu, 1994
Sericopimpla Kriechbaumer, 1895
Serphidae Kieffer, 1907
Serphus Stål, 1862
Sesbania Scop. [Plantae]
    Sesbania grandiflora (L.) Poiret
Seticornuta Morley, 1913
Shawiana van Achterberg, 1983

Shelfordia Cameron, 1902
Shorea Roxb. ex C.F.Gaertn [Plantae]
Sierolomorphidae Brues & Melander, 1932
Sigalphinae Haliday, 1833
Sigalphus Latreille, 1802
Signiphora Ashmead, 1880
    Signiphora bifasciata Ashmead, 1900
Signiphoridae Howard, 1894
Simplicibracon Quicke, 1988
Sinarachna Townes, 1960
Sinotilla Lelej, 1995
    Sinotilla boheana (Chen, 1957)
    Sinotilla calopoda Okayasu, 2017
Siphimedia Cameron, 1902
    Siphimedia nigriscuta Sheng and Sun, 2010
Sirex Linnaeus, 1760
    Sirex noctilio Fabricius, 1793
Sisyrostolinae Seyrig, 1932
Sitophilus Schoenherr, 1838 [Coleoptera]
    Sitophilus granaries (Linnaeus, 1758)
    Sitophilus oryzae (Linnaeus, 1763)
    Sitophilus zeamais Motschulsky, 1855
Skiapus Morley, 1917
Slonopotamus Khalaim, 2011
Smicromorpha Girault, 1913
Smicromyrme Thomson, 1870
    Smicromyrme dardanus (Smith, 1857)
Sogatella Fennah, 1956 [Hemiptera]
Solenofigites Díaz, 1979
Solenura Westwood, 1868
    Solenura ania (Walker, 1846)
Spalangia Latreille, 1805
Spalangiidae Haliday, 1833
Sparasion Latreille, 1802
Sparasionidae Dahlbom, 1858
Sparasionini (Dahlbom, 1858)
Spathius Nees, 1818
Sphaeropthalminae Latreille, 1825
Sphecidae (Latreille, 1802)
Sphecophaga Westwood, 1840
    Sphecophaga orientalis Donovan, 2002
Sphegigaster Spinola, 1811
Sphegigastrinae Thomson, 1876
Sphinctus Gravenhorst, 1829
    Sphinctus chinensis Uchida, 1930
    Sphinctus serotinus Gravenhorst, 1829
Spilomicrini Ashmead 1893
Spilomicrus Westwood, 1832
    Spilomicrus wallacei (Dodd, 1920)
Spilopteron Townes, 1965
Spinaria Brullé, 1846

Spinariella Szépligeti, 1906
Spinariina van Achterberg, 1988
Spodoptera Guenée, 1852 [Lepidoptera]
    Spodoptera frugiperda (Smith, 1797)
Standfussidia Lelej, 2005
Stantonia Ashmead, 1904
Staphylinidae Latreille, 1802 [Coleoptera]
Stauropoctonus Brauns, 1889
    Stauorpoctonus torresi Gauld, 1977
    Stauropoctonus townesorum Gauld & Mitchell, 1978
Stenarella Szépligeti, 1916
    Stenarella insidiator (Smith 1859)
Steninvreia Bouček, 1988
Stenobracon Szépligeti, 1901
Stenomacrus Förster, 1869
Stephanoidea Leach, 1815
Stephanus Jurine, 1807
    Stephanus borneensis (de Saussure, 1901)
    Stephanus serrator (Fabricius, 1798)
    Stephanus soror van Achterberg, 2002
Sternaulopius Fischer, 1965
Sthulapada Narendran, 1989
Stictolissonota Cameron, 1907
Stictopisthus Thomson, 1886
Stilbopinae Townes & Townes, 1949
Stilbops Förster, 1869
    Stilbops acicularis Kasparyan, 1998
    Stilbops gorokhovi Kasparyan, 1999
Stilbula Spinola, 1811
Stiropiini van Achterberg, 1993
Storeya Bouček, 1988
Storeyinae Bouček, 1988
Storozhenkotilla Lelej, 2005
Stratiomyiidae Latreille, 1802 [Diptera]
Streblocera Westwood, 1833
Strongylopsis Brauns, 1896
Stylopria Kieffer, 1914
Sussaba Cameron, 1909
    Sussaba sugiharai (Uchida, 1957)
Sychnostigma Baltazar, 1961
Sycobia Walker, 1871
Sycoecinae Hill, 1967
Sycophaginae Walker, 1975
Sycophila Walker, 1871
Sycoryctinae Wiebes, 1966
Sycophaga Westwood, 1840
Sycoscapter Saunders, 1883
Symplecis Gravenhorst, 1829
Symplocaceae Jacq. [Plantae]
Symplocos Jacq. [Plantae]
    Symplocos stellaris Brand

Synacra Förster, 1856
    Synacra paupera Macek, 1995
Synergini Ashmead, 1896
Synopeas Förster, 1856
Syntomernus Enderlein, 1920
Syntomopus Walker, 1833
Syrphidae Latreille, 1802 [Diptera]
Systasidae Bouček, 1988
Syzeuctus Förster, 1868
Syzygium P. Browne ex Gaertn [Plantae]
    Syzygium samarangense (Blume) Merr. & L.M.Perry
Tachinidae Bigot, 1853 [Diptera]
Tachynomyia Guérin-Méneville, 1842
Tachyphron Brown, 1995
Taeniogonalos Schultz, 1906
    Taeniogonalos fasciata (Strand, 1913)
Taimyrmosa Lelej, 2005
    Taimyrmosa nigrofasciata (Yasumatsu, 1931)
Tainaniella Masi, 1929
Tainitermini van Achterberg, 2001
Takyomyia Kimsey, 1996
Tamdaona Belokobylskij, 1993
Tamdaonini van Achterberg, 2020
Tanaodytes Masner, 1972
Tanaostigmatidae Ashmead, 1904
Tanycoryphus Cameron, 1905
Tanythorax Gibson, 1989
Tanyxiphium Huber, 2015
    Tanyxiphium harriet Zeya, 2015
Tatogaster Townes, 1971
Tatogastrinae Wahl, 1990
Tebennotoma Enderlein, 1912
Teleasinae Walker, Ashmead, 1902
Telegaiinae Tobias, 1962
Telenomus Haliday, 1833
    Telenomus remus Nixon, 1937
Teleutaea Förster, 1869
Temelucha Förster, 1869
Tersilochinae Schmiedeknecht, 1910
Tenebrionidae Latreille, 1802 [Coleoptera]
Tersilochus Holmgren, 1859
Tessaratoma Berthold, 1827 [Hemiptera]
Testudobracon Quicke, 1986
Tetracampidae Förster, 1856
Tetrastichinae Graham, 1987
Tettigoniidae Krauss, 1902 [Orthoptera]
Thaumatodryininae Perkins, 1905
Thaumatodryinus Perkins, 1905
    Thaumatodryinus alienus Olmi, 1987
    Thaumatodryinus asiaticus Olmi, 1984

    Thaumatodryinus beckeri Olmi, 1991
    Thaumatodryinus malayanus Olmi, 1984
    Thaumatodryinus noyesi Olmi, 1992
    Thaumatodryinus pasohensis Olmi, 1991
    Thaumatodryinus philippinus Olmi, 1984
    Thaumatodryinus sharkeyi Olmi, 2011
Theocolax Westwood, 1874
    Theocolax elegans Westwood, 1874
Theronia Holmgren, 1859
Therophilus Wesmael, 1837
Thoron Haliday, 1833
Thresiaella Narendran, 1989
Thrybius Townes, 1965
Thynnidea Shuckard, 1841
Thynnus Fabricius, 1908
Thyreodon Brullé, 1846
Thyreodonini Rousse, Quicke, Matthee, Lefeuvre, and Van Noort, 2016
Thysanus Duncan, 1863
Ticoplinae Nagy, 1970
Timberlakiella Compere, 1936
Tinea Linnaeus, 1758 [Lepidoptera]
Tineidae Latreille, 1810 [Lepidoptera]
Tiphiidae Leach, 1815
Tiphodytes Bradley, 1902
Tipula Linnaeus, 1758 [Diptera]
    Tipula irrorata Macquart, 1826
Tirathaba Walker, 1864 [Lepidoptera]
Tissahamia Huber, 2018
    Tissahamia gombak (Huber, 2011)
Tobiason Belokobylskij, 2004
Tobleronius Fernandez-Triana & Boudreault, 2018
Tomicobomorpha Girault, 1915
Torymidae Walker, 1833
Torymoides Walker, 1871
Torymus Dalman, 1820
    Torymus kovaci Narendran and Girish Kumar, 2005
Tossinola Viktorov, 1958
Townesion Kasparyan, 1993
Townesioninae Kasparyan, 1993
Toxoneuron (Say, 1836)
    Toxoneuron nigriceps Viereck
Toya Distant, 1906 [Hemiptera]
Trachypetidae Schulz, 1911
Trachyserphus Kolyada, 2017
Trachysphyrus Haliday, 1836
Trassedia Dessart, 1975
    Trassedia yanegai Miko and Trietsch, 2018
Trathala Cameron, 1899

Triancyra Baltazar, 1961
Triaspis Haliday, 1838
Triatoma Laporte, 1832 [Hemiptera]
    Triatoma rubrofasciata (De Geer, 1773)
Trichoferus Wollaston, 1854 [Coleoptera]
    Trichoferus campestris (Faldermann, 1835)
Trichogramma Westwood, 1833
    Trichogramma dendrolimi Matsumura, 1926
Trichogrammatidae Haliday, 1851
Trichogrammatoidea Girault, 1911
Trichomalopsis Crawford, 1913
Trichoplusia McDunnough, 1944
    Trichoplusia ni (Hübner, 1800–1803)
Trichopria Ashmead, 1893
    Trichopria clavatipes (Kieffer, 1911)
    Trichopria drosophilae (Perkins, 1910)
Trichrysis Lichtenstein, 1876
Triclistus Förster, 1869
    Triclistus aitkeni (Cameron, 1897)
Trigastrotheca Cameron, 1906
Trigonalidae Cresson, 1887
Trigonalyoidea Cresson, 1887
Trigonoderinae Bouček, 1964
Trigonoderus Westwood, 1832
Trigonura Sichel, 1866
Trimorus Förster, 1856
Trioxini Ashmead, 1901
Trioxys Haliday, 1833
Triraphis Ruthe, 1855
Trisecodes Delvare and LaSalle, 2000
Trispinaria Quicke, 1986
Trissolcus Ashmead, 1893
    Trissolcus basalis (Wollaston, 1858)
    Trissolcus japonicus (Ashmead, 1904)
Triteleia Kieffer, 1906
Tropobraconini Quicke *et al.*, 2023
Trogaspidia Ashmead, 1899
    Trogaspidia doricha (Smith, 1860)
Trogini Foerster, 1868
Tromatobia Förster, 1869
    Tromatobia hutacharerni Kusigemati, 1988
Tropimeris Steffan, 1948
Tropobracon Cameron, 1905
Troticus Brullé, 1846
Tryphoninae Shuckard, 1840
Typhoctinae Schuster, 1949
Ufens Girault, 1911
Uga Girault, 1930
Ungunicus Fernandez-Triana & Boudreault, 2018

Uniclypea Bouček, 1976
Urosigalphus Ashmead, 1889
Ussurohelcon Belokobylskij, 1989
Utetes Foerster, 1863
Vanhornia Crawford, 1909
    Vanhornia eucnemidarum Crawford, 1909
    Vanhornia leileri Hedqvist, 1976
    Vanhornia quizhouensis (He & Chu, 1990)
    Vanhornia yurii Timokhov & Belokobylskij, 2020
Vanhorniidae Crawford, 1909
Vespoidea Laicharting, 1781
Vespula Thomson, 1869
    Vespula germanica (Fabricius, 1793)
    Vespula vulgaris (Linnaeus, 1758)
Vidlinus Herdty, 2010
Vietcolastes Belokobylskij, 1992
Vipio Latreille, 1804
Walkerella Westwood, 1883
Wallaceaphytis Polaszek and Fusu, 2014
Waterstoniella Grandi, 1921
Watshamia Bouček, 1974
Watshamiella Wiebes, 1981
Westwoodiini Townes, 1970
Wilkinsonellus Mason, 1981
Wolbachia Hertig, 1936 [bacteria]
Wroughtonia Cameron, 1899
Wuda Cheesman, 1936
Xanthopimpla Saussure, 1892
    Xanthopimpla stemmator (Thunberg, 1822)
Xenomerus Walker, 1836
Xiphozele Cameron, 1906
Xiphozelinae van Achterberg, 1979
Xiphydriidae Leach, 1815 (Hymenoptera: 'Symphyta')
Xiphypronia He & Chen, 1991
Xoridinae Shuckard, 1840
Xyelidae Newman, 1834
Xyalaspis Hartig, 1843
Xyalophora Kieffer, 1901
Xylocopa Latreille, 1802
Xylophrurus Förster, 1869
Ycaploca Nagy, 1975
Yelicones Cameron, 1887
Yeliconini van Achterberg, 1991
Yeppoona Gauld, 1984
Yezoceryx Uchida, 1928
    Yezoceryx shengi Pham, Broad, Matsumoto, van Achterberg, 2017
Ypsolophidae Guenée, 1845 [Lepidoptera]

Zachterbergius Fernandez-Triana & Boudreault, 2018
Zaglyptogastra Ashmead, 1900
    Zaglyptogastra plumiseta Enderlein, 1920
Zagrammosoma Ashmead, 1904
Zaischnopsis Ashmead, 1904
Zatypota Förster, 1869
    Zatypota albicoxa (Walker, 1874)
    Zatypota yambar Matsumoto, 2011
Zavatilla Tsuneki, 1993

Zealaromma Gibson, Read and Huber, 2007
Zebe La Salle, 2005
Zelodia van Achterberg & Long, 2010
Zelomorpha Ashmead, 1900
Zeugomutilla Chen, 1957
Zolotarewskya Risbec, 1956
Zombrus Marshall, 1897
Zosteragathis Sharkey, 2017
Zygaenidae Latreille, 1809 [Lepidoptera]

Index

1rs-m 68

Ablerus clisiocampae 350
Ablerus sp. 350–351, *352*
abscissa 68
Acaenitellus sp. 189
Acaenitinae 157, 160, 191–193
Acaenitus dubitator 192
Acampsohelconinae 100, 118
Acanthocoris scaber 298
Acanthoplus discocidalis 293
Aceratapsis sp. *183*
acetabulum *see* epicnemium
Aclista sp. 329
Acmopolynema sp. 373, *374*
Acrias varicornis 365
Acroclisoides luzonensis 385
Acroclisoides sinicus 385
Acroclisoides sp. 385
Acrodactyla sp. 201
Acropimpla sp. 202
acropleuron *see* subalar region
Aculeata 47, 83–84
aculeate wasps 81
Adelencyrtus sp. *356*
Adeliinae 128
adhesives and gluing techniques 425–426
Aenasius arizonensis 357
aerial and butterfly nets 411–412
Afromegischus gigas 220
Afrostilba sp. *311*, 314
Aganaspis daci 314
Aganaspis sp. 314
Agaonidae (fig wasps) 344, 346–349
  key to 346
agastoparasitism *see* cleptoparasitism

Agathidinae *97*, 98, 101, 116–118, *117*
Ageniaspis sp. 354
Agriotypinae 167
agroforestry 3
Agrothereutes group 168
Agupta sp. *131*
Aleiodes 38, 109–110
Aleiodini 110
Aleurodicus rugioperculatus 357
alitrunk *see* mesosoma
Allanagrus sp. 374
Allobethylus sp. *251*, 252
Allophrys sp. 188
Alloxysta sp. *311*, 313
Alomyinae 167
Alysiinae *97*,112–113, *112*
alysioid subcomplex 111–115
Amauromorpha sp. 169
Ambocybe petiolate 365
Ambositrinae 328, 329
Amiseginae 253, *254*
Amitus sp. 293
Ampulicomorpha collinsi 263
Ampulicomorpha sp. 263
amyl acetate 423
Amylostereum sp. 315
Amyosoma 103
Anacharatinae 309
Anacharis sp. 311
Anacharitinae 311–312
Anagrus nilaparvatae 373
Anagrus optibalis 373
Anagyrus lopezi 4, 356
anal lobe 69
Anaphes kailashchandrai 374

Anaphes sp. 374
Anastatus mantoidae 368
Anastatus sp. 368
Anastrepha spp. 314
ancestral biology 130
Ancylotropus montanus 341, 359
Andricus mukaigawae 309
Anicetus sp. 355
Anisopteromalus calandrae 5, 386–387
Anomaloninae 157, 158, 173–174, *174*
Anomalon sp. 173–174, *174*
anomalous diversity 40–41
Anselmella malacia 366
antennae 58
antennal hammers 25
Anteon sp. 260
anterior subalar depression 95
Antrocephalus decipiens 353
Apanteles sensu lato 130–131
Aphanogmus hakonensis 235
Aphanogmus sp. 235
Aphanomerus sp. 294
Aphelinidae 345
     Calesidae and 349–350
Aphelinus sp. 313
Aphelopus sp. 259, 262
Aphidiinae 101–103
Aphidiinae–Ephedrini 100
aphidioid subcomplex 101–103
Aphytis chionaspis 350
apical wing membrane 272
Apiomorpha frenchi 380
Aplorhinus sp. 353
Apocharips sp. 313
Apocrita 47, 48
Apocrypta sp. 386, 387
Apoderus sp. 187
Apoidea 82
Apophua sp. 175, *176*
Aporus sp. 278
Aprostocetus sp. 341, 364, 366
apterous *Gelis* sp. 173
Aptesini 168
Arachnospila sp. 278
Archaphidius greenideae 102
Area de Conservation Guanacaste (Costa Rica) 39
areolet 152–154, 158, 167, 181, 197
Arhytis sp. 170
Ariphron sp, 281
Aroplectrus dimerus 365
arrhenotoky 15
Arthula sp. 170
Asaphesidae (Chalcidoidea *incertae sedis*) 378
Asaphes sp. 378
Aspicera sp. 312
Aspicerinae 309, 312–313
Aspidimorpha miliarus 391

Aspidimorpha sp. 391
associative learning 24
Astiphromma sp. 181, *182*
Astomaspis sp. 173
Astymachus sp. 355
Ateleutinae 167–168, *168*
Atoposega sp. 254
Atopotrophos sp. 190
Atrophini 175
augmentative biocontrol 4
Aulacaspis tubercularis 350
Aulacaspis yasumatsui 235, 391
Aulacidae 225–226
Aulacocentrum philippinense 121
Australolaelaps sp. 379
Austrocardiochiles sp. *127*
AXA method, of chemical drying 423
Axemopsis sp. 369
axillae 362, 364
axullae 59
Azotidae 350–351

Bactrocera dorsalis 314
Bactrocera sp. 353
Baeini 295, 297
Baeus sp. 297
Bairamlia sp. 378, 385
Bakeronymus sp. 245
Balcha sp. 367.
Balimer *et al.* (2023) 48–50
Banchinae 159, 160, 174–175, *176*
Banchini 175
Banchus sp. 175
barcode index numbers (BINs) 39
Bareogonalos sp. 245
Barronia sp. 161
Barycnemis sp. 188
basal coxa 61
Basalys sp. 330–331
Bates, Henry 5
behaviour 21
     host location and assessment
          and 23–26
     sex, courtship, and mating and 21–23
Belisana khaosok 297
Belyta sp. 331
Belytinae 328, 329–330
Bephrata sp. 369
Bethylidae (flat wasps) 85, 248–252, *250*
     key to subfamilies of 249
Bethylinae 249, 250
Betylobraconini 110
biological control, types of 3–4
biology 9
     developmental features and 12–15
     fig ecosystem and 17–18

host manipulation by venoms and viruses
  and 16–17
idiobiont/koinobiont spectrum and 10–12
life history strategies and 9–10
sex determination and mating system and 15–16
tropical parasitoid diversity and 39–41
Bischoffitilla lamellate 274
Bischoffitilla sp. 274, 277
Blastophaga psenes (Agaonidae) 348
Blastophaga sp. 347, 348
Bocchus sp. 260
Bootania sp. 371
Bootanomyia sp. 371, 372
Brachistinae 99, 100, 118–120, *120*
Brachycyrtiformes 161–164
Brachycyrtinae 158, 161
Brachycyrtus hawaii 161
Brachycyrtus sp. 164
Brachygaster sp. 227, *228*
Brachymeria sp. 352, *353*
Brachymeriinae 352
Brachyscleroma apoderi 187
Brachyscleroma sp. *188*
Brachyserphus sp. 322
Bracon genus 103
Braconidae 30, 87, 92
  cyclostomes and 101–115
  morphological terminology of 94–96
  non-cyclostomes and 115–133
  phylogenetic studies of 92–94
  subfamily identification of 96–101
  wing of 96
braconids 2, 25
Braconinae 98, 103–104
bracoviruses 17
Bradynobaenidae (extralimital) 279
BugDorm design, of Malaise traps *415*
Bupon sp. 254

Caenoscleroglibba longiceps 264
Caenoscleroglibba rossi 264
Caenoscleroglibba sp. 264
Calesidae 345
  Aphelinidae and 349–350
Cales noacki 350
Cales sp. 350
Callajoppa group 171
Callaspidia sp. 311
Callimomoides ovivorus 381
Callimomoides sp. 381
Calliscelio elegans 298
Calliscelio sp. *296*, 298
Callocleonomus sp. 379
Calosotinae 367–368
Calyoza sp. 251
Cambodia 33

Cameron, Peter 6
Campoletis chlorideae 177
Campopleginae 159, 175–177
Campsomeris sp. 280
Camptopteroides sp. 373
Camptotypus sp. 201, *202*
Canadian National Collection of Insects 423
Canalicephalus spp. 119
canopy fogging 32, 419–420
Carabunia sp. *356*
Cardiochilinae 101, 126–128
card points and rectangles 425
Cassidocida aspidomorphae 391
Cassidocida sp. 391
Cedriini 106–107
Cenocoliinae 100, 123
Centistidea sp. 132
Centrodora sp. *350*, 351
Cephaleta sp. 381
Cephalonomia tarsalis 249
Cephalonomia waterstoni 249
Ceraphron (*Allomicrops*) sp. 236
Ceraphronidae 235–236
Ceraphron manila 236
Ceraphronoidea 47, 85, 233–235
  Ceraphronidae and 235–236
  keys to 235
  Megaspilidae 237
Ceraphron sp. 236
Ceratobaeus sp. 297
Ceratobaeus-type ovipositor system 296
Ceratomansa group 169
Ceratosolen pilipes 349
Ceratosolen sp. 347, *348*, 349
Cerocephala sp. 378
Cerocephalidae 378
Cerrena sp. 315
Certonotus sp. *172*
Chalcedectidae 378
Chalcedectus sp. 378
Chalcididae 344, 351–354
Chalcidinae 352
Chalcidoidea 30, 31, 59, 70, 84, 86, 335–336,
  339–340
  Agaonidae (fig wasps) and 346–349
  antenna of 338
  Aphelinidae (and Calesidae) and 349–350
  Azotidae and 350–351
  Chalcididae and 351–354
  Chrysolampidae and 354
  Cynipencyrtidae and 354
  Encyrtidae and 354–357
  Eucharitidae and 358–360
  Eulophidae and 360–367
  Eupelmidae and *Eopelma* and 367–370
  Eurytomidae and 370
  families of 337–338

Chalcidoidea (continued)
    family key to female fully winged species of 343–346
    Former Pteromalidae and 376–390
    identification of 342
    Leucospidae and 370–371
    Megastigmidae and 371
    Metapelmatidae and 371–372
    metasoma of 342
    morphological terminology of 336
    Mymaridae (fairy flies) and 372–374
    Mymarommatoidea (false fairy wasps) and 407–408
    Neanastatidae and 374–375
    Ormyridae and 375
    Perilampidae and 375–376
    Signiphoridae and 390–391
    Tanaostigmatidae and 391
    Tetracampidae and 391
    thoracic morphology of 338, 341–342
    Torymidae and 392–393
    Trichogrammatidae and 393–395
chalcidoids 2, 70
Chalcis sp. 352
Charipinae 309, 313
charmontinae 120–121
Charops sp. 177
Chartocerus javensis 390
Chartocerus sp. 391
Cheiloneurus nankingensis 357
Cheiloneurus sp. 355, 357
Cheloninae 100, 101, 128, *129*
Chelonus formosanus 128
chemical cues 24
chemical drying 422, 423
Chiasmia sp. 366
Chilo 103
Chlorocryptus purpuratus 170
Chlorocryptus sp. 170
Chriodes sp. 184, 185
Chrysididae 248, 252–258
    key to subfamilies of 253
Chrysidinae 253–256
Chrysidoidea 84, 85, 247
    Bethylidae (flat wasps) 248–252
    Chrysididae 252–258
    Dryinidae (pincer wasps) 258–263
    Embolemidae 263–264
    key to families of 248
    Sclerogibbidae 264–265
    Scolebythidae 265
Chrysis sp. 255
Chrysobothris succedanea 382
Chrysochalcissa sp. 392
Chrysolampidae 354
Chrysolampus sp. 354
Cibdela janthina 133

Cidaphus sp. 181
Cirrospilus sp. 367
Cisaris sp. 173
Cladobethylus sp. 253, *254*
classical biological control 3–4
classification and phylogeny 46–47
    Hymenoptera phylogeny research and 47–50
    simplified total-evidence tree 50
Clatha sp. 174
claval lobe 69
claws 62, 116, *120*, 158, 159, 160, 181, *182*, 190, *191*, 196–198, *200*, 256, 259, *260*
Cleonomus sp. 379
Cleonymidae 378–379
Cleptes asianus 256
Cleptes humeralis 256
Cleptes sp. 256
Cleptes thaiensis 256
Cleptidea sp. 256
Cleptinae 253, 256
cleptoparasitism 310
Clinocentrini 110
Closterocerus sp. 363
Clypeodromus sp. 171
clypeus 57, 94–95, 98, 158–160, *166*, *177*, 178, 181, 182, 195–197
Clytocera chinospila 382
Cnaphalocrocis medinalis 183
Coccophagus sp. 349
Coelopencyrtus sp. 355
Coelopencyrtus sp. 355
collecting permit problem 33–34
collection 410–411
    lesser-used methods of 420
    prominent methods of 411–420
Colotrechninae 384
Colotrechnus sp. 384
Colpa sp. 281
Comperiella bifasciata 356
complementary sex determination (CSD) 15, 21
Comstock–Needham system 66, 67
concealed versus exposed hosts 11
concurrent host feeding 11
Conopomorpha cramerella 357
Conostigmus sp. 236
Convention on Biological Diversity (CBD) and permitting issues 32–36
Copidosoma floridanum 12
Copidosoma sp. 354, 355
Copidosomyia ambiguous 357
Coptera haywardi 331
Coptera sp. 331, *332*
Coraebus cavifrons 372
Corcyra sp. 353
Cosmocomoidea sp. 373, 374
Cosmocomopsis sp. 373
Cosmophorus sp. 125, *126*

Costa Rica 39
Cotesia glomerata 131
coxae 61, 83, 227
Cratocentrinae 352
Cratocentrus sp. 352
Cratolabus sp. 172
Cremastinae 159, 177–178
Crinibracon 103
critical point drying (CPD) 422–423
cross-veins 67, 68
Crovettia sp. 262
Cryptinae 168–170, *170*
Cryptini 168
Cryptoserphus sp. 322
Cryptus group 169
Ctenopelatinae *180*
Ctenopelmatinae 159, 178–180
culturing, of parasitoids 429
Cyanotiphia sp. 283
Cybocephalus nipponicus 235
cyclostomes 94–95
    alysioid subcomplex 111–115
    aphidioid subcomplex 101–103
    mainline 103–111
Cylloceria melancholica 193
Cylloceria sp. 193
Cylloceriinae 193
Cynipencyrtidae 354
Cynipencyrtus sp. 354
Cynipidae *307*, 309–310
Cynipini 309–310
Cynipoidea (gall wasps) 86, 87, 306
    Cynipidae and 309–310
    Figitidae and 310–314
    Ibaliidae and 314–315
    key to major groups of 308–309
    Liopteridae and 315
    morphological terminology of 306–307
    phylogeny of 308
Cyphacolus sp. 297
Cyrtorhyssa moellerii 22, 204

Dacnusini 112
Dallatorrella sp. 315
Dallatorrellinae 315
Darna pallivitta 365
deeply concealed hosts, locating 24–25
deforestation 30
Deilagion sp. 348
Delomeristini 197, 199
Dendrocerus carpenter 237
Dendrocerus sp. *236*, 237
Dendrolimus sp. 368
Derris elliptica 391
destructive host feeding 11
Diacamma scalpratum 358

Diachasmimorpha longicaudata 13, 114
Diadegma sp. 177
Diaparsis sp. 188
Diapriidae 327, 328–332, *331, 332*
    key to subfamilies of 328
Diapriinae 328, 330–331, *332*
    key to tribes of 330
Diapriini 331
Diaprioidea 84, 85, 327
    Diapriidae and 328–332
    Ismaridae and 332
    key to families of 327
    Monomachidae and 332–333
Diaziella bizzarea 387
Diaziella sp. 385, 386
Dicamptus sp. 186
Dicroscelio sp. 296
Dictyomotus sp. 186
Digonogastra sp. 429
Dilyta sp. 313
Dinocampus coccinellae 125
Dinocryptus niger 169
Dinumma sp. 366
Diolcogaster sp. 131
Diomorus aiolomorphi 393
Diparidae 379, *380*
Diplazon laetatorius 194
Diplazontinae 160, 193–194, *194*
diptera 80
direct pinning method 424
Dirhininae 352–353
Dirhinus sp. 353
Dirrhopinae 128
discrimen 61
dispersal and phoretic copulation 23
Ditropinotella sp. 379–380
Ditropinotellinae (Chalcidoidea *incertae sedis*) 379–380
diversity, in S.E. Asia 29–30
    Convention on Biological Diversity (CBD) and permitting issues and 32–36
    DNA barcoding and 36–39
    surveying 30–32
    tropical, and biology 39–41
DNA barcoding 36–39
Doddifoenus wallacei 383
dorsal carina 95
dorsope 96
Doryctinae 98, *99*, 104–105, *106*
Drepanoctonus spp. 184
Drosophila melanogaster 22
Dryinidae 248
Dryinus sp. *260*
Dryinidae (pincer wasps) 258–263
Dryocosmus okajimai 309, 310
Dryocosmus sp. *310*
D-Vac 421
Dvivarnus agamades 299

Eccinetus sp. 329
echolocation 25–26, 169
Echthromorpha sp. 197, *200*
Ecnomius sp. 125
ecological role and importance
    pest management utility and 3–5
ectoparasitoids 9–10
egg parasitism 14–15
Eileeniella sp. 314
Elaeis guineensis 32
Elasmus sp. 360, *363*, 384
Eleonoria 122
Elysoceraphron sp. *236*
Emargininae 308, 313
Embidobiini 295
Embolemidae 248, 263–264
Embolemus krombeini 263
Embolemus pecki 263
Embolemus sp. 263
Encarsia flavoscutellum 350
Encarsia sp. 350, *351*
Encyrtidae 344, 354–357, *356*
Encyrtinae 357
Encyrtoscelio sp. 298
Encyrtus lecaniorum 355
Enderlein, Günther 6
endoparasitoids 9, 10, 14
Enicospilini 186
Enicospilus sp. 185, 186
Entedoninae 362, *363*, 365
Entiinae 363, 365
Entomacis sp. *331*
entomological pin *426*
entomophytophagy 336
Eopelma sp. 369–370, 374
Ephialtini 197, 199–202, *202*
Epichrysomallidae *see* Former Pteromalidae
epicnemial carina 95, 151
epicnemium 61
Epirhyssa sp. 204, *205*
episternal scrobe 152
Epitraninae 353
Epitranis sp. *353*
Epitranus sp. 353
epomia 150–151, *152*
Epyrinae 249, 250–251
Eretmocerus sp. 350
Eriaporidae *see* Former Pteromalidae
Eriborus sp. *177*
Erimyrmosa burmanensis 277
Erotolepsiinae 389
Ettchellsia sp. *240*
Eucalyptus sp. 366, 371
Euceros albitarsus 161
Euceros frigidus 161
Euceros gilvus 164
Euceros pruinosus 161

Euceros sp. 161–164, *165*
Euceros unifasciatus 161
Eucerotinae 157, 160, 161–164
Eucharitidae 341, 345, 357–358, *358*
Eucharitinae 358
Eucoilidae 307
Eucoilinae 309, 313–314
Eucoilini spp. *312*
Euderinae 359, 365
Euderus sp. *363*
Eugalta sp. *203*
Eulophidae 345, 360–367
    key to groups of 361–363
Eulophinae 361, 362, *364*, 365–366, *366*
Eumenes arcuate 199
Eunotidae 380–381
Eunotiscus sp. *384*
Eupelmatinae 367
Eupelmidae 341, 345, 367–369
    key to subfamilies of 367
Eupelminae 368–369
Eupelmus sp. 368
Euperilampus sp. *375*
Euphorinae 100, 101, 123–125
Euphoroid complex 123–125, *126*
Euplectrini 365, *366*
Euplectrus sp. *361*
Eupristina sp. *348*, 386
Eurytoma sp. 371
Eurytomidae 345, 370
Eusterinx sp. 196
Eutrichosomellini (Aphelininae) *351*
Evania appendigaster 227
Evania postfurcalis 228
Evania sp. 227, *228*
Evaniidae (ensign wasps) 226–227
Evanioidea 83, 224
    Aulacidae and 225–226
    Evaniidae (ensign wasps) and 226–227, 228
    Gasteruptiidae (carrot wasps) and 227–230
    keys to families of 225, 227
Exallonyx sp. *324*
Exetastes sp. 175, *176*
Exochus sp. *183*
Exothecinae 99, 113
exposed hosts 11–12

Facitorini 110
female pheromones 22
femur 62
Ficobracon 104
Ficomila sp. *369*
Ficus araneosa 381
Ficus carica 347, 386
Ficus cereicarpa 349, 388
Ficus crassiramea 387

Ficus hirta 386
Ficus ixoroides 349
Ficus macrocarpa 408
fig ecosystem 17–18
Figitidae 308, 309, 310–314, *311, 312*
Figitinae 314
filum spinosum 342, 357
fine forceps 420–421, *422*
final instars 272
Fissicaudus hailandicus 102
flagellomeres 58
    basal 272
flagellum 361
Flavipimpla sp. 200
Flavopimpla sp. 202
Foenatopus sp. 219, *221*, 222
Foenatopus turcomanorum 220
Foenobethylus sp. 131
Foersterella sp. 391
fold and flexion lines 68
fore tibia 98, *99*, 159, *179*, 192
fore wing 65, 66, 67, 69, 84, 85, 100, 157, 158, 227,
    243, 272, 291, 308, 309, 320–321, *321, 322*,
    342, *343*, 362, 372
    vein nomenclature 155
Former Pteromalidae 376–390
former Sycoecinae 385–386
Formicidae 81, 272
Fornicia sp. 131
France 33
free veins 67
frenum and frenal area 342
Friona sp. 170
frons 57
fused veins 67, 68
Fustiserphus intrudens 322

Gabunia group 168
Gahanopsis sp. 373
Galodoxa sp. 252
Gasteruptiidae (carrot wasps) 225, 227–230
Gasteruption sp. 224, 228, *229*
Gastraspis sp. *311*
gastrocoeli and thyridea 155, *156*
Gastrogonatocerus sp. 373
genal carina 57
genitalia preparations, for specimen
    preparation 427, *428*
Gigantochloa scortechinii 393
Gilen orientalis 180
glymmae 155
Glyptini 159, 175
Gnathochorisis sp. 196
Gollumiella sp. 358–359
Gollumiellinae 358–359
Gonatocerus sp. 373

Gonatopus bonaerensis 262
Gonatopus flavifemur 262
Gonatopus javanus 259
Gonatopus nearcticus 259
Gonatypus sp. 263
Grasseiana sp. 386
Gravenhorstiini 174
Greenideidae 102
gregarious development 13
gregariousness 130
group identification, major 79–80
    keys to 80–88
Gryon ancinla 298
Gryon flavipes 298
Gryonini 298, 300
gula 59

Habrobracon brevicornis 103
Habrobracon hebetor 5, 103
Haltichella sp. 353
Haltichellinae 353–354
Halyomorpha halys 299
Hamuli and wing coupling and 63–64
haplodiploid insects 16
Hartemita sp. 127
Hayatosema initiator 360
Hayatosema sp. 359
head 57, *58*
Helconinae 100, 120
Helconoid complex 118–123
Helicoverpa armigera 177
Heliothis virescens 127
Heloridae 320, 321
Helorus sp. 321
Helosepilachna vigintioc-tomaculata 323
Henryana sp. 364
Heptascelio sp. 296, 298
Herbertia indica 381
Herbertiidae 381
herbivore-induced plant volatiles (HIPVs) 23
'Hercules' Insect glue 425
Heresiarches sp. 171
Heteribalia confluens 314
heteronomy 350
Heterospilus 104
Hevea brasiliensis 31–32
hexamethylsilazane (HMDS) 423
Heydenia sp. 381
Heydeniidae 381
hind femur 196, 361
hind wing 65, 66, 69
    vein nomenclature 155
Histeromerus 108
Homalotylus sp. 354, *356, 357*
Homobolinae 121
Homolobinae 100

Homolobus spp. 122
Homonotus sp. 278
Homoporus sp 388
'honorary' parasitoid wasps 2
Hormiinae 105–106
Hormiinae–Hormiini 98
Hormiinae–Lysitermini 98
Hormoserphus sp. 323
host feeding 11
host location and assessment 23–26
host manipulation by venoms and viruses 16–17
Hyatosema sp. 359
Hyblaea puera 202
Hymenoptera 2, 3, 31, 46, 74, 411
    Bayesian phylogenetic tree for *49*
    dated phylogenetic hypothesis of *51*
    head, features of *58*
    maximum-likelihood phylogeny of *52*
    phylogeny research and 47–50
    wing venation ground plan of 64–65
    see also individual entries
Hymenoptera, The 335
Hymenoptera Anatomy Ontology Portal, The 55
Hymenoptera of the World, The 79, 247, 271, 335
Hyperimerus sp. 378
hypopygium 160
    male 273
hypostomal carina 57

Ibalia kalimantanica 314
Ibalia sp. 307
Ibaliidae 308, 314–315
Icerya purchase 382
Ichneumonidae 30, 40, 41, 62, 87
    subfamilies of *148–149*
Ichneumonidae–Darwin wasps 147–150
    Brachycyrtiformes and 161–164
Ichneumoniformes and 165–172
keys to subfamilies of 157–160, 166–167, 196–197
Labeniformes and 172
morphological terminology of 150–156
Ophioniformes and 172–191
phylogeny and informal subfamily groupings 160–161
Pimpliformes and 191–206
Xoridiformes and 206–207
ichneumonids 25, 30
Ichneumoniformes 159, 165–172
Ichneumoninae 170–171, *172*
Ichneumonoidea 35, 47, 48, 57, 64, *66*, 69, 71, 73, 87
Ichneutinae 100, 128
idiobionts 10–11
Idris sp. *296*, 297
Indaphidius curvicaudatus 102
Indonesia 33
Inostemmatinae 293
inquilinism 2

insect separators 412–413, *413*
integrative pest management (IPM) 4, 30
International Commission on Zoological
    Nomenclature (ICZN) 149
International Institute of Tropical Agriculture
    (IITA) Benin 4
inventory 34–35
Ipobracon parvispeculum 36
Ipomea batatus 391
Ishigakia sp. 193
Ishtarella thailandica 101
Ismaridae 327, 332
Ismarus sp. 332
Ismaya sp. 382
Italy 33
Itoplectis sp. 196, 197–198
Ixodiphagus hookeri 354

Japan 33
Jezarotes sp. 193
Jezarotes yanensis 192
Jimwhitfieldius sp. *131*
jugal lobe 69
Jurine–Cresson system 66

Karlidia peterseni 274
Kieffer, Jean-Jacques 6
kite-shaped butterfly nets *411*
Klutiana sp. 184, 185
koinobionts 10, 11, 47, 50, 73, 94, 101, 107, 113, 116,
    120, 121, 124, 125, 127, 128, 132, 173, 174,
    176, 178, 180, 182, 186, 188, 192, 195, 196,
    225, 242, 258, 265, 278, 293, 313, 314, 321,
    322, 328, 332, 365, 371, 428
Kristotomus sp. *190*
Krombeinella sp. 277
Krombeinius sp. 375
Kudakrumia sp. 277

Labeniformes 172
Labeninae 172
labium 59
Lagynodes sp. 237
Lambdobregma sp. 374
Laodelphax sp. 373
larval feeding and external feeding phase 13–14
larval instars 12
lateral carina 95
lateral ocelli 295
lateral pronope (subpronope) 95
laterope 96
Lathrolestes sp. *179*
Laxiareola ochracea 188
leaf-mining hosts 25
learning and associative learning 24

Leeuweniella sp. 381
legs 61–62, *63*
Leiophron sp. 126
Lepidopsocus sp. 408
Leptacis sp. 294
Leptocorisa oratorius 298
Leptocybe invasa 371
Leptofoeninae 383
Leptofoenus sp. 383
Leptopilina sp. 144
Leptopilina boulardi 313
Leptopilina heterotoma 313
Leucospidae 343, 370–371
Leucospis dorsigera 371
Leucospis sp. 370–371, *370*
light traps 418–419
linea calva 342, 357
Liopteridae 307, 308, 315
Liosphex trichopleurum 283
Lipolexis oregmae 101
Lipothymus sp. 385, 386
Liriomyza huidobrensis 367
Lissopimpla sp. 196
Listrodomus sp. 171
Lithosaphonecrus sp. 310
Loboscelidia sp. 257–258, *257*
Loboscelidiinae 253, 256–258
Loboscelidae 84
longitudinal veins 64, 67, 68
Losbanus sp. 359
Louricinae (Chalcidoidea *incertae sedis*) 381
Lustrina assamensis 256
Lustrina sp. 256
Lyciscidae 381–382
Lycogaster flavonigrata 244
Lycogaster rufiventris 244
Lycogaster sp. 245
Lycorina borneoensis 180
Lycorina sp. 180, *181*
Lycorininae 159, 180
Lymaenon sp. 373
Lysitermini 107
Lyteba sp. *331*
Lytoxysta sp. 313

Macrocentrinae 99, 121–123, *123*
macrocentroid subcomplex 120–123
Macromesidae 382
Macromesus sp. 382
Macroteleia sp. *296*, 298
Mahinda bo sp. 254
Mahinda sp. 253
Malaise traps 31, 34, 40, 116, 175, 199, 253, 255, 257, 323, 360, 375, 410, 411, 413–416, 420
    aerial *416*
    samples, sorting of *422*
    standard design of *415*

Malaysia 33
male ants, assorted *82*
male genitalia 73–74, *74*
Malinka sp. 379
Mallada desjardinsi 357
Malta 33
mandibles 10, 11, 13, 57–59, *97*, *109*, 112, 160, 161, 179, 194, 196, 197, *203*, 204, *205*, *206*, 207, 228, 243, 245, *263*, *332*, 346, 408
Mangostigmus sp. 371
Mansa sp. 170
Mantibaria sp. 295
Marietta leopardine 351
Masner sp. 235
masoninae 103
maxillae 59
Megachalcis sp. 352
Megalyroidea 83, 239
    Megalyridae and 239–241, *240*
Megaphragma sp. 36–37
Megarhyssa sp. 204, *205*
Megascolia azurea 279
Megascolia sp. 279
Megaspilidae 235, 237
Megastigma sp. 372
Megastigmidae 344, 371
Megastigmus nigrovariegatus 371
Megastigmus sp. 336, 371
Megischus sp. 221, *222*
Melanagromyza sojae 384
Melittobia sp. 364, 367
mercury vapor lamps, as traps 418–419
Merostenus (*Hirticauda*) sp. 368
mesepimeral sulcus 291
Mesitiinae 249, 251–252
Mesochorinae 157, 158, 180–182
mesochorines 24, *182*
Mesochorus curvulus 181
Mesochorus sp. 180–181, *182*
Mesocynips sp. 315
mesopleural suture 196
mesopleuron 249, 272
mesoscutum 59, 157, 345
mesosoma 56, 59–61, 81
    dorsal, features of *60*, *362*, *372*
    lateral, features of *61*, *292*, *341*
    of sphecid wasp *82*
    ventral, features of 61, *62*
Mesosomal dorsum 273
Mesostenus group 169
mesosternal+mesopleural complex 60, 61
mesothoracic spiracles 59
metanotum 60, 249
Metapelma sp. 371, *372*, *374*
Metapelmatidae 371–372
metapostnotum 60

metasoma 56, 69–71, *71*, 81, 83, 159, 295, 307, 320, 427
metasomal horn 291
metasomal laterotergites 328
metasomal tergite 70, 95, 98, 99, *156*, 159, 160, 176, 272, 273, 308, 363
metathoracic spiracles 59
Meteoridea hutsoni 133
Meteoridea sp. 133
Meteorideinae 100, 116, 133
Meteorinae 123–125
Meteorus pulchricornis 125
Meteorus sp. 124, 125, *126*
Methocha granulosa 282
Methocha malayana 282
Methocha sp. 282
Methochinae 273
Methocinae 282
Metopheltes sp. 179
Metopiinae 158, 160, 182–184
Metopius sp. 182, *183*
Mickelomyrme puttasoki 274
Micranisa sp. 387
Microgastrinae 101, 129–132
Microgastroid complex 125–133
'microhymenoptera' 413, *414*, 416, 417
Microleptinae 171–172
Microplitis sp. *131*
microscope slide preparations, for specimen preparation 427
Microterys sp. 355
Milletia sericea 391
Minagenia sp. 278
minimalist movement and barcoding 39
minuten pins 424–425
Miracinae 100, 132
Mirax sp. 132
Miscogasteriella sp. 388
Miscogasterinae 384–385
molecular studies 48–49
Monacon gawai 376
Monacon sp. 376
Monema flavescens 255
Monoblastus sp,. 190
Monodontomerinae 392
Monodontomerus dentipes 392
Monodontomerus sp. 392
Monomachidae 327, 332, *333*
Monomachus antipodalis 332, *333*
Monomachus sp. 332
Moranilidae 382
morphological studies 47–48
morphology 55
    antennae 58
    head 57, *58*
    legs 61–62, *63*
    male genitalia 73–74, *74*

    mesosoma 59–61
    mouthparts 58–59
    ovipositor 71–73, *72*
    sculpture 74
    wasp waist 55–57
    wings 63–71
    see also wasp waist
mounting, for specimen preparation 423–425
mouthparts 58–59
Mt Halimun National Park (W. Java) 39
multiparasitism 13
Museum für Naturkunde, Berlin 36
Mutillidae (velvet ants) 272, 273–277
Myllenyxis sp. 204–205, *204*
Mymaridae (fairy flies) 84, 344, 372–374
Mymaromma anomalum 408
Mymaromma menehune 407, 408
Mymaromma sp. 408
Mymarommatoidea (false fairy wasps) 84, 335, 407–408
Mymar sp. 374
Myrmosidae 273, 277
Myzininae 273, 281–282

Nagoya Protocol 32
nasty-host hypothesis 40
natural enemies, conservation of 3
Natural History Museum, London 425
Neanastatus grallarius 374–375
Neanastatus sp. 374
Neapterolaelaps sp. 379
Nefoenus sp. 383
Neocalosoter sp. 378
Neochalcis sp. 353–354
Neodiprion biremis 392
Neodiprion sertifer 179
Neolosbanus palgravei 359
Neorhacodinae 188
Neoxorides sp. 203
Nephotettix spp. 32
Neralsia sp. 314
Nesomesochorinae 159, 184–185
Netelia sp. 189, *190, 191*, 411–412
Neurocrassus sp. 105
Neurogenia sp. 179, 181
Neuroscelio doddi 292
Neuroscelionidae 291, 292
Neuroscelio sp. 292
Nezara viridula 299
Nilaparvata lugens 32, 262, 367
Nilaparvata sp. 373
Nipisa phyllicola 297
Nipponosega lineata 254
Nipponosega sp. 253
Nixonia krombeini 292
Nixonia sp. 292

Nixonia watshami 292
Nixoniidae 291, 292–293
Nomosphecia sp. 199
Nomosphecia zebroides 199
non-cyclostomes 91, 95, 115–116
    Euphoroid complex 123, 124
    Helconoid complex 118–123
    Microgastroid complex 125–133
    Sigalphoid complex 116–118
Nonnus sp. 184
Notanisomorphella sp. 366
Notanisus sp. 379
notauli 59, 361–362
Nothofagus sp. 322
Nothoserphus afissae 323
Nothoserphus sp. 322–323, 324
Notosemus sp. 171, 172
notum 59
Noyes-design sweep net 414
Nyleta sp. 298

occipital carina 57, 98, 114
occiput 57
Odontacolus sp. 297
Odontepyris sp. 250
Odontofroggatia sp. 380
Oecophylla smaragdina 354
Oedemopsini 189–190
Olenecamptus bilobus 382
olfactory system 22
Oligoneurus spp. 129, 130
Oligosita sp. 394
Omphale sp. 361, 363
Ooctonus sinensis 374
Ooctonus sp. 374
Oodera longicollis 382
Oodera sp. 382
Ooderidae 382
Ooencyrtus sp. 355, 356, 357
Opheliminae 360, 361, 366
Ophelimus eucalypti 366
Ophelimus maskelli 366
Ophelimus sp. 366
Ophelosia crawfordi 382
Ophioniformes 172–191
Ophioninae 157, 185–186
Ophionini 186
Ophion sp. 185, 186
Opiinae 99, 100, 114–115
Opisina arenosella 133
Opisthacantha sp. 296
Orasema sp. 359
Oraseminae 359–360
organic produce, demand for 4–5
Orgilinae 100
Orgilonia sp. 121, 122

Orgilus 122
Oriental bethylid 74
Orientilla vietnamica 275
Orientochorus sp. 181
Ormyridae 345
Ormyrus sp. 375
Orseolia oryzae 294, 375, 388
Orthocentrinae 159, 195–196
Orthocentrus sp. 182, 196
Orussoidea (Orussidae) 47, 81, 89–90, 221
Osphranteria coerulescens 378
Osphrynchotus group 169
Otitisellini 385–386, 387
ovipositor 71–73, 72, 98, 159, 160, 228, 230,
    296, 324, 354
ovoparasitism 14–15
Oxyserphus russian 323
Oxytorinae 159, 186–187
Oxytorus carinatus 187
Oxytorus rufopropodealis 187
Oxytorus sp. 186

Pachycrepoideus vindemmiae (Pachyneurinae) 384
Pachyneurinae 385
Palachia sp. 393
Pambolinae 98, 107, 109
Paniscomima sp. 283, 284
Pantoclis sp. 329
pan traps 416–417
Papua New Guinea 33
Papuopsia sp. 389
Parabioxys songbaiensis 102
Parablastothrix sp. 354
Paramblynotus grossus 315
Paramblynotus sp. 315
Paraplitis 131
Pararhabdepyris sp. 252
Parasaphodes sp. 383
Parasaphodes townsendi 382
Parasaphodinae 382
Parasapyga boschi 279
Parasapyga sp. 279
Parasapyga yvonnae 279
'Parasitica' 47
parasitoid food conversion efficiency 14
Parasteatoda merapiensis 201
Parastephanellus sp. 222
parastigma 342
Pareumenes quadrispinosus 255
Parevania sp. 227, 228
Paridris nephta 297
Paridris sp. 297
Peckidium enigmaticum 329
Pediobius sp. 361, 364
Pelecinellidae 383
Pentatermini 107

Perilampidae 341, 345, 357–357, *358*, 375–376
Perilampus auratus 358
Perilampus singaporensis 376
Perilampus sp. 375, 376
Periplaneta americana 227, 366
Perisphincter sp. 174
Peristenus sp. 181
Perthiola sp. 366
pest management utility 3–5
petiole 70
Phaenacoccus solenopsis 357
Phaenoglyphis sp. 313
Phalgea sp. 193
Phaneroserphus calcar 322
Phaneroserphus sp. 324
Phanerotoma sp. 129
Phanerotomella sp. 129
Phenacoccus manihoti 4, 161, 356
Phenacoccus solenopsis 384
Pherotilla rufitincta 276
Philippines 33
Philomides frater 376
Philoplitis 131
Philotrypesis caricae 386
Philotrypesis sp. 341, 386, 387
phoretic copulation 23
Phrudinae 187, 188
Phygadeuontinae 157, 167, 172
Phytodietini 190
Phytodietus sp. *190*
phytophagous sawflies and wood-wasps 46, 47
Phytophagy 336
Pilocutis sp. 252
Pimpla sp. 196, 198
Pimpliformes 191–206
Pimplinae 158, 160, 196–202, *200*
pionines 179
Pirenidae 383–384
Pison argentatum 275
Plagiotrochus indochinensis 309
Plastanoxus sp. 251
Plastazote® 426
Platygaster foersteri 294
Platygaster oryzae 294
Platygaster spp. 294
Platygastridae 291, 293–294
Platygastrinae 293
Platygastroidea 86, 290–291
    key to families of 291
    morphological terminology of 291
    Neuroscelionidae and 292
    Nixoniidae and 292–293
    Platygastridae and 293–294
    Scelionidae and 294–300
    Sparasionidae and 300–301
Playaspalangia sp. 389
Plectochorus sp. 181

Pleistodontes sp. 349
pleuron 59
Pleurotropis sp. *363*
Plutella xylostella 177
Podagrion ahlonei 393
Podagrioninae 392–393
Podagrion sp. 393
Podoschistus sp. 203
Poemeniinae 157, 202–203
Poland 33
polydnaviruses 17, 126, 177
polyembryony 12
Polysphincta group 196, 197, 199, *201*, 428
polystyrene 426
Pompilidae (spider-hunting wasps) 272, 277–278
Pompiloidea 271
Popillia japonica 283
postgena 57
postpectal carina 152
Praestochrysis shanghaiensis 255
Praestochrysis sp. 255
precoxal sulcus 95
prepectal carina 95
prepectus 60–61, 338, 341, 345, 346
Pristapenesia asiatica 265
Pristaulacus comptipennis 226
Pristaulacus sp., fore wing of 67
Pristepyris sp. 251
Pristocerinae 249, 252
Pristomerus sp. 178, *178*
Probles sp. 188
Proconura sp. 354
Proctotrupidae 85, 320, 321–323
Proctotrupoidea 85, 320
    Heloridae and 321
    key to families of 320–321
    Proctotrupidae and 321–323
    Roproniidae and 323
    Vanhorniidae and 323–325
Promecidia sp. 275
Promuscidea unafasciaventris 384
Pronkia sp. 133
pronope (median pronope) 95
pronotum 82, 86, 160, 345
Propalachia sp. 393
Propicroscytus (Pteromalini) sp. 384
Propicroscytus sp. 388
propleuron 61
propodeal carinae 152
propodeum 157, 158, 249
Prorops sp. 252
Prosevania sp. 227
Protanaostigma sp. 391
Proteropinae 132–133
Proterops borneoensis 132, 133
Proterops fumosus 132
Proterops sp. 129, 132–133

pro-thorax 59
Pseudencyrtus sp. 354, *356*
Pseudococcus jackbeardsleyi 4
Pseudoligosita sp. *394*
Pseudomegischus sp. 222
Pseudonomadina sp. 245
Pseudorhyssa sp. 203, 204
Pseudoshirakia 103
Psilini 331
Psilocharis hypaena 359
Psilocharis sp. 358
Psix group 300
Pteromalidae 344, 346
Pteromalidae *sensu stricto* 384–388
Pteromalinae 385–387
Pteromalini 386–387, *388*
pterostigma 66–67
Pumplini 197–199
Pycnetron sp. 384

Queen Sirikit Botanic Garden (Chiang Mai, Thailand) 38

radial sector 67
Ralstonia sp. 336
Rhadinoscelidia lixa 258
Rhadinoscelidia sp. *257*, 258
Rhopalosomatidae 87, 272, 283–284, *284*
Rhorus sp. 179
Rhynchoticida sp. 392
Rhysipolinae 99, 107–108
Rhysipolis sp. 107–108, *109*
Rhyssalinae 99, 108, *110*
Rhyssa persuasoria 25
Rhyssinae 157, 204–206
Rhytidaphora sp. 180
Rhyzopertha dominica 378
richogramma dendroliti 394
Robertsia sp. 386
Rogadinae 98, 99, 109–111, *111*
Rogadini 110
Roman, Per Abraham 6
Ropronia malaise 323
Ropronia sp. 323, *324*
Roproniidae 321, 323
Rothneyia sp. 173

Sachtlebenia 175
Sapyga luteomaculata 278
Sapygidae 272, 278–279
Scelionidae 291, 294–300
   key to clusters of 295
scelionid wasp *292*
Scelioninae 295
   sensu authors 296–298
Scelio pembertoni 296

Scelio sp. 296, 298
Sceliotrachelinae 293
Scelio-type ovipositor system 296
Schizaspidia aenea 359
Schizaspidia diacammae 358
Schizaspidia nasua planidium *358*
Schizopyga sp. 201
Schlettererius cinctipes 220
Schoenlandella sp. 127
Schoenobius 103
Scirpophaga 103
Scleroderminae 249, 252
Sclerogibba sp. 265
Sclerogibbidae 85, 248, 264–265
Scolebythidae 248, 265
Scolebythus sp. 265
Scolia sp. 279
Scoliidae (digger wasps, mammoth wasps) 272, 279–281
Scolioidea 271, 279–281
scrobes 57
Scudderopsis sp. 189
scutellar sulcus 307
scutellum 59, 307, 309, 362
Semiotellus sp. 389
Sericopimpla sp. 202
Seriola sp. 249
Sesbania grandiflora 371
sex, courtship, and mating 21–23
sex ratio 15–16
sex ratio distortion 16
sexual dimorphism 58, 247, 252, 253, 256, 258, 368, 379
Shelford, Robert Walter Campbell 6
short-termism 37
side gluing 425
Sierolomorphoidea 271
Sigalphinae 99, 118
Sigalphoid complex 116–118
Signiphora bifasciata 391
Signiphora sp. *390*, 391
Signiphoridae 343, 390–391
Sinotilla calopoda 275
Sinotilla decorate 275
Siphimedia nigriscuta 193
Sirex noctilio 315
siricid wood wasps 315
Sisyrostolinae 187, *188*
Sitophilus granaries 378
Sitophilus oryzae 378
Sitophilus zeamais 378
skaphion 291
Slonopotamus 188
Smicromorpha sp. 354
Smicromorphinae 354
Smicromyrme dardanus 274
sodium benzoate 416–417

Sogatella sp. 373
Solenura ania 379, 381–382
Solenura sp. 381
Spalangia cameroni 389
Spalangia sp. 389
Spalangiidae 389
Sparasionidae 291, 300–301, *301*
Sparasion sp. 301
specimen impediment 35–36
specimen preparation 420
    adhesives and gluing techniques for 425–426
    drying of small and fragile samples for 422–423
    equipment and consumables for 420–422
    genitalia preparations for 427
    microscope slide preparations for 427
    mounting for 423–425
    store boxes for 426–427
    wet samples sorting for 422
sphecid wasp 82
Sphecophaga orientalis 169
Sphecophaga vesparum 169
Sphegigaster sp. 385
Sphinctini 190
Sphinctus chinensis 190
Sphinctus serotinus 190
Spilomicrini 330
Spilomicrus sp. *331*
Spilopteron sp. 193
Spodoptera frugiperda 299, *300*
stage pinning method 424–425
Stantonia 122
Stauropoctonus sp. 186
Stauropoctonus torresi 186
Stauropoctonus townesorum 186
Stenarella insidiator 170
Stenobracon 103
Stenomacrus sp. *195*
Stephanoidea (crown wasps) 83, 219–220
    identification of 222
    recognition of 220–221
    wing venation of 220
Stephanus sp. 222
sternaulus 151–152
Stilbopinae 187
Stilbops acicularis 187
Stilbops gorokhovi 187
Stiropiini 110
store boxes, for specimen preparation 426–427
Storeya sp. 389, *390*
Storeyinae (Chalcidoidea: *incertae sedis*) 389
Streblocera sp. 125, *126*
stripes 58
subalar region 61
submarginal vein 362–363
suction traps *see* vacuum cleaners

superparasitism 12–13
sweep nets 410, 412–413, *412, 414*, 416
Sychnostigma sp. 205
Sycobia sp. *381*
syconium 346, 348
Sycophaga sp. *388*
Sycophaginae 387–388
Sycophila sp. *369*
Sycoscapter sp. 386
'Symphyta' 46, 47, 81, 89
Symplocos stellaris 372
Synacra paupera 330
Synergini 310
Synopeas sp. 294
synovigenic idiobionts 11
Syntomernus 104
Systasidae 389
Systasis sp. 389
Syzeuctus sp. *176*
Syzygium samarangense 366
Szépligeti, Gyötö 6

Taeniogonalos fasciata 244
Taeniogonalos sp. 242, 245
Taeniogonalos tricolor 244
tagmata 55
Taimyrmosa nigrofasciata 277
Takyomyia sp. 281
Tanaostigmatidae 345, 391
Tanyxiphium harriet 373
tarsal segments 62
taxa
    with costal cell 84–85
    lacking enclosed fore wing costal cell 85–87
taxonomy
    early work of 6–7
    history, in S.E. Asia 5–7
    importance of 5
    need to speed up 37–39
'Teleasinae' 295, 298–299
Telengaiinae 99, 100, 113
'Telenominae' 295, 299–300
Telenomus remus 299, *300*
Telenomus sp. 299
Teleutaea sp. *175*
Temelucha sp. *178*
temple sculpture 157
teratocytes 12, 125
Tersilochinae 158, 187–188
Tessarotoma sp. 368
Testudobracon 104
Tetracampidae 391
Tetracneminae 357
Tetrastichinae *361, 362, 363*, 366
Thailand Insect Group for Entomological Research (TIGER) project 38, 116

thelytoky 16
Theocolax elegans 378
Theronia group 197, 198–199, *199*
Thoreauella sp. *311*
Thoron dayi 298
Thoron sp. 297–298
Thynnidae 271, 273
Thynninae 281
thynnine 281
Thynnoidea (flower wasps) 271, 281–282, *282*
Thyreodon sp. 186
thyridea 155
tibia 61
Tiphia intrudens 283
Tiphia sp. *282*, 283
Tiphiidae 271, 273, 283
Tiphiinae 273, *282*
Tiphioidea 271, 281–283, *282*
Tiphodytes sp. 298
Tipula irrorate 193
Tirathaba sp. 353
Tissahamia gombak 297
Tomicobomorpha sp. 382
Torymidae 344, 345, 392–393
Toryminae 393
Torymus kovaci 393
Townes, Henry 40, 55, 69, 149–150, 152, 168, 186, 322
Townes-design traps 414, *415*
Townesionini 175
Toxoneuron nigriceps 127
Toya sp. 373
Trachyserphus sp. 323
transverse carina 61, 95
Trassedia sp. 233
Trassedia yanegai 233
Trichoferus campestris 382
Trichogramma dendrolimi 394
Trichogrammatidae 343, 393–395
Trichoplusia ni 12
Trichopria clavatipes 330
Trichopria drosophilae 22
Trichopria sp. 330, *331*
Triclistus aitkini 183
Triclistus sp. 183, 184
Trigastrotheca 104
Trigonalyoidea 83, 242–243, *244*
    key to genera of 245
    morphology of 243
    S.E Asian fauna of 243
Trigonoderinae 388
Trigonoderus sp. 388
Trimorus sp., 299
Trisecodes sp. 360
Trissolcus basalis 299
Trissolcus japonicus 299, *300*
Trissolcus sp. 299

Triteleia sp., 298–299
tritrophic interactions 24
trochantellus 62
trochanter 61, 62
Trogaspidia doricha 274
Tromatobia sp. 201
tropical parasitoid diversity and biology 39–41
Tropobracon 103
Tryphoninae 150, 169, 188–191
Tryphonini 190–191
turbotaxonomy 38–39
tyloids 150, *151*

ultraconserved elements (UCE) 270, *271*, 273, 280, 360
USA National Science Foundation (NSF) 38

vacuum cleaners 421
Vanhornia eucnemidarum 325
Vanhornia sp. 323–324
Vanhorniidae 320, 323–324, *325*
vein naming systems 65–68
Vein RS&M 307
ventral mesopectus *see* mesosternal+mesopleural complex
ventral sternum 59
Vespidae 272
Vespina 47
Vespoidea 81, 83, 85, 87
Vespoidea *sensu lato* 270–272
    key to 272–273
    Pompilloidea 273–279
    Rhopalosomatidae 283–284
    Scolioidea 279–281
    Thynnoidea (Thynnidae) 281–282
    Thynnoidea and Tiphioidea (flower wasps) 281
    Tiphioidea (Tiphiidae) 282–283
Vespula germanica 169
Vespula vulgaris 169
vibrational sounding 25–26
vital sac 350
volsella 73

Wallace, Alfred Russel 5–6
wasp waist
    body regions of 55–56, *56*
    importance of 56–57
    see also morphology
wasp-waisted wasps 47
Waterston's evaporatorium 233, *234*
wild-collected hosts, rearing from 427–428
Wilkinsonellus sp. *131*
wing-fanning 22

wings 152–155
    bullae 154–155, *156*
    cell naming systems of 68–69
    fold and flexion lines and 68
    fore and hind *65, 66, 69*
    Hamuli and wing coupling and 63–64
    Hymenopteran wing venation ground plan and 64–65
    metasoma of 69–71, *71*
    vein naming systems and 65–68
Wolbachia sp. 16, 194, 259
Wroughtonia sp. *121*

Xanthocampoplex sp. *177*
Xanthopimpla sp. *197*, 198, *200*
Xanthopimpla stemmator 198
Xenomerus sp. 299
Xiphozele sp. *123*
xiphozelinae 123
Xiphyropronia sp. *323*
Xorides sp. *206*, 207
Xoridiformes 206–207
Xoridinae 158, 206–207

Xyalophora sp. 314
Xyelidae 47
Xylocopa sp. 355
xylophagous, locating 24–25

Ycaploca sp. 265
Yeliconini 110
yellow pan traps 410, 416–417, *417*
Yezoceryx shengi 192
Yezoceryx sp. 192

Zaglyptogastra sp. 104
Zaglyptus sp. *201, 202*
Zagrammosoma sp. 367
Zaischnopsis sp. *368*
Zatypota sp. *201*
Zatypota yambar 202
Zebe sp. 384
Zelomorpha sp. 39
Zoosphere project 36
Zatypota albicoxa 200, *201*